New Technologies for Energy Transition Based on Sustainable Development Goals

Kaviyarasu Kasinathan ·
Rasiah Ladchumananandasivam ·
S. Beer Mohamed
Editors

New Technologies for Energy Transition Based on Sustainable Development Goals

Factors Contributing to Global Warming

Springer

Editors
Kaviyarasu Kasinathan
School of Interdisciplinary Research
and Graduate Studies
The University of South Africa
Johannesburg, South Africa

S. Beer Mohamed
School of Technology
Central University of Tamil Nadu
Thiruvarur, India

Rasiah Ladchumananandasivam
Department of Textile Engineering
and Mechanical Engineering
Universidade Federal do Rio Grande do Norte
Natal, Rio Grande do Norte, Brazil

ISBN 978-981-97-2526-7 ISBN 978-981-97-2527-4 (eBook)
https://doi.org/10.1007/978-981-97-2527-4

This Springer imprint is published by the registered company Springer Nature Singapore Pte Ltd.
The registered company address is: 152 Beach Road, #21-01/04 Gateway East, Singapore 189721, Singapore

Preface

This volume entitled *New Technologies for Energy Transition Based on Sustainable Development Goals* highlights the several Sustainable Development Goals (SDGs) outlined by the United Nations (UN). The individual chapters explain the different SDGs which will be accessible to affordable and clean energy. Energy storage technologies will offer increased access to reliable energy. Advanced metering infrastructure improved reliability and stability of renewable energy sources, as well increased access to reliable energy. The advanced metering infrastructure offers enhanced grid management and reduced losses and better control over energy consumption and costs. Energy-efficiency technologies reduce consumption in households and lower energy bills and a reduced environmental footprint. Electric vehicles use electricity instead of traditional fuels, thus reducing air pollution and dependence on fossil fuels, shifting toward cleaner and sustainable mobility. Capturing and storing carbon dioxide emissions from industrial processes allows for continued industrial activities with reduced environmental impact and mitigates the impact of greenhouse gas emissions on climate change. Decentralized energy systems, where affordable and clean energy with increased access to electricity, improved resilience against grid failures. Overall, this volume provides the expertise knowledge about the new technologies for energy transition in achieving the Sustainable Development Goals, with various techniques, that will find abundant application as tools for the energy transitions. This volume will also be extremely beneficial to students, scientists, and technical sectors, providing a broad framework for interdisciplinary research, creating opportunities for academics to explore innovative solutions to complex global challenges, and, As a matter of fact, providing a better insight on the subject concerning the innovative application of various techniques on energy transition based on sustainable development.

Prof. Rasiah Ladchumananandasivam
University of Rio Grande do Norte (UFRN)
Natal, Brazil

In the spectrum of challenges facing our planet, few are as urgent and consequential as the quest for sustainable energy. The pressing need to transition from conventional, often environmentally taxing, energy sources to cleaner alternatives is not just a scientific or technological pursuit—it is a moral imperative, a commitment to the well-being of our planet and its inhabitants. This book is a humble contribution to the ongoing dialogue on sustainable energy. It is a tapestry woven with threads of insight, research findings, and visionary ideas from experts across the globe. As we stand at a critical juncture in history, where the choices we make today will reverberate for generations, the need for a sustainable energy revolution has never been more apparent. To me, one of the most exciting aspects of the areas covered in this book is that they transcend the barriers between the mainstream scientific disciplines of Physics, Chemistry, Biology, and Engineering. Thus, they provide new insights into the energy storage, microbiology, agriculture, electrocatalysis, photocatalysis, energy from waste, and dazzling possibilities for new technology. This book also gives an insight of the development of advanced materials, for low carbon emission plays a crucial role in climate change mitigation. Few topics in the book demonstrate recycling of e-waste, circular economy, protection of the environment, and improvement of agricultural productivity by using advanced technology. Of course, the entire sweep of the topic, including as it does, all the mainstream sciences and engineering, is incredibly broad, and the intention was to cover it with a relatively to give the reader the "*feel*" of what is exciting about the areas. This book was written in mind that the readers could easily understand the technologies involved in the characterization and fabrication of advanced devices. In the book chapters, the readers will embark on a journey through the dynamic landscape of interdisciplinary science and technology—a realm where boundaries blur and collaborative exploration gives rise to new frontiers. This book can be used by industries and university students studying interdisciplinary subjects and involved in interdisciplinary research. To the authors who have dedicated their expertise and passion to these pages, we extend our heartfelt appreciation. Your work represents a collective effort to illuminate the path toward a more sustainable and resilient energy.

Prof. S. Beer Mohamed
Central University of Tamil Nadu (CUTN)
Thiruvarur, India

Low- and middle-income people (individuals) contribute significantly to global warming fueled by the energy demand by fulfilling five critical roles in society. It is the objective of this perspective to examine the results of the study and to suggest how these results might be reduced in terms of energy and climate impacts. As part of the theme of this book, we aim to reach a broad audience across media, science, and popular culture activities. In addition to the recent IPCC mitigation report, SDGs have been addressed at other conferences that have elevated things to new heights. One of the most interesting things about our Springer Publishing was that they attracted the curiosity of our readers. The partnership with the United Nations sustainable goals counted us in organizing their climate action actions, with our five-role framework accompanying each category on a periodic basis. An analysis of how our findings from a scholarly chapter reached a wider audience led us to take five steps to effectively communicate our findings.

- An idea of broader societal interest is presented with a few hints, focusing on elites' disproportionate climate sensitivity.
- The team is working hard to write take-home messages that are clear and concise.
- Contributing to the development and communication of key messages within and beyond the paper through collaboration with talented authors.
- This book contains dozens of chapters that we directly pitched to.
- The benefit of working with a publishing house like Springer, which tends to attract more audiences and publishes more with a high impact through recognition.

A book such as this can be classified into five different parts.

(i) **Goal 3: Good Health and Well-Being:** (*a healthy lifestyle should be promoted for everyone—promoting well-being*). As a result of their ability to reduce diagnostic time, produce rapid results, be portable, affordable, highly sensitive, highly specific, and have a low detection limit, biosensors play a crucial role in clinical and biomedical settings. As a result, they are able to be used in any situation where there is a natural disaster or emergence.

(ii) **Goal 6: Clean Water and Sanitation:** (*it is imperative that water and sanitation are made accessible to all and that they are managed sustainably*). As well as protecting and restoring river systems, aquifers, wetlands, mountains, and forests, it also involves restoring ecosystems that are related to water. We will see changes in the extent of water-related ecosystems over time as we assess the changes in their habitats. By 2030, we will expand our international cooperation and capacity-building support to assist developing countries with water- and sanitation-related activities and programs.

(iii) **Goal 7: Affordable Clean Energy:** (*in order to achieve this goal, it is essential to ensure that everyone has access to clean, affordable, sustainable, and modern energy sources*). The reason for this is that energy fuels human progress since it is an integral part of all universal processes. About 80% of earth's energy comes from the sun. In general, fossil fuels are responsible for high energy consumption. In order to keep up with technological advancements, sustainable energy solutions are needed. While we are still a long way off from reaching that level despite the support we have from modern technology, we are close to it. The following part consists of seven chapters arranged according to the objectives.

(iv) **Goal 13: Climate Action:** (*it is imperative that we act urgently to combat the effects of climate change*). Due to the establishment of a single framework of regularity from multiple countries, energy source supply can be distributed similarly to traditional energy supply. In fact, biomass is an important renewable energy resource that can be used for biofuel production, power generation, and heating. It will recognize that bioenergy is an important economic force and has great potential in developed countries. Wastewater disposal, municipal waste disposal, and medicinal waste disposal pose major problems for cities. There remains a lack of fundamentals that can lead to a more sustainable bio-body food standard based on renewable energy. It has been detailed in this part how hydrogen is generated thermochemically and biologically. Each technique has its own advantages when it comes to generating hydrogen from biowaste. A cleaner environment is created by using biomass waste, and the natural system is supported to be sustainable by utilizing biomass waste.

(v) **Goal 15: Life on Land:** (*agricultural practices must conserve, restore, and utilize biological diversity sustainably*). As far as promoting farming is concerned, it is favorable compared to its goals. There is a discussion of the detrimental effects of seed germination, plant growth, and crop yield on crops in this part. As a result of the existing situation, population growth leads to a greater demand for agriculture-related products worldwide. The overconsumption of natural resources requires an interesting factor in food products and other essentials.

In *New Technologies for Energy Transition Based on Sustainable Development Goals—Factors Contributing to Global Warming*, we ensure the author concisely formulates his or her argument, captures the reader's attention, includes moral ambiguities, and engages the reader. The purpose of this series is to reach a large audience with research and development by identifying all appropriate topics. It is true that the main objective of the proposed book is to address several issues, therefore providing a platform for future technology development in energy conversion and environmental management. The book is especially relevant to undergraduates, postgraduates, and doctoral students and researchers working in interdisciplinary fields who are interested in materials science. Following biology's efficient and selective mechanisms, materials synthesis and energy harvesting have raised many concerns.

Dr. Kaviyarasu Kasinathan
UNESCO—Unisa Africa Chair in Nanoscience and Nanotechnology
School of Interdisciplinary Research and Graduate Studies
The University of South Africa
Johannesburg, South Africa

Contents

Editors and Contributors

About the Editors

Dr. Kaviyarasu Kasinathan completed master's in Physics at Loyola College (*private Catholic higher education institution run by the Society of Jesus in Chennai, Tamil Nadu, India*) and Ph.D. in Physics at Manonmaniam Sundaranar University, India, in 2014. I continued my academic career as an assistant professor at Sri Sankara Arts and Science College (*it is supported by the Shri Kanchi Kamakoti Peetham Charitable Trust, Kanchipuram*) affiliated with University of Madras, India, before moving to South Africa in 2015. During my role as an assistant professor, I have handled undergraduate physics subjects for undergraduate students conducting and hosting laboratory tours across nanoscience research facilities. I have served as a student coordinator to guide newly arrived newcomers (*entry level*) students to fit into the colleges and urban community. Dr. Kaviyarasu is an emerging early career researcher, has vast experience in materials science and nanotechnology, and discovered new mechanism of advanced materials for low carbon emission technologies which plays a key role to mitigate climate change. My goal is to understand how to design and control the nanoscale organization of macromolecular nanomaterials and their nanocomposites to achieve improved structure, properties, and functionality. As well as developing hydrogen-based materials, I am also developing materials for the energy sector. Finally, I believe the contribution of sustainable materials to achieving the

Sustainable Development Goals (*SDGs*) for a sustainable future for humanity. Research activities that are multidisciplinary are necessary in order to reach SDGs due to sustainability's multidimensional characteristics.

Research institutions in South Africa and the National Research Foundation (NRF) serve as intermediaries between government policies and strategies. Therefore, the NRF awarded me the C2 rating (*which means established researcher*) a year ago, which is valid for five years. Moreover, Stanford University named me one of the top 2% of South African scientists in 2023 as well. Moreover, I've been named to the list of Highly Cited Researchers in 2020 and 2021 by Clarivate Web of Science. According to my understanding, I am among the top cited persons in the materials science field at University of South Africa (UNISA). In my current position, I am a material scientist working at the University of South Africa (UNISA) under the category of UNESCO Africa Chair in Nanosciences and Nanotechnology Laboratories.

Prof. Rasiah Ladchumananandasivam graduated from the University of Manchester, England, UK, (B.Sc. Textile Chemistry (1974), Postgraduation Polymer and Fibre Science (1975)). He joined the Intex Yarns Limited in the textile technology division and in 1976 joined the English Numbering Machines in London. In 1977, he joined the Federal University of Rio Grande do Norte (UFRN), Natal, Brazil, as a visiting professor (1977–1979). He became a permanent member and an assistant professor in 1979. He was promoted to the adjunct professor in 1989 and to the associate professor in 2004. In 2010, through national public exam, he became a full professor (Professor Titular). He taught many subjects to the students from Textile, Chemical, and Production Engineering courses; supervised several M.Sc. and Ph.D. students from different courses; and served as a visiting professor at UFCE, Ceará, Brazil, and an affiliate professor at B. S. Abdul Rahman University in India. He published many papers in local and international in scientific journals and presented many papers in national and international conferences. He hosted and presided over many international conferences in Brazil, and participated in many bilateral projects with universities from countries such as South Africa, India, England,

and Scotland. He joined the Textile Institute in 1972. He is a chartered textile technologist, a chartered colorist, a fellow of the Textile Institute, and a fellow of the Society of Dyers and Colorists and of the Royal Society of Arts.

Prof. S. Beer Mohamed born in the district of Pudukkottai, Tamil Nadu, India. He did his schooling in the same town and moved to Tiruchirappalli for higher studies in Jamal Mohamed College and NIT. He did his Ph.D. from University Putra Malaysia. He is currently working as a professor in the Central University of Tamil Nadu, Thiruvarur. He has worked as a professor in the department of Mechanical Engineering in B. S. Abdur Rahman Crescent University, Chennai. He worked for 21 years in abroad research laboratories (AIST, Japan) and universities, Nagoya University in Japan, Okalahoma University in USA, National University Singapore, University Putra Malaysia, and King Faisal University, Saudi Arabia. He has guided 3 Ph.D. students and 25 M.Tech. students. He has won many national and international fellowship and research awards. Mentioning a few NEDO fellowship (Highest Fellowship award) in Japan, VBL fellowship Japan, Best Researcher Award, Best Scientist Award, and Best Thesis Award. He has published more than 120 research papers in peer-reviewed international journals. His research area of interest includes Nanotechnology, III-V Semiconductors, High Tc Superconductors, Li-Ion Batteries, Solar cells, Alloys, and Composites.

Contributors

Ali Afkham Department of Environmental Health Engineering, School of Health, Mashhad University of Medical Sciences, Mashhad, Iran

Rathi Muthaiyan Ahalliya Department of Biochemistry, Karpagam Academy of Higher Education, Coimbatore, Tamil Nadu, India

Ishaq Ahmad NPU-NCP Joint International Research Center on Advanced Nanomaterials and Defects Engineering, Northwestern Polytechnical University, Xi'an, China;
UNESCO-UNISA Africa Chair in Nanosciences/Nanotechnology, iThemba LABS, Somerset West, South Africa

N. M. I. Alhaji Department of Chemistry, Khadir Mohideen College (Affiliated to Bharathidasan University, Tiruchirappalli), Adirampattinam, Tamil Nadu, India

M. V. Arularasu Department of Electronic Engineering, Chang Gung University, Taoyuan, Taiwan

A. Ayeshamariam Department of Physics, Khadir Mohideen College (Affiliated to Bharathidasan University, Thiruchirappalli), Adirampattinam, Tiruchirappalli, Tamilnadu, India

R. Babujanarthanam Department of Biotechnology, Thiruvalluvar University, Vellore, Tamil Nadu, India

P. Baraneedharan Department of Electronics and Communication Engineering, Saveetha Engineering College, Thandalam, Chennai, Tamilnadu, India

M. Benhaliliba Film Device Fabrication Characterization and Application (FDFCA), Research Group USTOMB, Oran, Algeria;
USTOMB University, Oran, Algeria

Majid Bonyadi Manesh Department of Oral and Maxillofacial Prosthodontics, School of Dentistry, Mashhad University of Medical Sciences, Mashhad, Iran

J. Bosco Franklin Department of Physics, Sacred Heart College (Autonomous), Tirupattur, Tamil Nadu, India

Meenakshi Kaniyur Chandrasekaran Department of Biochemistry, Karpagam Academy of Higher Education, Coimbatore, Tamil Nadu, India

Pandian Chithamparam Department of Biotechnology, Kalasalingam Academy of Research and Education, Krishnankoil, Tamil Nadu, India

G. Damodaran Department of Physics, Sacred Heart College (Autonomous), Tirupattur, Tamil Nadu, India

A. Dhayal Raj Department of Physics, Sacred Heart College (Autonomous), Tirupattur, Tamil Nadu, India

R. Ganesamoorthy Department of Chemistry, Vinayaga Missions Kirupanandar Variyar Arts and Science College, Salem, Tamil Nadu, India

Suriyakala Gunasekaran Department of Pharmacognosy and Pharmaceutical Botany, Faculty of Pharmaceutical Sciences, Chulalongkorn University, Bangkok, Thailand;
Center of Excellence in DNA Barcoding of Thai Medicinal Plants, Faculty of Pharmaceutical Sciences, Chulalongkorn University, Bangkok, Thailand

R. Harikrishnan Department of Physics, St. Joseph College of Engineering, Sriperumbudur, Chennai, Tamil Nadu, India

M. Ismail Fathima Department of Physics, Sethu Institute of Technology, Kariyapatti, Tamil Nadu, India

S. Janani Iswarya Department of Biotechnology, Thiruvalluvar University, Vellore, Tamil Nadu, India

M. Jayachandran Department of Physics, Sethu Institute of Technology, Pullor, Kariyapatti, Tamil Nadu, India

Subhaschandrabose Jeyabharathi Department of Biotechnology, Kalasalingam Academy of Research and Education, Krishnankoil, Tamil Nadu, India;
Department of Microbiology, The Standard Fireworks Rajaratnam College for Women, Sivakasi, Tamil Nadu, India

S. John Sundaram Department of Physics, Sacred Heart College (Autonomous), Tirupattur, Tamil Nadu, India

K. Kanimozhi Department of Chemistry, Science and Humanities, Loyola Institute of Technology (LIT), Kuthambakkam, Tamil Nadu, India

Ayesha Kausar NPU-NCP Joint International Research Center on Advanced Nanomaterials and Defects Engineering, Northwestern Polytechnical University, Xi'an, China;
UNESCO-UNISA Africa Chair in Nanosciences/Nanotechnology, iThemba LABS, Somerset West, South Africa

K. Kaviyarasu UNESCO-UNISA Africa Chair in Nanosciences/Nanotechnology Laboratories, College of Graduate Studies, University of South Africa (UNISA), Pretoria, South Africa

M. Kumar Department of Physics, Arignar Anna Govt. Arts College, Cheyyar, Tamil Nadu, India

A. Lavanya Department of Physics, Sacred Heart College (Autonomous), Tirupattur, Tamil Nadu, India

F. Liakath Ali Khan Department of Physics, Islamiah College (Autonomous), Vaniyambadi, Tamil Nadu, India

Majid Bonyadi Manesh Department of Oral and Maxillofacial Prosthodontics, School of Dentistry, Mashhad University of Medical Sciences, Mashhad, Iran

M. Mani Department of Physics, Kalaignar Karunanidhi Govt. Arts College, Tiruvannamalai, Tamil Nadu, India

Valan Arasu Mariadhas Department of Botany and Microbiology, King Saud University, Riyadh, Saudi Arabia

S. Beer Mohamed Department of Material Science, Central University of Tamil Nadu, Thiruvarur, Tamilnadu, India

Azhaguchamy Muthukumaran Department of Biotechnology, Kalasalingam Academy of Research and Education, Krishnankoil, Tamil Nadu, India

Suresh Naveenkumar Department of Biotechnology, Kalasalingam Academy of Research and Education, Krishnankoil, Tamil Nadu, India

Y. S. Ocak Smart-Lab, Diyarbakir, Turkey

R. Parameswari Department of Biotechnology, Thiruvalluvar University, Vellore, Tamil Nadu, India

Hari Babu Pengonda Department of Chemical Engineering, Rajiv Gandhi University of Knowledge Technologies, Nirmal District, Basar, Telangana, India

A. Periya Nayagi Shilpa Department of Physics, Sacred Heart College (Autonomous), Tirupattur, Tamil Nadu, India

Sampath Kumar Puttapati Department of Chemical Engineering, National Institute of Technology Warangal, Warangal, Telangana, India

K. Rafath Fathima Department of Physics, Sacred Heart College (Autonomous), Tirupattur, Tamil Nadu, India

M. Ragamathunnisa PG & Research Department of Physics, Periyar EVR College (Autonomous), Tirchy, Tamil Nadu, India

Meenakshi Sundari Rajendran Department of Biochemistry, Karpagam Academy of Higher Education, Coimbatore, Tamil Nadu, India

Manikandan Vani Raju Department of Biochemistry, Karpagam Academy of Higher Education, Coimbatore, Tamil Nadu, India

R. Ramesh Department of Physics, Sacred Heart College (Autonomous), Tirupattur, Tamil Nadu, India;
Department of Physics, Islamiah College (Autonomous), Vaniyambadi, Tamil Nadu, India

Naresh Kumar Rotte M/S. Regal Enterprises, Eluru District, Kamavarapukota, Andhra Pradesh, India

S. Sachin Department of Physics, Sacred Heart College (Autonomous), Tirupattur, Tamil Nadu, India

A. Saran Department of Physics, Sacred Heart College (Autonomous), Tirupattur, Tamil Nadu, India

Maryam Sarkhosh Department of Environmental Health Engineering, School of Health, Mashhad University of Medical Sciences, Mashhad, Iran;
UNESCO-UNISA Africa Chair in Nanosciences and Nanotechnology, College of Graduate Studies, University of South Africa, Pretoria, South Africa

Mohammad Mahdi Sarkhoshkalat Department of Mechanical Engineering, Islamic Azad University, Mashhad, Iran

D. Shankari Department of Electronics and Communication Engineering, Saveetha Engineering College, Thandalam, Chennai, Tamilnadu, India

M. Sivabharathy Department of Physics, Sethu Institute of Technology, Kariyapatti, Tamil Nadu, India

Sathiyaraj Sivaji Nano and Energy Bioscience Laboratory, Department of Biotechnology, Thiruvalluvar University, Serkkadu, Vellore, Tamilnadu, India

A. Sudha Department of Chemistry, Khadir Mohideen College (Affiliated to Bharathidasan University, Tiruchirappalli), Adirampattinam, Tamil Nadu, India; Department of Chemistry, Kunthavai Nachiyar Government Arts College for Women (Affiliated to Bharathidasan University, Tiruchirappalli), Thanjavur, Tamil Nadu, India

Suchada Sukrong Department of Pharmacognosy and Pharmaceutical Botany, Faculty of Pharmaceutical Sciences, Chulalongkorn University, Bangkok, Thailand;
Center of Excellence in DNA Barcoding of Thai Medicinal Plants, Faculty of Pharmaceutical Sciences, Chulalongkorn University, Bangkok, Thailand

K. M. Tajun Meera Begum Department of Chemistry, Arignar Anna Government Arts College, Karaikal, Puducherry, India

Force Tefo Thema Department of Bioeconomy and Material Science, Botswana University of Agriculture and Natural Resources, Gaborone, Botswana

Narayanan Venkateshan Department of Pharmaceutical Chemistry, Arulmigu Kalasalingam College of Pharmacy, Krishnankoil, Tamil Nadu, India

Subbareddy Yerramala Department of Chemistry, Andhra Loyola College (Autonomous), Vijayawada, Andhra Pradesh, India

T. Yogesh Department of Biotechnology, Thiruvalluvar University, Vellore, Tamil Nadu, India

Abbreviations

•OH	Hydroxyl radicals
3DG	Three-dimensional graphene
3D-PER	Photoelectrocatalytic reactor
A_{bs}	Aqueous batteries
AE	Alkaline electrolysis
AFC	Alkaline fuel cell
AG	Agarose image
AGG	Acryloyl guar gum
AI	Anaerobic ingestion
AI	Artificial intelligence
ALAB	Advanced Lead Acid batteries
ALP	Alkaline phosphatase activity
AOP	Advanced oxidation processes
APTS	3-aminopropyltriethoxysilane
ARFBs	Aqueous redox flow batteries
ASC	Asymmetric supercapacitor
ATR-FTIR	Attenuated total reflectance-fourier-transform infrared spectroscopy
AuNPs	Gold nanoparticles
BC	Bacterial Cellulose
BDD	Boron-doped diamond
BG	Brilliant green
$BiVO_4$	Bismuth vanadate
BLAB	Bipolar Lead Acid Battery
BNF	Bacterial nanofiber
BTE	Behind the ear
BTUs	British thermal units
CA	Carbonic anhydrase
CAD	Computer-assisted diagnosis
CAES	Compressed Air Energy Storage
CAGR	Compound annual growth rate

CAM	Complementary and alternative medicine
CBD	Chemical bath deposition
CBG	Compressed biogas
CC-film	Collagen-Chitosan hybrid film's
CCG	Chemically converted graphene
CEA	Central Electricity Authority
CERS	Carbon-containing composite electrode with renewable surface
CES	Chemical Storage
CH_4	Methane
ChE	Cholesterol esterase
ChO_x	Cholesterol oxidase
CI Reactive red 2	Photoelectrocatalytic oxidation of X-3B azo dye
CMG	Chemically modified graphene
CNBs	Conductive nano biomaterials
CNF	Cellulose nanofibril
CNFs	Cellulose nanofibers
CNG	Compressed natural gas
CNT	Carbon nanotubes
CO	Carbon monoxide
CO_2	Carbon dioxide
Co_3O_4	Cobalt oxide
COD	Chemical oxygen demand
CPC	Conductive polymer composite
CPs	Conducting polymers
CRT	Cathode Ray Tubes
CS	Chitosan
CS	Compressed air
CSBP	Council on sustainable biomass production
Csp	Specific capacitance
CT	Computed technology
CT	Computed tomography
CTAB	Cetyl-tri-methyl ammonium bromide
CV	Cyclic voltammetry
CVD	Cardiovascular disease
CVD	Chemical vapor deposition process
DDS	Drug delivery systems
DENV	Dengue virus
DFIM	Double-fed inductor machine
DFSCs	Dental flesh stem cells
DM	Diabetes mellitus
DME	Dimethyl ether
DMFC	Direct methanol fuel cell
DNA	Deoxyribonucleic acid
DNOM	Decomposition of Natural Organic Materials
DSSC	Dye-sensitized solar cells

E.C.	Electrochemical biosensors
E-beam	Electron beam evaporation technique
EC	Electrical conductivity
EC	Electrocatalysis
EC	Electrochromic material
ECD	Electrochromic devices
ECM	Extracellular matrix
ECM	Natural extracellular matrix
ECSE	Electrochemical Storage Energy
EDAX	Electronic Dispersive X-ray Spectroscopy
EDLC	Electric Double-Layer Capacitance
EEE	Electrical and electronic equipment
EIS	Electrochemical impedance spectroscopy
ELIZA	Enzyme-linked immunosorbent assay
ELP	Elastin like protein
EMP	Embden-Meyerhof pathway
EMV	Equilibrium in the moisture value
EPA	Environmental Protection Agency
EPR	Enhanced permeability and retention
EPR	Extended producer responsibility
EPR	Permeability and retention
EPS	Human encouraged pluripotent stem cells
ES	Energy storage
ESS	Energy storage system
EST	Electrical storage of energy
ETBE	Ethyl-tertiary butyl-ether
EU	European Union's
E-waste	Electronic waste
FAD	Flavin-adenine dinucleotide
FC	Fuel Cells
FC	Stacked fuel cell
FCC	Face centered cubic
FDA	Food and Drug Administration
FECS	Federation of European Chemical Sciences
FES	Flywheel Energy Storage
FET	Field-effect transistor
FNC	Fiber nanocrystal
FNF	Fiber nanofibril
FW	Flywheels
GAGs	Glycosaminoglycans
GBR	Guided bone regeneration
GCD	Galvanostatic charge-discharge
GD	Glycoluril derivatives
GDP	Gross domestic product
GG	Guar Gum

GHG	Greenhouse
GHG	Greenhouse gas
GO	Graphene oxide
GO_x	Glucose-oxidase
GW	Giga Watts
H_2	Hydrogen
H_2O_2	Hydrogen peroxide
H_2S	Hydrogen sulphide
HA	Hyaluronic acid
HA	Hydroxyapatite
HATC	Humanoid adipose-resulting trunk cells
HBCSC	Human bone core mesenchymal stem cells
HDL	Cholesterol and high-density lipoprotein
HEC	Hydroxyethyl cellulose
HHP	Hydraulic holding period
HIPSC	Human induced pluripotent stem cells
HNSCs	Human neural stem cells
HPDEC	Human progenitor-derived endothelial cells
HRP	Horseradish peroxidase
HS	Humic substances
HTS	High-Temperature Sodium
ICT	Information, and communication technology
IDF	The International Diabetes Federation
IE-SEM	Inverse scanning electron microscope
IHC	Immunohistochemistry
IoT	Internet of Things
IPCC	International Panel on Climate Change
ISGF	Intelligent Grid India Forum
ITO	Indium Tin Oxide
IUPAC	International Union of Pure and Applied Chemistry
KE	Kinetic energy
LAB	Lead Acid batteries
LCA	Life Cycle Analysis
LCAO	Linear combination atomic orbitals
LCD	Liquid crystal displays
LCO	Lithium Cobalt Oxide
LDL	Low-Density Lipoprotein
LED	Light emitting diode
LFP	$LiFePO_4$
LFP	Li-iron phosphate
LFP	Lithium Iron Phosphate
LHS	Latent Heat Storage
LIB	Li-ion batteries
LiB	Lithium-Ion Batteries
LIBs	Laser induced breakdown spectroscopy

LMO	$LiMn_2O_4$
LMO	Lithium Manganese Oxide
LSPR	Localized surface plasmon resonance
LTO	Lithium Titanate Oxide
MBs	Streptavidin-coated magnetic beads
MCFC	Molten carbonate fuel cell
MCSE	Chemical storage of energy
MEC	Microbial Electrohydrogenesis Cell
MFC	Microbial fuel cells
MFF	Micro-fibrillated fiber
MGH	Membranes made of graphene hydrogel
MI	Myocardial Infarction
MIP-2D	Molecularly imprinted 2D
MIPs	Molecularly imprinted polymers
miR	MicroRNAs
MMPC	Mouse mesenchymal progenitor cells
MMT	Million metric tons
MNF	Microbial nanofiber
MnO_2	Manganese oxide
MNPs	Metallic nanoparticles
MoO_2	Molybdenum
MRI	Magnetic resonance imaging
MS	Metal-semiconductor
MSC	Mesenchymal stem cells
MSE	Mechanical storage of energy
MSN	Mesoporous Silica Nanoparticles
MSW	Municipal solid waste
MT	Monotherapy
MTB	Mycobacterium tuberculosis
MW's	Metal wires
MWCNTs	Multi-walled carbon nanotubes
N_2	Nitrogen
NAANI	N-acetylaniline
Na-$NiCl_2$	Sodium Nickel Chloride
NaS	Sodium Sulphur
NCA	$LiNi_{0.8}\ Co_{0.15}Al_{0.05}O_2$
NCA	Lithium Nickel Cobalt Aluminum Oxide
NF	Period nanofiber
NFF	Nano-fibrillated fiber
NG	Nicollian and Goetzberger approximation
NiO	Nickel oxide
NIR	Near-infrared
NMC	Layered $LiNi_{1-x-y}Mn_y\ Co_zO_2$
NMC	Lithium Nickel Manganese Cobalt Oxide
NSAIDs	Non-steroidal Anti-inflammatory Drugs

NSCs	Neural stem cells
O_2	Superoxide radicals
OECD	Organization for Economic Cooperation and Development
OEM	Original equipment manufacturers
OI	Optical imaging
OLED	Organic light-emitting diodes
ORR	Oxygen reduction reaction
OVA	Antigen ovalbumin
P	Phosphorus
PAFC	Phosphoric acid fuel cell
PAM	Polyacrylamide
PANI	Polyaniline
PAT	Photo ablation therapy
PAT	Photoacoustic therapy
PBI	Polybenzimidazole
PC	Photocatalysis
PCBs	Printed Circuit Boards
PCL	Polycaprolactone
PCR	Polymerase chain reaction
PDA	Polydopamine
PDDA	Polydiallyldimethylammonium chloride
PDMS	Polydimethylsiloxane
PDT	Photodynamic therapy
PE	Potential energy
PEC	Photoelectrocatalysis
PECO	Photoelectrocatalytic oxidation
PEDOT	Poly(3,4 ethylenedioxythiophene)
PEF	Photoelectro-Fenton
PEG	Polyethylene glycol
PEMC	Proton exchange membrane electrolysis
PEMFC	High temperature
PEMFC	Polymer electrolyte membrane
PFU	Plaque forming units
PGA	Poly glycolic acid
PHS	Pumped hydro storage
PHSN	Porous hollow silica nanoparticles
PL	Photoluminescence
PLA	Poly lactic acid
PLGA	Poly(lactide-co-glycolide acid)
PMSM	Permanent magnet synchronous machine
PPP	Pentose phosphate pathway
PPT	Plasmonic photothermal
PPy	Polypyrrole
PRF	Platelet-rich fibrin
PSA	Prostate-specific antigen

PSB	Photosynthetic bacteria
PtG	Power to Gas
PTh	Polythiophene
PTT	Partial thromboplastintime
PTT	Photothermal therapy
PTX	Paclitaxel
PVA	Poly vinyl alcohol
PVDF	Polyvinylidene fluoride
Q.C.	Quality control
QC	Quartz crystal
QCM	Quartz crystal microbalance
QSAR	Quantitative structure-activity relationship
QUS	Quantitative ultrasonic imaging
RCRA	Resource Conservation and Recovery Act
rGO	Reduced graphene oxide
RIA	Radioimmunoassay
RNA	Ribonucleic acid
RoHS	Restriction of Hazardous Substances regulations
ROS	Reactive oxygen species
RT-PCR	Reverse transcription-polymerase chain reaction
RuO_2	Ruthenium oxide
RZBs	Rechargeable zinc-based chemistry
SAED	Selected area electron diffraction
SC	Supercapacitors
SCIM	Squirrel cage inductor machine
SCLC	Space charge limited current
SCP	Spent coffee ground
SCs	Electrical Storage Supercapacitors
SDGs	Sustainable Development Goals
SELP	Silk Elastin like protein
SEM	Scanner Electron Microscopy
SERS	Raman scattering surface enhancer
SGH	Self-supporting graphene hydrogel
SHED	Human exfoliated deciduous teeth
SHT	Sensible Heat Storage
SMES	Superconducting Magnetic Energy Systems
SMM	Sustainable Materials Management
$SmVO_4$	Samarium vanadate
SOD	Superoxide dismutase
SOE	Solid oxide electrolysis
SOFC	Solid oxide fuel cell
SPEF	Solar photoelectro-Fenton
SPR	Surface plasmon resonance
SSA	Specific surface area
$_S$SGH	Self-supporting graphene hydrogel

STZ	Streptozotocin
SWCNTs	Single-walled carbon nanotubes
T1	Longitudinal relaxation
T1DM	Type 1 diabetes
T2	Transverse relaxation
TCS	Thermochemical Storage
TE	Tissue engineering
TEM	Transmission electron microscope
TES	Thermal energy storage
TMDs	Transition metal dichalcogenides
TMOs	Transition metal oxides
TOC	Total organic carbon
TRIS	Tris-hydroxymethyl amino-methane
TSE	Thermal storage of energy
U	Internal energy
UCNPs	Up-conversion inorganic nanoparticle
UNU	United Nations University
UST	Ultrasound scanning technology
V_2O_5	Vanadium oxide
VOCs	Volatile organic compounds
VONc	Nano-Vanadyl 2,3-naphthalocyanine
VRB	Vanadium Redox battery
VRFBs	Vanadium redox flow batteries
WCO	Waste cooking oil
WEEE	Waste electrical and electronic equipment
WHO	World Health Organization
WO_3	Tungsten trioxide
XRD	X-ray diffraction
ZAFs	Zinc Oxide Fuel Cells
ZnO	Zinc oxide
ZnT	New zinc transporter

Symbols

%	Percentage
$^{\bullet}OH$	Hydroxyl
$O_2^{\bullet-}$	Superoxide radicals
μm	Micrometer
α	Accumulation
Φb	Barrier height
K	Boltzmann constant
C-V	Capacitance–Voltage
μ	Charge carriers
q	Charge of electron
G-V	Conductance–Voltage
I-V	Current–Voltage
dd	Deep depletion
N_v	Density of states in valence band
d	Depletion
ϵ_s	Dielectric constant of silicon
E_g	Bandgap energy
H_c	Coercivity
hν	Light quantum
n	Ideality factor
D_{it}	Interface state density
mg	Milligram
mL	Milliliter
mM	Milli molar
M_r	Retentivity
M_s	Saturation magnetization
nm	Nanometer
V_{OC}	Open-circuit voltage
ϵ_0	Permittivity of vacuum
e^-/h^+	Photo-generated electrons/holes
R_S	Series resistance

$I/_{SC}$	Short circuit current
T	Temperature
v/e	Volume-to-energy ratio
w/e	Weight-to-energy ratio
λ	Wavelength of light

Goal 3: Good Health and Well-Being

Overview of the chapter:

Goal 3: Good Health and Well-Being

(Promoting well-being and a healthy lifestyle for every individual)

A description of advances in nanomaterials and biotechnology will be present in this part, which are aimed at achieving goal 3 in terms of satisfying human health-related and safety requirements, including diagnosing a variety of harmful and infectious diseases, and treating them using highly sensitive and technically sound equipment. The use of conducting nanobiomaterials (CNBs) in various biomedical applications has shown tremendous promise, including, regenerative medicine, and implantable and biosensor technology. A wide range of diseases and disorders can be diagnosed, managed, and treated better with CNBs. Indeed, CNBs are an intriguing area of biomedical research. These materials are undoubtedly going to be used in new and inventive ways by researchers who will ultimately provide patients with better healthcare. Furthermore, nanocellulose scaffolds participated in the differentiation and creation of new tissues along with the variation of stem cells. There are now numerous construction methods resembling 3D production that have remained very helpful in the process of creating bioactive nanocellulose scaffolds in the new century. Through the 3D manufacturing process, polymer scaffolds can be regulated near their microstructure, creating an effective microenvironment for cells to communicate and use energy effectively. Nanocellulose scaffolds with crosslinked properties therefore offer a wide range of potential applications in tissue engineering. Further improvements of nano-based scaffolds are also required for the application of cardiac tissue engineering in this review, which reveals several unsolved questions. Finally, the use of two-dimensional (2D) materials in biological sensors has brought exciting advances and innovations to the field. The use of 2D materials such as carbon materials, transition metal dichalcogenides (TMDs), and black phosphorus will enable us to find a variety of potential applications in clinical and analytical research. This may change how we detect, monitor, and treat diseases for the better. To begin, 2D materials have shown to be very sensitive, which has led to the creation of very accurate and quick biosensors. There is no question that biosensors are crucial in clinical and biomedical settings since they reduce diagnostic time, produce rapid results, are portable, are affordable, have high sensitivity and specificity, and have a low detection limit, allowing them to be used at any location where there is an emergence or natural disaster. In spite of the fact that clinical biosensors are prominent and enormous during COVID-19, industry forecasters expect the market to be worth more than $36.7 billion by 2036. Biosensors and their clinical applications will be discussed in this part.

Pioneering Role of Two-Dimensional Materials in Revolutionizing Biomedical Sensing

P. Baraneedharan, D. Shankari, and S. Beer Mohamed

Abstract A lot of people are interested in functional nanomaterials because of their unique sensing and optoelectronic capabilities, biocompatibility, chemical and environmental stability, and controllable form across several research domains, particularly within the healthcare sector. Biosensors possessing exceptional sensitivity, specificity, and a minimal detection threshold, enabling the quantification of biomolecules at nano/picomolar concentrations, hold significant relevance in the realm of medical sciences and the healthcare sector. The emergence of 2D materials has significantly expedited the investigation into producing cost-effective electrode materials because of its distinctive physical characteristics, such as its notable specific surface area, exceptional carrier mobility, elevated electrical conductivity, flexibility, and optical transparency. The hydrophilicity of 2D nanosheets is significantly boosted by the presence of surface functional groups, which in turn enables the chemical modification necessary for the development of a variety of 2D-based composites. These composites exhibit heightened sensitivity, making them particularly suitable for biosensor applications. This chapter provides a summary of the advancements in cholesterol sensing using 2D materials, as well as their combinations with organic, inorganic, and molecularly imprinted polymers (MIPs) for prospective detection purposes.

P. Baraneedharan · D. Shankari
Department of Electronics and Communication Engineering, Saveetha Engineering College, Thandalam, Chennai, Tamilnadu 602105, India

S. B. Mohamed (✉)
Department of Material Science, Central University of Tamil Nadu, Thiruvarur, Tamilnadu 610005, India
e-mail: sbmohammed@cutn.ac.in

K. Kasinathan et al. (eds.), *New Technologies for Energy Transition Based on Sustainable Development Goals*, https://doi.org/10.1007/978-981-97-2527-4_1

1 Introduction

Cholesterol is a waxy, fat-like substance that is essential for the proper functioning of the body. It can be derived from specific foods or is mostly created by the liver; yet, it is present in all cells. Hormones, vitamin D, and lipid-digestive assistance chemicals are all made possible by cholesterol. Cholesterol comes in two main forms: LDL and HDL. Each plays an important role in the body's lipid profile. The risk of cardiovascular disease and stroke is increased when LDL cholesterol levels are high, which is why it is commonly known as "bad" cholesterol. The "good" cholesterol, or HDL, helps filter out the harmful LDL cholesterol and lowers the risk of cardiovascular disease. When there is an imbalance between the levels of LDL and HDL cholesterol or when LDL cholesterol levels are excessively high, it can lead to the formation of plaque in the arteries. This condition, known as atherosclerosis, narrows and hardens the arteries, limiting the blood flow and increasing the risk of cardiovascular diseases. A bad diet high in saturated and trans fats, lack of exercise, smoking, obesity, and certain medical disorders can raise cholesterol. Genetics also affect cholesterol levels, as some people are predisposed to higher cholesterol. Maintaining good cholesterol levels helps cardiovascular health. Lifestyle adjustments and medicines can minimize the risk of heart disease and enhance health by regulating cholesterol.

Cholesterol, which is frequently vilified due to its association with cardiovascular illness, is, in fact, a remarkably diverse chemical with profound effects on the human body. It acts as a foundational building element for the formation of cell membranes, offering both solidity and flexibility, much like the pillars of a strong fortification. Aside from its function as a structural cornerstone, cholesterol converts into a flexible molecular workhorse, acting as a precursor for the manufacture of various important chemicals. Hormone production is one critical role. Cholesterol serves as the raw material for the manufacture of steroid hormones such as estrogen, testosterone, and cortisol, which regulate a variety of physiological processes ranging from reproduction to stress response. It is as if cholesterol is the brilliant composer orchestrating our bodies' hormonal symphony. Likewise, cholesterol is essential for the synthesis of vitamin D, which is important for bone health, immunological function, and overall well-being. Under the sun's beneficent gaze, cholesterol transforms into active vitamin D, much like an artist drawing inspiration from the sun to create a work of wellness art. Cholesterol is converted into bile acids, the unsung heroes of digestion, in the inner sanctuary of the liver. These acids are essential for emulsifying dietary fats, allowing fat-soluble vitamins to be absorbed, and breaking down fats into absorbable components. In addition, cholesterol plays an important role in brain function. It is abundant in neuronal cell membranes and helps to form synapses and release neurotransmitters, which are essential for cognitive processes, memory, and neurological health. It is like cholesterol is the conductor of a symphony, directing the beautiful interplay of neurons in the brain. However, as necessary as cholesterol is, the danger emerges when its levels in the bloodstream exceed safe levels. The recommended maximum level for total cholesterol is 5.5 mmol L^{-1}, assuming there are no additional risk factors present. If you possess cardiovascular risk factors such as hypertension, pre-existing cardiovascular ailment, diabetes, or engage in smoking, the target for LDL cholesterol levels should be below 2 mmol L^{-1}. Excess cholesterol is a significant risk factor for atherosclerosis, a disorder defined by the accumulation of cholesterol and other lipids in artery walls, resulting in plaques that restrict arteries and block blood flow. The implications of arterial congestion are severe. The presence of elevated cholesterol levels is a primary contributor to the development of cardiovascular diseases, including coronary artery disease, heart attacks, and strokes. As cholesterol accumulates in the arteries, it causes inflammation, draws white blood cells, and promotes the creation of plaques, which can rupture, resulting in blood clots that can block key blood passages. Furthermore, excessive cholesterol levels can worsen high blood pressure, which is another major risk factor for heart disease and stroke. As arteries narrow due to cholesterol buildup, resistance to blood flow increases, raising pressure against artery walls. Cholesterol blockages are not limited to the heart; they can affect arteries all over the body, potentially leading to illnesses like peripheral artery disease. Pain, suffering, and tissue damage result from this condition's reduction of blood flow to the extremities. Consequently, cholesterol sensing is a cornerstone of modern healthcare, tackling the need to closely monitor and expertly manage cardiovascular health.

In an epoch where cardiovascular diseases persist as a foremost global health exigency, the importance of cholesterol surveillance assumes paramount significance. Cholesterol, specifically the low-density lipoprotein (LDL) variant, assumes a pivotal role in the genesis of atherosclerosis, a pathological condition characterized by the accretion of atheromatous plaque within arterial lumina. This plaque constricts arterial passages, obfuscating hemodynamic flow, and gravely escalating the proclivity towards myocardial infarctions and cerebrovascular accidents. Thus, routine cholesterol monitoring becomes an intrinsic cog in the healthcare machinery, allowing medical practitioners to holistically appraise an individual's cardiovascular jeopardy. Through the judicious assessment of cholesterol levels, healthcare providers can adroitly tailor interventions, encompassing lifestyle adaptations, and pharmacological regimens, to preemptively mitigate the peril of cardiovascular afflictions. The technological milieu underpinning cholesterol sensing has borne witness to a staggering confluence of advancements, bequeathing precise, expeditious, and accessible modalities for cholesterol quantification. While conventional methodologies like enzymatic assays and electrochemical biosensors retain their utility for routine lipid profiling, recent strides have profoundly diversified the landscape of cholesterol detection. Portable, point-of-care cholesterol sensing apparatuses have emerged as a disruptive force within the domain, capitalizing on microfluidic architectures and cutting-edge biosensor technologies to confer swift outcomes with nominal blood samples. These instrumentalities empower individuals with the convenience of recurrent cholesterol scrutiny, thereby galvanizing proactive stewardship of cardiovascular health and endowing accessibility to a broader spectrum of clientele.

Concomitant with the pivotal role of cholesterol sensing, the acuity of 2D materials in scientific and technological domains has galvanized substantive interest. Exemplified by graphene and transition metal dichalcogenides, these materials exhibit an array of extraordinary properties, encompassing prodigious mechanical robustness, augmented electrical conductivity, and superlative optical transparency. In the sphere of cholesterol sensing, investigators are fervently assimilating 2D materials into biosensors to enhance their discernment and precision. Leveraging these materials' distinctive attributes, the biosensors attain heightened acumen in the rapid and precise identification of cholesterol moieties within biological substrates. Consequently, the amalgamation of 2D materials augments the fidelity of cholesterol profiling, endowing it with a more potent capacity to underpin efficacious healthcare interventions. The significance of two-dimensional materials in cholesterol sensing cannot be emphasized. These exceptional materials, such as graphene and transition metal dichalcogenides, have a distinct combination of characteristics that improve the accuracy and efficiency of cholesterol detection dramatically. Thus, the use of cholesterol sensors has a substantial influence on the realm of preventative and therapeutic healthcare, enabling the timely identification of diseases and customized interventions. Nevertheless, it is vital to acknowledge the significance of tackling the obstacles associated with precision, availability, and technological progress in order to augment their efficacy in the management of cardiovascular ailments and enhance patient results. As shown in Fig. 1, because of their great mechanical strength, electrical conductivity, and other significant properties of high surface area, excellent mechanical stability they are suitable candidates for enhancing the sensitivity and specificity of cholesterol sensors. Researchers are paving the door for more precise and quick identification of cholesterol molecules in biological samples by adding 2D materials into biosensors. This development not only improves the accuracy of cholesterol profiling, but it also has the potential to change healthcare interventions, promoting better cardiovascular health and overall well-being.

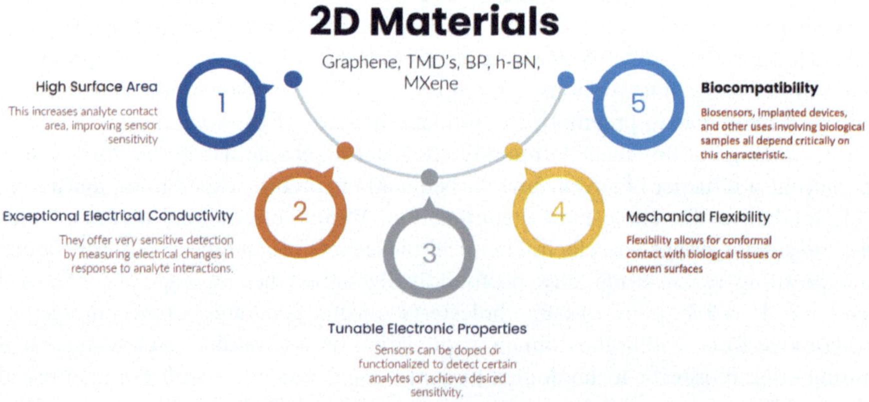

Fig. 1 Properties of 2D materials that are far better and provide a significant boost to the sensor design

2 Cholesterol Sensing Mechanism

Through the use of bioreceptors, it is possible to acquire quantitative analytical data from these sensing devices. There are a number of ways to detect changes in electrical signals caused by an analyte's redox reaction. By comparing the analyte's properties to a reference electrode, electrochemical sensors can identify changes brought about by electron gain or loss. Instead of using direct redox reactions, some electrochemical devices use a different method called indirect or assisted electron transfer. Mediators help move electrons from the surface of the electrodes to the location of the reaction. Electrochemical cholesterol sensors are severely limited in their operating performance by electrochemical reactions taking place at the electrode surface. Therefore, the electrochemical sensing capabilities for cholesterol are highly dependent on the electrode material properties [1]. Biosensors for cholesterol may be broken down into three broad categories: enzymatic, nonenzymatic, and redox mediator. Electron transfer occurs between the enzyme and the working electrode in enzymatic biosensors when hydrogen peroxide is produced during the enzyme reaction when analyte is present. Cholesterol oxidase (ChOx) is an enzyme that contains flavin-adenine dinucleotide (FAD) and is responsible for catalyzing the oxidation of cholesterol using molecular oxygen. Cholesterol esterase (ChE) is an enzymatic protein that initiates the hydrolysis of many fatty acid esters derived from cholesterol, hence playing a crucial role in the quantification of total cholesterol levels. Cholesterol is oxidized by ChOx, resulting in the production of cholest-4-en-3-one and hydrogen peroxide (H_2O_2). Typically, cholesterol biosensors rely on the electrochemical processes of hydrogen peroxide (H_2O_2) reduction or oxidation, which has the advantage of obviating the need for horseradish peroxidase (HRP) or redox mediators. The scientific community has shown significant interest in nonenzymatic electrochemical sensors due to the drawbacks associated with traditional enzymatic sensors, such as their low stability and storage issues. Electroactive compounds, namely MB, FC, and HQ, are employed as mediators in the advancement of mediator-based enzymatic or nonenzymatic electrodes. The redox species function as electron mediators and establish adducts with cholesterol. The species that undergo reduction in this process have the capability to be regenerated by electrochemical means, as illustrated in the subsequent mechanism.

$$M_{red} \rightarrow M_{ox} + ne^-$$

$$M_{ox} + \text{Cholesterol} \rightarrow M_{ox} \cdot \text{Cholesterol}$$

$$M_{ox} \cdot \text{Cholesterol} \rightarrow M_{red} + \text{Product},$$

where M_{red} and M_{ox} are the redox species in the reduced and oxidized form, respectively. The electrochemical signal may be detected using electrochemical methods, as it is influenced by the displacement between the analyte and the mediator. Since they are sensitive and selective, cholesterol oxidase (ChOx) and cholesterol esterase (CE) are commonly used in enzymatic cholesterol sensing. Enzymes often exhibit superior catalytic activity compared to chemical catalysts. Moreover, enzymes possess the advantageous ability to be readily immobilized and effectively used in the process of biosensor design.

3 Classification of 2D Materials

There exist several layered materials characterized by robust in-plane chemical bonding and little interlayer coupling, as seen in Fig. 2. The presence of these stratified arrangements affords the potential for the separation of discrete autonomous atomic layers. These layers with one dimension strictly restricted to a single layer are called two-dimensional (2D) material. The density of states for 2D materials exhibits a quasi-continuous step-like increase with increasing excitation energy. These changes in the density of states for exciting electrons alter the way excitation energy interacts with the valence electrons, thus changing the fundamental properties with size as well as shape in quantum-confined nanomaterials.

Specific geometries give these materials shape-dependent properties and successful sensing device use. Two-dimensional materials with single-atom or single-polyhedral layers are listed below.

(i) *Layered van der Waals solids*

These are the most common exfoliating materials for mono- and few-layered nanosheets. These materials have strong in-plane covalent or ionic bonding and weak van der Waals or hydrogen bonding. Common graphite exfoliation solvents have weak van der Waals bonding (40–70 meV) and surface tension (60–90 mJ m^{-2}), allowing micromechanical or liquid exfoliation of bulk parent crystals into nanosheets [2, 3]. Exfoliated nanosheets feature high aspect ratios and surface areas. Exfoliated nanosheets have lateral dimensions of a few lm and a thickness of less

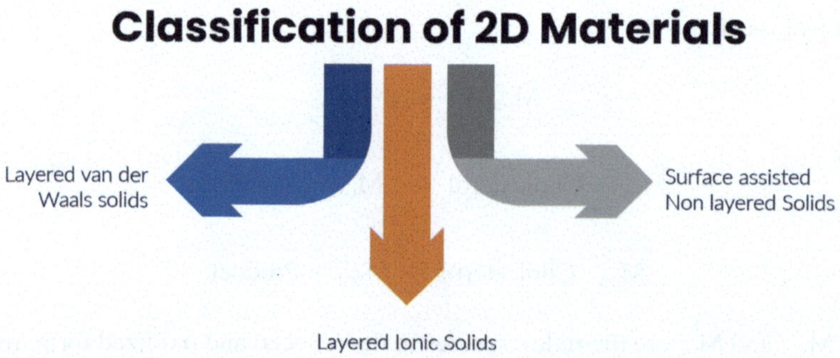

Fig. 2 Categorization of 2D materials is contingent upon their structural characteristics and surface properties

than 1 nm. MX2 (X-M-X layer) where M = Mo, Bi, Ta, V, Ti, Nb, W, Hf, Zr and X = S, Se Te) is a well-studied van der Waals solid layering system, especially for MoS_2, $MoSe_2$, and WS2. A stack of over 40 TMDs has been documented [4]. Single and few layer derivatives' exotic direct and indirect bandgap, Dirac point, electrochemical, and mechanical properties make them twenty-first-century materials for semiconductors, hydrogen production, thermoelectric devices, and more. Transition metal occupies octahedral (d0, d3, and some d1 metals) and trigonal prismatic (d1 and d2) coordinates in IT and 2H crystal structures. Sharing edges with neighbours creates a hexagonal honeycomb-like structure within each layer [3, 4]. H-BN, SiC, vanadium oxide, and Sb 2Te3 are also in this class.

(b) *Layered ionic solids*

Another important family of stacked 2D materials that is gaining interest is layered ionic compounds. Electrostatic forces sandwich a charged polyhedral layer between hydroxide or halide layers in these materials. Ion intercalation or ion exchange liquid exfoliation can easily exfoliate single and few-layered materials. The adsorption of bulkier foreign molecules, as well as the exchange of parent ions for heavier ions, increases the gap between the basal planes and reduces the basal plane bonding. Layered ionic solids are well represented by exfoliated europium hydroxide [5]. Perovskite-type oxides such as $La_{0.90}Eu_{0.05}Nb_2O_7$ [6], $K_{1.5}Eu_{0.5}Ta_3O_{10}$ [7], $KLnNb_2O_7$, $K_2Ln_2Ti_3O_{10}$ and $RbLnTa_2O_7$ (Ln: Lanthanide Ion) [8], $KCa_2Nb_3O_{10}$ [9, 10], metal hydroxides such as $Eu(OH)_{2.5}(DS)_{0.5}$ [6], $CO_2/_3Fe_{1/3}(OH)_2 1/3^+$ [11] are some examples of layered materials exfoliated from the ion exchange method. Because of the adsorption of solvent and bulkier transferred ions, the observed thickness of monolayers is frequently greater than the crystallographic thickness.

(iii) *Surface-assisted nonlayered solids*

Chemical vapour deposition and epitaxial growth are used to create this class of atomically thin layered material on a substrate. Silicene, a graphene counterpart, is a prime example of surface-assisted nonlayered solids. Si atoms with a low buckled hexagonal shape have been placed on Ag (111) [12], Ir(111) [13], and a thin film of ZrB_2 on silicon wafer [14]. In contrast to graphene, the instability of silicene at room temperature has limited its promise in electrical and optoelectronic applications goes into great depth about silicone. Al_2O/SiO_2 [15], Cu_2N/Cu (100) [16], MgO/Mo (001) [17], TiO_2 [18], and Ge/Ag (100) and Ag (111) [1, 19–22] nanosheets are created for use in energy storage, field effect transistors (FET), and modulating the charge density of nanoparticles.

4 2D Materials in Cholesterol Sensing

2D materials exhibit unique electrical properties that organic compounds often lack. Graphene, for example, demonstrates high electron mobility and exceptional electrical conductivity. These properties facilitate rapid and efficient charge transfer between the cholesterol molecules and the sensor surface. The efficient charge transfer not only improves the sensitivity of the sensor but also leads to a faster response time. This means that cholesterol levels can be detected and measured more quickly, enabling real-time monitoring and timely intervention. Moreover, the electrical conductivity of 2D materials allows for direct electrical readout, eliminating the need for complex chemical reactions or signal transduction mechanisms. This simplifies the sensing process, making it more straightforward and reliable. In contrast, organic compounds often require additional steps or chemical modifications to enable efficient detection and measurement of cholesterol levels. Another advantage of 2D materials is their exceptional mechanical and chemical stability. They are more robust, and durable compared to organic compounds, which can be prone to degradation or instability over time. The stability of 2D materials ensures the longevity and reliability of the cholesterol sensing platform, enabling long-term use and accurate monitoring of cholesterol levels.

Graphene and MXenes are two-dimensional materials with remarkable properties and distinctive features that make them very valuable in sensing applications. Graphene, a single layer of carbon atoms arranged in a hexagonal lattice, exhibits remarkable electrical conductivity, high mechanical strength, and large surface area. These properties make it an ideal material for sensing. Graphene-based sensors have shown excellent sensitivity and selectivity in detecting a wide range of analytes, including gases, biomolecules, and pollutants. Its large surface area allows for efficient analyte interaction, while its electrical conductivity enables the transduction of sensing signals. MXenes, a class of 2D transition metal carbides/nitrides, also offer great potential in sensing applications. MXenes possess a layered structure with metal atoms sandwiched between carbon or nitrogen layers. This structure gives them unique electrical, mechanical, and chemical properties. Several analytes, such as gases, ions, and proteins, have been sensed by MXenes. Their massive surface area, adjustable surface chemistry, and high electrical conductivity make them ideal for selective and sensitive detection. MXene-based sensors have demonstrated excellent performance in terms of response time, sensitivity, and stability. The importance of 2D materials in sensing lies in their ability to enhance sensitivity, selectivity, and performance compared to traditional sensing materials. Their unique properties allow for efficient interaction with analytes and the transduction of sensing signals. Furthermore, 2D materials offer opportunities for functionalization and integration into diverse sensing platforms. With ongoing research and development, the integration of 2D materials into sensing devices holds great promise for various applications, including environmental monitoring, healthcare, and wearable sensing technology Furthermore, 2D materials offer the possibility of functionalization with various biomolecules, enzymes, or aptamers, enabling selective cholesterol recognition. By modifying the surface of 2D materials, specific interactions between cholesterol and the sensing platform can be achieved, enhancing the selectivity of the sensor. This selectivity ensures that the sensor responds specifically to cholesterol molecules, minimizing interference from other substances present in complex biological samples. 2D materials outperform organic compounds as cholesterol recognizers in cholesterol sensing applications due to their high surface area, unique electrical properties, mechanical and chemical stability, and the potential for selective functionalization. The enhanced sensitivity, faster response time, simplified sensing process, and improved selectivity provided by 2D materials make them superior choices for cholesterol sensing platforms, leading to more accurate and reliable cholesterol level measurements.

A number of chemicals have been shown to be very useful as recognition factors for systems that detect cholesterol. For instance, materials based on β-cyclodextrin (²-CD) could be used well because ²-CD itself could interact with cholesterol as a host or a guest [23, 24]. β-CD has a hydrophobic inner pocket that helps guest molecules stick to it and a hydrophilic outer surface that makes it easier for water to dissolve. Huang et al. [24] suggested thiolates β-CD as the main part of a cholesterol sensor system, methylene blue (MB) as a way to move electrons around, and an Au NPs-functionalized gold electrode as a part that conducts electricity. In order for the electrode to work, the MB that was in the β-CD host was specifically replaced by cholesterol when it was put into a cholesterol solution. MB moved from the electrode surface to the fluid and made the redox signal stronger enough. It could measure a wide range of concentrations, from 2×10^{-8} to 5×10^{-5} mol L^{-1}, and had a low error rate of 2×10^{-9} mol L^{-1}. Interfering substances that are normally found in human blood had no effect. The normal addition method was used to analyse real samples with DPV. For three blood samples, the recovery for figuring out cholesterol ranged from 96.5 to 103.6%. As a result, the sensor was said to be suitable for tracking cholesterol levels in living things. As an extra general method for finding cholesterol using ²-CD, Gr can be used to better conductivity. Papers [25–27] showed how this method works. Agnihotri et al. [25] made it clear that Gr helps to make the making process easier and improves how electrons are transferred by MB. The DPV response was straight from 0.001 mmol to 0.10 mmol L^{-1}, with a limit of detection (LOD) below 1 mol L^{-1} and a sensitivity of 0.01 A mol^{-1} L^{-1} cm^{-1}. Interference was found to be very small, but cholesterol detection in real items was not done. Yang et al. [27] used a *N*-acetylaniline (NAANI) film to keep β-CD in place and stop MB from sticking to Gr without being specific by using a π–π stacking interaction. So, in [25], the host–guest interaction happened in the electrochemical cell, not on the electrode surface. This is different from [27], where the interaction happened directly on the electrode surface. The DPV reaction of the sensor was straight from 1 to 50 mol L^{-1}, and the low-level detection (LOD) was 0.5 mol L^{-1}. Species that got in the way, like ascorbic acid, glucose, sodium dodecyl sulphate, and different chlorides (sodium, potassium, and magnesium), were not found to have any effect. The suggested sensing method was also successfully used for tests on human blood samples. In this way, the monitor was described as a useful medical tool. The work detailed above was logically built upon in [26]. MWCNTs that had been functionalized with benzoic acid were deposited on SPE. The electrode that was made was then submerged in β-CD water solution to make a layer that was uniform and immobilized. With a LOD of 0.5 nmol L^{-1}, the electrode that was made responded to DPV over a wide linear range, from 1 nmol L^{-1} to 3 mol L^{-1}. More research into how the most common interferents affect the sensor showed that it is very selective. Something that could work like an enzyme was proven to be possible. Because metal active cores work like enzymes, these kinds of mimic materials are mostly made up of metal-containing materials [28–31]. However, flavine adenine dinucleotide (FAD) is an active centre in cholesterol oxidase, which is used to identify cholesterol enzymatically. It was proven that some glycoluril derivatives (GD) with FAD-like structures can help electrolyze cholesterol. Both projects are based on the electrochemical deposition of a certain GD on a carbon-based hybrid electrode with a surface that can be used again and again (CERS). It was used to recognize cholesterol in [32], and diphosphonic acid of 2,6-diacetylglycoluril (DPADGU) was used as a detecting core in [33]. The major benefit of this method is that it is easy to make and the surface can be renewed quickly, which is important for the electrode's antifouling properties. The main problem with the suggested electrodes is that they need a fairly high overvoltage (+ 0.98 V in [32] and + 1.1 V in [33]), which is an advantage when it comes to construction. Also, a phosphate buffer with a pH of 6.86 was used as an electrolyte support, which meant that extra detergent had to be added to dissolve the cholesterol. Even so, the electrodes we've talked about have good linear ranges from 0.1 to 50 mol L^{-1} [32] and from 1 to 200 mol L^{-1} [33]. They also have very good sensitivity (20 A mol^{-1} L cm^{-2} [33]) and high selectivity, meaning they don't get messed up by ascorbic acid, uric acid, lactic acid, lactose, albumin, retinyl, or copherol acetates. The LODs were 0.01 [32] and 1.5 mol L^{-1} [32]. The electrode described in [32] was used to test how well it could measure cholesterol in human blood serum, and the electrode described in [33] was used to find cholesterol in dairy goods. However, 2D materials are clearly better than organic compounds when it comes to recognizing cholesterol in cholesterol sensing uses. This means that they are better in many ways. Their large surface area is a big plus. Graphene, transition metal dichalcogenides (TMDs), and black phosphorus are all 2D materials that have a lot of surface area compared to organic products. The sensor is more sensitive because it can deal with more cholesterol molecules because its surface area has grown. The sensor is much better at finding things because cholesterol and the 2D material surface interact and connect with each other more. A number of processes in the detecting mechanism of 2D materials on cholesterol make it possible to find and measure cholesterol molecules. To start, a sensing base or device is used to make the 2D material. This could be graphene or transition metal dichalcogenides (TMDs). A lot of the time, the surfaces of 2D materials are changed or functionalized to make them work better with cholesterol molecules [34–36]. When a sample with cholesterol is put on the detecting device, the molecules of cholesterol interact with the surface of the 2D material. It's possible that the surface of the 2D substance and the cholesterol combine in more than one way. Physical adsorption is one way. Weak van der Waals forces make cholesterol molecules stick to the surface of the 2D material. Chemical processes are another type of mechanism. Because of this, some functional groups in cholesterol molecules might connect with functional groups on the surface of 2D objects to create hydrogen bonds or electrostatic interactions. It's easier for cholesterol to stick to the surface of the 2D material because of these interactions. It is possible for cholesterol to bind to 2D objects and change their electrical, optical, or mechanical properties. It is possible to measure these changes and link them to the amount of cholesterol in the body. The sense process is different depending on the type of 2D material and the detection method used. Adsorption of cholesterol changes the conductivity or resistance of the 2D material when used with electronic devices. For example, graphene-based devices use the fact that when cholesterol molecules bind to the surface of graphene, they change how electricity flows through it. By keeping an eye on the voltage or current flowing through the sensor, one can measure the change in conductivity and figure out how much cholesterol is present. Optical sensing methods depend on changes in the 2D material's optical properties that happen when cholesterol binds to it. These changes can be in how light is absorbed, emitted, or scattered. For example, TMDs have a lot of photoluminescence, and cholesterol can either weaken or boost the light that is emitted by the TMD surface, based on how it interacts with it. This change in light output can be measured and named, which tells us something about the cholesterol level. Mechanical sense is the process of keeping an eye on how the 2D material's mechanical properties change as cholesterol binds to it. Cholesterol binding can cause changes in the way strain or stress is distributed in graphene because it is very flexible and stretchable. To find out if cholesterol is present and how much of it there is, piezoresistive sensors or microcantilever-based devices can be used to measure these changes [37, 38].

Molecularly imprinted 2D (MIP-2D) materials are able to offer specific recognition sites for target molecules because they integrate molecular imprinting principles with the two-dimensional structure of the material. There are mainly three types of imprinting procedures: As illustrated in Fig. 3, there are three types of imprinting: (a) covalent and semi-covalent, (b) metal-ion exchange, and (c) non-covalent. A multi-step procedure involving template molecule selection, monomer selection, polymerization, and template removal is used to manufacture MIP-2D materials. Selecting the template molecule—the starting point for imprinting the material—is the first step in creating MIP-2D materials. The application dictates the choice of the template molecule, which can be anything from a small organic molecule to a large biomolecule or macromolecule. Afterwards, a suitable monomer or mixture of monomers is selected according to its ability to engage with the template molecule through non-covalent interactions, like electrostatic interactions, hydrophobic interactions, or hydrogen bonding. It is common practice to use monomers because they include functional groups that can bind to the template molecule. The next step, after choosing the monomers, is to polymerize them while the template molecule is present.

Polymerization can be accomplished by a variety of methods, including bulk polymerization, precipitation polymerization, and surface-initiated polymerization. As a result, a polymer matrix containing imprinted cavities with the shape, size, and chemical functionality required for selective binding to the template molecule is formed. Following polymerization, the template molecule is removed from the MIP-2D material, leaving gaps or imprinted sites that retain the template's molecular memory. This step is critical for obtaining selectivity because it develops recognition sites capable of binding to the template molecule or molecules with similar structural properties. The final MIP-2D material has a high affinity and selectivity for the template molecule. Chemical sensing, separation and purification processes, medication delivery systems, and catalysis are only a few of the uses. Because these

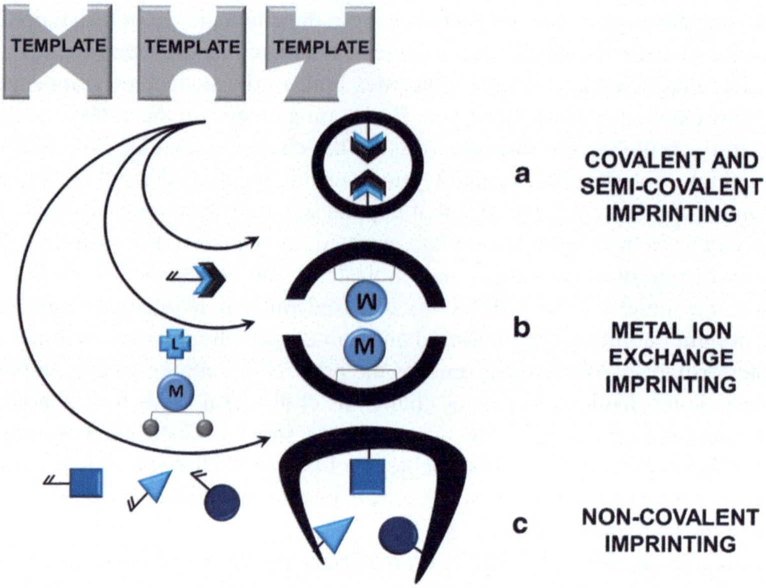

Fig. 3 Classification of Basic imprinting processes: **a** covalent and semi-covalent imprinting, **b** metal-ion exchange imprinting, and **c** non-covalent imprinting

materials are two-dimensional, they have advantages such as greater surface area, accessibility to target molecules, and potential incorporation into devices or coatings. In terms of cholesterol sensing, MIP-2D materials have several important advantages. Due to the establishment of specialized recognition sites via molecular imprinting, they demonstrate extraordinary selectivity towards cholesterol, even in the presence of structurally identical compounds. By reducing interference from other substances, this selectivity ensures reliable readings in complicated biological samples. Additionally, MIP-2D materials' specialized recognition sites and higher binding affinities make it possible to detect cholesterol accurately, even at low concentrations [39, 40].

For the early detection and monitoring of illnesses linked to high cholesterol, this sensitivity is essential. Additionally, MIP-2D materials have good stability, holding onto their binding qualities over time. Because of the carefully chosen monomers and polymerization processes employed during production, the final product is structurally sound and may be used again without suffering considerable degradation. They are more durable and have a longer shelf life because of their resistance to environmental influences. Due to their adaptability, MIP-2D materials can be used with a variety of cholesterol sensing platforms. They can be combined with other transduction mechanisms, including electrochemical, optical, or mass-based ones, to create a variety of cholesterol detecting devices. The two-dimensional structure of MIP-2D materials also enables miniaturization, making them appropriate for small and portable cholesterol sensing systems, such as wearable or implanted ones.

5 Device-Based Measurement of Cholesterol

The industry has been revolutionized by the development of miniaturization of cholesterol sensing devices since it has enabled the manufacture of compact, portable devices that may be utilized for point-of-care testing. In contrast to traditional laboratory-based testing methodologies, these little devices provide several advantages. The mobility of the subject under discussion is a significant benefit. Portable cholesterol measurement devices have been developed, which exhibit reduced dimensions and weight. These devices excel in conducting measurements on-site within diverse healthcare environments, including clinics, medical practices, and even remote areas with little laboratory infrastructure. Their portability facilitates seamless relocation between different sites. The mobility of cholesterol testing allows for simple and expeditious procedures, eliminating the necessity of transporting samples to a centralized laboratory and reducing the duration required to obtain test results. In contrast to traditional laboratory-based testing methodologies, miniaturized devices provide several advantages. These devices are suitable for fingerstick blood samples and other minimally invasive collection procedures due to their ability to accommodate lower sample volumes. Furthermore, the process of miniaturization enables the integration of several components, including sample preparation, detection, and data processing, onto a singular device. This amalgamation facilitates the acceleration of testing procedures and obviates the necessity for arduous and time-intensive laboratory protocols. An important advancement in the field of cholesterol detection is the integration of sensors into wearable technologies. The advent of wearable sensors has enabled the continuous monitoring of cholesterol levels, providing users with real-time data and comprehensive analysis of daily fluctuations and long-term patterns. Individuals are able to enhance their understanding of their cholesterol profiles due to the ongoing monitoring capabilities, so empowering them to assume control over their health. Wearable cholesterol sensors have the capability to be integrated into various forms like wristbands, patches, or smart garments, among other potential applications. These devices are designed to possess compact dimensions, lightweight construction, and inconspicuous appearance, therefore enabling individuals to comfortably wear them throughout the day without impeding their regular pursuits. Incorporation of molecularly imprinted 2D (MIP-2D) materials, which possess a targeted affinity for cholesterol molecules, is a prevalent practice in the integration of sensors. The wearable sensor provides continuous monitoring of cholesterol levels by detecting fluctuations in cholesterol content within sweat or interstitial fluid. The sensor employs MIP-2D materials that include specific recognition sites exclusively designed to interact with cholesterol molecules present in the fluid. The sensor is capable of monitoring fluctuations in cholesterol levels and generating an electrical or optical signal in response. The signal undergoes processing and is then shown as real-time data on the wearable device, or alternatively, it can be transmitted to a smartphone or other interconnected devices. Individuals can monitor their cholesterol levels continuously throughout the day and identify patterns or trends, owing to the availability of real-time data provided by wearable cholesterol sensors.

Individuals may utilize this acquired information to enhance their ability to regulate their cholesterol levels by informed decision-making regarding their lifestyle choices, including dietary patterns and physical activity. For example, in the event of a significant increase in cholesterol levels, individuals have the option to promptly modify their dietary habits or see healthcare professionals for assistance. Furthermore, the utilization of wearable cholesterol monitors enables the implementation of individualized healthcare measures. The data collected by the sensors may be examined by healthcare practitioners, who can afterwards provide recommendations or therapies according to the individual's cholesterol profile. The use of a personalized approach to managing cholesterol levels has the potential to provide more effective therapeutic outcomes and improved overall health outcomes [39, 41, 42].

In addition to cholesterol monitoring, wearable technology often encompasses supplementary functionalities such as heart rate monitoring, activity tracking, and sleep monitoring. A wearable device with many health monitoring capabilities empowers individuals to oversee their entire health and well-being conveniently and comprehensively. The use of portable electronics in the form of smartphone-based cholesterol sensor devices has emerged as a significant advancement in the field. These systems offer a simplified and convenient approach to cholesterol monitoring. The detection of cholesterol is commonly facilitated by systems that employ either a smartphone attachment or an external sensor to interface with molecularly imprinted 2D (MIP-2D) materials. The smartphone serves as the primary component responsible for processing, analysing, and displaying information inside smartphone-based systems. The incorporated external sensor or attachment of MIP-2D materials functions as the cholesterol sensing element. The sensor can identify and capturing cholesterol molecules, afterwards transmitting the resulting signal through either a cable or wireless connection to the smartphone. Smartphone-based solutions leverage the connection and computational capabilities of smartphones to provide clients with instantaneous outcomes. Specialized mobile applications or software are utilized to process the data, resulting in the generation of cholesterol concentration readings. The findings are shown on the screen of the smartphone, providing individuals with immediate accessibility to their cholesterol data. Smartphone systems facilitate the simplification of data sharing and administration processes. To facilitate further investigation and ensure continuous surveillance, the obtained cholesterol data may be securely kept on the mobile device or transmitted to cloud-based platforms. The application of lab-on-a-chip technology has significantly impacted cholesterol detection by the consolidation of several components onto a single microfluidic chip, hence enabling miniaturization and integration. The integration of sample preparation, mixing, and analysis inside a compact and portable equipment has revolutionized the conventional approach to cholesterol testing. The integration of lab-on-a-chip technologies with molecularly imprinted 2D (MIP-2D) materials offers several advantages in the assessment of cholesterol. The miniaturized architecture of lab-on-a-chip technology allows for reduced sample amounts and accelerated analysis times. The microfluidic channels of the chip have been designed to provide efficient fluidic manipulation, hence enabling rapid and precise measurements of cholesterol levels. The integration of several functional elements onto a single microchip is a fundamental aspect of lab-on-a-chip technology. Some of the components include the modules for sample loading and preparation, mixing chambers, reaction chambers, and detecting elements. Lab-on-a-chip technologies streamline the process of cholesterol sensing by integrating these functionalities onto a single chip, eliminating the requirement for supplementary apparatus, and reducing the complexity of analysis. The integration of MIP-2D materials into the lab-on-a-chip platform holds promise for enhancing cholesterol detection due to their distinctive binding properties. The components are rendered immobile within the reaction chambers or microfluidic channels, whereby they selectively extract the cholesterol molecules from the given sample [43].

The combination of cholesterol sensors with wireless connection Internet of Things (IoT) systems has made a big difference in how cholesterol is tracked and managed as shown in Fig. 4. This makes it possible to remotely watch, send data, and track cholesterol levels in real time. This means that healthcare providers can move quickly and individually based on each patient's cholesterol profile.

By adding cellular connection to the sensors, the data they collect can be sent to a central system or a platform in the cloud. So, doctors and nurses can access and watch real-time cholesterol data from afar without having to be there in person or manually pull the data. The digital link allows for perfect and constant data transfer, so doctors always have the most up-to-date information about the patient's cholesterol levels. Internet of Things (IoT) systems are used a lot for wireless communication and to connect cholesterol monitors. These devices have the space and tools needed to collect, store, and examine cholesterol data in a safe way. IoT platforms make it easy to connect different devices, so cholesterol gauges can talk to different healthcare systems and apps. With real-time tracking of cholesterol levels through Wi-Fi and

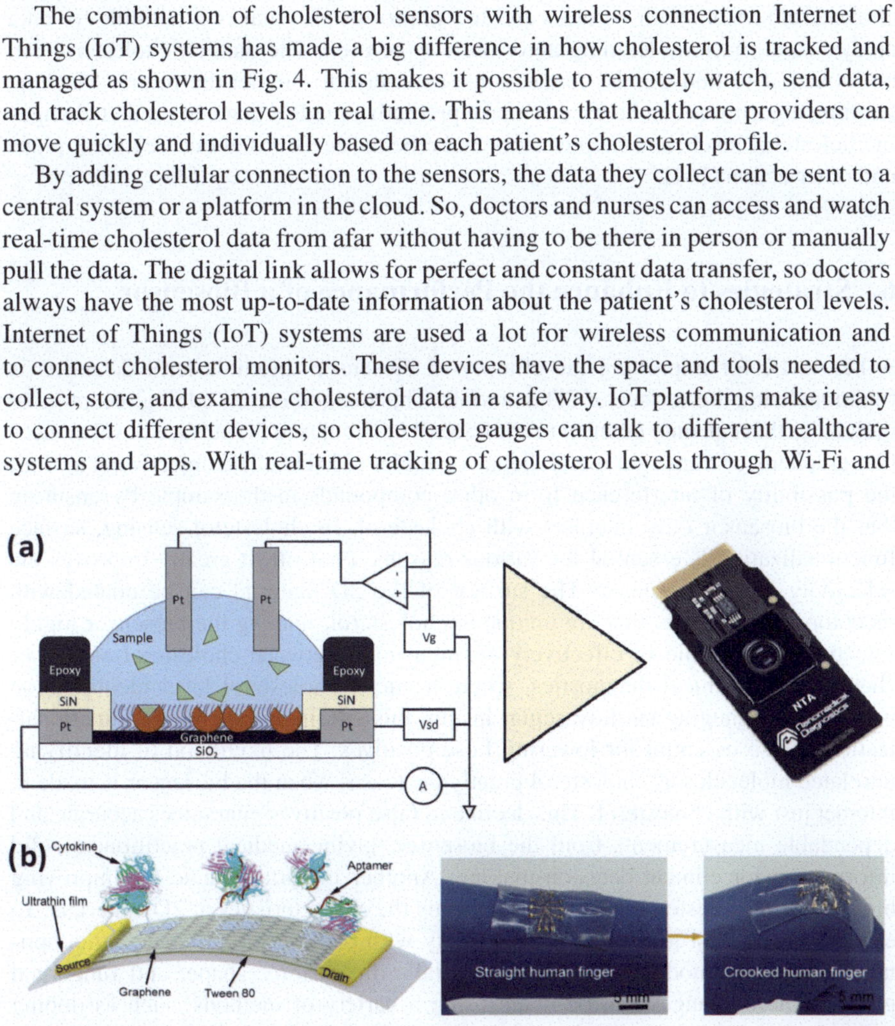

Fig. 4 Biosensors that are wearable, remote, and portable. **a** Design, production, and integration of a FEB biosensor into a small biosensor. **b** Creation of a flexible FET support for a skin biosensor

IoT connections, doctors can quickly spot trends, patterns, and changes in a patient's cholesterol levels. By keeping an eye on cholesterol data daily, healthcare workers can catch changes or oddities in cholesterol readings and act quickly. This makes it possible to take care of diseases linked to high cholesterol in a cautious way and to tailor treatments to each patient's cholesterol profile. The ability for wireless and IoT to work together to allow remote tracking also gives people convenience and freedom. Patients can use cholesterol sensors that they can wear or watch at home with gadgets that send data directly to the system of their healthcare provider. This means that people do not have to go to the hospital as often, and they can still get the care and monitoring they need while going about their daily lives. The integration of cholesterol sensors with wireless and IoT platforms also makes it possible for data analysis and machine learning algorithms to be used. The data on cholesterol can be looked at to find trends, connections, and ways to predict the future. Machine learning algorithms can be used to make predictive models that can predict changes in cholesterol levels and help doctors make better decisions about treatment plans and other actions [44].

6 Strategies to Enhance the Performance of a Biosensor

A key tactic for improving the efficacy of biosensors in cholesterol detection is functionalizing the surface of 2D materials. This method entails adding receptors or molecules that can only connect to cholesterol molecules to the surface of substances like graphene or transition metal dichalcogenides (TMDs). The objective is to limit the possibility of interference from other compounds in the sample by ensuring that the biosensor only interacts with cholesterol. In cholesterol sensing, surface functionalization is essential for various reasons. First off, it greatly improves the selectivity of the biosensor. The surface of the 2D material can be coated with receptors or molecules that are unique to cholesterol, making the biosensor highly selective and capable of effectively distinguishing between cholesterol and other chemicals. In clinical diagnostics, where accurate cholesterol level identification is crucial for gauging cardiovascular health, this specificity is crucial. Functionalization is also essential for lowering false positives. The likelihood of identifying unrelated molecules as cholesterol greatly decreases when the biosensor is made to interact just with cholesterol. This decline in false positives guarantees accurate and dependable measurements from the biosensor, giving medical practitioners solid information for clinical decision-making. Another important tactic for improving biosensors in cholesterol detection is tailoring the characteristics of 2D materials. To achieve the highest level of binding affinity with cholesterol molecules, this optimization entails modifying the 2D material's dimensions, shape, and functional groups. These changes can be made using a variety of methods, such as doping and chemical alteration. The improvement of the substance's affinity for binding cholesterol is a key component of this method. The 2D material can capture and hold cholesterol molecules more efficiently thanks to careful design. This higher affinity assures that cholesterol molecules will more easily adhere to the surface of the biosensor, enhancing its sensitivity and accuracy. The stability and lifespan of the material can also be affected by adjusting its qualities. Long-term cholesterol monitoring can benefit from the use of biosensors that incorporate 2D materials with improved characteristics since they are more durable and can sustain their function for longer periods of time [45]. Signal amplification in biosensors for cholesterol sensing can be effectively accomplished by taking use of the remarkable electrical conductivity of 2D materials. These substances, especially graphene, have exceptional electrical conductivity, which makes them the best choice for increasing electrochemical signals produced by cholesterol binding or related events. Electrical signals are produced when cholesterol molecules interact with the surface of the biosensor or go through processes. These signals might not be very strong, particularly when cholesterol levels are low. However, these signals can be further amplified by using 2D materials with extraordinary conductivity. The biosensor can precisely detect even minute levels of cholesterol because to the increased sensitivity brought on by signal amplification. This is crucial in clinical settings where early diagnosis of high cholesterol levels can result in prompt treatment and better patient outcomes. A complex method for improving biosensors in cholesterol detection is the incorporation of 2D materials into nanostructures, such as nanowires or nanosheets. This strategy takes advantage of the special qualities of nanostructured materials as well as the larger surface area they offer for cholesterol interaction. Since nanostructures have more surface area than flat surfaces, more cholesterol molecules can bind and engage with the biosensor at the same time. The biosensor's sensitivity is increased by this increased binding capacity, which also makes it more efficient in capturing cholesterol. In addition, the distinctive electrical and electrochemical properties that nanostructures frequently have can enhance signal transduction. A biosensor with greater performance, able to precisely detect cholesterol over a wide range of concentrations, is created by combining expanded surface area and improved electrochemical characteristics [46].

Another technique that can considerably improve biosensor performance in cholesterol detection is to stack numerous layers of 2D materials to generate heterostructures. The electrical capabilities of various 2D materials are harnessed by these layered structures, which also make charge transfer easier. As a result, signal transduction is improved, and cholesterol levels are detected more accurately. Each 2D material layer in a layered biosensor can perform a particular function. As an illustration, one layer might be designed to capture cholesterol molecules while a different layer might promote quick signal amplification. The design and functioning of biosensors may be precisely controlled thanks to this modular approach. In heterostructures, interactions between adjacent layers can also improve charge transfer and electron mobility, leading to quicker and more effective signal transduction. As a result, the biosensor's reaction time and sensitivity are increased, making it particularly useful for applications involving the detection of cholesterol. A technique that can considerably improve biosensor performance in cholesterol detection is the combination of 2D materials with functional nanoparticles, such as gold or silver nanoparticles. These nanoparticles give the biosensor special catalytic capabilities

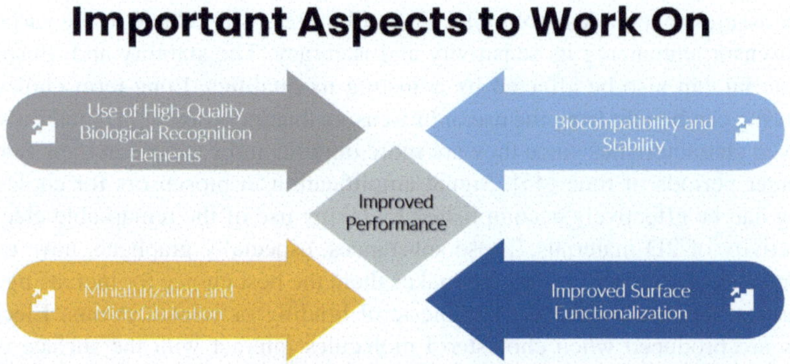

Fig. 5 Key areas that need to be worked on in order to improve the sensor's performance

that can be used to boost signals and increase sensitivity. The signal is amplified, making it stronger and easier to detect, increasing the biosensor's sensitivity in the process. The Fig. 5 illustrates the key areas for potential improvement in enhancing the overall performance of the sensor.

When incorporated into the biosensor, functional nanoparticles can accelerate electrochemical reactions involving the binding of cholesterol or other processes. The rate and effectiveness of these reactions are increased by this catalysis, producing stronger electrical impulses. As a result, the biosensor becomes more sensitive and can identify even extremely low cholesterol amounts. The high electrical conductivity of 2D materials and the catalytic capabilities of nanoparticles combine to produce a synergistic effect that significantly enhances cholesterol detection [47]. The biosensor is a flexible instrument for both clinical and research applications since it enables precise assessment of cholesterol levels across a broad dynamic range. A sophisticated technique that improves biosensor performance in cholesterol detection is the use of microfluidic channels into the biosensor architecture. By allowing for precise sample flow control and reducing interference, microfluidics ensures that the biosensor receives a consistent and carefully monitored sample supply. In biosensors with integrated microfluidics, the microchannels precisely direct the sample to the 2D material-based detecting surface. By minimizing fluctuations in sample delivery and response times, this regulated sample handling increases measurement accuracy. Additionally, it reduces the possibility of sample contamination or interference, ensuring the biosensor's accuracy.

7 Summary

In conclusion, the use of two-dimensional (2D) materials in biological sensors has started a very exciting time of progress and innovation. Because of how they are made, 2D materials like graphene, transition metal dichalcogenides (TMDs), and black phosphorus have made a lot of new medical and analytical options possible. This revolution could change how we find diseases, keep an eye on them, and treat them. To begin, 2D materials have shown to be very sensitive, which has led to the creation of very accurate and quick biosensors. Because they are very electrically conductive and have a lot of surface area, they have been used to make tools that can identify biomolecules, pathogens, and even single molecules. This awareness has made it possible to find diseases earlier, which can greatly improve the health of patients. The second thing is that 2D materials are very biocompatible, which means they can easily connect with living things. This compatibility has made it easier to make implantable sensors and wearable devices that can track bodily data in real time, deliver drugs, and provide personalized medicine. Now, patients have more power over their health, and doctors can view important data from afar. Also, the adaptability and plasticity of 2D materials have made it possible to create sensors that can identify more than one analyte at the same time. This new idea should make testing easier, cut costs, and make healthcare service more efficient. But there are still problems, like how to make it scalable, stable over time, and cost-effective for mass production. These problems need to be solved before 2D materials can be widely used in biological detection. To sum up, two-dimensional materials have already made big steps towards changing the way biological sensing is done. Their special features make it possible to find diseases earlier, keep an eye on them, and treat them better than ever before. As research and development continue, we may expect additional innovative uses that will improve health and the planet. Adding 2D materials to biological sensing is a huge step towards combining cutting-edge technology with healthcare, giving people hope for a brighter future.

References

1. A.G. Cullis, G.R. Booker, The epitaxial growth of silicon and germanium films on (1 1 1) silicon surfaces using UHV sublimation and evaporation techniques. J. Cryst. Growth **9**, 132–138 (1971)
2. S.Z. Butler, S.M. Hollen, L. Cao, Y. Cui, J.A. Gupta, H.R. Gutiérrez et al., Progress, challenges, and opportunities in two dimensional materials beyond graphene. ACS Nano **7**, 2898–2926 (2013)
3. J.A. Wilson, A.D. Yoffe, Transition metal dichalcogenides discussion and interpretation of observed optical, electrical, and structural properties. Adv. Phys. **18**, 193–335 (1969)
4. Q.H. Wang, K. Kalantar-Zadeh, A. Kis, J.N. Coleman, M.S. Strano, Electronics and optoelectronics of two-dimensional transition metal dichalcogenides. Nat. Nanotechnol. **7**, 699–712 (2012)
5. T.C. Ozawa, K. Fukuda, K. Akatsuka, Y. Ebina, T. Sasaki, Preparation and characterization of the Eu3+ doped perovskite nanosheet phosphor: $La_{0.9}Eu_{0.05}Nb_2O_7$. Chem. Mater. **19**, 6575–6580 (2007)
6. T.C. Ozawa, K. Fukuda, K. Akatsuka, Y. Ebina, T. Sasaki, K. Kurashima et al., $(K_{1.5}Eu_{0.5})Ta_3O_{10}$: a far-red luminescent nanosheet phosphor with the double perovskite structure. J. Phys. Chem. C **112**, 17115–17120 (2008)

7. S. Ida, C. Ogata, M. Eguchi, W.J. Youngblood, T.E. Mallouk, Y. Matsumoto, Photoluminescence of perovskite nanosheets prepared by exfoliation of layered oxides, $K_2Ln_2Ti_3O_{10}$, $KLnNb_2O_7$, and $RbLnTa_2O_7$ (Ln: lanthanide ion). J. Am. Chem. Soc. **130**, 7052–7059 (2008)
8. Y. Ebina, T. Sasaki, M. Harada, M. Watanabe, Restacked perovskite nanosheets and their Pt-loaded materials as photocatalysts. Chem. Mater. **14**, 4390–4395 (2002)
9. M. Osada, K. Akatsuka, Y. Ebina, H. Funakubo, K. Ono, K. Takada et al., Robust high-j response in molecularly thin perovskite nanosheets. ACS Nano **4**, 5225–5232 (2010)
10. R. Ma, Z. Liu, K. Takada, N. Iyi, Y. Bando, T. Sasaki, Synthesis and exfoliation of Co^{2+}–Fe^{3+} layered double hydroxides: an innovative topochemical approach. J. Am. Chem. Soc. **129**, 5257–5263 (2007)
11. B. Lalmi, H. Oughaddou, H. Enriquez, A. Kara, S. Vizzini, B. Ealet et al., Epitaxial growth of a silicene sheet. Appl. Phys. Lett. **97**, 223109 (2010)
12. L. Meng, Y. Wang, L. Zhang, S. Du, R. Wu, L. Li et al., Buckled silicene formation on Ir(1 1 1). Nano Lett. **13**, 685–690 (2013)
13. A. Fleurence, R. Friedlein, T. Ozaki, H. Kawai, Y. Wang, Y. Yamada-Takamura, Experimental evidence for epitaxial silicene on diboride thin films. Phys. Rev. Lett. **108**, 245501 (2012)
14. H. Jung, J. Park, I.-K. Oh, T. Choi, S. Lee, J. Hong et al., Fabrication of transferable Al_2O_3 nanosheet by atomic layer deposition for graphene FET. ACS Appl. Mater. Interf. **6**, 2764–2769 (2014)
15. C.D. Ruggiero, M. Badal, T. Choi, D. Gohlke, D. Stroud, J.A. Gupta, Emergence of surface states in nanoscale Cu 2N islands. Phys. Rev. B **83**, 245430 (2011)
16. M. Sterrer, T. Risse, U.M. Pozzoni, L. Giordano, M. Heyde, H.-P. Rust et al., Control of the charge state of metal atoms on thin MgO films. Phys. Rev. Lett. **98**, 096107 (2007)
17. B. Hao, Y. Yan, X. Wang, G. Chen, Synthesis of anatase TiO_2 nanosheets with enhanced pseudocapacitive contribution for fast lithium storage. ACS Appl. Mater. Interf. **5**, 6285–6291 (2013)
18. H. Oughaddou, B. Aufray, J.P. Bibérian, J.Y. Hoarau, Growth mode and dissolution kinetics of germanium thin films on Ag(0 0 1) surface: an AES–LEED investigation. Surf. Sci. **429**, 320–326 (1999)
19. H. Oughaddou, J.M. Gay, B. Aufray, L. Lapena, G.L. Lay, O. Bunk et al., Ge tetramer structure of the p(2p2 4p2)R45° surface reconstruction of Ge/Ag(0 0 1): a surface X-ray diffraction and STM study. Phys. Rev. B **61**, 5692 (2000)
20. E. Golias, E. Xenogiannopoulou, D. Tsoutsou, P. Tsipas, S.A. Giamini, A. Dimoulas, Surface electronic bands of submonolayer Ge on Ag(1 1 1). Phys. Rev. B **88**, 075403 (2013)
21. H. Oughaddou, A. Mayne, B. Aufray, J.P. Bibérian, G.L. Lay, B. Ealet et al., Germanium adsorption on Ag(1 1 1): an AES-LEED and STM study. J. Nanosci. Nanotechnol. **7**, 3189–3192 (2007)
22. H.M. Yadav, J.-D. Park, H.-C. Kang, J.-J. Lee, Recent development in nanomaterial-based electrochemical sensors for cholesterol detection. Chemosensors **9**, 98 (2021)
23. G. Zhu, Y. Yi, J. Chen, Recent advances for cyclodextrine-based materials in electrochemical sensing. Trends Anal. Chem. **80**, 232–241 (2016)
24. J. Huang, Z. Feng, L. Yang, Y. Qian, Q. Zhang, F. Li, A sensitive and selective non-enzyme cholesterol amperometric biosensor based on host–guest effect of β-cyclodextrine for cholesterol. Anal. Methods **4**, 4264 (2012)
25. N. Agnihotri, A.D. Chowdhury, A. De, Non-enzymatic electrochemical detection of cholesterol using β-cyclodextrin functionalized graphene. Biosens. Bioelectron. **63**, 212–217 (2015)
26. L. Yang, H. Zhao, G. Sh Fan, X. Zhao, C.-P. Ran, Electrochemical detection of cholesterol based on competitive host–guest recognition using β-cyclodextrin/poly(N-acetyl-aniline)/graphene-modified electrode. RSC Adv. **5**, 64146 (2015)
27. M.A.H. Nawaz, M. Majdinasab, U. Latif, M. Nasir, G. Gokce, M.W. Anwar, A. Hayat, Development of a disposable electrochemical sensor for detection of cholesterol using differential pulse voltammetry. J. Pharm. Biomed. Anal. **159**, 398–405 (2018)
28. J. De-La-Cuesta, I. Asenjo-Sanz, A. Latorre-Sanchez, E. Gonzalez, D.E. Martinez-Tong, J.A. Pomposo, Enzyme-mimetic synthesis of PEDOT from self-folded iron-containing single chain nanoparticles. Eur. Polym. J. **109**, 447–452 (2018)

29. Z. Wang, W. Yu Chen, J. Dong, B. Zhou, J. Han, L. Jiao, P. Lan, Q.C. Miao, Copper (II)-ploy-L-histidine functionalized multi walled carbon nanotubes as efficient mimetic enzyme for sensitive electrochemical detection of salvianic acid A. Biosens. Bioelectron. **121**, 257–264 (2018)
30. N. Chaibakhsh, Z. Moradi-Shoeili, Enzyme mimetic activities of spinel substituted nanoferrites (MFe_2O_4): a review of synthesis, mechanism and potential applications. Mater. Sci. Eng. C **99**, 1424–1447 (2019)
31. J. Xie, X. Zhang, H. Wang, Y. Huang, Analytical and environmental applications of nanoparticles as enzyme mimetics. Trends Anal. Chem. **39**, 114–129 (2012)
32. A.K. Dutta, S.K. Maji, A. Mondal, B. Karmakar, P. Biswas, B. Adhikary, Iron selenide thin film: peroxidase-like behavior, glucose detection and amperometric sensing of hydrogen peroxide. Sens. Actuators B **173**, 724–731 (2012)
33. K. Derina, E. Korotkova, Y. Taishibekova, L. Salkeeva, B. Kratochvil, J. Barek, Electrochemical non-enzymatic sensor for cholesterol determination in food. Anal. Bioanal. Chem. **410**, 5085–5092 (2018)
34. T. Peng, H. Li, R. Lu, Direct measurement of cholesterol based on anodic stripping voltammetry. Anal. Chim. Acta **257**, 15–19 (1992)
35. U. Saxena, A.B. Das, Nanomaterials towards fabrication of cholesterol biosensors: key roles and design approaches. Biosensors Bioelectron. **75**, 196–205 (2016)
36. J.N. Coleman, M. Lotya, A. O'Neill, S.D. Bergin, P.J. King, U. Khan et al., Two-dimensional nanosheets produced by liquid exfoliation of layered materials. Science **331**, 568–571 (2011)
37. Y. Li, H. Wang, L. Xie, Y. Liang, G. Hong, H. Dai, MoS_2 nanoparticles grown on graphene: an advanced catalyst for the hydrogen evolution reaction. J. Am. Chem. Soc. **133**, 7296–7299 (2011)
38. T. Wang, L. Liu, Z. Zhu, P. Papakonstantinou, J. Hu, H. Liu et al., Enhanced electrocatalytic activity for hydrogen evolution reaction from self-assembled monodispersed molybdenum sulfide nanoparticles on an Au electrode. Energy Environ. Sci. **6**, 625–633 (2013)
39. B. Poudel, Q. Hao, Y. Ma, Y. Lan, A. Minnich, B. Yu et al., High thermoelectric performance of nanostructured bismuth antimony telluride bulk alloys. Science **320**, 634–638 (2008)
40. J. Taha-Tijerina, T.N. Narayanan, G. Gao, M. Rohde, D.A. Tsentalovich, M. Pasquali et al., Electrically insulating thermal nanooils using 2D fillers. ACS Nano **6**, 1214–1220 (2012)
41. S.S. Lin, Light-emitting two-dimensional ultrathin silicon carbide. J. Phys. Chem. C **116**, 3951–3955 (2012)
42. S. Yang, Y. Gong, Z. Liu, L. Zhan, D.P. Hashim, L. Ma et al., Bottom-up approach toward single-crystalline VO_2-graphene ribbons as cathodes for ultrafast lithium storage. Nano Lett. **13**, 1596–1601 (2013)
43. D. Kong, W. Dang, J.J. Cha, H. Li, S. Meister, H. Peng et al., Few-layer nanoplates of Bi_2Se_3 and Bi_2Te_3 with highly tunable chemical potential. Nano Lett. **10**, 2245–2250 (2010)
44. L. Hu, R. Ma, T.C. Ozawa, T. Sasaki, Exfoliation of layered europium hydroxide into unilamellar nanosheets. Chem. Asian J. **5**, 248–251 (2010)
45. U. Saxena, A.B. Das, Nanomaterials towards fabrication of cholesterol biosensors: Key roles and design approaches. Biosens. Bioelectron. **75**, 196–205 (2016)
46. A. Gahlaut, V. Hooda, V. Dhull, V. Hooda, Recent approaches to ameliorate selectivity and sensitivity of enzyme based cholesterol biosensors: a review. Artif. Cells Nanomed. Biotechnol. **46**(3), 472–481 (2018)
47. J. Kowalski, J. Płoszyńska, A. Sobkowiak, J.W. Morzycki, A.Z. Wilczewska, Direct electrochemical acetoxylation of cholesterol at the allylic position. J. Electroanal. Chem. **585**, 275–280 (2005)

An Overview of Drug Delivery for Wellbeing Based on Polysaccharides

K. Kanimozhi and M. V. Arularasu

Abstract Biopolymers are an essential near grow an ecologically welcoming, non-toxic, recyclable polymer that can substitute natural and synthetic based polymers. They need supreme rank trendy medicinal requests then merged biopolymers must exposed talented possessions against separate complements. An excellent presentation of biopolymers merged with combined films derived from chitosan is now being regarded as major. A salt leaching technique allowed guar gum and polyvinyl alcohol (CS/GG/PVA) to remain ready. In this process, films were manufactured on CS, GG, and PVA continuously. Attenuated total reflectance-Fourier-transform infrared spectroscopy (ATR-FTIR) revealed a resolute understanding of polymer surface morphology, construction, and interface, as well as water fascinating volume characteristics. Inverse scanning electron microscope (IE-SEM), X-ray diffraction (XRD), and swelling characteristics are examined. FTIR showed intermolecular attachment among the biopolymer's charity aimed at films groundwork. FESEM films displays a framework displayed good absorbent constructions. The swollen designs stayed importantly enhanced by great bulge share. Consequences exposed that the chitosan, guar gum besides polyvinyl alcohol of ternary mergers through need latent aimed at regulator painkiller announcement.

1 Introduction

The normal polymers remain recyclable, biocompatible, besides then obviously nearby and their possessions can remain improved through alteration [1–3]. Natural polymers like chitosan, gums, In the list of polysaccharide ingredients are: gellans, grewias, acacias, locust beans, khayas, tragacanths, okras, terminalia catappa, karayas, guars, carrageenans, resins, and pectins [4–7]. It is equally important for

K. Kanimozhi (✉)
Department of Chemistry, Science and Humanities, Loyola Institute of Technology (LIT), Kuthambakkam, Tamil Nadu 600123, India
e-mail: kanivlr8@gmail.com

M. V. Arularasu
Department of Electronic Engineering, Chang Gung University, Taoyuan 33302, Taiwan

K. Kasinathan et al. (eds.), *New Technologies for Energy Transition Based on Sustainable Development Goals*, https://doi.org/10.1007/978-981-97-2527-4_2

mucilage, alginate, and chitosan to be adapted through a variety of methods so that they can meet the demands aimed at drug delivery, aquatic cleaning, etc., [8–11]. Currently, the tendency of the normal herbal medicinal creation remains cumulative meanwhile herbal products remain inoffensive and safe near usage [12, 13]. The resources founded happening usual polymers stay well-organized toward announcement the medication cutting-edge skillful way [14]. The medication announcement design is contingent upon the bodily and biochemical possessions of the medicine bringing go-between. Now this respect, biopolymer mixtures by dissimilar moieties need revealed talented competence. Polysaccharides contain of monosaccharide components starting great net of polymer [15]. Polysaccharides remain obviously rich resources, steady, recyclable, harmless near usage, non-toxic and hydrophilic trendy environment owing near the attendance dissimilar useful collections similar amino group, hydroxyl group, which remain talented near communicating through organic tissues [16, 17]. Normal resin contains of honey monomer that remain related composed near procedure large particles [18]. Ordinary secretions stay price actual, non-toxic besides simply available [19]. Guar gum (GG) consumes great swelling properties then stays the finest dishonorable aimed at the development of social colonic bacterium bactericides ovatus [20]. GG is inexplicable trendy greatest of the carbon-based thinners [21]. On little heat, it is accomplished of starting aqueous solution of tall thickness for of the situation great molecular weight (300,000–400,000 Da). The situation remains steady trendy the pH range of 6–7 conversely it destroys on developed heat besides trendy basic pH. It stands charity now nourishment then drinks by way of thickener, folder, and cash reformer then aimed at the discount of lubricant ingesting of cooked food [22, 23].

As for chitin, it is composed of some (1–4) dependent glucosamine subunits, while chitosan is composed of D-glucosamine and N-acetyl-Dglucosamine subunits. By using artificial polymer, chitosan consumes the high-quality amalgamation material and utilizes dormant requests for the production of trendy aliments [24]. Chitosan remains non-toxic, recyclable, biocompatible then hemostatic trendy environment therefore can remain rummage-sale now medicinal manufacturing [25]. Polyvinyl alcohol (PVA) remains aquatic solvable a artificial polymer. PVA system great excellence layers consuming outstanding ductile control then resistance then consumes varied requests trendy dissimilar arenas [26]. PVA mixtures through numerous biomaterials remain rummage-sale aimed at diverse requests. For example, PVA, starch, LDPE remains rummage-sale aimed at nourishment wrapping by way of provisional dampness of nutrition. Merged layers of chitosan-PVA remain charity by way of health implement then drug delivery [27, 28]. PVA must remain recycled aimed at biochemical alteration of extra polymer aimed at biomedical field [29]. Usual besides artificial polymer mergers system progressive resources aimed at dissimilar requests [30]. Polymer amalgamation impression remained interested by consciousness that novel particles remain not continuously enough near happen the obligation of progressive physical besides combination remains feasible towards achieve the wanted possessions somewhat than creating novel resources [31]. Artificial polymers stay varied by usual polymers near enhance the possessions of usual polymers. Trendy overall, definitive ternary mixture films stay complete of any binary contributors varied by unique acceptor before unique contributor varied through binary acceptors trendy the occurrence of certain stabilizers [32]. The situation customs a measurement association between polymers, monomers, and annoying linkers; the situation is like several dissimilar methods of charity aimed at irritating polymer linking. After observing the amount of certain monomers by admiration near the total amount of monomers in a polymer, describe the relationship [33] the quantity of monomer components then unique method near create the repeat element of polymer by way of single monomer component then/before the monosaccharide equally single component besides care trendy attention that the monosaccharide remain measured then now opinion of care besides present situation of atmosphere, near remains essential towards accept eco-kind resources [34–40]. A review of guar gum and chitosan-founded physical groundwork then requests remain obtainable now Table 1. Established happening aforementioned truths, the chief impartial of this education were near make an original sequence of merged films founded happening chitosan, guar gum and polyvinyl alcohol (CS/GG/PVA) through salt leaching technique. The construction, surface morphology, contact of polymers and marine fascinating volume remained strong-minded by FTIR, XRD, SEM and swelling properties.

2 Materials and Method

2.1 Reagent and Chemicals

Consequently, all chemicals and substances used were of a logical score and were purchased from Sigma-Aldrich, India. S&D fine chemicals, Mumbai, India, provided the nutritional score guar gum consuming a thickness of 5000 cps in addition to chitosan.

2.2 An Approach for Preparing Mixtures of Guar Gum, Chitosan, and PVA

Chitosan and Guar Gum (GG) scaffolds were prepared by salt leaching using Poly (vinyl alcohol) PVA (CS-GG&PVA). An acetic acid solution of 0.1 M was stirred with 1 g of CS powder into 100 ml of 0.1 M acetic acid to form a chitosan solution containing 1 wt% CS powder. The 1% clear PVA solution was prepared by mixing 1 g PVA powder with 100 ml purified water at 800 °C while stirring. The mixtures were then homogenized by 6 h of magnetic stirring at room temperature after 1 g of GG powder was added to the CS-PVA solution. An equal amount of polymer solutions was homogeneously mixed with sodium chloride 200–500 m granules (NaCl 200–500 m). To remove salt crystals, the solution was allowed to gel for three days at

Table 1 The review also claims that guar gum might be able to improve physical performance besides chitosan-based training

S. No	Biopolymer/derivatives	Applications	References
1	Hydroxylated guar gum	Chromate (IV) adsorption	[41]
2	Cross-linked guar gum dialdehyde polymer	Hg ions are removed by adsorbent	[42]
3	Poly(acrylic)-guar gum graft it is acidic)	Waste water with acidic pH adsorbs Pd(II)	[43]
4	Hydrogel derived from chitosan	In 12 h, cefixime released 81.6% of its active ingredient	[44]
5	Guar gum—alginate hydrogel	Delivering protein drugs in a controlled manner	[45]
6	Modified glycerin biochitosan	For low cholesterol, glycyrrhetinic acid is delivered	[46]
7	Nanogels made from chitosan	Cancer treatment with pravastatin	[47]
8	Sodium alginate and guar gum	In 10 h, 97% of the nateglinide has been delivered	[48]
9	Drug based on chitosan taking a stand on nanocarriers	TNF-α siRNA for gene carrier	[49]
10	Chitosan carboxymethylated	Anticancer therapy with doxorubicin	[50]
11	Guar gum derived from carboxymethyl guar	Treatment of wounds	[51]
12	Guar gum made from Acryloyl Guar	Dopamine and tyrosine release slowly	[52]
13	N, N′-dimethylacrylamide derived from guar gum	Removing lead, nickel, and zinc ions	[53]
14	Poly(2-dimethylaminoethylmethacrylate) Guar gum	Within 48 h, 98% of the 5-fluorouracil was released	[54]
15	Where do we stand with chitosan-based drug nanocarriers	Simulated TNF-siRNA	[49]

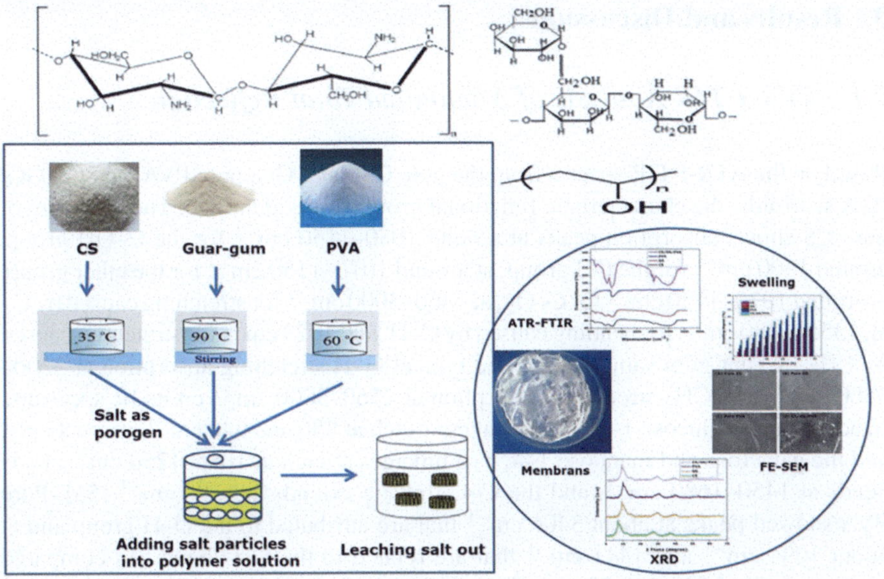

Fig. 1 Structures of Chitosan, guar gum and PVA and belnds (membranes) preparation schematic presentation

room temperature before being washed with deionized water. CS/GG/PVA scaffolds with salt-leached pores were obtained after 3 days of drying. It can be seen in Fig. 1 how chitosan, guar gum, and PVA are blended and their structures.

2.3 *Characterization Studies*

Attenuated total reflectance Fourier transform infrared spectrophotometer (ATR-FTIR) analysis of scaffolds was used in the biochemical construction of the scaffold (Shimadzu IR affinity—1S). In addition to a wavenumber range of 4000–500 cm^{-1}, each band was learned using Quest ATR ZnSe quartz cells in a makeup of 250 scans. We used a Japanese XRD Rigaku Co, model DMAX 2200, to perform a X-ray diffraction analysis. With a Hitachi SR-4700 operating at 1 kV, Germany at a magnification of 100 kx, a surface morphology analysis of prepared membranes was performed. The swelling of scaffolds remained measured aimed at numerous period intermissions. The heaviness of petri dish then dry scaffold's part stayed measured and formerly, the scaffolds stayed waterlogged now the purified water and later exact period the aquatic stayed detached besides weighed it once more. The similar process remained frequent aimed at every prepared scaffold later several exact intermissions of period.

3 Results and Discussion

3.1 ATR-FTIR Analysis of Attenuated Total Reflection

Based on the ATR-FTIR spectra from the pure CS, pure GG, pure PVA, and CS/GG/PVA scaffolds, the characteristic functional groups were identified. The spectrum of pure CS shows absorption peaks at around 1050–1150 cm^{-1} for the C–OH group, around 1600 cm^{-1} for the OH group, at around 1070–1150 cm^{-1} for the ether group, at around 1070–1150 cm^{-1} for C–O–, at 2860–3000 cm^{-1} for stretching caused by C–H, 1355–1480 cm^{-1} for bending caused by C–H, and 1023 cm^{-1} for stretching caused by CH_2. A guar gum sample exhibited typical O–H stretching absorption at 3200–3600 cm^{-1}, and CH_2 stretching absorption at 2560–3000 cm^{-1} on its IR spectrum. In addition, the glucose 1–4, 1–6 linkages stretch at 850 and 940 cm^{-1}, respectively, and the galactose and mannose 1–4, 1–6 linkages stretch at 1055–1250 cm^{-1}, C–H bends at 1450–1680 cm^{-1}, and the C–C linkage extends at 1200 cm^{-1} [55]. Pure PVA showed peaks at about 3406 cm^{-1} that are attributed to the O–H group and at about 1094 cm^{-1} and 1441 cm^{-1} that are related to the C–O bond. As compared to the spectra of CS, GG, PVA, and CS/GG/PVA scaffolds, the bands at 1460 cm^{-1} correspond to galactose linkage and the bands at 2900–2935 cm^{-1} correspond to O–H stretching. Strong C–O absorption is observed between 1000 and 1030 cm^{-1}, C=C stretching is observed between 1610 and 1690 cm^{-1}, and C–N stretching is observed between 1200 and 1370 cm^{-1}. Three-member ternary scaffolds based on PVA, GG, and chitosan are characterized by IR spectral analysis as shown in Fig. 2.

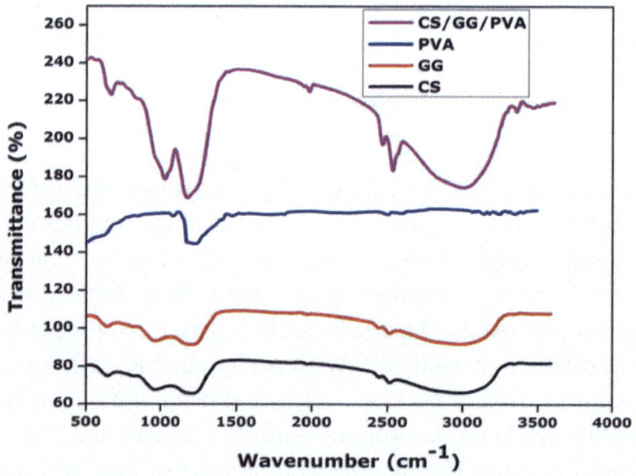

Fig. 2 IR spectrum of prepared blends (membranes) of pure CS, GG, PVA and combined CS-GG-PVA

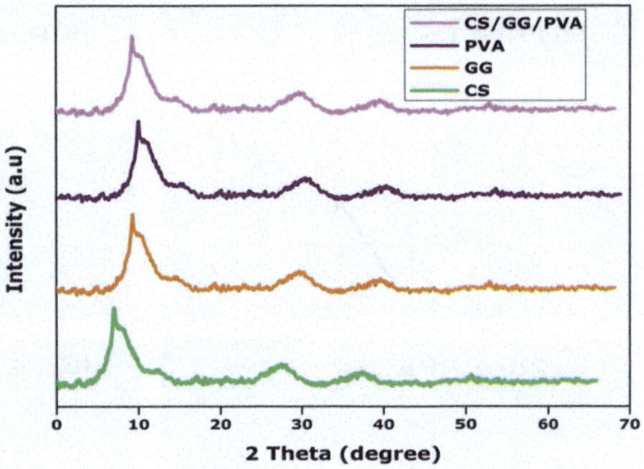

Fig. 3 XRD analysis of prepared blends (membranes) of pure CS, GG, PVA and combined CS-GG-PVA

3.2 *Diffraction Analysis Using X-rays*

This Fig. 3 shows the X-ray diffraction patterns of pure CS and GG scaffolds, pure PVA scaffolds, and CS/GG/PVA scaffolds. In this method, crystalline materials are analyzed to determine their structure. Through learning the quartz assembly, it recognizes the glassy stages besides biochemical arrangement evidence [56]. Pure CS comprises amino group besides responds by PVA useful collections near system novel promises, consequently a minor change in the ultimate of 2θ remained experiential. This diagram demonstrations the glassy ultimate on overweight near 14° which remained not existing trendy separate mechanisms i.e., chitosan, guar gum and PVA. In the glassy peak of chitosan cellulose construction on 24.2°, this displays the positive attachment among chitosan, PVA and guar gum [57].

3.3 *An Electro-Scanning Field Emission Microscope Analysis*

FESEM examination remains beneficial near training superficial morphology, scope, crystallinity besides the locations of stages trendy the ready factual [58, 59]. In the chitosan and GG founded ternary mergers by PVA remained too examined aimed at morphology examination then consequences of ready scaffolds, while the pictures of ready merged scaffolds pure CS, pure GG, pure PVA and CS/GG/PVA as shown in Fig. 4. Moreover, in Fig. 5 shows photographs of blended membranes (CS/GG/PVA, CS/GG/CS) prepared using pure CS, GG, and PVA. Microscopy images showed uniformly distributed pores on the membranes. The FESEM descriptions discovered that the scaffolds remained permeable besides holes remained regularly dispersed [40]. The permeability permits the earlier bulge that stays valuable aimed at medication filling then announcement [38, 60].

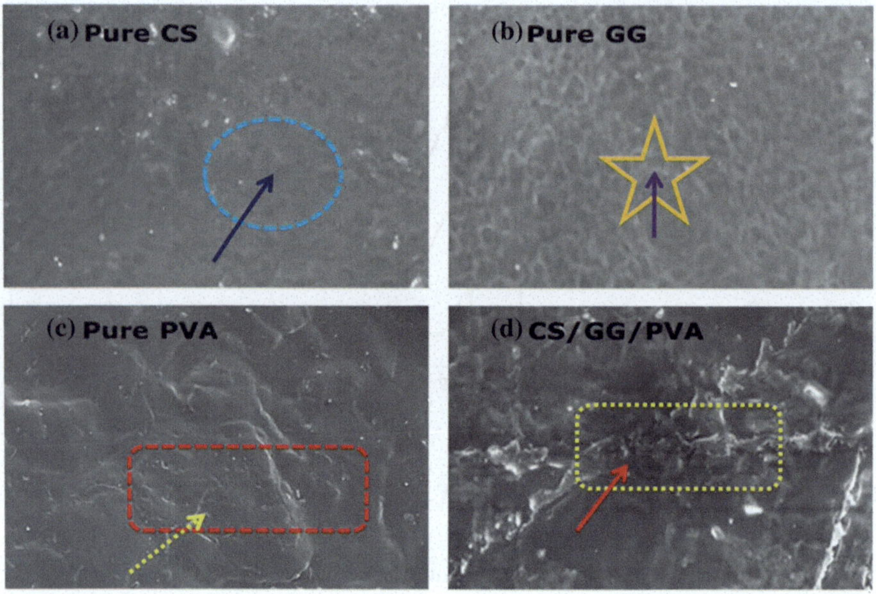

Fig. 4 SEM images of prepared blends (membranes) of pure CS, GG, PVA and combined CS-GG-PVA

3.4 *Swelling Characteristics*

Swelling performance of natural substantial be contingent happening the average, i.e., salts, acids and bases touch the bulge features mixed films. In the films ready by CS, GG and PVA remained verified aimed at bulge features. The bulge of ready films remained check through gauging the heaviness of augmented films on dissimilar period intermission then consequences remain exposed in Fig. 6. The films remained located trendy aquatic average then afterward 5 min the aquatic remained detached then film stayed considered. The film by 60% CS/GG/PVA, 20% PVA, 15% guar gum and 5% CS presented decent bulge stuff. It remained experiential that by way of the guar gum attentiveness remained a bridged, the bulge remained abridged, which designates the film consuming advanced attentiveness of CS/GG/PVA presented advanced bulge. These answers remain fashionable streak by preceding educations that chitosan founded physical presented decent bulge possessions trendy dissimilar media. As a result, a variety of bulge properties were observed in aquatic, buffer, non-buffer, and ionic solutions composed of polyvinyl pyrrolidone, chitosan, and aminopropyletriethoxy silane (cross linker). In the case of a supreme bulge (4386%) that remained experientially necessary to hydrogelize using 330 L of aminopropyletriethoxy silane (cross linker), 0.4 g of chitosan, and polyvinylpyrrolidone, the conduct of the bulge remained significantly altered [61]. Hydrophilic countryside is

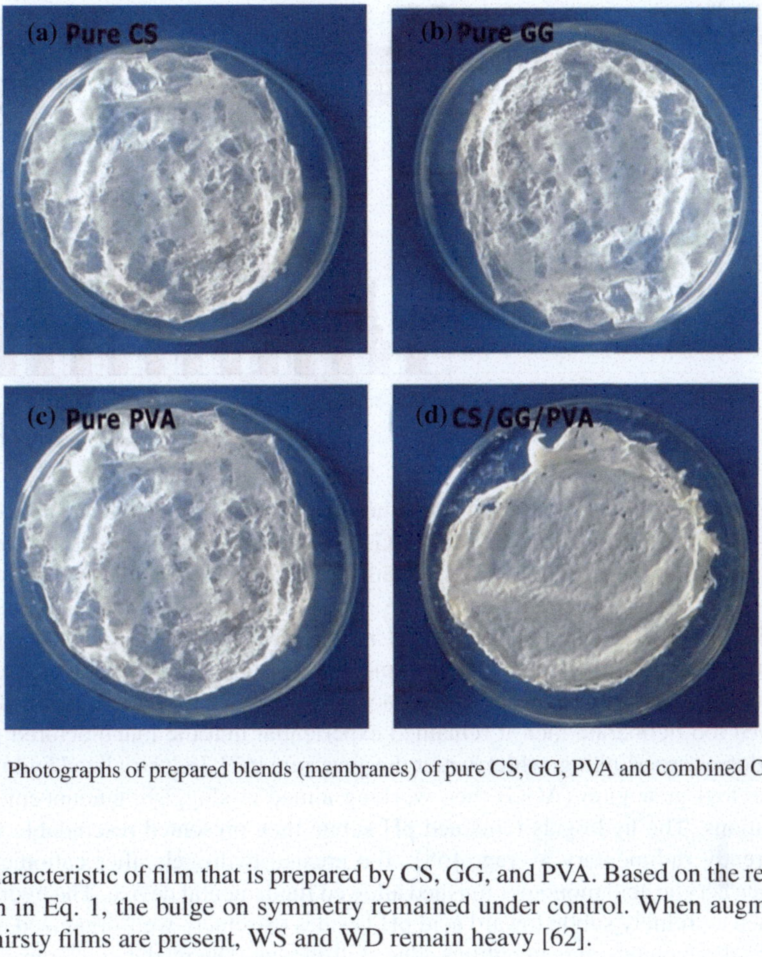

Fig. 5 Photographs of prepared blends (membranes) of pure CS, GG, PVA and combined CS-GG-PVA

the characteristic of film that is prepared by CS, GG, and PVA. Based on the relation shown in Eq. 1, the bulge on symmetry remained under control. When augmented and thirsty films are present, WS and WD remain heavy [62].

$$\text{Equilibrium swelling}\,(\%) = \text{WS} - \text{WD}/\text{WD} \times 100 \qquad (1)$$

The chitosan and guar gum founded physical must remained lengthily ready aimed at organic request i.e., carboxymethyl guar gum (CMG) remained ready through couture it by aqueous nanosilica sol then rummage-sale aimed at skillful medication announcement. Owed toward the tall viscid countryside of hydrogel nanocomposites, the medication announcement remained sluggish by way of likened toward well-ordered CMG [63]. Likewise, alginate-guar gum hydrogel irritated related by glutaraldehyde remained working aimed at skillful announcement of medications, which remained ready custody cutting-edge attention the pH of average meanwhile the medication leakage remains a main subject on advanced acid equal. Guar gum remained comprised fashionable the alginate medium aimed at supreme competence trendy the procedure of measured medication announcement. Glutaraldehyde and guar gum crosslink to improve encapsulation skills, thus improving medicine

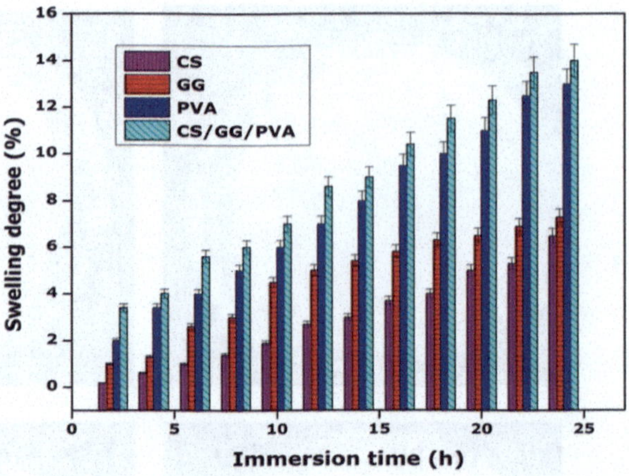

Fig. 6 Swelling pattern of prepared blends (membranes) of pure CS, GG, PVA and combined CS-GG-PVA

announcements [64]. Also, CMG after the innate guar gum remained spliced by ethylenediamine toward improve the CMG possessions. In vitro educations showed outstanding biocompatibility owing toward cross-linking then likewise presented development reserve in contradiction of *S. aureus and P. aeruginosa* bacteria [65]. Likewise, guar gum adapted below heat radioactivity presented talented competence aimed at skillful announcement of medication [66]. The meaning of collagen adapted guar gum aimed at flesh manufacturing requests by the corrosion device remained too deliberate then it remained experiential that the manufactured physical endorses crosslinking, which better the guar gum [67]. Industrialized hydrogels, i.e., acryloyl guar gum (AGG) then working aimed at sluggish announcement of medications. The hydrogels remained pH subtle then presented rescindable variations trendy rudimentary average [68]. Too created hydrogels after cationic guar gum then acrylic acid monomer finished allowed fundamental device. The hydrogels remained extremely subtle toward acid pH besides extremely well-organized aimed at medication announcement cutting-edge skillful way. The medication conveyance remained prejudiced through sharpness of the average. The GA hydrogels too presented possible aimed at colon-exact medication [69]. Expressed Nate glinide microspheres by ionic gelation technique, which remained rummage-sale aimed at continued announcement of medicine [70]. As well, suspension techniques were used to fabricate chitosan and guar gum microgels. Originally, the guar gum remained spliced then formerly mixed by chitosan aimed at the groundwork of microgels of dissimilar arrangements besides ready design presented talented competence by way of medicine transporters [71]. Assessed the possible of mini elements in contradiction of the illness of tuberculosis. As a result, a multitude of chitosan, guar gum, mannan and guar gum-covered chitosan preparations remained available. Outcomes presented that guar gum lots must exposed sensible variability owing toward their talented physical chemical possessions. The advantaged receipt of guar gum covered arrangements remained related by mannose devoted on the cellular equal [72]. Advanced an original medication distribution physical toward improve the board medicine distribution throughout the sequence of chemotherapy. Aimed at this drive, chitosan resulting solitary enclosed carbon nanotubes remained beaten fashionable pH subtle nanogels, which remained rummage-sale by way of talented nano-transporters aimed at the chemotherapy [73]. Manufactured nanogel founded happening the supra-molecular connections among silica, polyethyleneimine then chitosan by way of a steady genetic factor transporter substantial. In the nobbling consumes augmented the forte, constancy besides medicine distribution aptitude of the ready physical [74]. Also, rummage-sale chunk copolymer to procedure nanogels through the wanted possessions toward strengthens the request of nanogels aimed at the chemotherapy [75]. Extra lately, ready chitosan and polyvinyl pyrrolidone by crosslinking marvel-founded hydrogels besides verified aimed at exact medicine issue [76]. By way of manufactured physical presented talented competence aimed at skillful medication announcement cutting-edge stomach pH average besides ready polymeric film by curcumin cross-related by chitosan-PVA [77]. The coiled remedial possible of the ready polymeric film remained verified besides competence remained talented of chitosan-PVA mixtures. Henceforth, it remained experiential that the chitosan and guar gum founded preparations remained extremely steady besides real aimed at switch medication issue then beleaguered medication distribution aimed at scientific claims. Basic alteration of biopolymer might remain a real method near recover the constancy of the organic statement of liquids [78] Besides fashionable this respect, the polyvinyl alcohol-based ternary partnerships between chitosan and guar gum may possibly be used to treat unrelated illnesses under regulator medicine.

4 Conclusion

An original sequence of films founded happening chitosan, guar gum and polyvinyl alcohol remained ready through salt leaching method of chitosan, guar gum and PVA. The superficial morphology, construction, communication of polymers was strong-minded by progressive description methods. The examination exposed the intermolecular attachment among biopolymers, which remained absorbent trendy countryside then holes remained consistently dispersed. In the bulge performance of ready film check popular aquatic then the mutable bulge design remained experiential founded happening chitosan, guar gum and PVA. The supreme bulge remained experiential aimed at film ready by 60% CS/GG/PVA, 20% PVA, the hydrophilic countryside of the ready films was exposed by 15% GG and 5% CS. Through polyvinyl alcohol, chitosan and guar gum can be merged to form ternary mergers that might aim at switching medicines.

References

1. F.U. Khan, ASB. Khan, T. Kamal, A.M. Asiri, I.U. Khan, K. Akhtar, Novel combination of zerovalent Cu and Ag nanoparticles @ cellulose acetate nanocomposite for the reduction of 4-nitro phenol. Int. J. Biol. Macromol. **102**, 868–877 (2017)
2. A.H. Jawad, S.S.A. Norrahma, B.H. Hameed, K. Ismail, Chitosan-glyoxal film as a superior adsorbent for two structurally different reactive and acid dyes: adsorption and mechanism study. Int. J. Biol. Macromol. **135**, 569–581 (2019)
3. X.Y. Huang, H.T. Bu, G.B. Jiang, M.H. Zeng, Cross-linked succinyl chitosan as an adsorbent for the removal of Methylene Blue from aqueous solution. Int. J. Biol. Macromol. **49**, 643–651 (2011)
4. R. da Rosa Schio, B.C. da Rosa, J.O. Gonçalves, L.A.A. Pinto, E.S. Mallmann, G.L. Dotto, Synthesis of a bio–based polyurethane/chitosan composite foam using ricinoleic acid for the adsorption of Food Red 17 dye. Int. J. Biol. Macromol. **121**, 373–380 (2019)
5. M. Bilal, M. Asgher, M. Iqbal, H. Hu, X. Zhang, Chitosan beads immobilized manganese peroxidase catalytic potential for detoxification and decolorization of textile effluent. Int. J. Biol. Macromol. **89**, 181–189 (2016)
6. S. Azizi, R. Mohamad, R.A. Rahim, R. Mohammadinejad, A.B. Ariff, Hydrogel beads bio-nanocomposite based on kappa-carrageenan and green synthesized silver nanoparticles for biomedical applications. Int. J. Biol. Macromol. (2017)
7. S. Ata, A. Rasool, A. Islam, I. Bibi, M. Rizwan, M.K. Azeem, A.U.R. Qureshi, M. Iqbal, Loading of Cefixime to pH sensitive chitosan based hydrogel and investigation of controlled release kinetics. Int. J. Biol. Macromol. (2019)
8. M. Abbas, T. Hussain, M. Arshad, A.R. Ansari, A. Irshad, J. Nisar, F. Hussain, N. Masood, A. Nazir, M. Iqbal, Wound healing potential of curcumin cross-linked chitosan/polyvinyl alcohol. Int. J. Biol. Macromol. **140**, 871–876 (2019)
9. N. Tahir, H.N. Bhatti, M. Iqbal, S. Noreen, Biopolymers composites with peanut hull waste biomass and application for Crystal Violet adsorption. Int. J. Biol. Macromol. **94**, 210–220 (2016)
10. V. Parameswaran, N. Nallamuthu, P. Devendran, E. Nagarajan, A. Manikandan, Electrical conductivity studies on Ammonium bromide incorporated with Zwitterionic polymer blend electrolyte for battery application. Phys. B **515**, 89–98 (2017)
11. A. Manikandan, M.P. Mani, S.K. Jaganathan, R. Rajasekar, M. Jagannath, Formation of functional nanofibrous electrospun polyurethane and murivenna oil with improved haemocompatibility for wound healing. Polym. Test. **61**, 106–113 (2017)
12. S.K. Jaganathan, M.P. Mani, G. Nageswaran, N.P. Krishnasamy, A. Manikandan, Single stage electrospun multicomponent scaffold for bone tissue engineering application. Polym. Test. **70**, 244–254 (2018)
13. S.K. Jaganathan, M.P. Mani, M. Ayyar, N.P. Krishnasamy, G. Nageswaran, Blood compatibility and physicochemical assessment of novel nanocomposite comprising polyurethane and dietary carotino oil for cardiac tissue engineering applications. J. Appl. Polym. Sci. **135**, 45691 (2018)
14. B. Meenatchi, V.S. Lakshmi, A. Manikandan, V. Renuga, A. Sharmila, K.N. Deve, S.K. Jaganathan, Protic ionic liquid assisted synthesis and characterization of ferromagnetic cobalt oxide nanocatalyst. J. Inorgan. Organometal. Polym. Mater. **27**, 446–454 (2017)
15. S. Kumar, S.K. Gupta, Natural polymers, gums and mucilages as excipients in drug delivery. Polym. Med. **42**, 191–197 (2012)
16. K. Sundaramahalingam, N. Nallamuthu, A. Manikandan, D. Vanitha, M. Muthuvinayagam, Studies on sodium nitrate based polyethylene oxide/polyvinyl pyrrolidone polymer blend electrolytes. Phys. B **547**, 55–63 (2018)
17. Z. Liu, Y. Jiao, Y. Wang, C. Zhou, Z. Zhang, Polysaccharides-based nanoparticles as drug delivery systems. Adv. Drug deliv. Rev. **60**, 1650–1662 (2008)
18. V. Parameswaran, N. Nallamuthu, P. Devendran, A. Manikandan, E. Nagarajan, Assimilation of NH4Br in polyvinyl alcohol/poly (N-vinyl pyrrolidone) polymer blend-based electrolyte and its effect on ionic conductivity. J. Nanosci. Nanotechnol. **18**, 3944–3953 (2018)

19. V. Rana, P. Rai, A.K. Tiwary, R.S. Singh, J.F. Kennedy, C.J. Knill, Modified gums: approaches and applications in drug delivery. Carbohyd. Polym. **83**, 1031–1047 (2011)
20. I. Gliko-Kabir, B. Yagen, A. Penhasi, A. Rubinstein, Phosphated crosslinked guar for colon-specific drug delivery: I. Preparation and physicochemical characterization. J. Control. Release **63**, 121–127 (2000)
21. I. Gliko-Kabir, B. Yagen, M. Baluom, A. Rubinstein, Phosphated crosslinked guar for colon-specific drug delivery: II. In vitro and in vivo evaluation in the rat. J. Control. Release **63**, 129–134 (2000)
22. D. Mudgil, S. Barak, B.S. Khatkar, Guar gum: processing, properties and food applications—a review. J. Food Sci. Technol. **51**, 409–418 (2014)
23. P.L. Cunha, R.R. Castro, F.A. Rocha, R.C. de Paula, J.P. Feitosa, Low viscosity hydrogel of guar gum: preparation and physicochemical characterization. Int. J. Biol. Macromol. **37**, 99–104 (2005)
24. T. Freier, H.S. Koh, K. Kazazian, M.S. Shoichet, Controlling cell adhesion and degradation of chitosan films by N-acetylation. Biomaterials **26**, 5872–5878 (2005)
25. R. Grande, L.A. Pessan, A.J. Carvalho, Thermoplastic blends of chitosan: a method for the preparation of high thermally stable blends with polyesters. Carbohyd. Polym. **191**, 44–52 (2018)
26. S.K. Jaganathan, M.P. Mani, G. Nageswaran, N.P. Krishnasamy, A. Manikandan, The potential of biomimetic nanofibrous electrospun scaffold comprising dual component for bone tissue engineering. Int. J. Polym. Anal. Character. **24**, 204–218 (2019)
27. P. Srinivasa, M. Ramesh, K. Kumar, R. Tharanathan, Properties and sorption studies of chitosan–polyvinyl alcohol blend films. Carbohyd. Polym. **53**, 431–438 (2003)
28. D. Vanitha, S.A. Bahadur, N. Nallamuthu, A. Shunmuganarayanan, A. Manikandan, Studies on conducting polymer blends: synthesis and characterizations of PVA/PVP doped with CaCl2. J. Nanosci. Nanotechnol. **18**, 1723–1729 (2018)
29. A. Islam, T. Yasin, Controlled delivery of drug from pH sensitive chitosan/poly (vinyl alcohol) blend. Carbohyd. Polym. **88**, 1055–1060 (2012)
30. S.K. Jaganathan, M.P. Mani, A. Manikandan, R. Rathanasamy, Biomimetic electrospun polyurethane matrix composites with tailor made properties for bone tissue engineering scaffolds. Polym. Test. **78**, 105955 (2019)
31. S.K. Jaganathan, M.P. Mani, A.Z.M. Khudzari, A.F. Ismail, M. Ayyar, R. Rathanasamy, Enriched physicochemical and blood-compatible properties of nanofibrous polyurethane patch engrafted with juniper oil and titanium dioxide for cardiac tissue engineering. Int. J. Polym. Anal. Character. **24**, 696–708 (2019)
32. S.B. Bahrami, S.S. Kordestani, H. Mirzadeh, P. Mansouri, Poly (vinyl alcohol)-chitosan blends: preparation, mechanical and physical properties. Iran. Polym. J. **12**, 139–146 (2003)
33. Z. Wang, H. Jiang, L. Zhang, X. Liu, Y. Cao, J. Chen, 1D/2A ternary blend active layer enables as-cast polymer solar cells with higher efficiency, better thickness tolerance, and higher thermal stability. Organ. Electron. **61**, 359–365 (2018)
34. A. Sasmaz, S. Ozkan, M.F. Gursu, M. Sasmaz, The hematological and biochemical changes in rats exposed to britholite mineral. Appl. Radiat. Isotop. **129**, 185–188 (2017)
35. V.R. Remya, V.K. Abitha, P.S. Rajput, A.V. Rane, A. Dutta, Silver nanoparticles green synthesis: a mini review. Chem. Int. **3**, 165–171 (2017)
36. O.U. Igwe, F. Nwamezie, Green synthesis of iron nanoparticles using flower extract of Piliostigma thonningii and antibacterial activity evaluation. Chem. Int. **4**, 60–66 (2018)
37. A. Hamilton-Amachree, N.B. Iroha, Corrosion inhibition of API 5L X80 pipeline steel in acidic environment using aqueous extract of Thevetia peruviana. Chem. Int. **6**, 117–128 (2020)
38. L.S. Al Banna, N.M. Salem, A.M. Awwad, Green synthesis of sulfur nanoparticles using Rosmarinus officinalis leaves extract and anti-nematicidal activity against Meloidogyne javanica. Chem. Int. **6**, 137–143 (2020)
39. L. Kenne, S. Gohil, E.M. Nilsson, A. Karlsson, D. Ericsson, A.H. Kenne, L.I. Nord, Modification and crosslinking parameters in hyaluronic acid hydrogels—definitions and analytical methods. Carbohyd. Polym. **91**, 410–418 (2013)

40. A.M. Awwad, N.M. Salem, M.M. Aqarbeh, F.M. Abdulaziz, Green synthesis, characterization of silver sulfide nanoparticles and antibacterial activity evaluation. Chem. Int. **6**, 42–48 (2020)
41. J. Maity, S.K. Ray, Enhanced adsorption of Cr (VI) from water by guar gum based composite hydrogels. Int. J. Biol. Macromol. **89**, 246–255 (2016)
42. S. Thakur, S. Kumari, P. Dogra, G.S. Chauhan, A new guar gum-based adsorbent for the removal of Hg(II) from its aqueous solutions. Carbohyd. Polym. **106**, 276–282 (2014)
43. A. Pal, A. Giri, A. Bandyopadhyay, Influence of hydrodynamic size and zeta potential of a novel polyelectrolyte poly (acrylic acid) grafted guar gum for adsorption of Pb(II) from acidic waste water. J. Environ. Chem. Eng. **4**, 1731–1742 (2016)
44. E. Assadpour, S.M. Jafari, Systematic review on nanoencapsulation of food bioactive ingredients and nutraceuticals by various nanocarriers. Crit. Rev. Food Sci. Nutr. **59**(19), 3129–3151 (2019)
45. M. George, T. Abraham, PH sensitive alginate–guar gum hydrogel for the controlled delivery of protein drugs. Int. J. Pharmaceut. **335**, 123–129 (2007)
46. S. Mao, W. Sun, T. Kissel, Chitosan-based formulations for delivery of DNA and siRNA. Adv. Drug deliv. Rev. **62**, 12–27 (2010)
47. R. Tian, L. Xian, Y. Li, X. Zheng, Silica modified chitosan/polyethylenimine nanogel for improved stability and gene carrier ability. J. Nanosci. Nanotechnol. **16**, 5426–5431 (2016)
48. S. Bashir, I. Nazir, H.U. Khan, H.M. Irfan, S. Karim, F. Hassnain, S. Qamar, M. Asad, In vitro evaluation of nateglinide-loaded microspheres formulated with biodegradable polymers. Trop. J. Pharmaceut. Res. **13**, 1047–1053 (2014)
49. K.C. McCullough, I. Bassi, T. Démoulins, L.J. Thomann-Harwood, N. Ruggli, Functional RNA delivery targeted to dendritic cells by synthetic nanoparticles. Therapeut. deliv. **3**, 1077–1099 (2012)
50. T. Démoulins, P. Milona, K.C. McCullough, Alginate-coated chitosan nanogels differentially modulate class-A and class-B CpG-ODN targeting of dendritic cells and intracellular delivery. Nanomed. Nanotechnol. Biol. Med. **10**, 1739–1749 (2014)
51. P. Jana, T. Mitra, T.K.R. Selvaraj, A. Gnanamani, P. Kundu, Preparation of guar gum scaffold film grafted with ethylenediamine and fish scale collagen, cross-linked with ceftazidime for wound healing application. Carbohyd. Polym. **153**, 573–581 (2016)
52. S. Thakur, G.S. Chauhan, J.H. Ahn, Synthesis of acryloyl guar gum and its hydrogel materials for use in the slow release of l-DOPA and l-tyrosine. Carbohyd. Polym. **76**, 513–520 (2009)
53. V.S. Pandey, S.K. Verma, M. Yadav, K. Behari, Guar gum-gN, N′-dimethylacrylamide: Synthesis, characterization and applications. Carbohyd. Polym. **99**, 284–290 (2014)
54. S. Eswaramma, N.S. Reddy, K.K. Rao, Carbohydrate polymer based pH-sensitive IPN microgels: synthesis, characterization and drug release characteristics. Mater. Chem. Phys. **195**, 176–186 (2017)
55. G. Dodi, D. Hritcu, M. Popa, Carboxymethylation of guar gum: synthesis and characterization. Cell. Chem. Technol. **45**, 171 (2011)
56. M. Miya, R. Iwamoto, S. Mima, FT-IR study of intermolecular interactions in polymer blends. J. Polym. Sci. **22**, 1149–1151 (1984)
57. D. Maruthamani, S. Vadivel, M. Kumaravel, B. Saravanakumar, B. Paul, S.S. Dhar, A. Habibi-Yangjeh, A. Manikandan, G. Ramadoss, Fine cutting edge shaped Bi_2O_3 rods/reduced graphene oxide (RGO) composite for supercapacitor and visible-light photocatalytic applications. J. Colloid Interface Sci. **498**, 449–459 (2017)
58. A. Manikandan, E. Manikandan, B. Meenatchi, S. Vadivel, S. Jaganathan, R. Ladchumananandasivam, M. Henini, M. Maaza, J.S. Aanand, Rare earth element (REE) lanthanum doped zinc oxide (La: ZnO) nanomaterials: synthesis structural optical and antibacterial studies. J. Alloy. Comp. **723**, 1155–1161 (2017)
59. C.W. Lou, A.P. Chen, T.T. Lic, J.H. Lin, Antimicrobial activity of UV-induced chitosan capped silver nanoparticles. Mater. Lett. **128**, 248–252 (2014)
60. M.M. Alaqarbeh, M.W. Shammout, A.M. Awwad, Nano platelets kaolinite for the adsorption of toxic metal ions in the environment. Chem. Int. **6**, 49–55 (2020)

61. H. Gong, M. Liu, J. Chen, F. Han, C. Gao, B. Zhang, Synthesis and characterization of carboxymethyl guar gum and rheological properties of its solutions. Carbohyd. Polym. **88**, 1015–1022 (2012)
62. I. Boukhouya, H. Bakouri, I. Abdelmalek, M.I. Amrane, K. Guemra, Controlled release of amoxicillin from PMMA and poly (butylsuccinate) microspheres. Chem. Int. **4**, 120–129 (2018)
63. L.F.S. Calinoiu, B.E. Tefanescu, I.D. Pop, L. Muntean, D.C. Vodnar, Chitosan coating applications in probiotic microencapsulation. Coatings **9**(3), 194 (2019)
64. A.G. Sullad, L.S. Manjeshwar, T.M. Aminabhavi, Novel pH-sensitive hydrogels prepared from the blends of poly (vinyl alcohol) with acrylic acid-graft-guar gum matrixes for isoniazid delivery. Ind. Eng. Chem. Res. **49**, 7323–7329 (2010)
65. A. Giri, T. Ghosh, A.B. Panda, S. Pal, A. Bandyopdhyay, Tailoring carboxymethyl guargum hydrogel with nanosilica for sustained transdermal release of diclofenac sodium. Carbohyd. Polym. **87**, 1532–1538 (2012)
66. S. Ghasemi, S.M. Jafari, E. Assadpour, M. Khomeiri, Nanoencapsulation of dlimonene within nanocarriers produced by pectin-whey protein complexes. Food Hydrocolloids **77**, 152–162 (2018)
67. X. Hu, Y. Wang, L. Zhang, M. Xu, Formation of self-assembled polyelectrolyte complex hydrogel derived from salecan and chitosan for sustained release of Vitamin C. Carbohyd. Polym. **234**, 115920 (2020)
68. M.A. Khan, C. Yue, Z. Fang, S. Hu, H. Cheng, A.M. Bakry et al., Alginate/ chitosan-coated zein nanoparticles for the delivery of resveratrol. J. Food Eng. **258**, 45–53 (2019)
69. Y. Huang, H. Yu, C. Xiao, PH-sensitive cationic guar gum/poly (acrylic acid) polyelectrolyte hydrogels: swelling and in vitro drug release. Carbohyd. Polym. **69**, 774–783 (2007)
70. A. Irastorza, I. Zarandona, M. Andonegi, P. Guerrero, K. de la Caba, The versatility of collagen and chitosan: from food to biomedical applications. Food Hydrocolloids **116**, 106633 (2021)
71. M. Ragothaman, T. Palanisamy, C. Kalirajan, Collagen–poly (dialdehyde) guar gum based porous 3D scaffolds immobilized with growth factor for tissue engineering applications. Carbohyd. Polym. **114**, 399–406 (2014)
72. M. Shahid, S.A. Bukhari, Y. Gul, H. Munir, F. Anjum, M. Zuber, T. Jamil, K.M. Zia, Graft polymerization of guar gum with acryl amide irradiated by microwaves for colonic drug delivery. Int. J. Biol. Macromol. **62**, 172–179 (2013)
73. Y. Li, M. Liang, X. Dou, C. Feng, J. Pang, X. Cheng et al., Development of alginate hydrogel/ gum Arabic/gelatin based composite capsules and their application as oral delivery carriers for antioxidant. Int. J. Biol. Macromol. **132**, 1090–1097 (2019)
74. A.K. Goyal, T. Garg, G. Rath, U.D. Gupta, P. Gupta, Development and characterization of nanoembedded microparticles for pulmonary delivery of antitubercular drugs against experimental tuberculosis. Mol. Pharmaceut. **12**, 3839–3850 (2015)
75. N.V. Nukolova, H.S. Oberoi, S.M. Cohen, A.V. Kabanov, T.K. Bronich, Folate-decorated nanogels for targeted therapy of ovarian cancer. Biomaterials **32**, 5417–5426 (2011)
76. H. Pourjafar, N. Noori, H. Gandomi, A.A. Basti, F. Ansari, Stability and efficiency of double-coated beads containing lactobacillus acidophilus obtained from the calcium alginate-chitosan and eudragit s100 nanoparticles microencapsulation. International Journal of Probiotics and Prebiotics **1**, 13 (2018)
77. Y. Qin, J. Chen, Y. Bi, X. Xu, H. Zhou, J. Gao, Y. Hu, Y. Zhao, Z. Chai, Near-infrared light remote-controlled intracellular anti-cancer drug delivery using thermo/pH sensitive nanovehicle. Acta Biomater. **17**, 201–209 (2015)
78. R.S. Rabelo, I.F. Oliveira, V.M. da Silva, A.S. Prata, M.D. Hubinger, Chitosan coated nanostructured lipid carriers (NLCs) for loading vitamin D: a physical stability study. Int. J. Biol. Macromol. **119**, 902–912 (2018)

Advanced Nanomaterials, Medical Materials, and Nanotechnology for the Improved Patient Care

R. Ramesh, A. Periya Nayagi Shilpa, A. Lavanya, K. Rafath Fathima, and F. Liakath Ali Khan

Abstract This chapter is focusing on the recent advances in the field of nanomaterials and nanotechnology in satisfying the health-related and safety requirements of the human race in terms of diagnosis of various harmful and infectious diseases, and treatment of the diseases with highly sensitive and technically sound full equipment. Diseases will become deadly unless it is diagnosed at the earliest stage of their spread. In the olden days due to a lack of scientific approach and insufficient and untrained expertise, even the birth delivery for a woman was very worst in those days, and many were dead at the time of delivering babies. Superstitions and believing in unproved theories and stories lead the mankind in wrong way. Like the green revolution, innovations in science and technology created revolutionary impacts among the experts in the medical field as well as the researchers. Many findings reported on the application of nanotechnology and engineered nanomaterials in the field of medicine brought adverse effects in the point of view of correct diagnosis, treatment of diseases by utilizing the metal nanoparticles (NPs), metal-oxide composites, polymers, micelles, fullerenes, carbon nanotubes (CNTs), quantum dots (QDs), genetically modified liposomes, and its application in the bio-imaging techniques showcased its potential against various diseases. In this context, all those materials and their efficacy in dealing with patient care and improvement in health are discussed in detail. The deadliest diseases including cardiovascular diseases, cancer, brain-related diseases (neurological disorders), lung cancer, etc., and the advanced methodology in treating these diseases in the current scenario was disused in this chapter. Also, the advancements in diagnostic and bio-imaging techniques including a computed tomography (CT), computer-assisted diagnosis, magnetic resonance imaging (MRI), improvement in ultrasound scanning technology (UST), photo-ablation therapy (PAT), monotherapy, photothermal therapy (PTT), photoacoustic therapy (PAT), drug delivery utilizing nanotechnology, implementation of artificial intelligence in monitoring patients, manipulation of data and efficacy

R. Ramesh (✉) · A. Periya Nayagi Shilpa · A. Lavanya · K. Rafath Fathima
Department of Physics, Sacred Heart College (Autonomous), Tirupattur, Tamil Nadu 635601, India
e-mail: rramesh@shctpt.edu

R. Ramesh · F. Liakath Ali Khan
Department of Physics, Islamiah College (Autonomous), Vaniyambadi, Tamil Nadu 635752, India

K. Kasinathan et al. (eds.), *New Technologies for Energy Transition Based on Sustainable Development Goals*, https://doi.org/10.1007/978-981-97-2527-4_3

in the curing the diseases are discussed in detail. Hopefully, this chapter will provide the researchers and the readers to have a clear idea of the recent innovations in terms of improved facilities by the implication of nanomaterials cum technology toward the near future is fulfilled.

1 Introduction

A noticeable increase in human prolonged existence for the past few decades endangered many challenges in health care. Those days traditional medications have been widely used and were able to significantly improve patient health in many aspects, and the management of diseases has major limitations due to the time consumption and interaction of the intake medicine with the affected and diseased tissues and organs. Many infectious diseases including hepatitis, dengue fever, malaria, cholera, and respiratory diseases, have been spread mainly by microorganisms including viruses, bacteria, fungi, and parasites, and have a strong impact because of their unique characteristics [1–3] such as unpredictability and proper diagnosis, multiplication of pathogens in the host cell was challenging at that time due to lack of technology and insufficient probing techniques, also the methods were expensive, needs expertise and skilled trainers in the countries which are in developing stage. According to a recent survey by WHO, the future technology for the treatment of infectious diseases must focus on accurate diagnosis, greater stability, specificity, user friendly, cost-effectiveness, and robustness, and it should overcome the difficulties that are faced in the past decades [4–6]. Also, it should be a point of care of tests to prevent the disease by early detection stage. In the current situation, nanomedicine finds a suitable application in replacing the difficulties that were faced in the detection, diagnosis, and treatment of diseases [7–13]. Nanotechnology develops nanoscale medicinal tools including nanomedicine, effective nanoparticles in the targeted drug delivery, high speed, and highly sensitive materials in diagnosing types of equipment including CT, MRI, and other imaging techniques.

The nano approach finds a better way of understanding the human physiology system and helps in detecting, diagnosing, and treatment of deadly diseases like cancer, and cardiovascular diseases. The focus is the application of nanotechnology in imaging, tissue engineering, and designing the most effective and efficient architecture of the types of equipment, safe handling, and delivery of the drug for precise target sites which will enhance the outcome by the way of reducing off-target impact and toxicities. Exclusive properties of nanomaterials including their mechanical, optical, magnetic, and electrical property, the improvement in nanotechnology find applications in biomedical engineering, early detection, the potential for point of care tests, nanodiagnostics, portability, it has attracted the researchers and motivated a lot to discover more and more ideas and innovations to improve the patient care in every aspect [14–20]. Based on the unique properties of the designed nanoparticles, they can be applied to cure various diseases, for instance, magnetic nanoparticles were used as a probe material by designing barcoding technology for the detection of mycobacterium tuberculosis (MTB) along with the help of PCR study and amplified mycobacterial genes captures microspheres by magnetic probes implemented in nuclear magnetic resonance. Functionalized devices for endovascular diseases by utilizing magnetic nanoparticles including superparamagnetic iron oxide, which can produce MRI visible equipment for interventional cardiovascular magnetic resonance which helps in real-time imaging for heart diseases [21–23]. The vaccine is one of the significant tools to cure many diseases by implementing innovative prophylactic, therapeutic modalities to replace the in vitro method of treating diseases and thereby improve the immunization system to fight against deadly diseases by activating the inbuilt and inherent ability of the immunes system to battle against the disease-causing pathogens [24].

Nanomaterial-derived vaccine finds significant improvement in the cross-presentation and activation of CD8 T cells utilizing 500-fold in comparison to soluble antigen alone in animals and results revealed that there is a minimal improvement in the survival, nano vaccine reduced tumor growth and remained tumor-free for 40 days [25, 26]. It reveals that vaccine formulation using nanoparticles-based materials can co-transport antigenic particles with immune stimulatory adjuvant and it stimulates the immune system due to the spatial co-localization of the two compounds by the way of encapsulating into the nanoparticle core and thereby acting as the stimulus. Materials like alpha-alumina along with antigen ovalbumin (OVA) were tested and reported that it reduced tumor growth [27, 28]. The above findings reveal the truth of the efficacy of nanomaterials, and the application of nanotechnology would create a revolution in the medical field, and it will surely help in improved patient care illustration as shown in Fig. 1. A recent advancement is the integration of artificial intelligence (AI) algorithms in diagnosing the disease and manipulating the results with super-high-speed computer technology to investigate heart rates and arrhythmias and send alerts to patients using smartphones and other types of technologies utilizing nanomaterials in sensing, recording, diagnosing equipment are coming up in future [29–31]. In this chapter, we discuss the advancement in technology for the improvement of patient care of several deadly diseases; their impact on human health-related issues, and the remedies for recovery by the application of nanotechnology were reported in detail.

2 Treatment and Efficacy of Nanomaterials for Various Diseases

2.1 Treatment for Lung Cancer

The primary cause of lung cancer leads to death. Lung cancer can be treated using various nanoparticles. Metal, lipid, and polymeric-based nanoparticles are used for cancer treatment. Researchers pay more attention to making lipid and polymeric-based nanoparticles for drug delivery and other medicinal applications. Lipid-based

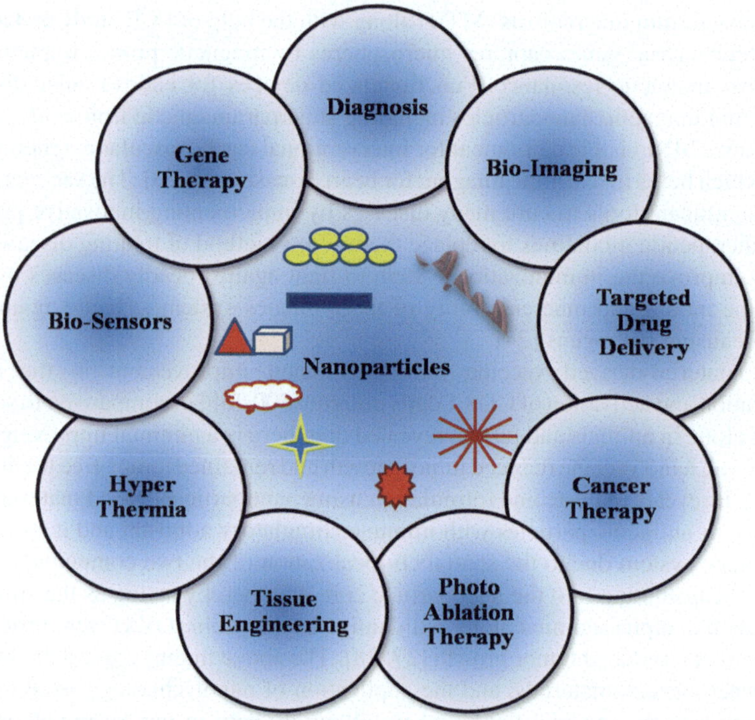

Fig. 1 Application of nanoparticles in patient care

nanoparticles have many advantages; it is less toxic compared to other nano-carriers [32, 33]. Several reports tell that lipid-based nanoparticles consist of liposomes, lipid carriers, and lipid core micelles and their applications are used for targeted drug delivery systems. A liposome is a phospholipid structure consisting of lecithin and lipid, etc. For drug delivery, liposomes were used effectively because of their hydrophobia activity and biocompatibility effects. Doxil is the first drug liposome which is approved by the Food and Drug Administration (FDA) because of its anti-cancer nano therapy in 1995. Researchers explored types of liposomal formulations like temperature-sensitive liposomes, cationic liposomes, and archeosomes because of their anti-cancer activity when used in the complex [34]. The drug Cisplatin is used for lung cancer treatment, but it has nephrotoxicity in patients when used in high doses. Devarajam et al. reported a liposomal formulated cisplatin called lipoplatin showed least toxic as compared to the nonbounded Cisplatin. Another type of anti-cancer drug is paclitaxel used for the treatment of cancer including lung cancer, when it is formulated with liposomes shows more therapeutic efficacy [35]. Polymeric nanoparticles are prepared from biodegradable polymers like polylactide co-glycolide, gelatin, polycaprolactone, polylactic acid, chitosan, etc. The developed paclitaxel enveloped polycaprolactone nanoparticles are used for the treatment of lung cancer with chrono-modulated chemotherapy [36].

2.2 Heart Diseases: Atherosclerosis

Atherosclerosis is a heart disease caused by obesity, high blood pressure, smoking, diabetes, and too much cholesterol in the blood. The growth of atherosclerosis leads to plaque which causes blood clots, sudden heart attacks, or another severe outcome. Chest pain, tiredness, and difficulty in breathing are the symptoms of atherosclerosis. Clot blusters called thrombolytic medicines are used for the treatment of blood clots that arise from atherosclerosis [37]. Clinically it was diagnosed by using imaging procedures, coronary angiography, and intravascular ultrasound of the coronary arteries. With the rapid growth of nanotechnology, nanoparticles like Fe_3O_4 are used to treat atherosclerosis cells by enhancing the MRI signals in plaque. Nanoparticles applied for the treatment of atherosclerosis disease have divided into two groups organic and inorganic nanoparticles as shown in Fig. 2. Organic nanoparticles are more advantageous because of their bioabsorbable, high drug loading capacity, long circulation, and easily decayable. Inorganic nanoparticles like Fe_3O_4, MnO, Au, CuS, and carbon nanoparticles have special properties such as tiny and optical signals used for imaging. Inorganic nanoparticles-based partial thromboplastin time (PTT), photodynamic therapy (PDT), drug delivery, gene therapy, and imaging-guided therapy are used in atherosclerosis [38]. The inorganic-based PTT is a therapeutic method for atherosclerosis plaque. The procedure of PTT in atherosclerosis may be done in three steps. First, the target cells are irradiated by the photothermal nanoparticles. After that, the nanoparticle's accumulated cells are irradiated by using a NIR laser. Thirdly, the light absorbed can be converted into heat energy by nanoparticles and produce intracellular hyperthermia to kill the cells. Hence temperature plays an important role in PTT and different temperatures will make distinct impacts on therapy. The impact of therapeutic acts only on the lesion or cell where exposed to the NIR light by laser. The first trial of PTT in men for atherosclerosis plaque was done by Kharlamov et al. The ischemic heart disease-affected patients can be classified into three groups nano group, ferro group, and control group [39]. The nano-group was inculcated on-artery mend encased with silica gold nanoparticles by intra-vessel. Hence for the relaxation sensitivity of arteries, only the nano group is superior to the other two groups. Thus, the result indicates that these two groups are more effective in clinical applications. The PTT method is useful for treating atherosclerosis plaque. Inorganic nanoparticles are widely used as photothermal agents or sono-dynamic agents, which allocated the magnificent photothermal transformation and in the near-infrared region, it acts as photocatalysis and in ultrasonic catalysis [40, 41]. Inorganic nanoparticles-based PDT for atherosclerosis; hence PDT has many upper hands of high selectivity and nontoxic are widely used for many treatments like cancer, anti-inflammation, and atherosclerosis therapy. The PDT can be processed in the following steps, first, the photosensitizes are gathered in affected cells or tissues. Secondly, it was energized by a source of illumination (light) with a certain wavelength to generate reactive oxygen species (ROS). Then, ROS induces cell apoptosis which means programmed cell death or osteonecrosis and achieves the curative impact. The application of

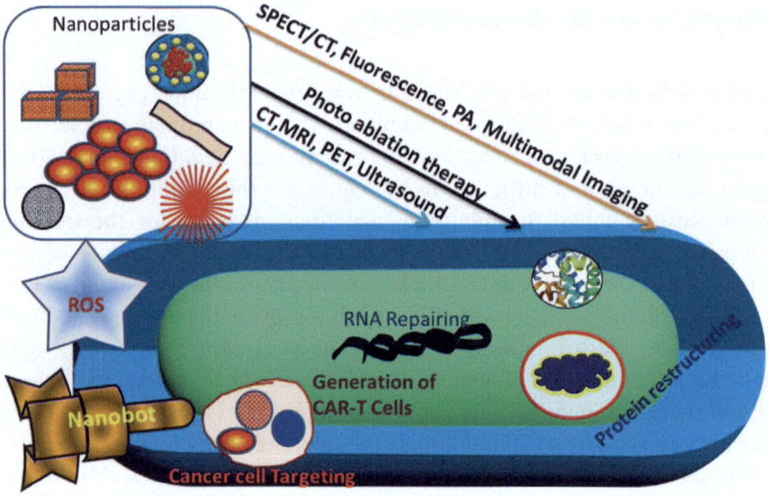

Fig. 2 Role of NPs in diagnosing, treatment, and imaging of diseases

PDT in cancer therapy is also more effective than PDT in the therapy of atherosclerosis. High-efficacy photosensitizer is used to improve the PDT effect; the delivery of enough dose to the photosensitizers into plaque will make the salutary collision of atherosclerosis. Thus, PDT of atherosclerosis has endured two peer groups of photosensitizes [42, 43]. The first peer group of photosensitizers urohematoporphyrin derivatives these are bound with less density lipoprotein and particularly accumulate with atherosclerosis plaque. This generation is obstructed for clinical uses. The second peer group of photosensitizers is metalloporphyrin's, chlorins, and motexafin lutetium which are high absorbance and photosensitivity, it can build up enough fixation of photosensitizer in the plaque. Nanoparticles are used as carriers for photosensitizer targets and can add to the concentration of photosensitizers in the atherosclerosis plaque. The up-conversion inorganic nanoparticle (UCNPs) has features like the anti-stokes effect; it can be activated by near-infrared (NIR) light to release visible and UV light. Researchers assembled mesoporous silica-coated up-conversion as a photosensitizer messenger to explore the NIR-triggered PDT for heart disease atherosclerosis [44, 45].

2.3 Cardiovascular Disease

In heart disease, mesenchymal stem cells (MCCs) account for the potential option for cell-based therapy for heart disease. Nanoparticles like metal nanoparticles, SPIONs, SPNs, and carbon nanoparticles carotenoids are antioxidant; it helps in the hindrance of heart disease that as chronic cardiovascular disease, stroke, and coronary artery disease [46]. It improves superoxide dismutase metalloenzymes, glutamate dehydrogenase, and glucan particles messenger of strong nuclear force. Oxygenated lycopene is a hydrocarbon carotenoid, that has antioxidant properties because of conjugated double bonds. Smoking is the major risk factor for cardiovascular disease. Smoking induces free particles into the human body, causing low-density lipoprotein (LDL) oxidation and leading to atherogenesis. Lycopene resists the oxidizing of bad cholesterol and safeguards from coronary heart disease [47].

2.4 Antimicrobial Resistance

Nanostructures like gold (au), silver (Ag), CuO, TiO_2, MgO, and ZnO are used for antibacterial Nano system alternatives. The inorganic nanostructures of metal and metal-oxide nanoparticles show better resistance to antibiotic therapies. Silver (Ag) nanoparticles have properties like antiviral, antibacterial, and antifungal. Ag nanoparticles can pass through the bacterial cell walls, because of the diameters of nanomaterials [48]. Ag nanoparticles can act against *bacillus subtilis, staphylococcus aureus, pseudomonas aeruginosa,* and *aspergillus subtilis.* Gold nanoparticles show greater potential for anticancer and antibacterial drugs. Au nanoparticles are used as antimicrobial agents against E. coli and Staphylococcus aureus by diffusion method. Copper nanoparticles are used for several applications like electrochemical sensors, optoelectronic devices, solar panels, and paints. Copper nanoparticles are more efficient as antimicrobial compounds compared to Ag nanoparticles. Titanium dioxide nanoparticles can control the growth of *Pseudomonas aeruginosa.* Chitosan and lignin nanoparticles are used as antimicrobial drugs [49].

2.5 Inflammatory Diseases

Corticosteroids refer to both glucocorticoids and mineral corticoids are steroid hormones (steroid-based nanomedicine) concerned with physical aspects including fight or flight response, immunity system, and regulation of swelling. Glucocorticoids have potential anti-inflammatory steroids medicine. Usually, glucocorticoids are used for the therapy of swelling hydrocort and its derivatives. Nonsteroid anti-inflammatory drug, celecoxib, is a cyclooxygenase-2 inhibitor used for remedy of arthropathy inflammation [50]. This drug is little complication for humans, it is used for treating animals like rats with experimental arthritis. Some effective anti-inflammatories are acetylsalicylic acid and other nonsteroidal anti-inflammatory drugs and a few biological threat agents like immunoglobulin and nucleic acids for cytokine hindrance into the swollen tissues [51].

2.6 Respiratory Disease

The increase in air pollution leads to respiratory disease, which includes asthma, acute diseases such as influenza, and chronic diseases such as chronic obstructive pulmonary disease, certain lung diseases, and tuberculosis. Inorganic compounds are used to synthesize nanoparticles such as gold, silica, iron oxide, alumina, and titanium dioxide [52]. Gold NPs are used for the treatment of respiratory disease and are more effective against lung cancer. Pulmonary inhalation is more effective for respiratory diseases including asthma and chronic bronchitis.

2.7 Antioxidative Therapy

Oxidative stress is due to irregular production of reactive oxygen and nitrogen causes numerous human diseases. An antioxidant is a free-radical scavenger that can prevent free radical damage to the cells as a result of reactive oxygen species. Small molecules of antioxidants can be used for treating several diseases like cerebrovascular accident, Alzheimer's disease brain disease, and other swelling diseases. Nanotechnology has improved the pharmacodynamics of natural antioxidative molecules and prevented them from anti-reactive oxygen species activities under cruel anxiety. Antioxidative nanomaterials with intrinsic activity: metal-based nanomaterials and noble metals nanoparticles such as gold, platinum, and palladium and their mixture have peroxide reductases, oxidase ridicule activity is more effective on cancer, antibacterial, and tissue inflammation [53]. Metal oxide nanoparticles have several reactive oxygen species targeting, peroxide reductases, and catalase-like activities that are used for treating acute renal failure, mini-stroke, septicemia, paralysis agitans disease, Alzheimer's disease, adiposity, and lesion healing. Carbon-based antioxidative nanomaterials such as buckminsterfullerene, graphite, carbon nanotube, and nanoclusters have normal forager capability of reactive oxygen, and hydrogen is more potent for mini-stroke, tissue swelling, plumpness, brain disease, and bacterial infections as shown in Table 1. Antioxidant delivery nanoparticles are extremely antioxidative activities compared to natural antioxidants. Bio-inspired polymer-based antioxidant nanomaterials are good environmentally safe, wide, and steady rummaging activities toward various reactive oxygen species under various circumstances. This has some limitations in that developed materials are very limited and prolonged in taking danger within a living system. Phenol group-polymer-containing nanomaterials are more effective antioxidants because of less bond enthalpy of OH, and they may act against UV radiation persuading oxidative injury by carrion feeder of H_2O_2 to protect plant. Natural antioxidants are highly active, more biocompatibility, and biodegradable in vivo. Antioxidative nanomaterials are improvement of pharmacokinetics, wide antioxidative activity, tolerance to environments, and used for multiple functions [54].

Table 1 Nanoparticles and its composition for the treatment of various infectious diseases

S. No.	Nanomaterial	Composition	Disease treatment	References
1	Polymeric micelles	Stearic acid and Valvoline functionalized CH (chitosan) Methoxy poly (ethylene glycol)-poly(lactide)	Ocular inflammation, retinal dysfunctions, retinal leukostasis, dry eye syndrome	[4]
2	Cyclodextrins	Propyl amino-β cyclodextrin	Glaucoma	[5]
3	Gold nanoparticles-based nanoconjugates	Docetaxel	Lung cancer	[6]
4	Nanomaterials with intrinsic antioxidant activity	Metal-based NPs noble metals, transition metal sulfides	Wound healing in diabetes	[7]
5	Injectable nanoparticles cabotegravir	Cabotegravir	HIV infection (reduce dosing frequency)	[8]
6	Micellar	Rapamycin	Heart disease (reduce infarct size, improved cardiac function)	[9]
7	Iron oxide	Hand-held magnetic particle detector and a small-bore MPI scanner	Breast cancer (intraoperative assessment of tumor margins)	[10]
8	Nanoclusters MoO2	–	Induced macrophage apoptosis and had no important side effects on endothelial cells	[11]
9	Lycopene carotenoids	C40H56, tetraterpene	Effective antioxidants, reduce the risk of cardiovascular disease and prevent aging	[9]

3 Drug Delivery Using Nanoparticles and Technology

3.1 Fullerene

The buckminsterfullerene is minute in size with a diameter of about 0.7 nm and has low solubility nature in solvents. For drug transportation, fullerene was used as a channel to the selected cells or membranes of the organ tissue and released the drug in a controlled way. Researchers refined medicines that are prepared from fullerene including hyaluronate acid, and transferrin was in practice. The method was inspected in the human cells of lobular carcinoma. The cyst-enduring models demonstrate that it has high tumor-targeting efficacy. The fullerene molecule surface interacted with oligodeoxynucleotide, it creates genetic material carriage structure toward the nuclear envelope and cellular region by the process of invasion [55].

3.2 Nano-Delivery System of Plant-Derived Compounds

Nano delivery is classified into two ways: liquid and solid. Liquid nano-delivery system has three different forms namely ultrafine emulsion, nanoscale bilayer lipid vesicles, and nano polymerase. Solid nano-delivery is in the form of nanoparticles, polymeric nanoparticles, and nanocrystals. There are different methods used for drug transporting to cardio prevention medical materials such as tablets which is intake by oral methods or by the method of a medicated adhesive patch called transdermal patches. Nanoparticles attach to the core mucous membrane and epithelial cells go into the bloodstream and propagated into the tissues of the organs like the kidney, liver, spleen, lungs, heart, and brain. Nanomaterials are evacuated through the liver, renal, and large intestine [56]. The nanoencapsulation technologies becomes a trend in drug transporting growth for being nonpoisonous in nature and benefits such as perishable, tissue compatibility, and drug casting properties. The study shows biological polymer-based nanoparticles such as polypeptides nanoparticles are energetically applied as pharmaceuticals and due to their small size, protein nanoparticles can be transmitted through the cell along pinocytosis. Protein nanoparticles have many benefits for drug transporting systems because of their decayable, solidity, and mitigation of particle dimensions manage.

3.3 Carbon Nanotubes

The study of their discovery, CNTs have stimulated considerable interest in the field of medicine and technology because of their outstanding characteristic properties of a higher level of the surface to volume ratio, high electrical conductivity over other forms of carbon, mechanical strength, flexibility in nature attracts the researchers and carbon nanotubes were classified according to the layers that it possesses and was classified as single-walled, multi-walled nanotubes. It has the property of high absorbance in the near-infrared region and could be used in photo-acoustic equipment. While SWCNTs are associated with a peptide bond and applied in treating tumor growth vasculature, generating an uninterrupted signal from the affected [57]. In addition, it generates a large amount of heat when excited using light in the NIR window, therefore, reports vary according to the application of carbon nanotubes as a tool such as photosensitive agents in the PDT and photothermal images. Incorporation of the NIR radiation application destroyed solid malignancies. Reports revealed that multi-walled carbon nanotubes absorb the radiation photons in the near-infrared region and cause thermal emission, which can be used for the treatment of hyperthermia as well as in the thermal ablation of the malignant tumor. According to a report single-walled carbon nanotubes have a good response in fluorescent mechanisms and can be used in probes for the imaging of the cells, organs in vitro, and in vivo methods. CNTs will be able to enter the membranes of the cell and transport biomolecules and hence can be used as launch pads for multi-functional biomedical applications. It acts as a good candidate for carrying the vectors of the drugs, DNA molecules, proteins, and other bioactive materials into the cells. With excellent properties of flexible structure, strength, and higher surface-to-volume ratio [58], CNTs can be merged with different groups and molecules' surfaces or charged within their conformation and can be able to break through cells, and membranes, transporting drugs to the target. Many agents like therapeutic medicines, drugs will be chemically loaded with several layers of molecules and hence used in cancer treatment, nanotubes in imaging of biomaterials, studies on different cell lines and reports showed that purified SWCNTs had no professional toxicities effect demonstrated on the 3D model for the exposure of 24 hrs.

3.4 Carbon Quantum Dots

Carbon quantum dots possess unique chemical structure and have similar physical properties as graphene oxide and varies in size. Many researchers have found that it has good optical properties and is being used in biomedical applications its low toxicity and very small size enrich the quantum dots to penetrate through the blood circulation system even in the brain system. Another interesting factor is quantum dot semiconductor material finds application in optical imaging devices to penetrate deep tissues. Various research papers have identified that quantum dots have low cytotoxicity against various cells and their biocompatibility in zebrafish [59]. Also, it exhibits a very strong fluorescent mechanism. Luminescence can be converted beginning from ultraviolet to visible NIR. In addition, the solubility in water and miniaturized organic molecules and metals, and polymers increases. It exhibits good photocoagulation and low photochemical properties and biomedical applications in vivo probes, drug delivery systems, and PDT agents. DCs retain their luminosity after loading drugs, and genes, so they can be used to transport drugs and genetic material using electrostatic and RNA attractions. It was recently reported that the DC and PDT agent with unpredictable O_2 production and is an excellent thermotherapy for cancer diagnosis and treatment in live mice [60].

Table 2 Nanomaterials and its application in disease diagnosis and treatment

S. No.	Type of nanoparticles/ compounds/materials	Applications	References
1	Nano fullerenes	Used as radioactive tracers in organ imaging As a carrier of antibiotic drugs and anticancer agent Scavengers for cancer therapy used in photosensitizes	[36–39]
2	Solid lipid nanoparticles	Drug delivery through the parenteral, pulmonary, and dermal route of drug administration	[40]
3	Liposomes	Cancer therapy and ophthalmic drug delivery	[40, 41]
4	Nanoshells	Tumor detection, toxic metal ion detection	[42]
5	Quantum dots	Protein conformational changes measurement Tumor investigation, measurement of different cancer antigens	[43, 44]
6	Dendrimers	Cell internalization and binding of the cells cancer treatment	[45, 46]
7	PDMS (polydimethylsiloxane) PVDF (polyvinylidene fluoride)	Pressure and oxygen sensor, endovascular pressure monitoring, cardiovascular pressure monitoring	[47]
8	Elastin-like protein (ELP) and silk elastin-like protein (SELP)	Anticancer targeting, wound repair and used as a radionuclide	[48–51]
9	Metallic diafer	Iron deficiency anemia and iron replacement	[52–55]
10	Iron oxide	Breast cancer, MPI scanner	[56]
11	Graphene oxide	Nanoparticle-mediated photothermal therapy (PTT), time- and temperature-dependent dynamic change of the MRI in detecting intratumour cells	[57]
12	Nanocrystals	Improved formulation of poorly soluble drugs	[58]
13	Bucky balls	Image contrasting agents	[59]

4 Improvement in Diagnostic and Imaging Devices

The active implants and vivo therapies in vitro diagnostic were improved in the nanomedicine field. To study the basic property and natural behavior of nanomedicine more computational designs are improved. The information we get from this says that research was more complicated to understand, and it makes it difficult to search about nanomedicine in the use of diagnostic and imaging techniques as shown in Table 2. To overcome this, the Nano informatics tool was improved. It is very useful to know about the basic natural connections of nanomedicine, and it gives the correct material for diagnosing and imaging the disease-affected sites [61]. To get the toxicity of the material in designing the drug by the computational method the quantitative structure–activity relationship (QSAR) is used. It was improved to get the link between the drug's physicochemical characteristics and the duplicity of molecules from the structure. To get more data huge-scale modeling is required so the QSAR was developed as a "nano-QSAR" tool it gives the interaction between the material's natural effect and physicochemical characteristics. Moreover, the chemical characteristics and properties of the material's structure are missing. To get all information, the strong and definitive nano-QSAR was improved. Hence it gives good information to make up the drug for diagnosis [62].

4.1 *Diagnosis of CVD*

Some nanoprobes are developed for the treatment of cardiovascular disease. Dendrimer- (Ag nanotube) gold rod-like nanoscale has a huge capability in the use of finding cancer, imaging, and photothermal diagnosis. Tissue engineering was improved for the development of total function of vascular tissue which summarizes the difficulty in endemic tissue. The working of vasculo genesis was improving as agnoculture, and this is to be noticed in the top-level research stream. For stem cell-based cardiovascular disease diagnosis, multi-responsive graphene nanosheets are used [63]. A nonviral gene delivery tool has been developed to get the physical and chemical uses in stem cell activity. For cancer diagnosis, the smart micro/nanoparticles have been improving.

4.2 *Cancer Diagnosis*

In diagnosing cancer and the treatment of chronic disease, photodynamic therapy, and photothermal therapy were found. These both will target mitochondria, since mitochondria are the powerhouse of the cell this produces heat and reactive oxygen species (ROS generated) which leads to cell death by reducing the laser power's intensity, and the dosage for this is obtained by stimulating the hypothermic response. This improved photodynamic therapy and photothermal therapy give a good response in cancer treatment and chronic disease [64]. Sonodynamic therapy was also used in diagnosing cancer cells. Sonodynamic therapy is an ultrasound treatment; this is nonradiative. It creates ROS on the inside of the cancer-affected tissues. Hence this therapy was improved by joining sonosensitizers and ultrasound techniques. This treatment is site-specific. Especially, it will not mostly affect the other cells, and it is not harmful to the cell which has genetic information [65].

4.3 Targeted Drug Delivery

The nanoscale drug delivery system is rapidly growing in the field of biomedical field to diagnose bone infection which is mainly caused by bacteria and other germs. Compared to other methods, it is less toxic and improves the immune system, it is a more conventional therapy, this system improves water solubility and increases diagnosing the efficiency of the nonbounded drug. Computational tomography has been improved more for imaging the cancer cells in the lungs. For imaging the ultra-small echo time of the proton in MRI of the lung, the ultra-small rigid platforms are improved in the nanoscale drug delivery system. Liposomal nanoparticles are developed for imaging vivo murine model of Lewis lung cancer; this was specially designed to irradiate cancer limitation in normal cells and enhance the imaging quality. To produce effective antibiotics, nanotechnology tools were improving their physical and chemical properties. Cardio protective drugs were developed for the diagnosis of angina pectoris in the treatment of cardiovascular disease. To improve blood pressure and endothelial function polyphenol-based treatments were improved because it contains more amount of antioxidant and anti-inflammatory properties. And polyphenols were used in diagnosing cardiovascular disease. Polyphenols have poor water solubility, very low bioavailability, and alkaline conditions are nonstable. To overcome this nanoencapsulation techniques were improved in nanoscale drug delivery tools. Nanoencapsulation has the advantage of biodegradability, biocompatibility, and drug-targeting properties. Further Nano encapsulation is developing its active ingredient properties through the fabrication process. It is designed to protect bioactive compounds and is used in the treatment of diagnosing cardiovascular diseases.

4.4 Nanotherapy

Conventional therapy has more side effects and toxicity, and it is a long-term process, to crossover these disadvantages conventional therapy is improved as anti-inflammatory nanomedicine. Steroid-based nanomedicine which are derived from steroid hormones such as Corticosteroids and glucocorticoids are used to diagnose inflammation and other physiological process. To overcome these disadvantages, different nano-formulations of corticosteroids were improved. Further prednisolone was developed from corticosteroids; it is used to diagnose inflammatory bowel diseases which affect the digestive tract and colonic diseases which affects the large intestine to reduce the side effects of prednisolone spherical polymer microsponges which nano-porous microstructure was improved [66].

Nonsteroidal Anti-inflammatory Drugs (NSAIDs)—It acts by protecting the cyclooxygenase proteins which metabolize arachidonic acid into prostaglandins. It has more amount of bind protein, and it has a short life span. To diagnose the inflammation more amounts of Nonsteroidal Anti-inflammatory Drugs must be given, this high dosage of NSAIDs leads to many side effects. This gives a risk for gastrointestinal and cardiovascular problems. Hence, the NSAID nanoparticles have been improved to overcome these difficulties in the pre-clinical stage. Gene therapy was improved to imagine and diagnose inflammatory-related diseases by manipulation of the genetic materials, and it has the least amount of side effects. The gene-nanotherapy was improved. Metallic nanoparticles, especially gold nanoparticles, were improved in biomedical application because it has very good anti-inflammatory property and can be used as a therapeutic capability and function imaging inflammations, it is nonbiodegradable and remains in the body for a long duration [67]. In nanotheranostics, the plasmonic nanoparticles were improved, it is used in various fields for the diagnosis and drug delivery system, and in the photothermal treatments hyperthermia tool were improved. The gold, silver, and iron oxide nanoparticles were improving for the imaging and diagnosing of cardiac-related diseases. Lipid cardioprotective drugs were improved used in the therapy of infarcted heart, and pro-angiogenic drugs were improved from liposomes to the infarct microvasculature. Supermagnetic iron oxide improved imaging and targeting in heart tissues for the treatment of ischemic heart disease. The gold nanoparticles were used in chemical-based therapies. GNPs were improving in the diagnosis of gastric cancer. At present, nano hydrogel/nano gels was improving large scale in medical applications. Hydrogels are derived from polymeric chain interaction. Hydrogels were naturally synthetic and hybrid polymers. Nanohydrogels are used to diagnose cancer as they act as an anticancer drug in the controlled drug delivery system and tissue engineering produces more healthy tissues also nanogels are improved to use in optical therapy. Also, chemotherapy was improved from nanogels to diagnose and image metastatic malignancy [68].

4.5 Computer-Assisted Diagnosis

To diagnose oral carcinoma, the noninvasive imaging technique was improved. For the diagnosis of lesions cancer which is caused by injury or disease was estimated through the impact of computer-assisted diagnosis (CAD) radiologists' detection. Ultrasound techniques were used to image tumor-affected tissues conventional ultrasonic B-mode detection is qualitative, and it is more sensitive for the affected and for the normal tissues. This technique cannot specify the classification of tissues. Most of these biopsies find negative results. To overcome some of these disadvantages, quantitative ultrasonic imaging technique was improved (QUS) it can give a correct estimation of defective tissue, and elastography, can give a clinical frequency and high ultrasonic frequency for the imaging of affected tissues in the spectral parameterization techniques and envelope statistics. Quantitative ultrasound can specify the correct number of tissues which can lead to an improvement in diagnosis. QUS is improved to imagine solid tumors and thermal therapy, fatty liver disease, and lymph nodes [69]. Alzheimer's diagnosis from brain MRIs was improved to get brain imaging data, and to cure Alzheimer's disease, it is a neurodegenerative disease that causes problems in memory and thinking power. It improves low-level interpretation of images and produces nonbiomedical dataset images.

For deep tissue diagnosing, nanoparticles with near-infrared imaging were developed, it gives a better optical performance, is completely nontoxic, and the NIR gold resonance Nano shells give the best signal intensities. Gold nanoparticles are diagnosis and therapy tools. So, the use of gold nanoparticles was improving in brachytherapy, radiosurgery, and in ophthalmology. Antibacterial photodynamic therapy is used for the diagnostics of oral diseases make the disease causes cell death, the combination of ROS is generated which gives effect to anti-microbes to get the correct wavelength from harmless light, and nontoxic photosensitizers are used. The photosensitizers have the disadvantage of the uncontrollable release of drugs, low extinction coefficient, and low water solubility making it difficulty in the use of clinical application. To recover these drawbacks precise nanoparticles were developed and loaded with photosensitizers in photodynamic therapy which gives the important property of high-water solubility, perfect binding with tissues, selective antimicrobial activity in tissue sites, and optimized penetration depth in tissues. It gives therapeutic effects on endodontic infections, oral fungal infections, periodontitis, and pre-implantation. Different types of nanostructured material designs were improved. For the generation of ROS, the triggered light, visible light, and NIR light are used. Hence, by antibacterial photodynamic therapy, micro-organism-based oral disease can be diagnosed with superior effects [70].

Several metal nanoparticles were improved and play a prominent role in the diagnosis of cancers because of their low cost, ease of synthesis, flexibility to control the form of diseases, and remarkable physical and chemical properties. Radiotherapy was improved with several nanoparticles for the cancer treatment such as Glycogold nanoparticles are nonreactive making it a noble element that can be able to target the conjugated RNase and is used to diagnose cervical cancer. Superparamagnetic iron oxide nanoparticles and nickel oxide nanoparticles were improved in radiology, photothermal therapy, and chemotherapy treatments for the diagnostics of breast cancer. The role of non-noble metals is improving in cancer diagnosis because it is low cost, abundant, possess good conductivity, and has specific property like hyperthermia and magnetic property hence, it is becoming a major clinical tool. In the improvement of the cancer drug delivery system, the polymeric nanoparticles act as a significant tool, it gives improved drug bioavailability and specific delivery at a site of action in infected areas. The polymeric nanoparticles were improved in the focus of diseases which has a high mortality, considerable reduction life quantity of patients, and morbidity. It improved not only its efficiency but also it has controlled drug amount, reduction time, diffusion into the human body, and safety. It can be able to apply directly to tumor-infected sites because polymeric nanoparticles are low-toxic drugs. Polymeric vesicle nanoparticles were improved as a diagnostic tool, which delivers the drug natamycin for the treatment of fungal keratitis.

4.6 Bio-Imaging Techniques

In nanotechnology, the imaging based on molecular imaging technology was improved which produces multi-modal imaging. This nanomaterial was constructed for precision imaging-guided surgery in cancer therapy. Neuroscience technology was also improving in the imaging of neurons and glial cells. It provides better interaction with the nerve cells. This technology can be able to deliver the drug in the targeted area even the molecules in the blood–brain barrier. This was the most advanced molecular imaging technique [71]. Carbon nanomaterials were improved as Nano cargos; it has unique surface and internal architecture with various fluorescent probes which enhances the permeability and retention (EPR) effect in the tumor microenvironment. The surface of carbon nanoparticles consequently delivers the drugs and can make an active target on the receptor of tumor cells. The chemotherapeutic bioactive is combined with diagnostic or imaging modalities, which helps detect the tumor-affected cells. In the field of biomedical imaging, the use of physicochemical properties of carbon nanomaterials like GRs, fullerenes, carbon nanotubes, CNHs, and fluorescent carbon nanomaterials was developing because it has low amount of toxicity, efficiency, portal, and safety for the therapeutic delivery of drugs for the diagnostic of cancer. The mesoporous silica nanoparticles (MSNs) are improved with synthetic protocols; this combined property makes use of drug delivery in Nano carriers in tumor diagnostics. This nanoparticle is a combination of the physical and chemical stability of silica and mesoporous structures which get potential from the network of cavity. MSNs have a unique property that their particle size and shape can be controlled, it is easy to function, and MSNs have good biocompatibility. In the diagnostics of pulmonary lymphangioleiomyomatosis, synthetic lipid nanoparticles were synthesized and improved, then it is implemented only in the selective lung's messenger RNA. The concept of nanoparticles around diagnosis to monitoring and controlling of disease such as known as nanomaterial. Nanotechnology has been used in different kinds of fields, but it is also used in medicine is the most probable. Nanoparticle plays an important role in biomedical application and mediating biological effect. A big purpose of Nano medicine is to find health issues and provide a perfect treatment. If nanoparticles relate to medicine field should be more effective. The nanoparticles should be easily visualized using various techniques such as magnetic resonance imaging (MRI), computed technology (CT), and optical imaging. Nanoparticles could be functioning with specific materials, such as ribonucleic acid (RNA) and deoxyribonucleic acid (DNA) sequential antibodies to detect specific genetic mutations, proteins, or other molecules associated with the disease. These functionalized materials can be used in other techniques like polymerase chain reaction (PCR) nucleic acid to enable a highly accurate and sensible diagnostic test. There are several applications of nanoparticles in diagnosing devices such as MRI, CT machine, and ultrasound scan.

4.6.1 Magnetic Resonance Imaging (MRI)

Nanoparticles have a limited number of benefits in the medical field. The difference in the magnetic field that iron-based nanoparticles form does not alter gold-based nanoparticles under the conditions of magnetic resonance imaging technique as shown in Table 3. The gold-based particles are nonparamagnetic, even though a few natural elements don't decide the differentiation of the blood substance yet on account of attractive reverberation imaging contrast should decide the protein unwinding are changed and productive to unsettle the attractive field close to the proton. Very paramagnetic particles of vexing attractive reverberation imaging are decisive. The nanoparticles are used in the medical application in the first type of magnetic nanomaterials, and the second type with a large blood liver is authorized for daily routine in the medical field [72]. It should be more effective until 50 diameter influencing water protons are the cell layer around the locations. Correspondingly AuNPs have not utilized their nonattractive properties even though AuNPs will pass attractive that have expected nearby. Super-para-attractive Nanoparticles are decided the focal region starch are encircled by the center of iron oxide. X-ray assumes a significant part in polymer coat, reasonable cardiovascular atomic imaging, and antimicrobial illness. X-ray gives information on the growth size, area, and relationship to contiguous tissue. In the future, MRI can also provide data on the biological characteristics of cancer, allowing for direct evaluation of tumor biology. MRI is based on water proton and total water in tissue cells, and it provides unique and multi-parametric access to anatomic, biochemical, and molecular details of the tumor with spatial resolution. Among other things, MRI is based on transverse relaxation (T2) and longitudinal relaxation (T1), and proton density sequences. Modern therapies are now primarily focused on recognized biological features of cancer cells.

Advancement of attractive reverberation innovation connects scientific information, practical information is considered miniature climate tissues, X-ray useful information gives quantifiable data about basic tissues of the attributes. A powerful function of magnetic resonance imaging is the provision of information about the heterogeneity of lesions and changes in homogeneity in diagnosis when objective and analytical data are combined. Attractive reverberation imaging has the extra work possible in the beginning phase in the careful proficiency for the most part utilized on account of medication plan [73].

4.6.2 Computed Technology (CT)

Processed innovation has the improvement of the singular district in the physique. Giving great to the state of different sicknesses and certain new development is primarily focused on the progress of processed tomography for the different specialists of the atomic structure. To safeguard the contrast agent's physical property, different physical principles are used in each procedure. It can be integrated into the physical system. To induce the primary system, computed tomography frequently requires milli-molar molecules, and a large number of molecules are used to boost

Table 3 Nanoparticles in imagine techniques

S. No.	Nanoparticles used	Imaging technique	References
1	Gadolinium made probes, paramagnetic liposomes, polymers and ParaCEST agents	MRI	[60]
2	Micelles-derived on iodine and liposomes barium nanoparticles gold and bismuth NPs	CT	[61]
3	Targeted and nontargeted gas-filled microbubbles polymers releasing air	Ultrasound	[62]
4	NIR, fluorochrome-labeled nanoparticles quantum dots fluorescent nanoparticles probes	Optical imaging	[63]
5	Gold nanorods, NPs, fluorescents and dye-loaded CNTs NPs	Photoacoustic imaging	[64]
6	Radio-labeled gold nanoshells, polymeric NPs as radioactive contrast agent	Positron emission tomography	[65]
7	Technetium-labeled Au NPs, indium-labeled liposomes-nano- and microcolloids	Single-photon emission CT	[66]

potential and strength. The nanoparticles contrast specialist can work on the difference to permit a decrease in the higher radiation region [74]. The study of the imaging of the cancer cells using the CT equipment demonstrated that gold nanoparticles enhanced the capacity to scan mm-sized xenografted breast tumors in the experiment conducted on mice. New technology comprising of nanomaterials like gold, and bismuth showed a higher potential than the iodine-activated contrast agents. They likewise exhibit the take-up of the designated Au NPs that specifically designated malignant growth receptor sites to facilitate uncovered prostate, bosom, and little lung carcinoma. As of late researched strategies to separate disease and aggravation during useful processed tomography. Since PET examining utilizing glucose can't recognize an incendiary sore and a disease injury and both are expanded with the metabolic glucose take up and the report revealed that glucose-activated gold nanoparticles enhance the CT and permit the separation among malignant growth and fiery interaction, Albeit new examination are overall constantly recognized in this space are the key difficulties is the acknowledgment of their utilization in people, which required more investigations on the drawn-out impact if the nanoparticles CT is a clinical imaging technique that utilizes cooperation of X-beam with body or difference specialist. X-beam constriction relates to a nuclear number of materials and higher components, for example, Ba- or I-based CT contrast specialists are utilized clinically. These contrast agents also have problems with rapid clearance and toxicity. On the other hand, gold nanoparticles (GNPs) provide 1.8 times as much contrast per unit weight as I-base contrast. Godfrey Hounsfield and Allan Mc lead at Cormack developed computed tomography (CT).

The main parts of the CT machine are the source that generates X-rays and the detector arrangement. After passing through the patient, the X-rays come into contact with the detector arrangement, and the X-ray flux is recorded. The sequence of steps and setups involves synchronizing the rotation of the source and the detector arrays around the patient to create a 360-degree dataset that reveals the patient's X-ray absorption from all angles. Computers and algorithms use this dataset to reassemble CT images. Reconstructions based on filtered back projections currently dominate. It has been reported for use in biomedical applications like polymeric emulsions, lipoprotein viruses, micelles, nanoparticles comprising polymeric materials, metal nanoparticles, oxides of the metals, salts, CNTS, grapheme sheets, etc. These have been contemplated fundamentally using the structural material based on lipids. The solids are strong center based like metal–metal compounds, or metal salt or the blend of the two-lipid-based structure are especially engaging as nanoparticles for biomedical applications because of their simplicity of combination and consideration of various practical substances as well as their overall property [75], A rundown of lipid-based molecule is shown; liposome is a compound which is made up of a group of lipids framed on the bilayer which encases the fluid center. The at present endorsed registered tomography contrast specialists for intravenous infusion are iodine more modest particles, for instance, the benzene center has three iodine, three amide, and six liquor branches for water solvency and has very low osmolality and the iodine contrast gave little blood half-life, yet are brilliant for various medical applications including imaging of vascular tissues because of its quick filtering pace CT machine. Nevertheless, a large number of components give great CT contrast. The following are the materials used in CT.

- *Iodine-based specialists*: The figured innovation of nanoparticles for iodine-based contrast is at first detailed in the imaging procedure which is fundamentally utilized in the clinical fie and the nanomaterials of scattering ethane oil and liposomes.
- *Gold-based agent*: The nanomaterials of the gold-based drug have been approached for the biomedical applications field, likely photo-acoustic imaging, photothermal in the cutting and detection of DNA, on the field of gold-based nanomaterials are evolutions to approach surface chemicals and control of dimension.
- *Bismuth subsalicylate*: Computed tomography has the similar energy of *Z* element to make X-ray tube voltage, in bismuth subsalicylate, it is regarded as subsalicylate, in place of bismuth and having less dense as gold and the specimens obtained are very distinct.

4.6.3 Ultrasounds Imaging

Numerous studies reported the utilization of nanoparticles in the ultrasound imaging technique. Ultrasonic sound waves can refine the strong nanomaterials in the scale imaging tissues in the liver. In an ultrasonic framework, the silica nanosphere (100 nm) scattered in mass focus on a high goal. They likewise consider different sizes of polystyrene particles with fixation are related to scattered in an agarose image (AG). The possibility exists that, in a simulative oxidant using 0.5 w/cm ultrasound power and 10% obligation in megahertz spectrum, metal oxide particles are catalyzed to examine their potential role as superoxide dismutases (SOD). In such an instance of human lymphoma, u937 got to be a piece of DNA cell examination and cell counting are picture by cells demise in the state of 100 Hz platelet-rich fibrin (PRF) for 2 min., elevated apoptosis to hold passing of the cell. In the disclosure of the mitochondrial system following cell passing for the treatment ultrasounds in the presence or nonappearance of nanoparticles were additionally assessed *ca* subordinate and mitochondria. Their result was approved because it interfered with the reduction of ultrasound-induced block, and it plays a significant role in the increase in cell death [76–78]. In the research, the mechanical effect of ultrasounds on controlling pro-survival pathways and the improvement of cell death-causing blocks caused by nanoparticles were examined. Examination of the fetus during the pregnancy period is used to observe the growth and health of the baby. Estimate blood flow in the genitals and identify problems in figures. In the study of inflammation, the thyroid gland and the breast in the lumps are analyzed and ultrasounds tests known as transducers produce sound influxes of recurrence over 20 Hz that are put on the skin.

5 Summary and Conclusion

The entire discussion in this chapter on advanced nanotechnology and nanomaterials and their application in improved patient care the point of view in terms of biodetection of pathogens by utilizing nanoparticles, treatment of various infectious and deadly diseases including lung cancer, brain tumor, and heart diseases including CVD, and other diseases and its treatment by employing the advanced nanotechnology was reported. The long-term aim and focus of nanomaterials research are to implement the quantitative and qualitative and preventive measures in diagnosing, treating, and curing diseases by employing new innovative materials in protein synthesis, DNA repairing, tissue engineering, detection of tumors at their earlier stage, isolation, and cleaning of biological molecules and cells are most promising and challenging one in the current scenario. The importance of research lies in the improvement and development of technology especially nanotechnology will influence the imaging techniques including MRI, CT scan, photo- acoustic, photothermal, ultrasound and other techniques will become an emerging tool in future by enhancing its characteristic properties like sensitivity, accuracy, rapid result, manipulation, signaling ability by merging the machinery technology along with nano scaled engineered particles also there are some discoveries during the corona period which includes medical wearable's by the application of new and advanced technology helped the mankind in safeguarding lives of many peoples and now the recent research is on the application of AI algorithms and nanotechnology for the improved diagnosis, treatment, and curing of various diseases is emerging as one of the interesting key feature for the future generation and concluded that advanced nanomaterials, nanotechnology, fabrication of new devices is a multi-disciplinary scientific approach can be utilized as a revolutionary advances in human care and health.

References

1. L. Yan, F. Zhao, J. Wang, Y. Zu, Z. Gu, Y.A. Zhao, A safe-by-design strategy towards nanomaterials in nanomedicines. Adv. Mater. **31**(45), 1805391 (2019)
2. M. d'Amora, S. Giordani, Carbon nanomaterials for nanomedicine, in *Smart Nanoparticles for Biomedicine*, 7, 103–113 (2018)
3. T. Dai, W. He, C. Yao, X. Ma, W. Ren, Y. Mai, A. Wu, Applications of inorganic nanoparticles in the diagnosis and therapy of atherosclerosis. Biomatr. Sci. J. R. Soc. Chem. **8**, 3784–3799 (2020)
4. Y. Sun, Y. Lu, L. Yin, Z. Liu, The role of nanoparticles in stem cell-based therapy for cardiovascular disease. J. Front. Bioeng. Biotechnol. **8**, 947 (2020)
5. R. Scherließ, Future of nanomedicines for treating respiratory diseases. Expert Opin. Drug Del. **16**(1), 59–68 (2019)
6. P. Sahare, P.G. Alvarez, J.M.S. Yanez, J.G.L. Bárcenas, S. Chakraborty, S. Paul, M. Estevez, Engineered titania nanomaterials in advanced clinical applications. Beilstein J. Nanotechnol.Nanotechnol. **13**, 201–208 (2022)
7. J. Majumder, T. Minko, Targeted nano therapeutics for respiratory diseases: cancer, fibrosis, and coronavirus. Adv. Therap. **4**, 2000203 (2021)
8. S. Modi, G.K. Inwati, A. Gacem, S. Saquib Abullais, R. Prajapati, V.K. Yadav, R. Syed, M.S. Alqahtani, K.K. Yadav, S. Islam, Y. Ahn, B.H. Jeon, Nanostructured antibodies and their emerging medicinal applications: an overview of nanoantibiotics. Antibiotics **11**(6), 708 (2022)
9. R.K. Sindhu, A. Goyal, E. AlgınYapar, S. Cavalu, Bioactive compounds and nano delivery perspectives for treatment of cardiovascular diseases. Appl. Sci. **11**(22), 1103 (2021)
10. G. Modugno, C. Ménard-Moyon, M. Prato, A. Bianco, Carbon nanomaterial combined with metal nanoparticles for theranostic applications. Br. J. Pharmacol.Pharmacol. **172**, 975–991 (2015)
11. R. Brusini, M. Varna, P. Couvreur, Advanced nanomedicines for the treatment of inflammatory diseases. Adv. Drug Deliv. Rev. **157**, 161–178 (2020)
12. F. Bruno, V. Granata, F. Cobianchi Bellisari, F. Sgalambro, E. Tommasino, P. Palumbo, F. Arrigoni, D. Cozzi, F. Grassi, M.C. Brunese, S. Pradella, M.L.M.S. Stefano, C. Cutolo, E. Di Cesare, A. Splendiani, A. Giovagnoni, V. Miele, R. Grassi, C. Masciocchi, A. Barile, Advanced magnetic resonance imaging techniques: technical principles and applications in nanomedicine. Cancers **14**(7), 1626 (2022)
13. S.A. Bansal, V. Kumar, J. Karimi, A.P. Singh, S. Kumar, Role of gold nanoparticles in advanced biomedical applications. Nanoscale Adv. **2**, 3764–3777 (2020)
14. N. Kumar, P. Chamoli, M. Misra, M.K. Manoj, A. Sharma, Advanced metal and carbon nanostructures for medical, drug delivery and bio-imaging applications. Nanoscale **14**, 3987–4017 (2022)
15. H. Ehtesabi, Application of carbon nanomaterials in human virus detection. J. Sci. Adv. Mater. Dev. **5**(4), 436–450 (2020)
16. C. Fan, J. Joshi, F. Li, B. Xu, M. Khan, J. Yang, W. Zhu, Nanoparticles-mediated drug delivery for treatment of Ishemic heart disease. Front. Bioeng. Biotechnol. **8**, 687 (2020)
17. M.Z. Quazi, N. Park, Nanohydrogels: advanced polymeric nanomaterials in the era of nanotechnology for robust functionalization and cumulative applications. Int. J. Mol. Sci. **23**, 1943 (2022)

18. A.R. Kirtane, M. Verma, P. Karandikar, J. Furin, R. Langer, G. Traverso, Nanotechnology approach for global infectious diseases. Nat. Nanotechnol.Nanotechnol. **16**, 369–384 (2021)
19. Y. Liu, J. Shi, Antioxidant nanomaterials and biomedical applications. NanoToday **27**, 146–177 (2019)
20. D.H. Kim, T.N. Nguyen, Y.M. Han, P. Tran, J. Rho, J.Y. Lee, H.Y. Son, J. Park, Local drug delivery using poly(lactic-co-glycolic acid) nanoparticles in thermo sensitive gels for inner ear disease treatment. Drug Del. **28**, 2268–2277 (2021)
21. K.P. Loh, D. Ho, G.N.C. Chiu, D.T. Leong, G. Pastorin, E.K. Chow, Clinical applications of carbon nanomaterials in diagnostics and therapy. Adv. Mater. **30**(47), 1802368 (2018)
22. M. Manzano, V.R. Maria, Mesoporous Silica nanoparticles for drug delivery. Adv. Func. Mater. **30**(2), 1–13 (2020)
23. R. Kanaoujiya, D. Porwal, S. Srivastava, Applications of nanomaterials for gastrointestinal tumors: a review. Front. Med. Technolol. **4**, 997123 (2022)
24. V. Chaudhary, Prospects of green nanotechnology for efficient management of neurodegenerative diseases. Front. Nanotechnol. **4**, 1055708 (2022)
25. M.X. Luo, S. Hua, Q.Y. Shang, Application of nanotechnology in drug delivery system for respiratory disease. Mole. Med. Rep. **23**(5), 325 (2021)
26. C. Zhu, J. Ma, Z. Ji, J. Shen, Q. Wang, Recent advances of cell membrane coated nanoparticles in treating cardiovascular disorder. Molecules **26**, 3428 (2021)
27. B. Begines, T. Ortiz, M. Perez-Aranda, G. Martinez, M. Merinero, F. Arguelles-Arias, A. Alcudia, Polymeric nanoparticles for drug delivery: recent developments and future prospects. Nanomaterials **10**(7), 1403 (2020)
28. M. Qi, M. Chi, X. Sun, X. Xie, M.D. Weir, T.W. Oates, Y. Zhou, L. Wang, Y. Bai, H.H. Xu, Novel nanomaterial-based antibacterial photodynamic therapies to combat oral bacterial biofilms and infectious diseases. Int. J. Nanomed.Nanomed. **14**, 6937–6956 (2019)
29. S. Hong, D.W. Choi, H.N. Kim, C.G. Park, W. Lee, H.H. Park, Protein-based nanoparticles as drug delivery system. Pharmaceutics **12**(7), 12070604 (2020)
30. S. Lara-Ochoa, W. Ortega-Lara, C.E. Guerrero-Beltrán, Hydroxyapatite nanoparticles in drug delivery: physicochemistry and applications. Pharmaceutics **13**, 1–10 (2021)
31. M. Qiu, Y. Tang, J. Chen, R. Muriph, Z. Ye, C. Huang, J. Evans, E.P. Henske, Q. Xu, Lung-selective mRNA delivery of synthetic lipid nanoparticles for the treatment of pulmonary Lymphangioleiomyomatosis. PNAS **119**(8), 8e2116271119 (2022)
32. A. Garcia-Fernandez, F. Sancenon, R. Martinez-Manez, Mesoporous silica nanoparticles for pulmonary drug delivery. Adv. Drug Deliv. Rev. **177**, 113953 (2021)
33. M. Kaloumenou, E. Skotadis, N. Lagopati, E. Efstathopoulos, D. Tsoukalas, Breath analysis: a promising tool for disease diagnosis—the role of sensors. Sensors **1238**, 1–58 (2022)
34. H. Sun, J. Ren, X. Qu, Carbon nanomaterials and DNA: from molecular recognition to applications. Account. Chem. Res. **49**(3), 461–470 (2016)
35. V. Raji, K. Paul, T. Zaheer, N. Kalarikkal, S. Sabu, F.G. de souza, A. Si, Gold nanoparticles against respiratory diseases: oncogenic and viral pathogens review. **11**(8), 521–534 (2020)
36. K. Komatsu, M. Murata, Y. Murata, Encapsulation of molecular hydrogen in fullerene C_{60} by organic synthesis. Science **307**(5707), 238–240 (2005)
37. Z. Liu, Y. Jiao, Y. Wang, C. Zhou, Z. Zhang, Polysaccharides-based nanoparticles as drug delivery systems. Adv. Drug Deliv. Rev. **60**, 1650–1662 (2008)
38. Z. Markovic, V. Trajkovic, Biomedical potential of the reactive oxygen species generation and quenching by fullerenes (C_{60}). Biomaterials **29**, 3561–3573 (2008)
39. P. Mroz, G.P. Tegos, H. Gali, T. Wharton, T. Sarna, M.R. Hamblin, Photodynamic therapy with fullerenes photochemical & photo biological sciences. Official J. Euro. Photo Chem. Assoc. Euro. Soc. Photobiol. **6**, 1139–1149 (2007)
40. S. Bhatia, Nanoparticles types, classification, characterization, fabrication methods and drug delivery applications, in *Natural Polymer Drug Delivery Systems: Nanoparticles, Plants, and Algae* (Springer International Publishing. Cham, 2016), 33–93
41. J.W. Park, Liposome-based drug delivery in breast cancer treatment. Breast Cancer Res. (BCR) **4**, 95–99 (2002)

42. L.R. Hirsch, R.J. Stafford, J.A. Bankson, S.R. Sershen, B. Rivera, R.E. Price, J.D. Hazle, N.J. Halas, J.L. West, Nano shell-mediated near-infrared thermal therapy of tumors under magnetic resonance guidance. Proc. Natl. Acad. Sci. **100**(23), 13549–13554 (2003)
43. S. Kalele, S. Gosavi, J. Urban, S. Kulkarni, Nanoshell particles: synthesis, properties and applications. Curr. Sci. **91**(8), 1038–1052 (2006)
44. S. Hohng, T. Ha, Single-molecule quantum-dot fluorescence resonance energy transfer. ChemPhysChem Euro. J. Chem. Phys. Phys. Chem. **6**(5), 956–960 (2005)
45. H.S. Choi, Renal clearance of nanoparticles. Nat. Biotechnol.Biotechnol. **25**(10), 1165–1170 (2007)
46. A. Saovapakhiran, A. D'Emanuele, D. Attwood, J. Penny, Surface modification of PAMAM dendrimers modulates the mechanism of cellular internalization. Bioconjugate Chem. **20**(4), 693–701 (2009)
47. J.R. Baker Jr., Dendrimer-based nanoparticles for cancer therapy. Hematol. Am. Soc. Hematol. Educ. Progr. **1**, 708–719 (2009)
48. Y. Chen, Y.-S. Kim, B.W. Tillman, W.-H. Yeo, Y. Chun, Advances in materials for recent low-profile implantable bioelectronics. Adv. Mater. **11**(4), 522 (2018)
49. M.R. Dreher, A.J. Simnick, K. Fischer, R.J. Smith, A. Patel, M. Schmidt, A. Chilkoti, Temperature triggered self-assembly of polypeptides into multivalent spherical micelles. J. Am. Chem. Soc. **130**, 687–694 (2008)
50. W. Teng, J. Cappello, X. Wu, Physical crosslinking modulates sustained drug release from recombinant silk-elastinlike protein polymer for ophthalmic applications. J. Controlled Rel. **156**(2), 186–194 (2011)
51. K.J. Isaacson, M.M. Jensen, D.B. Steinhauff, J.E. Kirklow, R. Mohammadpour, J.W. Grunberger, J. Cappello, H. Ghandehari, Location of stimuli-responsive peptide sequences within silk-elastin like protein based polymers affects nanostructure assembly and drug polymer interactions. J. Drug Target. **28**, 766–779 (2020)
52. Y.N. Zhang, R.K. Avery, Q. Vallmajo-Martin, A. Assmann, A. Vegh, A. Memic, B.D. Olsen, N. Annabi, A. Khademhosseini, A highly and rapidly crosslinkable elastinlike polypeptide based hydrogel for biomedical applications. Adv. Funct. Mater.Funct. Mater. **25**, 4814–4826 (2015)
53. V. Limaye, G. Fortwengel, D. Limaye, Regulatory roadmap for nanotechnology based medicines. Int. J. Drug Regul. Aff. **2**, 33–41 (2014)
54. W. Nawaz, S. Xu, Y. Li, B. Huang, X. Wu, Z. Wu, Nanotechnology and immune engineering: how nanotechnology can boost CAR-T therapy. Acta Biomater.Biomater. **109**, 21–36 (2020)
55. D.J. Irvine, M.C. Hanson, K. Rakhra, T. Tokatlian, Synthetic nanoparticles for vaccines and immunotherapy. Chem. Rev. **115**, 11109–11146 (2015)
56. W. Poon, B.R. Kingston, B. Ouyang, W. Ngo, W.C.W. Chan, A framework for designing delivery systems. Nat. Nanotechnol.Nanotechnol. **15**, 819–829 (2020)
57. G.Y. Lee, W.P. Qian, L. Wang, Y.A. Wang, C.A. Staley, M. Satpathy, S. Nie, H. Mao, L. Yang, Theranostic nanoparticles with controlled release of gemcitabine for targeted therapy and MRI of pancreatic cancer. ACS Nano **2**, 2078–2089 (2013)
58. F. Zhang, J. Cao, X. Chen, K. Yang, L. Zhu, G. Fu, X. Huang, X. Chen, Noninvasive dynamic imaging of tumor early response to nanoparticle-mediated photothermal therapy. Theranostics **5**, 1444–1455 (2015)
59. W.L. Monsky, D.S. Vien, D.P. Link, Nanotechnology development and utilization: a primer for diagnostic and interventional radiologists. Radiographics **31**, 1449–1462 (2011)
60. Y. Liu, K. Ai, J. Liu, Q. Yuan, Y. He, L. Lu, Hybrid BaYbF(5). Nanoparticles: novel binary contrast agent for high-resolution in vivo X-ray computed tomography angiography. Adv. Healthc. Mater. **1**, 461–466 (2012)
61. A. Jakhmola, N. Anton, T.F. Vandamme, Inorganic nanoparticles based contrast agents for X-ray computed tomography. Adv. Heal. Mater **1**, 413–431 (2012)
62. F. Kiessling, J. Bzyl, S. Fokong, M. Siepmann, G. Schmitz, M. Palmowski, Targeted ultrasound imaging of cancer: an emerging technology on its way to clinics. Curr. Pharm. Des. **18**, 2184–2199 (2012)

63. X. Cai, W. Li, C.H. Kim, Y. Yuan, L.V. Wang, Y. Xia, In vivo quantitative evaluation of the transport kinetics of gold nanocagesin a lymphatic system by noninvasive photoacoustic tomography. ACS Nano **5**, 9658–9667 (2011)
64. A. Avasthi, C. Caro, E. Pozo-Torres, P.M. Leal, M.L. García-Martí, Magnetic nanoparticles as MRI contrast agents. Top. Curr. Chem. **378**, 1–43 (2020)
65. S. Caspani, R. Magalhães, J.P. Araújo, C.T. Sousa, Magnetic nanomaterials as contrast agents for MRI. Materials **13**, 2586 (2020)
66. J. Puértolas, S. Kurtz, Evaluation of carbon nanotubes and graphene as reinforcements for UHMWPE-based composites in arthroplastic applications: a review. J. Mech. Behav. Biomed. Mater.Behav. Biomed. Mater. **39**, 129–145 (2014)
67. K. Chen, Y.H. Huang, J.L. Chen, Understanding and targeting cancer stem cells: therapeutic implications and challenges. Acta Pharmacol. Sinica. **34**, 732–740 (2013)
68. M.M. Gottesman, Mechanisms of cancer drug resistance. Annu. Rev. Med. **53**, 615–627 (2002)
69. X. Li, M.T. Lewis, J. Huang, C. Gutierrez, C.K. Osborne, M.F. Wu, S.G. Hilsenbeck, A. Pavlick, X. Zhang, G.C. Chamness, H. Wong, J. Rosen, J.C. Chang, Intrinsic resistance of tumorigenic breast cancer cells to chemotherapy. J. Natl. Cancer Inst. **100**(9), 672–679 (2008)
70. M.J. Hawkins, P. Soon-Shiong, N. Desai, Protein nanoparticles as drug carriers in clinical medicine. Adv. Drug Delivery Rev. **60**(8), 876–885 (2008)
71. A.P. Ramos, M.A.E. Cruz, C.B. Tovan, P. Ciancaglini, Biomedical applications of nanotechnology. Biophys. Rev. **9**, 79–89 (2017)
72. I.-K. Yoon, J.-Y. Hwang, W.-C. Jang, Natural bone-like biomimetic surface modification of titanium. Appl. Surf. Sci. **301**, 401–409 (2014)
73. A. Hubler, O. Osuagwu, Digital quantum batteries: energy and information storage in nano vacuum tube arrays. Complexity **15**, 48–55 (2010)
74. S. Acharya, S.K. Sahoo, PLGA nanoparticles containing various anticancer agents and tumour delivery by EPR effect. Adv. Drug Deliv. Rev. **63**, 170–183 (2011)
75. C.E. Probst, P. Zrazhevskiy, V. Bagalkot, X. Gao, Quantum dots as a platform for nanoparticle drug delivery vehicle design. Adv. Drug Deliv. Rev. **65**, 703–718 (2013)
76. A. Chrastina, J.E. Schnitzer, Iodine-125 radiolabeling of silver nanoparticles for in vivo SPECT imaging. Int. J. Nanomedicine **5**, 653–659 (2010)
77. A.M. Grumezescu, C.M. Chifiriuc, Prevention of microbial biofilms—the contribution of micro and nanostructured materials. Curr. Med. Chem. **21**, 3311 (2014)
78. M. Gaur, C. Misra, A.B. Yadav, S. Swaroop, F.Ó. Maolmhuaidh, M. Bechelany, A. Barhoum, Biomedical applications of carbon nanomaterials: fullerenes, quantum dots, nanotubes, nanofibers, and graphene. Materials **14**, 5978 (2021)

Electroanalysis and Sensors for Biomedical and Clinical Applications

R. Parameswari, R. Babujanarthanam, T. Yogesh, S. Janani Iswarya, and R. Ganesamoorthy

Abstract Biosensor is a biorecognition element device that measures biological reactions via physical and chemical sensing. Enzymes, antibodies, nucleic acids, proteins, receptor molecules, or any other intercellular agent that acts as an input, recognizes, and transfers signals and outputs them as digitalization, color, absorbance, or odor could be the biorecognition element. Biosensors play an unquestionable role in clinical and biomedical settings because they reduce diagnostic time, produce rapid results, are portable to any location where there is an emergence or natural disaster, are small in size, economically affordable, and have high sensitivity, specificity with a low detection limit. According to forecasts, clinical biosensors will be worth more than $36.7 billion by 2036, even though the biosensor function is prominent and enormous during covid-19. This chapter will focus on biosensors and their clinical applications.

1 Introduction

Analytical chemistry, physical chemistry, materials science, clinical biochemistry, fabrication industries, electrical and electronic engineering, nanotechnology, solid-state devices, and some statistical fields are all associated with electrochemical analysis [1]. It also includes information on biophysics, chemical engineering, and other related fields. It is a rapidly growing field with numerous industries, research, and education applications. As a result, electrochemistry is used for various applications rather than just one [2]. Electroanalysis is another name for electroanalytical chemistry, located between analytical science and electrochemistry and was defined in

R. Parameswari · R. Babujanarthanam (✉) · T. Yogesh · S. Janani Iswarya
Department of Biotechnology, Thiruvalluvar University, Serkkadu, Vellore, Tamil Nadu 632115, India
e-mail: babukmg@gmail.com

R. Ganesamoorthy
Department of Chemistry, Vinayaga Missions Kirupanandar Variyar Arts and Science College, Salem, Tamil Nadu 636308, India

K. Kasinathan et al. (eds.), *New Technologies for Energy Transition Based on Sustainable Development Goals*, https://doi.org/10.1007/978-981-97-2527-4_4

1993 by the Federation of European Chemical Sciences (FECS) and later by International Union of Pure and Applied Chemistry (IUPAC). Creating a model strategy, then applying procedures with instrumentation and collecting results based on the composition, time, and nature of observed measurements. Specifically, indefinite validation and fundamental protocol crosscheck must be prioritized. Simply put, electroanalysis is developing a chemical method to characterize modern trends such as easy carrying and most diminutive form instead of bulk, resulting in quick analysis and results.

1.1 Development of Sensors and Biosensor

Gremer pioneered the field of electrochemical sensors in 1906 with a device similar to a glass electrode [3]. In 1909, Gremer's student Klwmensiewicz abandoned his invention with basic analytical applications. This glass electrode desired to be introduced under the name Haber electrode'. However, in later literature, the glass electrode invention is referred to as the Haber Electrode rather than the Cremer. Since then, this invention has been used in various fields, including medicine, food processing, and environmental monitoring. Klwmensiewicz's improvements to Cremer's invention enabled basic analytical applications, such as pH measurement, to be performed with glass electrodes, which had previously been regarded as unreliable. The Haber electrode became an invaluable tool in many industries because it was more reliable and accurate than other methods [4]. This sensor resulted in more precise and accurate readings, allowing the Haber electrode to be used for even more complex measurements. Because the electrode was designed to be used in various environments, it also allowed for readings to take much less time. Because of its versatility, it is widely used in various industries. It has transformed how analytical measurements are performed and is credited with saving countless lives. Meanwhile, electrochemical sensors and their importance have played a role as a monitor, conservative predictor, disaster management, disease diagnosis, prevention management, and industrial applications over the last two decades [5]. A typical chemical substance-packed sensor known as a device absorbs and transfers chemical messages with a selected mode and a reversible bay depending on the loaded concentration of the desired sample *vs* total concentration through its signal in the ancient make of sensor [6]. Following this fundamental feature, enormous researchers worked for several years to achieve success in advanced electrochemical sensors with a particular hand kit model, rapid analysis, low cost with best features and detection, high sensitivity and specificity response sensors launched in the medical field as well as the industry [2, 7–9]. More than 2500 peer-reviewed articles have been published to date, and concern that electrochemical sensors are showing significant output in a variety of fields, with more to come in Table 1. Sensing technology applies to both fields due to its broad applicability in chemistry, biochemistry, and the environment in which it is used. As such, it can potentially transform how we measure, monitor, and analyze the physical world [18]. This is due to sensing technology's ability to collect real-time data and monitor environmental changes. It can also be used to detect the presence of specific substances in the environment, such as toxins as shown in Fig. 1. Furthermore, it can be used to analyze data for patterns and trends that can be used to inform decisions. Electrochemistry sensors can be optical, mass (*molecular equivalent-based*), or thermal [19–21]. In recent years, these sensors have significantly advanced. This chapter discusses sensors and biosensors and their applications in biomedical fields.

Table 1 Historical development of biosensor

S. No.	Types of biosensors	Year	Developer
1.	Ion-sensitive field-effect transistor	1970	Bergveld [10]
2.	Fiber-optic sensor for detection of CO_2 and O_2	1975	Luppers and Optiz [11]
3.	Commercial glucose biosensor detector in history	1975	YSI [11, 12]
4.	Microbe-supported immunosensor introduction	1975	Suzuki [13]
5.	Glucose biosensor which is made by fiber optic shielded invention	1982	Schultz [14]
6.	Immunosensor made by surface plasmon resonance (SPR)	1983	Liedberg [15]
7.	Ferrocene-based glucose detection by glucose oxidase enzyme method	1984	Commercial industries [16]
8.	Pharmacia-biacore biosensor designed by SPR pattern	1990	Commercial industries
9.	Blood biosensor	1992	i-STAT tech
10.	O_2 meter—biosensor	2000	Clinical Tech invention [17]

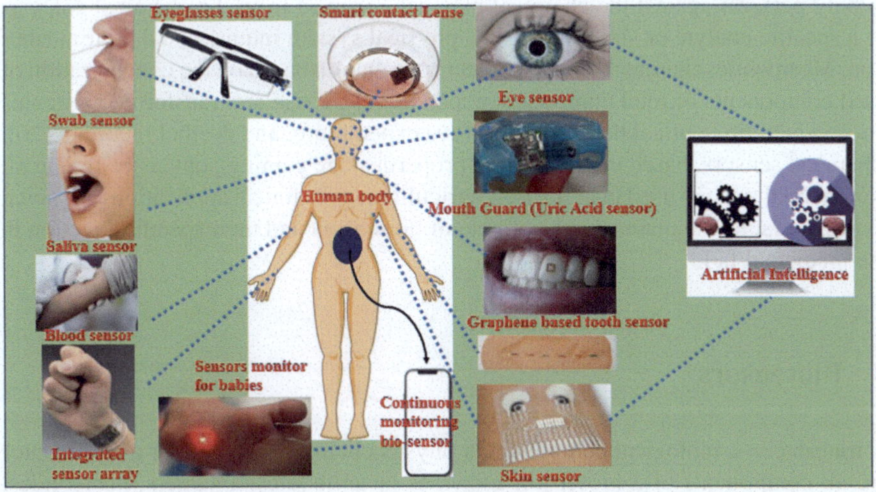

Fig. 1 Various categories of clinical electrochemical biosensor

2 Sensors

A *sensor* is a device that converts an input physical current into an output electric signal, which can then be read or detected by the human eye (sensation or color) or analytical instruments [18]. It is typically used to detect and quantify compounds associated with physical and chemical properties such as pH, odor, pressure of both gases and fluids, forced pressure, fluid flow during surgery, unique chemicals, and light intensity [22, 23]. Sensors are distinguished by the fact that they can be based on any of the following [24].

- When considering that it is not sensitive to all the other parameters encountered through its significant function or when considering that it is only sensitive to the chemical or physical matter of the total measured.
- When a sensor is processed, it cannot interfere with its input, such as chemical or physical matters. Although any changes or signal reflection in the measured physical or chemical quantity output indicate that the sensor's sensitivity or specificity diverges due to input selection. As a result, significant features such as accuracy, sensitivity, selectivity, ease of calibration, high resolution, reduced workforce, low-cost operation, and the ability to repeat if environmental conditions are required are considered when selecting a sensor.

The sensors are classified into physical and chemical types based on the type or attribute of the substance to be measured. In a nutshell, the physical type of sensors in the device was identified. Alternatively, measure pressure, force, optical density absorbance, RI, temperature, and atomic mass differences [25]. During the analysis, the chemical device has no function or interferes with the physical category. A chemical sensor of type 2 is composed primarily of a selective layer of chemically filled material that responds only to the appropriate analyte. These chemical sensors aim to only perform against the chemical message received from the chemical analysis of a specific analyte or substance or the physical system inquiries and their attribution. Meanwhile, chemical sensors collect information based on the concentration of used components or total component expenditures. These are transferred as signals via conductivity, optical density, light wave expenditure, and decibel recorded [26]. Chemical sensors, based on all of these properties, are gaining popularity among all commercial sensors used in industries, agriculture, and clinical applications as shown in Fig. 2. Table 2 explains that brief note of biosensors and their specific function in the medicinal field.

3 Biosensors

A transducer, a biological component, a physicochemical detector, or a bioreceptor are all components of a biosensor. Enzymes or antibodies, aptamers, nucleotides, or cells identified as biomolecules can specifically identify or detect the target

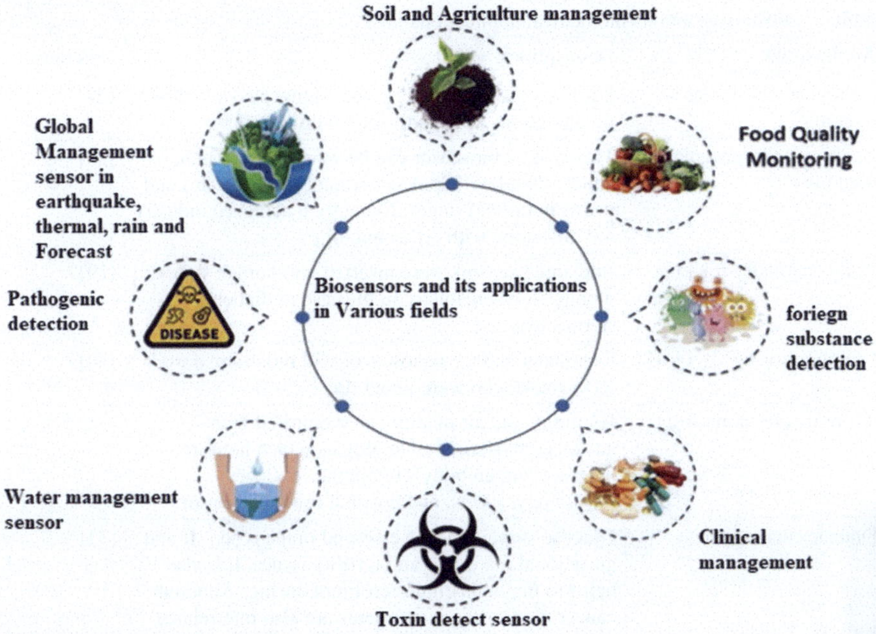

Fig. 2 Biosensor applications of various applications

and be commercially identified as receptors [26]. Because of biorecognition, the bioreceptor's specific structure will interact with its target analyte, which provides advantages, which is why receptors can be used as biosensors. The critical thing to remember is that receptors will avoid interference from other analytes due to specific interactions. Finally, the desired bioreceptor will interact, recognize, and transform into a metric signal via a transducer. A biosensor requires an immobilized bioreceptor on the surface of the transducer to produce a signal that can be measured as an output when it is in a stable state. A biosensor and its stable condition are pre-checked by its bioreceptor with immobilization conditions (reversible or irreversible methods) on the transducer's surface as shown in Fig. 3. To succeed in this state, various strategies such as cross-linking, entrapment with beads or fibers, bio-affinity, chelating ions, and absorption on the surface of the transducers of the desired sample can be used. The selection and challenges were kept track of and varied [35].

Signalization is an essential process in an element that converts energy produced by physical change into another reaction while the biosensor's transducer quantifies momentum. It was noted that different types of transducers are available depending on the element: optical, electrochemical, colorimetric, crystal piezoelectric, quartz, or thermal criteria [36]. Even though transducers can generate any of the optical or electrical signals that are directly proportional to the physical or chemical interactions between receptors and analytes, Clark and his team developed the first biosensor in 1962 for the detection of glucose in biological fluids using glucose oxidase (GO_x), an enzymatic analyte-bioreceptor processor [37]. Later, Rechnitz et al. [38] coined

Table 2 Biosensors and its biomedical applications

Applications	Descriptions	References
Biological anomalies tracking	Healthcare abnormalities, vital signs, physiological modifications and monitoring with low cost	[27]
Heart rate and pacemaker tracking	This type of biosensor can be wrist smart-watch, athletic band will tracks continuously heartrates and warns in critical stages. Recently it achieved industry 4.0 milestone with AI technology	[28]
Body chemistry tracking	Implanted sensors were inserted into human skin will monitor ions/chemistry in the case of diabetes/ sternus exercise	[29]
Diet monitoring	Dieticians' advice nanosensor will monitor calorie/ RDA those who were under diet	[30]
Air moisture monitoring	Humidity and air quality can monitored with advanced biosensor technologies which helps to measure human body's temperature, sleep, physiology, sports, and hospital used this sensor	[31, 32]
Diabetes management	Glucose surveillance is essential in the type I, II and gestational maternity and fertility issues. It is also helps to maintain cholesterol monitoring. Although cancer monitoring instruments are also interrelated with these biosensors	[33]
Heart rhythms tracking in cardiac patient	These biosensors were wearable and smaller in size which monitor tension levels in the workers of firefighters, law enforcement, miners, and service in border areas. Also helps in maintaining athletes' fitness during and after exercise	[34]

the term selective sensor for the first time to describe a selective arginine electrode used as a biorecognition material in the case of living organisms [38]. After two decades, a researcher named Scott discovered numerous advantages to the biosensor used to detect analytical and clinical diagnostic tools in various ways in 1988. Since its inception, all have recognized its ease of use, portability, low cost, easy and quick detection, and self-monitoring under medication such as cardiac issues, nephrotic syndromes, or diabetic complications known as dialysis [39]. On the other hand, biosensors are used in the food industry, quality control, and biodefense fields of various types. Meanwhile, well-planned commercialization is required to explore the biosensor with improved sensitivity, selectivity, consistency, and stability in various applications at a low cost. Various types of transducers for biosensors that are dependent on output signals have been introduced in the modern era, such as absorbance, chemiluminescence, luminescence, magnetoelectric, piezoelectric, electrochemical, and thermometric signals [40].

The electrochemical signal is the most used by users of all the above signals. Its popularity stems from its ease of use and low cost. It also enables quick and easy communication between various devices. It is also a secure method that is difficult

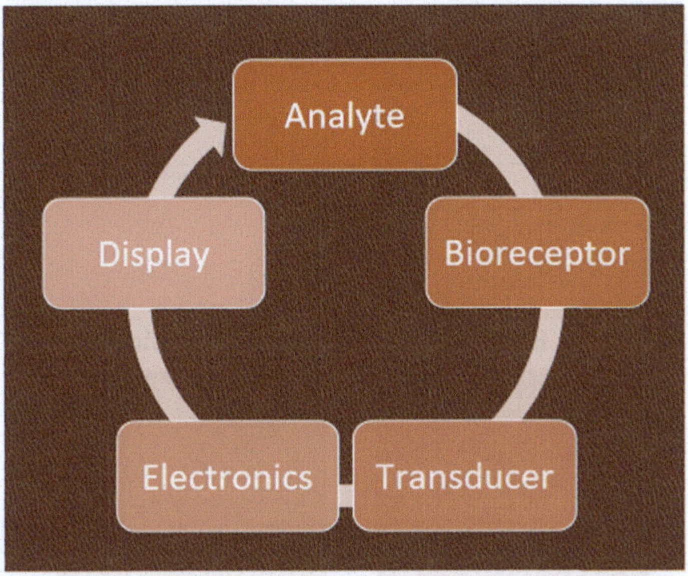

Fig. 3 Working protocol of biosensor applications

to intercept or hack. This is due to its user-friendliness, low cost, robustness, portability, and miniaturized size to carry wherever needed in emergencies or disasters [36]. Furthermore, electrochemical signals use little energy and are quickly returned, making them ideal for real-time applications. It is also resistant to interference from other signals and is unaffected by weather, making it dependable. Furthermore, superior monitoring and identification of any consequence in electrode signal-associated voltage fluctuation during the transfer of electro-potential capacitance determine the significance of various commercialized products in the current trend [41]. As a result, the electrochemical biosensor can be further classified based on its ability to monitor signals, as shown in Fig. 4.

4 Electrochemical Biosensor

As per the International Union of Pure and Applied Chemistry (IUPAC), an electrochemical biosensor is an interconnected and self-contained device that combines biological recognition elements (e.g., receptors, enzymes, and antibodies) with direct contact made by the transducer to produce quantitative or semi-quantitative electroanalytic-related information [35]. There was a list of commercial products or diagnostic kits that use electrochemical biosensors in various fields. Furthermore, maximum analytical and clinical instruments based on the electrochemical biosensor technique are in operation, such as various types of spectroscopies, colorimetric,

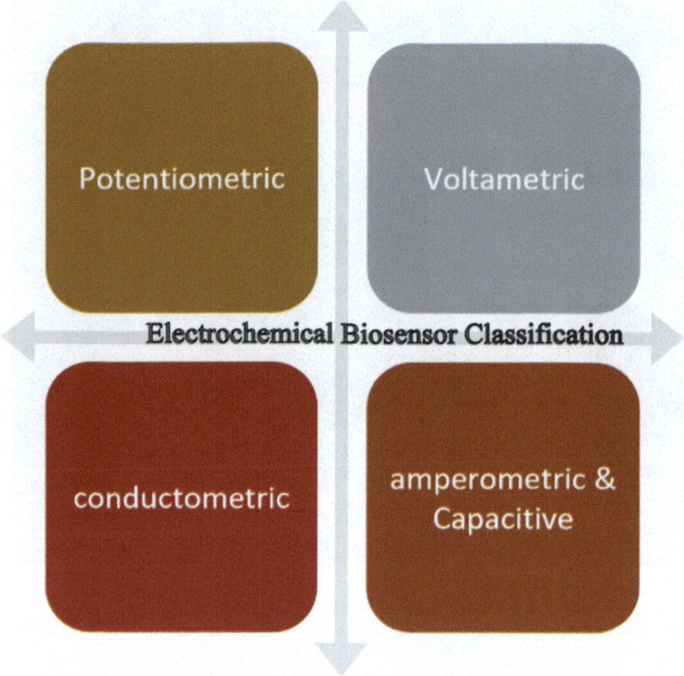

Fig. 4 Classification of electrochemical biosensor applications

UV–visible instruments, wave monitoring technology, electrochemistry, atomic force microscopy-combined electrochemistry, and crystallization by electro-quartz crystals [42]. The IUPAC guidelines defined functional and technical characteristics as shown in Fig. 5 to validate the execution of biosensors and their efficiency.

5 Biosensors in Biomedical and Clinical Applications

5.1 Biosensors in Diabetic

Maintaining blood glucose levels is critical in diabetic patients with Type I, II, and/or gestational conditions and their associated complications [43]. As a result, blood glucose biosensors were introduced in the 1990s for follow-up or sustained level maintenance as prescribed by clinicians [20, 44, 45]. There are more huge revelations in this sector available on the market, with enormous facilities and a lower price. Diabetes is caused by an insufficient release of endocrine hormones or the absence of islets of Langerhans, which results in poor glucose oxidation. Hyperglycemia (> 90–110 mg/dl) is caused by an increase in glucose in the blood. Diabetes is caused

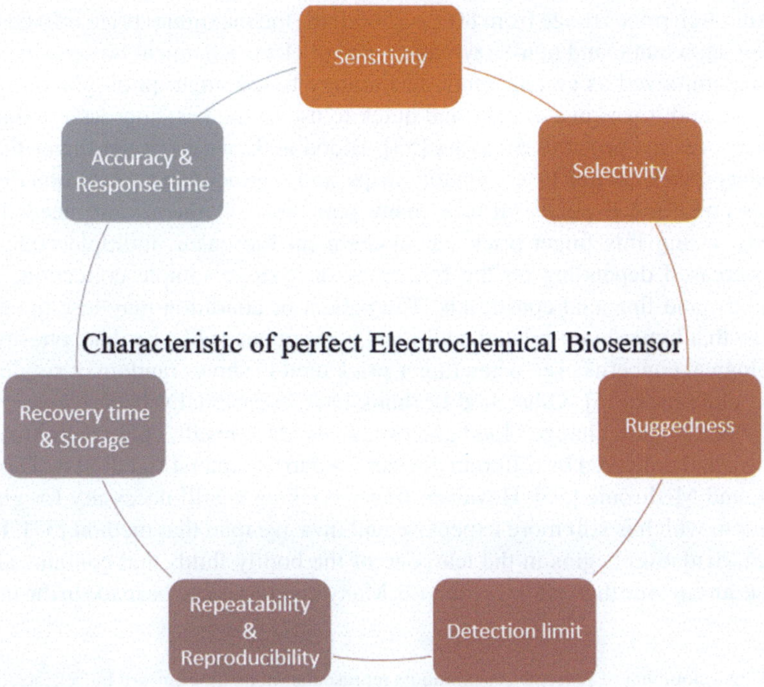

Fig. 5 Features of electrochemical biosensor applications

by a number of factors, and people ranging in age from infancy to middle age are at risk [46]. Although poor diabetes maintenance or inadequate knowledge causes kidney failure, retinopathy, neuropathy, abnormal blood pressure, and various types of cardiac issues. Furthermore, the ICMR reported in May 2023 that following the covid-19 pandemic, it was estimated that more than 70 million is increased during covid-19 of total diabetes cases increased, particularly in India (ICMR 2023 daily update). As a result, we are in a position to practice proper diabetic management. According to the clinician's or physician's advice, proper management of diabetic diet and medications is critical. It was designed as a hand kit blood glucose kit for routine check-ups in the home and to check AC levels in the early mornings for those who need to check AC or PC levels quickly. Furthermore, patients in OT or the emergency room were given this blood glucose sensor, which resulted in less than a minute, avoiding transportation and delayed laboratory reports. From 1962 to 2000, Table 3 summarizes the various types of blood glucose sensors and their commercialization as kits. There have been three generations of enzymatic blood glucose sensors developed and commercialized. The third generation created commercial kits, such as glucometers from brands like Contour Next One, Freestyle Libre, Dexcom G6, ever sense, Guardian Connect System, and Accu-Check. The data shows that more than half of the pharmaceutical industry profits from this diabetic blood glucose sensor

[10]. Although prices range from 800 to 2500 INR, the maximum price is fixed based on brand, enzymatic and non-enzymatic types of electrochemical biosensors.

It was employed as an enzymatic technology-based finger-prick glucose monitoring test, and it was just as easy and quick to use as the best approach to diabetes treatment over the previous 50 years [53]. Blood is drawn from the finger through a capillary prick to get to enzymatic strips and a glucometer. Additionally, the patient's condition depends on how many punctures or pinches are needed each day [54]. Using this finger-prick electrochemical biosensor, multiple collections were decreased depending on the frequency of syringe sample collection, social constraints, and financial constraints. The patient or condition may lose the ability to follow that hyper/hypo-related additional consequences like cardiac, obesity, and nephrological concerns even when finger-prick merely shows random or single-dose glucose outcomes [55]. Other bodily fluids have occasionally been recognized to have the potential to change blood glucose levels. As a result, commercial reagents are shown and delivered by different devices used to test interstitial fluid by Dexcom, Abbott, and Medtronic [56]. However, finger pricking is still necessary for glucose calibration, which is still more expensive and invasive than that method [57]. In this instance, oxidation begins in the tear, one of the bodily fluids that contain salivary amylase, an enzyme that oxidizes glucose. Many diagnoses can be made in the interim

Table 3 Development of glucose sensor and its representation history onward biosensors

S. No.	Inventors	Biosensor type	Year of development
1.	Clark and Lyons [37]	Prime description of the biosensor	1962
2.	Hicks and Updike [44]	At first enzymatic electrodes were for practical purpose	1967
3.	Guilbault and Lubrano [44, 47]	H_2O_2-based first enzyme electrode for the detection of glucose	1973
4.	Newman and Turner [45, 48]	As per the YSI analyzer, the glucose analyzer is restructured and introduced as the first commercialized kit	1975
5.	Miles [49]	History prime artificial beside pancreas	1976
6.	Shichiri [48]	Introduction of needle-based enzyme electrode for SC injection	1982
7.	Cass [16]	First sandwich typed amperometric glucose biosensor	1984
8.	Wang [50]	Blood glucose biosensor launched by Medisense Exac Tech	1987
9.	Milton [51]	First in vivo glucose biosensor	1999
10.	Wang [52]	Non-invasive wearable blood glucose biosensor	2000
11.	Clinical electrochemical sensors [43]	Accuacheck and strip-based commercial kits and gluconate	2007

using this [58]. Only some instruments were created to show glucose from tears, and few attempts were made with little success. Conversely, the majority of researchers accept contact lenses as wearable electronic devices. Effective measurement of tears will occur with the suitable sensor configuration for a specific localization, which is critical. Lacrimal glands and goblet cells in the eye's conjunctiva may collaborate to make tears, secreting 6–8 L of a thin film at a rate of 1–2 L/min [55, 59].

Based on their patented literature, NovioSense BV just began operations. According to that literature, the sensor was created with a micro-electrochemical cell coated in polysaccharide gel. These sensors are employed. It was reporting data on glucose monitoring while crying. Through tears, they are providing data on glucose monitoring. To feel or find extra macromolecules or foreign substances in blood fluids [56]. This polysaccharide-coated hydrogel was utilized to monitor glucose, a macromolecule. Therefore, glucose was diffused and evaluated. Following this, spray, spin-coated, and spray-type biosensors were created under patents and published literature, as illustrated in Fig. 6. The version 1 NovioSense sensor is shown to be a flexible, spring-shaped coil that can be quickly applied to the lower eyelid and detects glucose in tears. Later, it can be utilized continuously or once, producing findings comparable to finger-prick tests but non-invasive and less expensive than current systems. Biosensors on the exterior of lacrimal-secreting tears are utilized as an external sample to determine glucose levels based on the literature that is currently accessible and is testing. However, there have not been any studies showing that tear glucose can be used in place of blood glucose to monitor hypo- or hyperstates in the clinical setting of POCT [36].

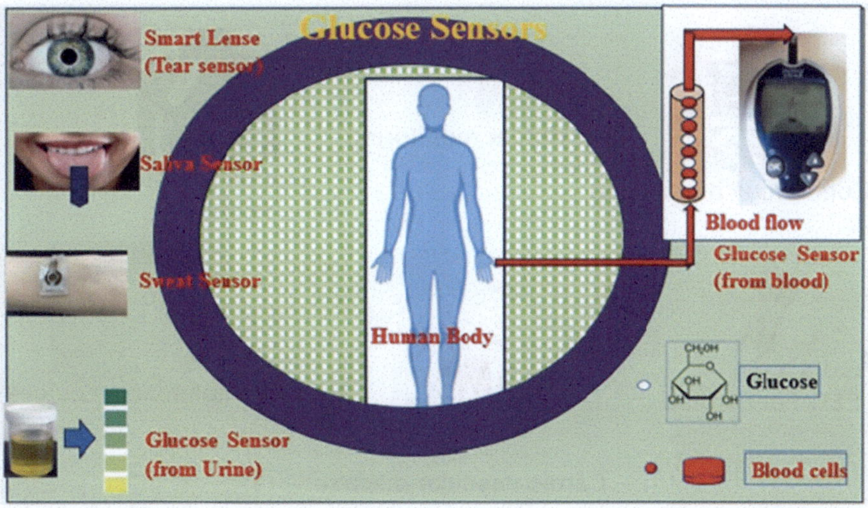

Fig. 6 Various body fluids secreting glucose and its detection by different kinds of electrochemical biosensor

6 CVD Detecting Biosensors

As a genetic or lifestyle factor, cardiovascular disease (CVD) is among the most common cause of death. Until 2015, more than 17.7 million people died due to CVD, accounting for 31% of all deaths worldwide. Furthermore, 7.4 million had coronary defects, and 6.7 million had strokes due to BP or CVD-related stroke [17]. With this in mind, early detection is critical to avoid unnecessary casualties. However, the technologies available to detect CVD are time-consuming procedures, and the patient must wait for further treatment for more than 7–8 h, and it may take up to 24 h [60, 61]. Furthermore, this arrangement is made by the WHO, and the CVD patient must follow any of the protocols, such as ECG peaks followed by blood biochemical markers. Even if an ECG is a critical parameter if the patient has CVD, it may be a poor diagnostic due to average cardiogram results, even if the patient is experiencing chest pain and its related causes [60]. At this point, introducing cardiac biosensors in the healthcare sector will provide quick assistance in determining the exact cause and treatment as shown in Fig. 7. Numerous biomarkers are available for the detection of CVD [62]. CVD is caused by blood flow, artery thickness, altered blood pressure, diabetes, an incorrect BMI, a hormonal defect, mental stress, psychological distress, a sedentary lifestyle, poor cholesterol management, and smoking [61]. As mentioned earlier, diagnosing CVD based on its causes is difficult. However, based on the diagnosis, it was divided into therapeutic and pathogenic classes. It was previously stated that early detection of CVD should follow the following rules [63].

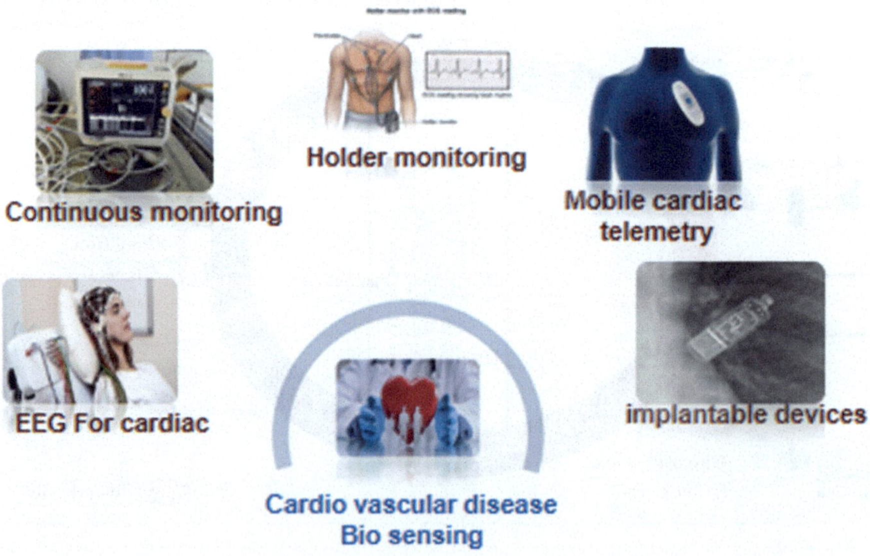

Fig. 7 Various cardiac diseases diagnosis and monitoring electrochemical biosensor

- Clinical aspects with the highest specificity and sensitivity
- Rapid biomolecule release and biorecognition in the bloodstream
- Biorecognition should be stable and consistent over time to pinpoint the exact cause of CVD
- Self-programming allows for spontaneous assessment and quantification.

According to studies, the most used CVD biomarkers are CRP, amyloid A, MMP 1, 2, 10, selectin, CD40L, PPAR receptors, COX-2, TGF beta, IL, LDL, myeloperoxidase, NADPH oxidases, and HGF [64]. Even though these biomarkers were assigned to any of the biosensors based on the type of biomolecule recognition, the fluorescence was determined by the

- CVD biosensor [65]
- Luminescence CVD biosensor [66]
- SPR CVD biosensor [67]
- CVD ELISA biosensor [68].

After all this, pacemakers are also one of the three types of electronic-based sensors. Single-chamber, dual-chamber, and biventricular pacemakers were developed for oxygen saturation, pressure control, and ventilation impedance biosensors [69]. However, there is a rumor that pacemaker sensors can be illegally hacked, but this has yet to be proven.

7 Cancer Biosensor

Cancer is the world's second most common disease and the leading cause of death. Before covid-19, it was reported that 19.1 million people were affected, with 9.6 million dying due to delayed diagnosis and poor treatment in developing and low-income countries compared to developed countries [70]. A cancerous tumor is formed by the uncontrolled growth of cells caused by genetic mutations and toxin exposure. The tumor can be benign or malignant. When cancer is discovered in an advanced stage, it is far more dangerous than benign cancer. As a result, determining the initial state is critical before they spread [71]. The available techniques for suspecting cancer are CT scan imaging, cell culture, cytological testing, and biopsy screening [72]. However, these are invasive and cost-effective diagnoses with severe side effects. On this note, innovative, low-cost diagnostics are required for low-income and developing countries. At this point, the World Health Organization (WHO) defines 'cancer biomarkers' as any molecule, structure, or action that can be quantified in the host to assess disease dominance. Over 50 different types of cancer have been identified worldwide. Circulating proteins, free radicals, oxidative stress, and the cytogenetic basis were chosen as biomarkers and are used as diagnostic tools in cancer research. These biomarkers, detected in body fluids early, are required for treatment and patient care [73]. PSA, found in sperm and blood, is used to detect prostate cancer biomarkers. P53, like PSA, is a well-known biomarker used to detect leukemia, a type of blood cancer, and other cancers of the lung, bones, and ovaries. Breast cancer is measured using CD44. Apolipoprotein A1 (Apo-A1) in urine is used to detect early-stage bladder cancer. Biomarkers such as AFP, CA125, CA15-3, and SOX-2, among others, have been discovered to detect various types of cancer [74]. However, a non-invasive cancer treatment protocol is required globally. Polymerase chain reaction (PCR), Enzyme-linked immunosorbent assay (ELIZA), radioimmunoassay (RIA), immunohistochemistry (IHC), the high-performance liquid, and Western blot analysis are some of the more common and expensive techniques used in laboratories. However, biosensors for cancer have been introduced in the last ten years in the form of electrochemical, optical, and mass-based devices.

Biosensors for cancer biomarkers are intended to detect proteins or miRNAs on the surface of the tumor cell membrane. Meanwhile, various other methods for detecting cancer biomarkers were available; however, among all electrochemical techniques, electrochemical is ideal due to its quick results, understandable SOP, easily quantifiable, adjustable cost for all economics, and high sensitivity [75]. It was reported that enzymes, synthetic molecules of aptamers, peptides, DNA fingerprints, nucleic acid sequencing, antibodies, and cell receptors are among the biorecognition materials identified as biomarkers to diagnose cancer in its early stages [73, 74]. They are classified based on the precise biorecognition of biosensors [76]:

- Apatasensors
- Immunosensors
- Enzymatic sensors
- Genbiosensors.

A change occurs in normal cells that cancer cells alter, resulting in anomalies. On the other hand, the anti-tumor gene becomes deactivated, as evidenced by chromosomal mutations and hypermethylation of the gene. Even though these cancer-causing suspects were all considered nucleic acids, cancer biomarkers include microRNAs (miR) and a p53 gene mutation. Boriachek et al. [77] reported that early detection of cancer is possible due to this cancer biomarker-based electrochemical sensor called streptavidin-coated magnetic beads (MBs), which has an almost 93% recovery rate. Another study found that a significant cancer marker called EGF 2(+) BT 474 can be detected in the very early stages with the help of HER2 antibody at a detection limit of 4.7×10^5 exosomes/L [78]. With current techniques, avoiding or improving the prognosis followed by cancer treatment is highly desired and cost-effective. After the introduction of biosensors as an alternative method to detect cancer in its very early stages with no side effects, new procedures, such as nanofabrication-associated clinical applications, which are required for lab-on-chips and NSG at a lower cost, are expected to be launched [79]. Nanostructured and nanocomposite biosensors are currently used in electrochemical sensors and are expected to be introduced soon [80].

8 Covid-19-Detecting Biosensor and Its Emergence

Covid-19 is the worst scenario, and the world has undergone it. When the diagnosis is accessible in the symptomatic subjects and, in the case of asymptomatic, it is challenging. The suspected subjects' respiratory swabs are collected and subjected to paper-based real-time PCR. This aids in detecting the presence of the disease-causing virus in the specimen. The test results are usually available within a few hours. The biosensors used in the tests are highly sensitive, which means they can detect the presence of the virus in the patient's sample at deficient concentrations [81]. Furthermore, the test results can be obtained quickly, allowing for an accurate diagnosis. Furthermore, the test is inexpensive, allowing many people to be tested without breaking the bank. All of this makes it an excellent choice for screening people for covid-19. The suspected subjects' respiratory swabs are collected and subjected to paper-based real-time PCR [82]. This aids in detecting the presence of the disease-causing virus in the specimen. The test results are usually available within a few hours. The biosensors used in the tests are highly sensitive, which means they can detect the presence of the virus in the patient's sample at deficient concentrations.

8.1 Asymptomatic Cases Versus Microbial Investigation Associated Biosensor in Covid-19 Cases

Antigens can be found by antibodies in early-stage patients who have not yet shown symptoms, especially in those who have been exposed to microbes (like covid-19 individuals) [83]. Using biomarkers or other indicators in this study delayed treatment rather than a cure in the early stages. Therefore, creating sensitive and speedy diagnostic techniques will focus on antigens and enable quick therapy [84]. The World Health Organization advised clinicians to develop more diagnostics during the covid-19 outbreak to stop the virus from spreading. Recognizing the location or epidemiology of the alien pathogen depends mainly on the diagnosis or kind of diagnosis. The afflicted's condition and kind of infection are also quickly identified and released, which is essential in a contagious outbreak. This will also aid the patient's isolation, lowering the communication disorder [85].

8.2 Sensor and Scan Already in Place for Covid-19 Diagnostic

Meanwhile, there is a considerable risk to the drug testing industry and front-line healthcare professionals whenever an outbreak occurs. Three different types of tests are utilized during covid-19 to identify, validate, and assess the severity of the infection: (i) a computed tomography (C.T.) scan; (ii) an ELIZA-based automatic chemiluminescence analysis that identifies antibodies; and (iii) reverse transcription-polymerase chain reaction (RT-PCR) for miRNA detection [86]. However, there are few places to get C.T. scans, and radiologists who can handle them are needed to review the C.T. results. The other approach, RT-PCR, is one of the longer processes and will take at least 4 h. However, there is a maximum risk of creating false-negative findings in patients who may transmit the viral infection to healthy, uninfected people. The antibodies cannot be a reliable source for primary detection in asymptomatic patients once the antigen–antibody response titer is attained on the tenth day following infection. However, other chemicals, such as proteins in other substances used in serology, may impact false-positive results. As a result, a more accurate technique, such as the pin-top biosensor type now in use, is needed to detect virus infections. At the same time, precise direct binding through the biorecognition molecule to virus antigens for identification is one strategy that must be used in conjunction with these biosensing technologies [87]. Nevertheless, this binding will produce quantifiable signals via transducers directly interacting with SPR or tagged enzymes [88]. For the highly sensitive electrochemical biosensors to be detected, quick detection of viral infections, and urgent therapy, the available biomarkers and bioindicators linked with the sample are still required. Electrochemical biosensors are a great instrument to help detect viral infections, according to the diagnostic sector quality control (Q.C.) report base. Even if the sample comes from different biological habitats, it will be rapidly and affordably discovered and reported [89, 90].

8.3 PCR, RT-PCR in Covid-19 Antigen Diagnosis

The possibility of increasing speedy detection in emergencies during Covid-19 will, however, improve due to the design of biosensors in new aspects. Currently, ssRNA, antigen, antibody, and A.I. sensing have all been utilized as biomarkers and indicators for Covid-19 detection. With the aid of the RT-PCR technique, single-stranded RNA is a crucial biomarker utilized in the SARS-CoV-2 viral infection. It examines the entire genome of the target's conserved region, compares it to specially created primers, and then discovers it. Because only the primers target only SARS, RT-PCR cannot be used to identify other viral illnesses like MERS and its variants. In addition, individual-specific primers are used for the ssRNA biomarker for all viral infections [91, 92]. However, compared to current biomarker analysis, RT-PCR's sensitivity and ssRNA detection are superior for detecting nasopharyngeal infections. The only negative is that it takes a lot of time and money. Antigens are another type of biomarker since they are structural proteins like S, M, N, and E, typically found on viruses' surfaces. More than 28 different SARS-CoV-2 viral protein types were also discovered, and M and E varieties predominated on the structure's surface. It was discovered that the CoV-2 virus's 'S' protein gains higher significance when it enters host cells, where it then rapidly attaches to the receptor domain and engages with ACE2 receptors. As a result, S and N proteins are the essential protein cum antigens that serve as biomarkers to identify SARS-CoV infections. The other virally associated proteins, however, could quickly evade or avoid being recognized by the proteins mentioned above or antigens at this time [90, 92]. On the other hand, a recent study by Jiang and his colleagues discovered that more than 25 proteins were created by the microarray proteome approach used to assess the antibody response against infection, demonstrating an increase in antibody synthesis in proteomes like N and S1 during the recovery state [93]. In addition, lateral flow analysis, rather than microarray, was used to identify the same Covid-19 proteome and proteins. Both of these methods were found to be efficient in terms of cost [94].

8.4 Immuno Antibodies as Sensors in Covid-19 Diagnosis

It also demonstrated that, when compared to other infections and chronic diseases, IgM responds swiftly to viral diseases. Serological antibodies such as IgM have more binding sites, which are elevated after infection and are predominantly found in covid-19 patients. Although a case study by Zhang et al. demonstrates that IgG and IgM of primary antibodies were detected after the fifth day of infection, the other study discovered that following covid-19 infection on the fourteenth, there is a maximal increase in IgM compared to other antibodies. Even if it may be helpful to assess the patient's level of recovery for additional treatment, there are more reliable approaches than covid-19 detection using antibodies [95]. While the pandemic was underway, AI-based biomarkers, as well as a few additional biomarkers, were also used to detect covid-19 from samples, including blood, urine, level of hemagglutination range, blood O_2/CO_2 gas index of saturation rate, viral load, and level of cytokines. RT-PCR was utilized to use all these indicators to find covid-19 infection, but this method is insufficient to prove the infection. As a result, advanced created technology known as artificial intelligence (A.I.) is employed to identify potential illnesses. While this happens, blood samples from covid-19-infected patients exhibit vital pro-inflammatory cytokines and lymphopenia, broadening the range of diagnoses that can be made using the current techniques [96].

8.5 Trending Biosensor Against Covid-19

Promising biosensors were chosen to diagnose viral infection depending on the patient's conditions, how the sample was obtained for the test, and the degree of severity (respiratory problems/O_2 demand). However, it was demonstrated in crises during the epidemic (19). When the patients were in critical condition, these biosensors continuously monitored the covid-19 biomarkers and indicators. In contrast to current techniques, biosensor-based covid-19 diagnostics provides fast, high-sensitivity results and can be detected using smartphones. As a result, unique biosensors are now available, including localized surface plasmon resonance (LSPR), field-effect transistor (FET), and Raman scattering surface enhancer (SERS). During covid-19, these electrochemical biosensors underwent improvements that allowed them to be used for diagnosis. These are now the most conserved biosensors for detecting covid-19, according to a diagnostic survey from the centers FET, SERS EC, and LSPR [97].

9 Plasmonic-Based Biosensor

The precise target analytes were analyzed using these SPRs, freely labeled biosensors with the highest sensitivity. A genetically created gold fusion polypeptide protein on the surface of the virus antigen is employed in SPR to diagnose SARS-CoV-2 antibodies because it has the highest sensitivity and specificity [98]. It was recently revealed that Masson and his colleagues discovered that the optimal sample to use when using biosensing SPR technology to identify nucleocapsid antibodies against SARS-CoV-2 correctly is an undiluted human blood sample. However, the tests are simple to carry out thanks to the portable SPR device. The SPR's operational protocol states that when the sensor opens to let the SARS-CoV-2 virus in, the immune responses of produced antibodies detect antigens in patients [99]. Additionally, SPR precisely identifies the antibodies produced by the SARS virus, providing a crucial hint for developing a vaccine for each SARS-CoV-2 suspect. Additionally, Wang and his team demonstrate how SPR working with dual plasmonic photothermal (PPT) architecture will aid in the fast diagnosis of the covid-19 virus. Additionally, the specific sequence of SARS-CoV-2 will be predicted using 2D-designed AuNI-coupled DNA receptors. The PPT sensor, which was manually created, may boost in situ hybridization thermal status while also producing heat and working to enable 2D sequencing genes automatically. It can detect up to 0.22 pM [100].

10 FET Biosensor Versus Covid-19 Diagnosis

Compared to alternative diagnostic sensing strategies, the FET biosensing platform offers the most promising advantages, including the capacity to demonstrate its sensitivity, selectivity, and specificity in presenting data. A limited number of targets can be rapidly found with FET. As a result, this kind of biosensor has a lot of potential applications in the clinical field, point-of-care testing, and on-site diagnosis, including diagnostics in the operating room. One of the greatest biosensors in covid-19 for diagnosing SARS-CoV-2 is the FET, which has hexagonal carbon atoms coated with graphene on the surface and is easy to conduct electrons with. In order to carry out immunological antibodies with higher sensitivity, perfect, unique sensing is built, and graphene-coated FET biosensors play a crucial role in that [101, 102]. As shown in Fig. 8, Seo and his colleagues successfully created a device utilizing FET-based technology to detect the SARS-CoV-2 virus using this hexagonal graphene-coated approach. In more detail, the FET biosensor observed an increase in virus entrance

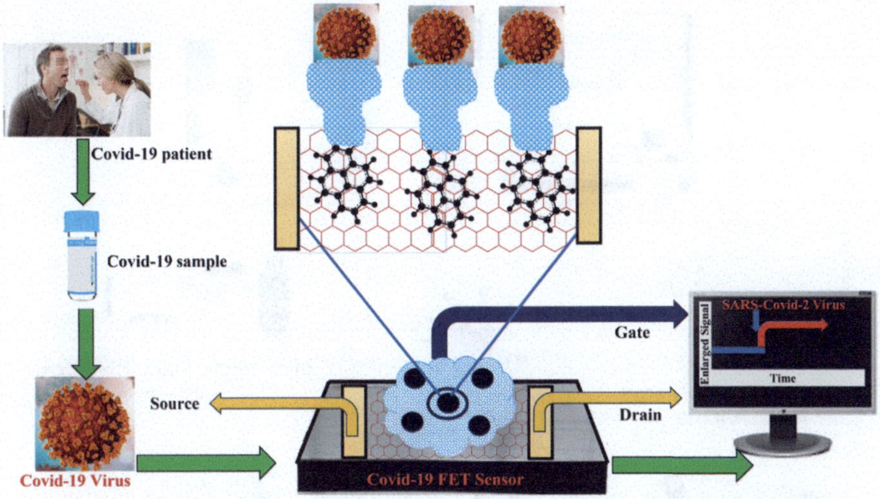

Fig. 8 Leading FET biosensor and its working principle in Covid-19 detection

into the host and the graphene sheets were coupled with antibodies against antigens. The FET biosensor's strength is that the self-cultured virus was obtained from covid-19 pneumonia patients utilizing nasal swabs and protein as an antigen for assessment. This FET can detect the SARS-CoV-2 spike protein at clinics at one fg/mL conc in PBS medium and 100 fg/mL in transportation media. Nevertheless, it was demonstrated that the FET biosensor could identify SARS-CoV-2 spike protein in the self-cultured medium with much lower detection thresholds, like 2.4×10^2 copies/mL. Interestingly, FET exclusively picks up the SARS-CoV-2 spike protein, does not react with MERS antigens, and does not give false-positive results [103].

Like FET, electrochemical biosensors (E.C.) attracted increased attention from analytical professionals because of their ease of use, low cost, accessibility to people of all income levels, sizes, and availability to large companies. Both at home and in clinics, it is applicable. Early or rapid diagnosis is the sole option in the early stages of the pandemic fight when there is no vaccination for covid-19 infections [102, 104] Eissa's team created a recombinant spike protein as a biomarker for MERS CoV infection. Later, this was reorganized by Abad-Valle et al., into a straightforward and affordable gold film coating to detect the SARS-CoV gene as a gene E.C. sensor. At 6 pM., the genomic DNA libraries reached the nanosensor's detection limit [105]. During the covid-19 epidemic, Gandhi and his research team recently tested an internal diagnostic biosensor known as an oven with commercial marketing tools. To create a potentiostatic sensor to detect the Covid-19 spike antigen, they made AuNPs coated with immunized monoclonal antibodies [106]. The design of the ovens, which is based on fluctuation in the chosen electrical conductivity, uses a carbon electrode with an immunological covid-19 antibody screen printed on the surface. Like viral, various microbial infections against biosensors in commercial market as shown in Fig. 9. In Table 4, you may find information about covid-19

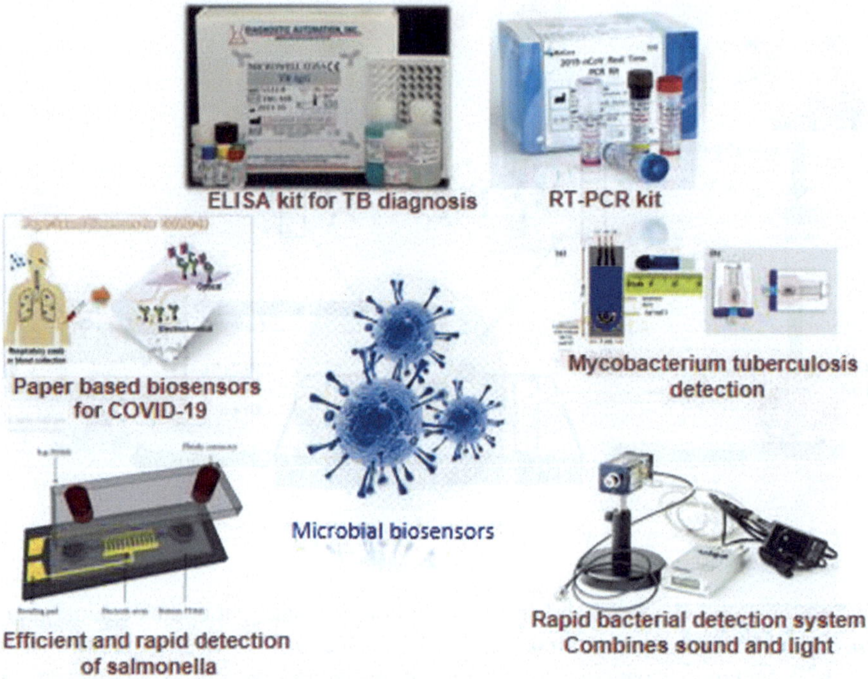

Fig. 9 Various microbial diagnosis of biosensor

detection, different biosensors and related approaches, processing time, detection ranges, and accuracy sensitivity.

11 AI in Covid-19 Diagnosis

Furthermore, the test results can be obtained quickly, allowing for an accurate diagnosis. Furthermore, the test is inexpensive, allowing many people to be tested without breaking the bank. All of this makes it an excellent choice for screening people for covid-19. ACE-2 and CRISPR are biomarkers for covid-19 detection. ACE-2 is a cell receptor to which the virus binds to infiltrate cells. To confirm the presence of SARS in a sample, CRISPR technology is used [113]. This technique is highly accurate and can detect the virus in minutes. This technology is faster, more accurate, and more straightforward than traditional virus detection methods like PCR tests. It is also less expensive and can quickly identify and isolate infected people. Philips began early detection of covid-19 as a wearable biosensor to detect covid-19 in its early stages; the system combines AI-driven imaging technologies, patient data, and analytics. It is promoted to assist healthcare professionals in quickly identifying and prioritizing patients with covid-19 [114]. Philips is also developing an

Table 4 Various techniques and its features in diagnosis of viral Covid-19

Method/ Apparatus used	Technique	Detection limit	Sensitivity	Time required	Proposed/ Invented by
PCR	RT-PCR		71%	4 h	[29, 107]
PCR	MNPs made RT-PCR	10 times copies	–	Less than 30 min	[53]
ELISA	ELISA reader	–	83.7%	2 h	[108]
POC strip	AuNPs coated colloid (IgM + IgG)	–	96.9%	Less than 15 min	[95]
2D LSPR	Plasmonic biosensor and FET	0.22 pM	3.7 RNA copies	–	[109]
CT	Chest CT scan	–	97%	–	[110]
Water bath	Colorimeter	–	97.7%	Exactly 30 min	[99]
Automatic analyzer/semi conduct analyzer	FET/ chemiluminescence	90 fM	–	–	[111]
EC station	EC biosensor	2.42×10^2 copies	–	Few min	[112]

AI-powered triage system to automate patient evaluation (online source software developer). The system detects covid-19 early, allowing healthcare professionals to treat patients before the virus spreads. The AI-powered triage system aids in the assessment process by automating and analyzing patient data, potentially shortening the time it takes to diagnose a patient.

12 Market Value/Net Growth of Biosensors

The global market for net growth biosensors is expected to reach $36.7 billion by 2027 [115, 116]. The increased demand for wearable medical devices and point-of-care diagnostics can be credited to this expansion. Furthermore, the growing demand for home healthcare propels the biosensor market forward. Additionally, the biosensor market is anticipated to expand due to the increased prevalence of long-term diseases like diabetes. Rapid technological advancements in biosensors are also propelling the growth of the biosensor market. Governments worldwide are investing heavily in biosensor research and development, expected to drive market growth. Furthermore, the growing acceptance of preventive healthcare is fueling the growth of the biosensor market as shown in Fig. 10.

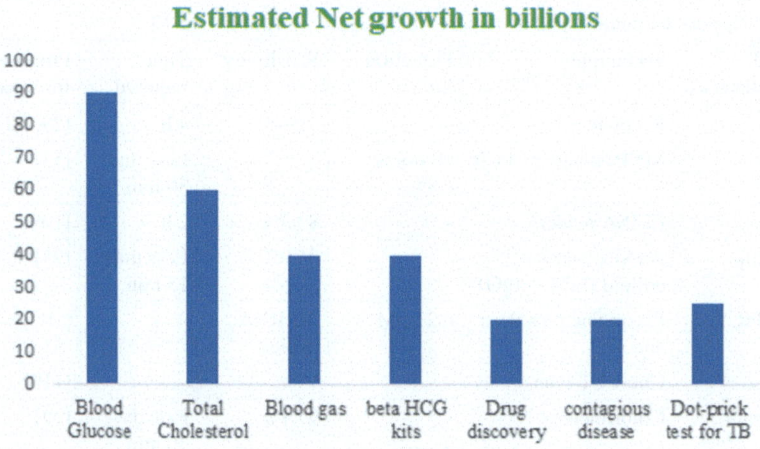

Fig. 10 Biosensors and its net growth in various clinical tools by 2036

13 Chat GPT AI Biosensor

Chat GPT is an innovative biosensor device that detects and measures biological signals in real-time. It is appropriate for various clinical applications, including vital sign monitoring, diagnostics, and drug delivery. Chat GPT can transform healthcare by enabling rapid and more precise diagnosis and treatment [117]. Chat GPT can also detect and measure physiological signals more precisely than traditional methods, such as heart rate, blood pressure, and respiration rate [118]. It can also monitor patients, improving care, and patient outcomes remotely. GPT can supplement clinical diagnosis by providing a thorough overview of a patient's medical history and condition. It can also detect subtle patient condition changes those traditional methods would miss. This can assist physicians in anticipating and preventing adverse health issues, allowing them to provide better care and improve patient outcomes [119, 120]. Furthermore, it can potentially lower healthcare costs by allowing for earlier diagnosis of conditions and reducing the need for costly medical procedures.

14 Conclusion

More than 50 years have passed since the introduction of biosensors. Their position in clinical and biomedical research is unrivaled. Clinical testing has a low cost, quick turnaround, and excellent sensitivity and specificity. Despite this, biosensors remain essential in clinical settings. Other kits, particularly cancer biosensors, are still struggling as commercial kits, except for glucose sensors. Glucose sensors are the only commercialized and successful biosensor kit. Researchers need more funding to be given more priority. Furthermore, due to a lack of research grants, it is taken for granted with essential science biosensors. Some people might use it for practical purposes or to teach students.

Acknowledgements The authors are thankful to Thiruvalluvar University, Tamil Nadu, India.

Conflict of Interest The authors declare that nothing to competing interest.

References

1. M.J. Lobo-Castanon, S. Campuzano, Electrochemical biosensors—driving personalized medicine. Anal. Bioanal. Chem. **415**(6), 1001–1002 (2023)
2. B. Can Demirdogen, A literature review of biosensors for multiple sclerosis: towards personalized medicine and point-of-care testing. Multiple Scler. Relat. Disord. **48**, 102675 (2021)
3. M.J.B. Cremer, Über die Ursache der elektromotorischen Eigenschaften der Gewebe, zugleich ein Beitrag zur Lehre von den polyphasischen. Elektrolytketten **47**(29), 562–608 (1906)
4. F. Haber, Z. Hlemensiewicz. Über elektrische phasengrenzkräfte. Aus der Zeitschrift Zeitschrift für Physikalische Chemie **67**(1), 385–431 (1909)
5. A.A. Belyustin, The centenary of glass electrode: from Max Cremer to F. G. K. Baucke. J. Solid State Electrochem. **15**(1), 47–65 (2011)
6. N. Lakshminarayanaiah, *Membrane Electrodes* (Elsevier, 2012)
7. E. Eroglu, W.F. Graier, R. Malli, The preamble to the free radical biology and medicine virtual special issue on "Targeting genetic biosensors to intracellular signaling pathways." Free Radic. Biol. Med. **187**, 202–203 (2022)
8. R.S. Gaster, D.A. Hall, C.H. Nielsen, S.J. Osterfeld, H. Yu, K.E. Mach, R.J. Wilson, B. Murmann, J.C. Liao, S.S. Gambhir, S.X. Wang, Matrix-insensitive protein assays push the limits of biosensors in medicine. Nat. Med. **15**(11), 1327–1332 (2009)
9. M. Gray, J. Meehan, C. Ward, S.P. Langdon, I.H. Kunkler, A. Murray, D. Argyle, Implantable biosensors and their contribution to the future of precision medicine. Vet. J. **239**, 21–29 (2018)
10. S.J. Reyes, Y. Durocher, P.L. Pham, O.J.P. Henry, Modern sensor tools and techniques for monitoring, controlling, and improving cell culture processes. Processes **10**(2), 189 (2022)
11. C.V. Mun'delanji, K. Kerman, I.-M. Hsing, E. Tamiya, *Nanobiosensors and Nanobioanalyses* (2015)
12. E.-H. Yoo, S.-Y. Lee, Glucose biosensors: an overview of use in clinical practice. Sensors **10**(5), 4558–4576 (2010)
13. S. Suzuki, F. Takahashi, I. Satoh, J. Sonobe, Ethanol and lactic acid sensors using electrodes coated with dehydrogenase—collagen membranes. Bull. Chem. Soc. Jpn. **48**(11), 3246–3249 (1975)
14. J.S. Schultz, Optical sensor of plasma constituents. Google Patents (1982)
15. B. Liedberg, C. Nylander, I. Lunström, Surface plasmon resonance for gas detection and biosensing. Actuators **4**, 299–304 (1983)
16. A.E.G. Cass, G. Davis, G.D. Francis, H.A.O. Hill, W.J. Aston, I.J. Higgins, E.V. Plotkin, L.D.L. Scott, A.P.F. Turner, Ferrocene-mediated enzyme electrode for amperometric determination of glucose. Anal. Chem. **56**(4), 667–671 (1984)
17. S.A. Tabish, Is diabetes becoming the biggest epidemic of the twenty-first century? Int. J. Health Sci. (Qassim) **1**(2), 12 (2007)
18. A.J. Bard, L.R. Faulkner, H.S. White, *Electrochemical Methods: Fundamentals and Applications* (Wiley, 2022)
19. J. Chen, H. Pu, M.C. Hersam, P. Westerhoff, Molecular engineering of 2D nanomaterial field-effect transistor sensors: fundamentals and translation across the innovation spectrum. Adv. Mater. **34**(3), e2106975 (2022)

20. U. Fischer, Fundamentals of glucose sensors. Diabet. Med. **8**(4), 309–321 (1991)
21. M. Heidari, B. Dabirmanesh, H. Mahdavi, K. Khajeh, Study break: glucose-sensitive holographic (bio)sensors: fundamentals and applications. Iran. Biomed. J. **21**(6), 347–348 (2017)
22. G. Gauglitz, Optical biochemical and chemical sensors–Europt(r)ode VIII. Anal. Bioanal. Chem. **386**(5), 1199–1200 (2006)
23. M. Hanko, N. Bruns, S. Rentmeister, J.C. Tiller, J. Heinze, Nanophase-separated amphiphilic conetworks as versatile matrixes for optical chemical and biochemical sensors. Anal. Chem. **78**(18), 6376–6383 (2006)
24. A.A. Ensafi, A. Kazemzadeh, Optical pH sensor based on chemical modification of polymer film. Microchem. J. **63**(3), 381–388 (1999)
25. D.G. Buerk, *Biosensors: Theory and Applications* (Technomic Publishing Co., 1993)
26. D.G. Buerk, *Biosensors: Theory and Applications* (CRC Press, 2014)
27. C.R. McMahon, I.C. Field, C.J. Bradshaw, G.C. White, B. Hindell, Tracking and data-logging devices attached to elephant seals do not affect individual mass gain or survival. Ecology **360**(2), 71–77 (2008)
28. N. Irawati, A.M. Hatta, Y.G.Y. Yhuwana, Sekartedjo, Heart rate monitoring sensor based on singlemode-multimode-singlemode fiber. Photon. Sens. **10**, 186–193 (2020)
29. S. Imani, A.J. Bandodkar, A.M. Vinu Mohan, R. Kumar, S. Yu, J. Wang, P.P. Mercier, A wearable chemical–electrophysiological hybrid biosensing system for real-time health and fitness monitoring. Nat. Commun. **7**(1), 11650 (2016)
30. F. Maduell, E. Cholbi, L. Morantes, V.J. Escudero-Saiz, J. Ollé, M. Martínez-Chillarón, L.M. Rodas, D. Rodriguez-Espinosa, M. Arias-Guillén, M. Vera, N. Fontseré J.J. Broseta, Results of salt intake restriction monitored with the new sodium control biosensor. Am. J. Nephrol. 1–1 (2023)
31. K.M. Rao, B.G. Sudarshan, Design and development of real time respiratory rate monitor using non-invasive biosensor. Int. J. Res. Eng. Technol. **4**(6), 437–442 (2015)
32. E. Eltzov, A.L. De Cesarea, Y.K.A. Low, R.S. Marks, Indoor air pollution and the contribution of biosensors. EuroBiotech J. **3**(1), 19–31 (2019)
33. S.M. Khor, J. Choi, P. Won, S.H. Ko, Challenges and strategies in developing an enzymatic wearable sweat glucose biosensor as a practical point-of-care monitoring tool for type II diabetes. Nanomaterials (Basel) **12**(2), 221 (2022)
34. K. Garikapati, The utility of handheld and wearable devices in the diagnosis of cardiac arrhythmias (2023)
35. C. Karunakaran, K. Bhargava, R. Benjamin, *Biosensors and Bioelectronics* (Elsevier, 2015)
36. A.J.R. Hobson, *Handbook of Modern Sensors*, ed. by J. Fraden (Springer/AIP Press, 2004), xvii+589 pp., with index, ISBN 0-387-00750-4. 295 ref. distributed chapter by chapter (Hardback£ 73) **23**(3), 397–397 (2005)
37. L.C. Clark Jr, C. Lyons, Electrode systems for continuous monitoring in cardiovascular surgery. Ann. N. Y. Acad. Sci. **102**(1), 29–45 (1962)
38. G.A. Rechnitz, R.K. Kobos, S.J. Riechel, C.R. Gebauer, A bio-selective membrane electrode prepared with living bacterial cells. Anal. Chim. Acta **94**(2), 357–365 (1977)
39. A.O. Scott, *Biosensors for Food Analysis* (1998)
40. Z. Taleat, A. Khoshroo, M.J.M.A. Mazloum-Ardakani, Screen-printed electrodes for biosensing: a review (2008–2013). Microchim. Acta **181**, 865–891 (2014)
41. J. Janata, *Principles of Chemical Sensors* (Springer Science & Business Media, 2010)
42. G. Selvolini, G. Marrazza, *Sensor Principles and Basic Designs. Fundamentals of Sensor Technology* (Elsevier, 2023), pp. 17–43
43. A. Harper, M.R. Anderson, Electrochemical glucose sensors—developments using electrostatic assembly and carbon nanotubes for biosensor construction. Sensors (Basel) **10**(9), 8248–8274 (2010)
44. S.J. Updike, G.P. Hicks, The enzyme electrode. Nature **214**(5092), 986–988 (1967)
45. J.D. Newman, A.P.F. Turner, Home blood glucose biosensors: a commercial perspective. Biosens. Bioelectron. **20**(12), 2435–2453 (2005)

46. W. Zhang, G. Li, Third-generation biosensors based on the direct electron transfer of proteins. Anal. Sci. **20**(4), 603–609 (2004)
47. G.G. Guilbault, G.J. Lubrano, An enzyme electrode for the amperometric determination of glucose. Anal. Chim. Acta **64**(3), 439–455 (1973)
48. J. Newman, S. Setford, Enzymatic biosensors. Mol. Biotechnol. **32**, 249–268 (2006)
49. W.V. Tamborlane, W. Bonfig, E. Boland, Recent advances in treatment of youth with type 1 diabetes: better care through technology. Diabet. Med. **18**(11), 864–870 (2001)
50. J. Wang, Glucose biosensors: 40 years of advances and challenges. Electroanalysis **13**(12), 983–988 (2001)
51. R.D. Milton, F. Giroud, A.E. Thumser, S.D. Minteer, R.C.T. Slade, Hydrogen peroxide produced by glucose oxidase affects the performance of laccase cathodes in glucose/oxygen fuel cells: FAD-dependent glucose dehydrogenase as a replacement. Phys. Chem. Chem. Phys. **15**(44), 19371–19379 (2013)
52. J. Wang, Electrochemical glucose biosensors. Chem. Rev. **108**(2), 814–825 (2008)
53. Y. Fang, H. Zhang, J. Xie, M. Lin, L. Ying, P. Pang, W. Ji, Sensitivity of chest CT for COVID-19: comparison to RT-PCR. Radiology **296**(2), E115–E117 (2020)
54. D. Bruen, C. Delaney, L. Florea, D. Diamond, Glucose sensing for diabetes monitoring: recent developments. Sensors (Basel, Switzerland) **17**(8) (2017)
55. J.D. Lane, D.M. Krumholz, R.A. Sack, C. Morris, Tear glucose dynamics in diabetes mellitus. Curr. Eye Res. **31**(11), 895–901 (2006)
56. M.I. Svechtarova, I. Buzzacchera, B.J. Toebes, J. Lauko, N. Anton, C.J. Wilson, Sensor devices inspired by the five senses: a review. Electroanalysis **28**(6), 1201–1241 (2016)
57. W.C. Lee, E. Smith, B. Chubb, M.L. Wolden, Frequency of blood glucose testing among insulin-treated diabetes mellitus patients in the United Kingdom. J. Med. Econ. **17**(3), 167–175 (2014)
58. I. Buzzacchera, M. Vorobii, N.Y. Kostina, A. de Los Santos Pereira, T. Riedel, M. Bruns, W. Ogieglo, M. Möller, C.J. Wilson, C. Rodriguez-Emmenegger, Polymer brush-functionalized chitosan hydrogels as antifouling implant coatings. Biomacromolecules **18**(6), 1983–1992 (2017)
59. T. Bailey, B.W. Bode, M.P. Christiansen, L.J. Klaff, S. Alva, The performance and usability of a factory-calibrated flash glucose monitoring system. Diabetes Technol. Ther. **17**(11), 787–794 (2015)
60. A.J. Al-Mawali, Non-communicable diseases: shining a light on cardiovascular disease. Oman's Biggest Killer **30**(4), 227 (2015)
61. P.W. Wilson, R.B. D'Agostino, D. Levy, A.M. Belanger, H. Silbershatz, W.B. Kannel, Prediction of coronary heart disease using risk factor categories. Circulation **97**(18), 1837–1847 (1998)
62. K. King, L.P. Grazette, D.N. Paltoo, J.T. McDevitt, S.K. Sia, P.M. Barrett, F.S. Apple, P.A. Gurbel, R. Weissleder, H. Leeds, E.J. Iturriaga, A. Rao, B. Adhikari, P. Desvigne-Nickens, Z.S. Galis, P. Libby, Point-of-care technologies for precision cardiovascular care and clinical research: national heart, lung, and blood institute working group. JACC Basic Transl. Sci. **1**(1), 73–86 (2016)
63. M. Wolf, D. Juncker, B. Michel, P. Hunziker, E. Delamarche, Simultaneous detection of C-reactive protein and other cardiac markers in human plasma using micromosaic immunoassays and self-regulating microfluidic networks. Biosens. Bioelectron. **19**(10), 1193–1202 (2004)
64. G.J. Kost, N.K. Tran, Point-of-care testing and cardiac biomarkers: the standard of care and vision for chest pain centers. Cardiol. Clin. **23**(4), 467–490 (2005)
65. A. Qureshi, Y. Gurbuz, J.H. Niazi, Biosensors for cardiac biomarkers detection: a review. Sens. Actuators B Chem.**171**, 62–76 (2012)
66. W. Leung, C.P. Chan, M. Leung, K. Lehmann, I. Renneberg, M. Lehmann, A. Hempel, J.F. Glatz, R. Renneberg, Novel "digital-style" rapid test simultaneously detecting heart attack and predicting cardiovascular disease risk. Anal. Lett. **38**(3), 423–439 (2005)
67. H.R. Jang, A.W. Wark, S.H. Baek, B.H. Chung, H.J. Lee, Ultrasensitive and ultrawide range detection of a cardiac biomarker on a surface plasmon resonance platform. Anal. Chem. **86**(1), 814–819 (2014)

68. I.H. Cho, E.-H. Paek, Y.-K. Kim, J.-H. Kim, S.-H. Paek, Chemiluminometric enzyme-linked immunosorbent assays (ELISA)-on-a-chip biosensor based on cross-flow chromatography. Anal. Chim. Acta **632**(2), 247–255 (2009)
69. M. Ibrahim, A. Alsheikh, A. Matar, Attack graph modeling for implantable pacemaker. Biosensors (Basel)**10**(2), 14 (2020)
70. B. Bohunicky, S.A. Mousa, Biosensors: the new wave in cancer diagnosis. Nanotechnol. Sci. Appl. 1–10 (2010)
71. S. Valastyan, R.A. Weinberg, Tumor metastasis: molecular insights and evolving paradigms. Cell **147**(2), 275–292 (2011)
72. C. Loo, A. Lowery, N. Halas, J. West, R. Drezek, Immunotargeted nanoshells for integrated cancer imaging and therapy. Nano Lett. **5**(4), 709–711 (2005)
73. A.B.R. Santosh, T. Jones, J. Harvey, A review on oral cancer biomarkers: Understanding the past and learning from the present. J. Cancer Res. Ther. **12**(2), 486–492 (2016)
74. T.N. Zamay, G.S. Zamay, O.S. Kolovskaya, R.A. Zukov, M.M. Petrova, A. Gargaun, M.V. Berezovski, A.S. Kichkailo, Current and prospective protein biomarkers of lung cancer. Cancers (Basel) **9**(11), 155 (2017)
75. P.G.C. Selwyna, P.R. Loganathan, K.H. Begam, Development of electrochemical biosensor for breast cancer detection using gold nanoparticle doped CA 15-3 antibody and antigen interaction, in *2013 International Conference on Signal Processing, Image Processing & Pattern Recognition* (IEEE, 2013), pp. 75–81
76. J. Wu, F. Yan, J. Tang, C. Zhai, H. Ju, A disposable multianalyte electrochemical immunosensor array for automated simultaneous determination of tumor markers. Clin. Chem. **53**(8), 1495–1502 (2007)
77. K. Boriachek, M. Umer, M.N. Islam, V. Gopalan, A.K. Lam, N.-T. Nguyen, M.J.A. Shiddiky, An amplification-free electrochemical detection of exosomal miRNA-21 in serum samples. Analyst **143**(7), 1662–1669 (2018)
78. K. Bravo, F.G. Ortega, G.A. Messina, M.I. Sanz, M.A. Fernández-Baldo, J. Raba, Integrated bio-affinity nano-platform into a microfluidic immunosensor based on monoclonal bispecific trifunctional antibodies for the electrochemical determination of epithelial cancer biomarker. Clin. Chim. Acta **464**, 64–71 (2017)
79. M.Ç. Canbaz, Ç.S. Şimşek, M.K. Sezgintürk, Electrochemical biosensor based on self-assembled monolayers modified with gold nanoparticles for detection of HER-3. Anal. Chim. Acta **814**, 31–38 (2014)
80. A.A. Saeed, J.L.A. Sánchez, C.K. O'Sullivan, M.N. Abbas, DNA biosensors based on gold nanoparticles-modified graphene oxide for the detection of breast cancer biomarkers for early diagnosis. Bioelectrochemistry**118**, 91–99 (2017)
81. H. Wei, C. Zhang, X. Du, Z. Zhang, Research progress of biosensors for detection of SARS-CoV-2 variants based on ACE2. Talanta 123813 (2022)
82. S.A. Abid, A.A. Muneer, I.M.S. Al-Kadmy, A.A. Sattar, A.M. Beshbishy, G.E.-S. Batiha, H.F. Hetta, Biosensors as a future diagnostic approach for COVID-19. Life Sci. **273**, 119117 (2021)
83. I.S. Mahmoud, Y.B. Jarrar, W. Alshaer, S. Ismail, SARS-CoV-2 entry in host cells-multiple targets for treatment and prevention. Biochimie **175**, 93–98 (2020)
84. M. Mascini, S. Tombelli, Biosensors for biomarkers in medical diagnostics. Biomarkers **13**(7–8), 637–657 (2008)
85. Z.S. Miripour, R. Sarrami-Forooshani, H. Sanati, J. Makarem, M.S. Taheri, F. Shojaeian, A.H. Eskafi, F. Abbasvandi, N. Namdar, H. Ghafari, Real-time diagnosis of reactive oxygen species (ROS) in fresh sputum by electrochemical tracing; correlation between COVID-19 and viral-induced ROS in lung/respiratory epithelium during this pandemic. Biosens. Bioelectron. **165**, 112435 (2020)
86. L. Zhang, H. Guo, Biomarkers of COVID-19 and technologies to combat SARS-CoV-2. Adv. Biomark. Sci. Technol. **2**, 1–23 (2020)
87. A. Lahav, M. Auslender, I. Abdulhalim, Sensitivity enhancement of guided-wave surface-plasmon resonance sensors. Opt. Lett. **33**(21), 2539–2541 (2008)

88. D.K.W. Chu, Y. Pan, S.M.S. Cheng, K.P.Y. Hui, P. Krishnan, Y. Liu, D.Y.M. Ng, C.K.C. Wan, P. Yang, Q. Wang, M. Peiris, L.L.M. Poon, Molecular diagnosis of a novel coronavirus (2019-nCoV) causing an outbreak of pneumonia. Clin. Chem. **66**(4), 549–555 (2020)
89. J.F.W. Chan, C.C.-Y. Yip, K.K.-W. To, T.H.-C. Tang, S.C.-Y. Wong, K.-H. Leung, A.Y.-F. Fung, A.C.-K. Ng, Z. Zou, H.-W. Tsoi, G.K.-Y. Choi, A.R. Tam, V.C.-C. Cheng, K.-H. Chan, O.T.-Y. Tsang, K.-Y. Yuen, Improved molecular diagnosis of COVID-19 by the novel, highly sensitive and specific COVID-19-RdRp/Hel real-time reverse transcription-PCR assay validated in vitro and with clinical specimens. J. Clin. Microbiol. **58**(5), 85 (2020)
90. A. Wu, Y. Peng, B. Huang, X. Ding, X. Wang, P. Niu, J. Meng, Z. Zhu, Z. Zhang, J. Wang, J. Sheng, L. Quan, Z. Xia, W. Tan, G. Cheng, T. Jiang, Genome composition and divergence of the novel coronavirus (2019-nCoV) originating in China. Cell Host Microbe **27**(3), 325–328 (2020)
91. J.L. Nieto-Torres, M.L. DeDiego, C. Verdiá-Báguena, J.M. Jimenez-Guardeño, J.A. Regla-Nava, R. Fernandez-Delgado, C. Castaño-Rodriguez, A. Alcaraz, J. Torres, V.M. Aguilella, L. Enjuanes, Severe acute respiratory syndrome coronavirus envelope protein ion channel activity promotes virus fitness and pathogenesis. PLoS Pathog. **10**(5), e1004077 (2014)
92. P.C.Y. Woo, S.K.P. Lau, B.H.L. Wong, H.-W. Tsoi, A.M.Y. Fung, R.Y.T. Kao, K.-H. Chan, J.S.M. Peiris, K.-Y. Yuen, Differential sensitivities of severe acute respiratory syndrome (SARS) coronavirus spike polypeptide enzyme-linked immunosorbent assay (ELISA) and SARS coronavirus nucleocapsid protein ELISA for serodiagnosis of SARS coronavirus pneumonia. J. Clin. Microbiol. **43**(7), 3054–3058 (2005)
93. H.-W. Jiang, Y. Li, H.-N. Zhang, W. Wang, D. Men, X. Yang, H. Qi, J. Zhou, S.-C. Tao, Global profiling of SARS-CoV-2 specific IgG/IgM responses of convalescents using a proteome microarray (preprint) (2020)
94. K.K.W. To, C.C.Y. Yip, C.Y.W. Lai, C.K.H. Wong, D.T.Y. Ho, P.K.P. Pang, A.C.K. Ng, K.-H. Leung, R.W.S. Poon, K.-H. Chan, V.C.C. Cheng, I.F.N. Hung, K.-Y. Yuen, Saliva as a diagnostic specimen for testing respiratory virus by a point-of-care molecular assay: a diagnostic validity study. Clin. Microbiol. Infect. **25**(3), 372–378 (2019)
95. W. Zhang, R.-H. Du, B. Li, X.-S. Zheng, X.-L. Yang, B. Hu, Y.-Y. Wang, G.-F. Xiao, B. Yan, Z.-L. Shi, P. Zhou, Molecular and serological investigation of 2019-nCoV infected patients: implication of multiple shedding routes. Emerg. Microbes Infect. **9**(1), 386–389 (2020)
96. Y. Pan, X. Li, G. Yang, J. Fan, Y. Tang, J. Zhao, X. Long, S. Guo, Z. Zhao, Y. Liu, H. Hu, H. Xue, Y. Li, Serological immunochromatographic approach in diagnosis with SARS-CoV-2 infected COVID-19 patients. J. Infect. **81**(1), e28–e32 (2020)
97. M. Asif, A. Aziz, M. Azeem, Z. Wang, G. Ashraf, F. Xiao, X. Chen, H. Liu, A review on electrochemical biosensing platform based on layered double hydroxides for small molecule biomarkers determination. Adv. Colloid Interface Sci.**262**, 21–38 (2018)
98. T.J. Park, M.S. Hyun, H.J. Lee, S.Y. Lee, S. Ko, A self-assembled fusion protein-based surface plasmon resonance biosensor for rapid diagnosis of severe acute respiratory syndrome. Talanta **79**(2), 295–301 (2009)
99. A. Djaileb, B. Charron, M.H. Jodaylami, V. Thibault, J. Coutu, K. Stevenson, S. Forest, L.S. Live, D. Boudreau, J.N. Pelletier, J.-F. Masson, A rapid and quantitative serum test for SARS-CoV-2 antibodies with portable surface plasmon resonance sensing. Anal. Chem. (2020)
100. M. Asif, M. Ajmal, G. Ashraf, N. Muhammad, A. Aziz, T. Iftikhar, J. Wang, H. Liu, The role of biosensors in coronavirus disease-2019 outbreak. Curr. Opin. Electrochem. **23**, 174–184 (2020)
101. J. Liu, X. Chen, Q. Wang, M. Xiao, D. Zhong, W. Sun, G. Zhang, Z. Zhang, Ultrasensitive monolayer MoS_2 field-effect transistor based DNA sensors for screening of down syndrome. Nano Lett. **19**(3), 1437–1444 (2019)
102. M. Asif, A. Aziz, H. Wang, Z. Wang, W. Wang, M. Ajmal, F. Xiao, X. Chen, H. Liu, Superlattice stacking by hybridizing layered double hydroxide nanosheets with layers of reduced graphene oxide for electrochemical simultaneous determination of dopamine, uric acid and ascorbic acid. Microchim. Acta **186**, 1–11 (2019)

103. M. Asif, A. Aziz, A.Q. Dao, A. Hakeem, H. Wang, S. Dong, G. Zhang, F. Xiao, H. Liu, Real-time tracking of hydrogen peroxide secreted by live cells using MnO2 nanoparticles intercalated layered doubled hydroxide nanohybrids. Anal. Chim. Acta **898**, 34–41 (2015)
104. M. Asif, A. Aziz, G. Ashraf, Z. Wang, J. Wang, M. Azeem, X. Chen, F. Xiao, H. Liu, Facet-inspired core–shell gold nanoislands on metal oxide octadecahedral heterostructures: high sensing performance toward sulfide in biotic fluids. ACS Appl. Mater. Interfaces **10**(43), 36675–36685 (2018)
105. L.A. Layqah, S. Eissa, An electrochemical immunosensor for the corona virus associated with the Middle East respiratory syndrome using an array of gold nanoparticle-modified carbon electrodes. Mikrochim. Acta **186**, 1–10 (2019)
106. S. Mahari, A. Roberts, D. Shahdeo, S. Gandhi, eCovSens-ultrasensitive novel in-house built printed circuit board based electrochemical device for rapid detection of nCovid-19 (2020). https://doi.org/10.1101/2020.04.24.059204
107. C. Wang, C. Wang, X. Wang, K. Wang, Y. Zhu, Z. Rong, W. Wang, R. Xiao, S. Wang, Magnetic SERS strip for sensitive and simultaneous detection of respiratory viruses. ACS Appl. Mater. Interfaces **11**(21), 19495–19505 (2019)
108. I. Santura, P. Kawalec, M. Furman, T. Bochenek, Chest computed tomography versus RT-PCR in early diagnostics of COVID-19—a systematic review with meta-analysis. Pol. J. Radiol. **86**, e518–e531 (2021)
109. Z. Zhao, H. Cui, W. Song, X. Ru, W. Zhou, X. Yu, A simple magnetic nanoparticles-based viral RNA extraction method for efficient detection of SARS-CoV-2 (2020) https://doi.org/10.1101/2020.02.22.961268
110. T. Ai, Z. Yang, H. Hou, C. Zhan, C. Chen, W. Lv, Q. Tao, Z. Sun, L. Xia, Correlation of chest CT and RT-PCR testing for coronavirus disease 2019 (COVID-19) in China: a report of 1014 cases. Radiology **296**(2), E32–E40 (2020)
111. L. Yu, S. Wu, X. Hao, X. Dong, L. Mao, V. Pelechano, W.-H. Chen, X. Yin, Rapid detection of COVID-19 coronavirus using a reverse transcriptional loop-mediated isothermal amplification (RT-LAMP) diagnostic platform. Clin. Chem. **66**(7), 975–977 (2020)
112. J. Xiang, M. Yan, H. Li, T. Liu, C. Lin, S. Huang, C. Shen, Evaluation of enzyme-linked immunoassay and colloidal gold-immunochromatographic assay kit for detection of novel coronavirus (SARS-Cov-2) causing an outbreak of pneumonia (COVID-19) (2020). https://doi.org/10.1101/2020.02.27.20028787
113. S. Liu, L. Han, J. Li, H. Li, Electrochemical detection of ACE2 as a biomarker for diagnosis of COVID-19 and potential male infertility. Biosens. Bioelectron. **198**, 113788 (2022)
114. X. Zhang, Y. Yang, J. Cao, Z. Qi, G. Li, Point-of-care CRISPR/Cas biosensing technology: a promising tool for preventing the possible COVID-19 resurgence caused by contaminated cold-chain food and packaging. Food Front. **4**(1), 207–232 (2023)
115. V. Chaudhary, S. Rustagi, A. Kaushik, Bio-derived smart nanostructures for efficient biosensors. Curr. Opin. Green Sustain. Chem. 100817 (2023)
116. A. Shinde, K. Illath, U. Kasiviswanathan, S. Nagabooshanam, P. Gupta, K. Dey, P. Chakrabarty, M. Nagai, S. Rao, S. Kar, T.S. Santra, Recent advances of biosensor-integrated organ-on-a-chip technologies for diagnostics and therapeutics. Anal. Chem. **95**(6), 3121–3146 (2023)
117. R. Qureshi, M. Irfan, H. Ali, A. Khan, A.S. Nittala, S. Ali, A. Shah, T.M. Gondal, F. Sadak, Z. Shah, Artificial intelligence and biosensors in healthcare and its clinical relevance: a review (2023)
118. J. Zhang, Z. Li, Y. Pang, Y. Fan, H. Ai, Genetically encoded boronolectin as a specific red fluorescent UDP-GlcNAc biosensor. ACS Sens. (2023)
119. Y. Tong, L. Zhang, Discovering the next decade's synthetic biology research trends with ChatGPT. Synth. Syst. Biotechnol. **8**(2), 220 (2023)
120. T. Marbun, Integration of artificial intelligence, ChatGPT, and critical thinking method for academic assignment at theological higher education (2023)

Biomedical Applications of Nanocellulose-Based Biomaterials: Recent Advancements

K. Kanimozhi, M. V. Arularasu, R. Babujanarthanam, and Force Tefo Thema

Abstract Nanocellulose, is a single besides auspicious natural polymer material removed from native cellulose, takes increased abundant care aimed at the situation habit by way of biomedical material for the reason that of its extraordinary superior surface chemistry, physical properties, and excellent biological properties. Nanocellulose comes in three types: nanocrystals, nanocrystalline cellulose, and nanocrystalline cellulose nanofibers. A cellulose nanofibril (CNF) is a very small piece of cellulose and bacterial cellulose (BC), are presented then associated trendy positions of manufacture, possessions, and biomedical claims fashionable this manuscript. Due to its elasticity, nanocellulose holds an extensive variety of potential requests owing near the situation energetic efficient groups and amino groups. The biodegradability and biocompatibility of nanocellulose also its products laterally by their numerous biochemical functionalities variety among the development and application of innovative technologies in the pharmaceutical, nutritional, medicinal, environmental, agricultural, and drug delivery industries. In the current paper, we goal near offer an inclusive and planned account reformed nanocellulose, and its derivatives used to build biomedical nanomaterials. It covered nanocomposite materials' biological and physicochemical properties, including the advantages of nanocellulose and its derivatives as well as better physical biological constancy then compartments flesh interface, painkiller relief control, and improved bio-obtainability, besides ability to perform medical exercises. In addition, this review does not fully explore the multitude of alteration methods and their practical uses.

K. Kanimozhi (✉)
Department of Chemistry, Science and Humanities, Loyola Institute of Technology (LIT), Kuthambakkam, Tamil Nadu 600123, India
e-mail: kanivlr8@gmail.com

M. V. Arularasu
Department of Electronic Engineering, Chang Gung University, Taoyuan 33302, Taiwan

R. Babujanarthanam
Department of Biotechnology, Thiruvalluvar University, Serkkadu, Vellore, Tamil Nadu 632115, India

F. T. Thema
Department of Bioeconomy and Material Science, Botswana University of Agriculture and Natural Resources, Private Bag 0027, Gaborone, Botswana

K. Kasinathan et al. (eds.), *New Technologies for Energy Transition Based on Sustainable Development Goals*, https://doi.org/10.1007/978-981-97-2527-4_5

1 Introduction

A natural homopolymer of glucose and glycosidic bonds, fiber stands collected of d-glucopyranose elements [1]. Despite being an extremely hydrophilic bio polymer, cellulose stays not answerable trendy aquatic owing towards the situation extremely glassy structure, making it unique of the greatest plentiful natural polymers happening ground. The situation remains possible to make cellulose soluble by chemically substituting its ordered crystalline regions. In many fields, including food, pharmaceuticals, textiles, and adhesives, cellulose derivatives are used. In addition to being informal near custom, HPMC consumes superior flick starting capabilities, decent bio compatibility besides biodegradability making it an important cellulose derivative. Pharmaceutical companies typically use it as drug delivery matrixes [2]. Plants, animals, bacteria, and some amoebas produce cellulose, which is an abundant and sustainable biopolymer. As agricultural waste is not harmful to food supplies, its use for industrial purposes is attractive [3]. The cellulose microfibers extracted from Gongura life have previously been shown to be effective reinforcement additives [4]. Alkaline washing, bleaching, and mechanical attrition were used to prepare the microfibers of diameter 10 nm and length 10 nm 100 nm. By further processing, cellulose nanocrystals eliminate these amorphous regions, enhancing their stiffness. By preparing nanocrystals from Gongura leaf, we removed these feebler fields since the fiber to create a great-worth-totaling supporting assistance on behalf of polymer atmosphere composites. Nanocrystals prepared from Gongura leaf will be used for the first time in plastic as a functional filler. Numerous foundations, such by way of wheat grass then tunicin, stood recycled to extract and isolate cellulose whiskers, which remained recycled equally cavalries now polymer conditions [5–8]. Numerous advantages can be derived from using cellulose crystallites in nanocomposites. Most renewable polymers originate in plant cells. Bacteria and animals can also synthesize cellulose, making it the world's most abundant polymer. Incorporating quartz voguish a medium, however, can involve glitches trendy regulatory the equal of dispersion since they must be isolated. For the elimination of formless districts of fiber, biting hydrolysis is well known. There have been several studies that have identified and separated cellulose crystallites from different sources. An acid hydrolysis method was reported by Ranby for preparing cellulose whiskers from microfibrils. Birefringence was reported by Marchessault and co-workers when acid hydrolysis was used to hydrolyze chitin and mercerized, native, and bacterial cellulose. There is intense research and development going on with cellulose, a fascinating biopolymer and a valuable raw material. Increasingly, cellulose is being considered as an environmentally friendly and economical strengthening mediator aimed at composites. In wide variety of products, counting biocomposites besides nanocomposites, can be produced using cellulose in polymer production. Nanofibers are single of the greatest good-looking courses of resources to use in nanoamalgams due to their outstanding properties [9]. In addition to whiskers, nanofibrils can also be made from cellulose [10]. A nanoparticle with a needlelike appearance is known as a whistle (or "nanowhiskers/nanocrystals"), while a nanofibril before "micro fibrillated fiber" remains an extended besides elastic nanoparticle collected of an alternating shapeless besides glassy domain. As a result of the dispersion of reinforcing nanosized particles into continuous polymers, Abdul Khalil et al. [11] obligate remained appealing great care now fresh ages aimed at their ability near improve bodily belongings by actual little stages of nanosized plasters [12]. There has been extensive use of nanosized reinforcements in polymer crowds towards improve polymers possessions, e.g., nanomuds, carbon nanotubes, graphite, besides inorganic nanoparticles. Several novel nanocomposite products have been proposed that incorporate cellulose nanocrystals for reinforcement. As mentioned above, nanocellulose have great potential, but the high concentration of -OH on their surfaces makes them highly attracting to each other [13]. The high polarity of cellulose surfaces makes similar spreading of nanofiber happening polymers problematic unpaid near a short interfacial compatibility through hydrophobic polymer matrixes [14]. An illustration of cellulose's structure can be found in Fig. 1. Among its properties are well constructions, great superficial part, decent motorized, rheological, fence, non-poisonous besides biocompatible nature, creation nanocellulose an perfect substantial aimed at tissue engineering. The biomedical applications of polymeric ternary blends have recently been studied [15]. In order to produce nanocomposites for tissue engineering applications, many types of polymers are combining with nanoparticles, including natural, synthetic, and combinations. This figure shows an overview of the number of journals on nanocellulose scaffolds in the SCOPUS catalogue beginning in 2002 and ending in 2023. A document containing unique trainings, appraisals, volume sections, meeting documents, discussion appraisals, in addition literatures is included trendy the document as shown in Fig. 2. There is a need for enlarged noticeably hip new ages in journals connecting to "*nanocellulose scaffolds*". In fashionable 2019, 804 articles were published in book form, whereas there were only 95 articles in fashionable 2005. It is likely that this tendency will endure in the centuries to come.

It has been found that the publications concerning "*Nanocellulose Scaffolds*" have enlarged significantly in the current century. A total of 804 articles were published in 2019, whereas 95 articles were published in 2005. In the years to come, this trend determination may continue.

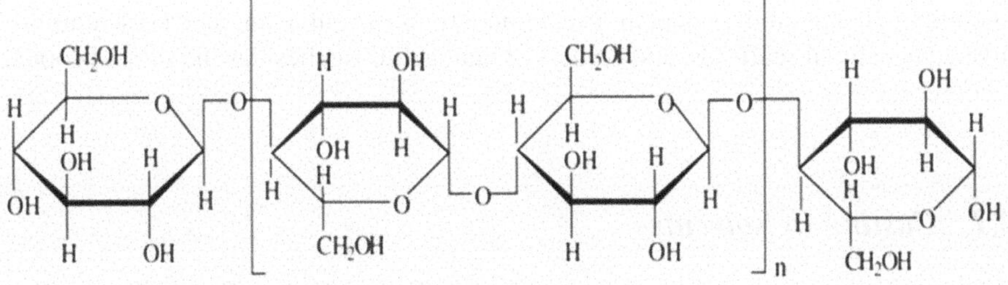

Fig. 1 Schematic structure of nanocellulose

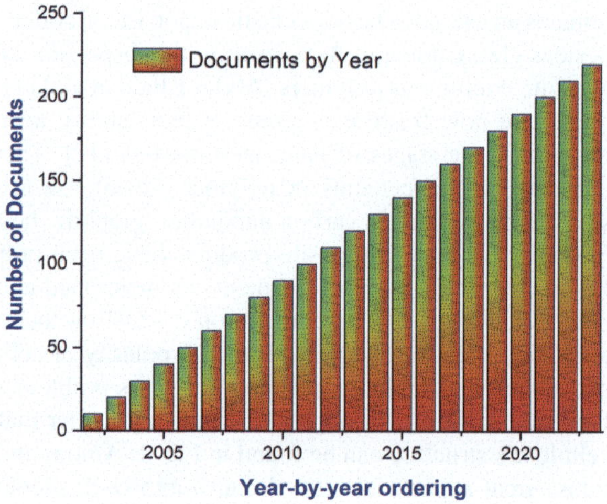

Fig. 2 Publications per year from 2002 to June 2023 in the Scopus database, including original; articles, reviews, book chapters, conference papers, conference reviews, and letters in all areas of knowledge, using the keyword "*nanocellulose scaffold*"

2 The Current Possibility of Effort

Normally, journal object stands a part of collected works those summaries the sympathetic of existing national fine painting improvements, current/future tasks on a specific issue trendy concurrence by the investigational truths of numerous classes of investigation nearby arranged that specific topic. The possibility of the current effort remains towards highlight dissimilar features of nanocellulose then the situation byproducts focused nanocomposites only by way of healthy as trendy mixture through numerous usual or artificial polymeric supplies fashionable varied efficient systems then toward commercial the investigation part that remain combining moderately significant modernizations fashionable the previous period in stimulating potential requests. Furthermore, the major unit of this appraisal object presents overall evidence and chemistry of nanocellulose specifying combination, properties, and altered ways of alterations, another unit journals the biodegradation development of nanocellulose before the situation derived focused ingredients laterally by the varied series of potential uses. At the last unit emphases happening the assumption and upcoming viewpoint of nanocellulose. Foundations of source, organic presentations, diverse beneficial scaffolds, and tenders of nanocellulose, besides the subject copied resources.

2.1 *Casting in Solvents*

Despite the current trend of polymer solution scaffold fabrication, solvent casting remains a humble and cost-effective method. While scaffolds must have a minimum thickness and physical properties, they must not be too thick.

2.2 In the Second Method, Freeze Drying Is Used

Anhydrous or nearly anhydrous 3D-polymer constructions are freeze-dried in a vacuum environment to remove freeze water/solvent. An adhesion between a spongy construction and a scaffold is caused by redirected freeze crystals [24]. Almost 99% of nanocellulose scaffolds are porosity-free and have pores between 30 and 300 μm in size [25]. As a result, this method does not require several hard porogens to be consumed, whereas porosity frequently remains uneven after prolonged handling.

2.3 Leaching of Particulates

As a result of particle leaching, aquatic-solvable salts (e.g., sodium chloride, sugar particles, then sodium citrate) are combined hooked on a polymer clarification. As a result, the explanation of polymer is coated onto the frame with a particular figure and then allowed to dry by means of the restriction ventilation process. A porous polymer structure is created by letting the salt particles materials rest filtered close to it [26].

2.4 The Fourth Method Is Electrospinning

Using the electrospinning method, polymer solutions are stretched by the force of electrostatic forces before being electrospun into sufficient scaffolds. There is an enormous amount of external space in these scaffolds intended for connecting chambers with restricted machine-driven properties [27].

2.5 Foaming of Gas

Carbon dioxide (CO_2) is rummaged sale in the direction of convincing porosity in this method of constructing highly porous scaffolds stripped of cleaners. It consequences now the nucleation besides development of air foams in the solid when polymer arrangement is exposed to high-density fume. The final result is they remain restriction dehydrated without a 3D-arrangement, so they have a hole extent of all over 100 mm and porosity of up to 95%. However, the scaffolds' interconnectivity and pore sizes remain challenging [28].

2.6 The 3D Printing Revolution

Currently, 3D-printed cells are varied by polymer solutions placed coating-through-coating besides irritated related near create a 3D-skeleton that imitators the flesh before organ [29]. A multifaceted 3D figure is generated by measuring pore sizes, resolution, morphology, and inside constructions. The scaffold can also be continuously filled with cells [30].

3 Nanocellulose

In plants, animals, and bacteria, nanofiber remains the creation before excerpt of natural fiber. A variety of characteristics distinguish nanocellulose from customary resources, with morphology besides geometric sizes that are different from those of traditional materials. A great exact superficial part, quartz that is high, rheological possessions, fluid glassy conduct, and arrangement besides location are all desirable. Biocompatibility, biodegradable, biodegradable, mechanical reinforcement, barrier properties [16]. There are three general parts to the properties of nanocellulose. Two properties of physical nature. Properties of chemicals and properties relating to biology. Nanocellulose's biological properties and mechanical reinforcement surface groups are mainly discussed in this article, which is associated with the topic biomedicine. Hydrogels, layers, nano/micro threads, pipes, micro compasses, nanoparticles before films can be fabricated from nanocellulose scaffolds with different shapes and sizes [17, 18]. A diversity of systems tin stands charity near generates these biomaterials, such as solvent casting, freeze drying, particulate leaching, electrospinning, gas foaming, and 3D printing [19–23] as shown in Fig. 3.

3.1 Nanocellulose Scaffolds as Physical Materials

Physical nanocellulose scaffolds can remain formed so long as binary serious situations: (a) inter chain connections of scaffolds essential be there robust sufficient near system semi-perpetual connection opinions fashionable the polymer grid and (b) the polymer grids of the scaffolds necessity remain bright towards conversation then rambling liquid particles over their hydrated construction [31]. Figure 4 displays the chief bodily connections elaborate trendy starting nanocellulose scaffolds. In the scaffolds construction can stand steadied through altering besides monitoring the outside incentives such as pH, ionic forte, heat, otherwise using the being there of exact particles or else ions [32].

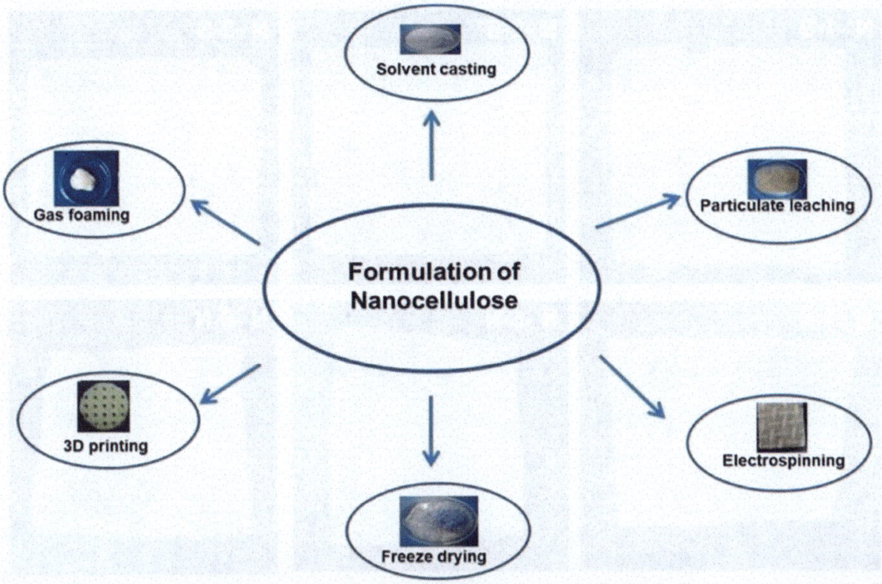

Fig. 3 Schematic illustration of nanocellulose scaffold in different methods

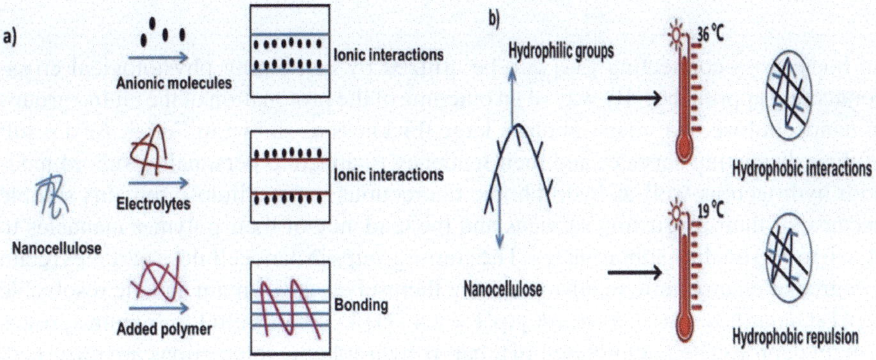

Fig. 4 Schematic representation of nanocellulose scaffolds grids resulting from different physical interactions

3.2 *Nanocellulose Scaffolds Are Prepared by Physical Crosslinking*

Alginate and carrageenan are other marine polysaccharides that are found as gel-like structures in nature. Nanocellulose is not one of these compounds [33]. An alternative advancing method for stable scaffolds aimed at tissue engineering applications based

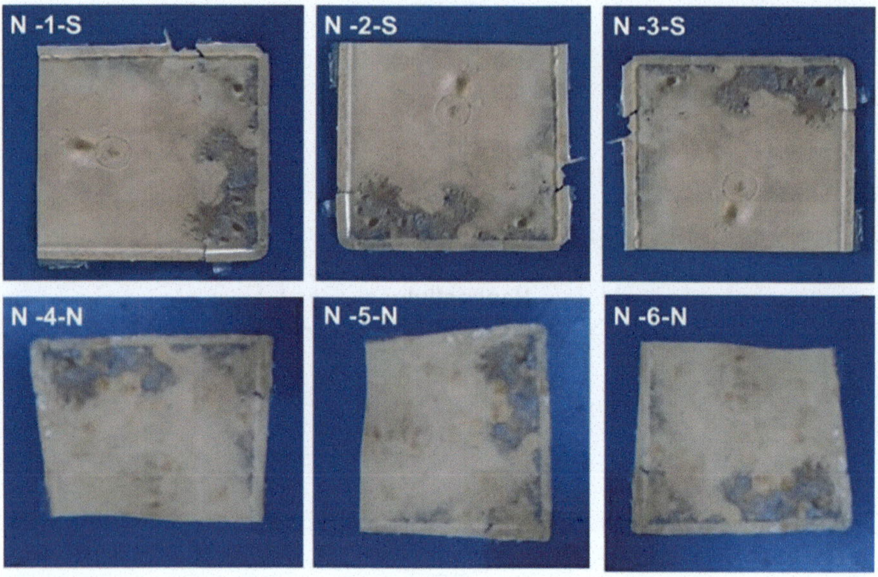

Fig. 5 Nanocellulose scaffolds fabricated using different polymer concentrations (5.0, 3.0, and 2.0% w/w) and two dissimilar gelation methods ("S" for the NaOH solution and "N" for NH_3 gaseous)

on body cross-connecting [34] can be utilized by combining physiological cross-connecting approaches. By way of an outcome of the protonation of the amino groups in nanocellulose, that material has a large thickness as shown in Fig. 5. As a result of their disgusting services and their tendency to attach to personality-subordinates with hydrogen as well as hydrophobic interactions, nanocellulose remains soluble because of their disgusting services and the tendency of their polymer manacles to attach to personality-subordinates. The amino groups in nanocellulose particles retain a degree of electrostatic repulsion among themselves and remain aquatic resolvable at pH standards lower than one pK_a (pH = 5.4–7) [35]. Nanocellulose's amino groups remain deprotonated at unbiased pH, but its hydrophobic interactions are refreshed, creating a gel-like hastiness. A nanocellulose scaffold can remain physically, chemically, and biologically customized after limitation. A range of methods have been employed, including polymer attention, DDA, solubilization (in later aqueous or alcoholic solutions), and crosslinking [36].

3.2.1 The Neutralization Course Irritates Nanocellulose

Using growing pH standards, nanocellulose scaffolds can be found minus consuming particles such as polyions or complexation mediators [37]. In the counteraction procedure persuades an entire responsibility thickness damage of amine groups of the nanocellulose now clarification (pK_a = ~ 5.5–7.0) using straight besides measured acquaintance near basic thinners similar $NaHCO_3$ [38], urea [39], NaOH [40–42], or gaseous ammonia [43]. Fashionable these situations, hydrophobic connections and hydrogen relationship remain heartened. Examined the creation kinetics of the nanocellulose scaffolds (nanocellulose meditation among 2 and 5% w/w) with the range of 1–8 M, NaOH in place of a bodily fractious linker. The carriage of NaOH particles over the nanocellulose scaffolds happened through circulation besides the gelation amount stood earlier by little nanocellulose attentions then great NaOH attentions [44] found fractious-connected nanocellulose scaffolds by the side of dissimilar nanocellulose meditations (5% and 8% w/w) consuming a thickening wash (6 M NaOH). Nanocellulose scaffolds remained similar besides showed plane sides [45]. By adding NaOH solution to nanocellulose solution of 1% w/w, nanocellulose scaffolds are formed. The scaffolds remained unchanging aimed at a lengthy dated next to area heat [46]. It is formed actually fractious-related nanocellulose scaffolds consuming aqueous NaOH (2.5 mol/L) then vaporous NH_3 (NH_4OH next to a meditation of 3.6 mol/L) by dissimilar nanocellulose meditations (Fig. 4) shows (0.5, 1, 2, 3, 4, 5, and 10%). Scaffolds organized by a nanocellulose meditation since 1.0–5.0% (w/w) remained extra similar than individual's scaffolds organized by a polymer meditation since 9% (w/w). In the scaffolds organized through 0.5% (w/w) remained delicate contravention addicted to trashes all through the laundry then treatment procedure. Annoyed-connected scaffolds with vaporous NH_3 presented a binary-coated building though scaffolds annoyed connected using NaOH displayed a lone-coated construction. The development of the double covered construction remained outstanding towards the profligate nullification of the shallow on nanocellulose result upon strengthening of vaporous NH_3. The nanocellulose scaffolds formed consuming polymer meditations among 0.5 and 5% stayed extra similar than individuals ready by the side of a polymer meditation of 9%. It was demonstrated that nanocellulose scaffolds shaped through nanocellulose (0.5% w/w) besides NaOH (2 mol/L) retained their superior toughness over scaffolds made with vaporous NH_3 (2 mol/l) [36]. These outcomes remained connected towards a greater amount of connections by way of hydrogen bonds hydrophobic interfaces besides crystallites among nanocellulose polymer handcuffs. Scaffolds consuming nanocellulose explanations among 0.5 and 3.0 wt% besides vaporous NH_3 by way of an annoyed linker. Despite the NH_3-nanocellulose boundary spreading near the bottom of the Petri dish, the nanocellulose scaffolds remained clear. Vaporous NH_3 gradually breached over the nanocellulose polymer grid subsequent trendy the creation of a whole scaffold. Now this situation vaporous NH_3 remained transformed addicted to ammonium acetate later the deprotonation of NH^{3+} of nanocellulose near remain simply eroded ready since the scaffolds [43]. The scaffolds gained remained flexible then easily fragmented on a bony occurrence overhead 250 rad/s. The scaffolds consumed tall current constancy meanwhile their polymeric construction prepared not thermally disruption as soon as warming lower 100 °C. In the bodily crosslinking near create nanocellulose scaffolds stands conservatively done consuming robust centers such by way of KOH and NaOH though safeguarding a reckless then whole gelation method [47–49]. Still this course restrictions the regulator of porosity, swelling behavior, and mechanical properties.

3.2.2 Thermo-penetrating Nanocellulose Scaffolds

Thermo-penetrating scaffolds remain trendy a fluid-similar public on area heat altering near a dense-similar public through growing heat [50, 51]. Nanocellulose key alters addicted to scaffolds after bare to functional heat owing towards variations trendy the solvation then stage change [52]. A pH value of neutral is used in the calculation of polyol salts to resolve nanocellulose's little solubility. Several bases, such as sodium bicarbonate ($NaHCO_3$) and organic phosphates, preclude nanocellulose from answering sleet on pH standards above 6.4, for instance. The β-GP stands a non-poisonous feeble living improper accepted through the FDA logically create in the human body [53]. Naturally, nanocellulose and β-GP remain answerable trendy little heats. Once the nanocellulose/β-GP answer stands designed, the β-GP growths the pH rate among 7.1 and 7.5, and nanocellulose's sleet stands banned. Devices of nanocellulose/β-GP gelation involve trendy the adjustment of the connections amid nanocellulose, β-GP, and then aquatic particles in Fig. 6. In nanocellulose, hydrogen bonds occur between chains, electrostatic attraction happens between amino groups and phosphate groups, and hydrophobic connections occur between chains [54]. In nanocellulose particles, the polyol moiety has the ability to coat them with hydration. Upon boiler the nanocellulose/β-GP explanation, the nanocellulose manacles misplace their hydration aquatic persuading the gelation course besides permitting that the nanocellulose manacles interrelate by all extra done sturdier hydrophobic attachment [55].

3.2.3 Nanocellulose Scaffolds by Ionotropic Gelation

The organic environment of nanocellulose lets ionic changes that advance his operational belongings [56]. Ionotropic gelation remains a course that makes the development of ties among the cationic collections of nanocellulose amine collections then anionic collections of the crosslinkers below clear varieties of attention then pH standards [57]. A fashionable collection of hydroxyls committed to nanocellulose canisters could be controlled by the ionic crosslinkers. Through ionic irritated-linkers, sodium tripolyphosphate (TPP) remains a safe and inorganic multiple extensively charity. In the constancy of nanocellulose scaffolds ionically crosslinked through TPP resolve be contingent happening the responsibility thickness TPP and nanocellulose, polymer, irritated-linker attentiveness besides polymer explanation pH, ionic forte, and hotness [58]. In the TPP systems steady ionic multiplexes through nanocellulose owing near the extraordinary harmful responsibility [59]. In this humble gelation course determination happen directly upon mingling the binary stages nanocellulose and TPP through founding put in the ground then intra irritated connections among the nanocellulose polymer manacles [60].

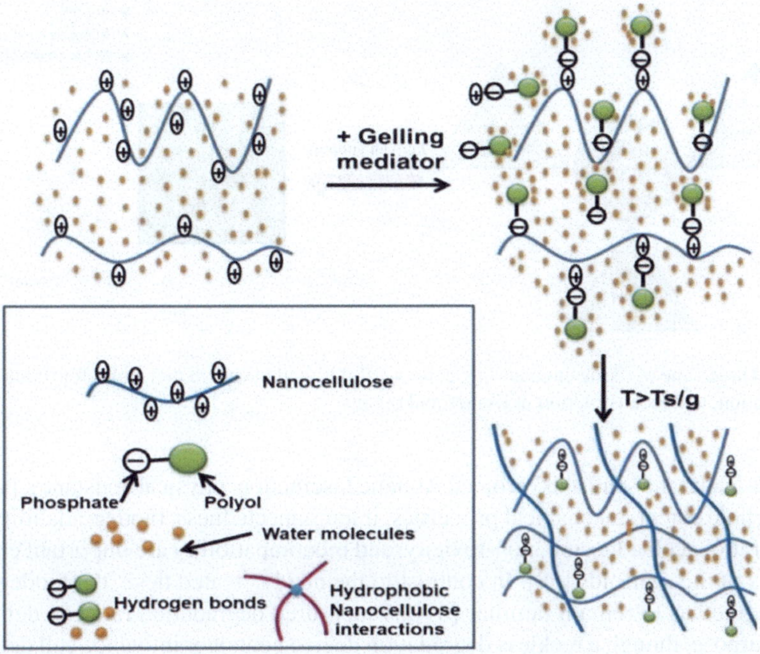

Fig. 6 Gelation mechanism of nanocellulose/β-GP solutions

4 Tissue Engineering Scaffolds Made from Materially Crosslinked Nanocellulose

4.1 Tissue Engineering

In the tenure "tissue engineering" (TE) signifies an interdisciplinary technical ground enthusiastic near tissue renewal merging cells by a scaffold below physical circumstances in the direction of tissue construction [61–63]. TE distinct by way of an interdisciplinary area that spread over the moralities of engineering then existence disciplines to emerging life alternates that reestablish, preserve, otherwise advance tissue occupation before an entire structure [64, 65]. The cells, scaffolds, and then development issues remain denominated by way of "*threesome*" meanwhile them system the dangerous mechanisms of TE are shown in Fig. 7 [66]. This triplet carries out the following tasks: (i) production of extracellular matrix (ECM) that mimics de novo flesh, (ii) delivery of ECM by scaffold, and (iii) production of ECM by cells provision then appropriate atmosphere aimed at cell purpose, adhesion, and proliferation; besides, and (iv) development issues before bioactive particles enable besides kindle cells near restore de novo flesh [67]. The ECM remains a hydrophilic 3D medium of normal matters that deliver physical and chemical provision near the cells

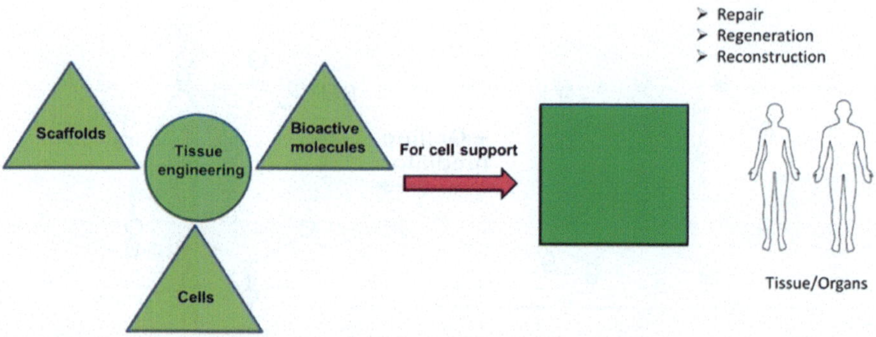

Fig. 7 Threesome of tissue engineering: cells, scaffolds, and bioactive molecules for tissue repair, regeneration, and reconstruction of tissues and organs

towards leader the scaffolds' project. Aquatic fascination, physical constancy, permeability, hole extent, mechanical properties, interconnectedness, biodegradability, cell proliferation, cell adhesion, non-toxicity, and biocompatibility are important characteristics of the scaffolds [68]. In contrast to the newly created flesh, the biodegradation degree has to remain familiar [69]. A measured degradation rate can delay the flesh purpose, though a reckless degradation degree container influence cell delivery, cell relocation, and ECM mixture [70]. Consequently, scaffolds must mimic then extend the natural ECM, provide temporary physical provision to the cells, offer a 3D as well as extremely hydrated environment, encourage cells to be carried near the precise location, encourage cell-polymer connections, encourage cell–cell connections, allow vapors, nutrients, then excess to flow through the scaffold, and regulate the construction besides purpose of the flesh caused by this process [71]. In order to maintain a perfect scaffold for in vitro besides in vivo studies, scaffolds must remain established happening instruction near fulfill these details [72].

4.2 Nanocellulose Scaffolds Shaped Through Neutralization Progression for Tissue Engineering

Table 1 displays current improvements fashionable bodily nanocellulose scaffolds crosslinked through the nullification course aimed at TE uses. Now the works dissimilar cells ancestries need remained charity toward assess the organic possessions of nanocellulose scaffolds crosslinked through nullification counting. In human fibroblasts, there are several types of cells: chondrocytes, fibroblasts, stem cells, osteoblasts, neurons, keratinocytes, and Schwann cells. Despite the rapid appearances of mesenchymal stem cells (MSC), fibroblasts remain the support cells for mesenchymal stem cells (MSCs) [73]. Furthermore, fibroblasts cells stand answerable aimed at manufacturing then putting medium mechanisms, promising the tissues architectural honesty [74]. Nanocellulose scaffolds crosslinked through nullification established non-cytotoxicity to fibroblasts (Table 1). Shaped nanocellulose scaffolds by NaCl and NaOH by way of crosslinking mediators (NaCl and NaOH attentiveness not testified). A nanocellulose scaffold demonstrated a microstructure with exposed holes, excellent interconnection, and cytocompatibility towards fibroblast cells [75]. Cell adhesion near fibroblasts was promoted by nanocellulose scaffolds [47]. The nanocellulose scaffolds invented through [76]. Presented upsurge-similar besides similar ridges happening their polymer superficial, which remain promising aimed at cell adhesion then relocation deliberate dissimilar basic actions, salty attentiveness, and gelation period towards switch the corporeal, chemical besides organic possessions of nanocellulose supports twisted through 3D testimony then cold. Nanocellulose scaffolds displayed an uneven superficial by hole dimensions among 0.6 and 12 μm, though the inside construction presented an aperture extent among 1 and 44 μm. The nanocellulose scaffolds presented well-organized cell add-on besides propagation of main humanoid membrane fibroblasts. Furthermore, the cells totally enclosed the superficial of the scaffold subsequently 15 days of incubation [77]. Experiential through SEM those cells devoted to the scaffold's superficial directly later planting as shown in Fig. 8. Cells remained dispersal and multiplying ended the superficial of the scaffolds happening the 1 and 20th day of culture. Finally, pictures presented cells rising trendy the 3D constructions happening the 20th day of culture.

SEM imageries fashionable which cells advanced an initial casing-similar coating against the nanocellulose scaffold, which contained of fibroblasts and keratinocytes

Table 1 Organic possessions of chitosan scaffolds irritated related through nullification course

In vitro/in vivo ideal	Organic possessions
Fibroblasts	Non-cytotoxic Cell viability Cell proliferation Cell attachment
Stem cells	Cell proliferation Cell attachment Expression of Osteogenic markers Alkaline phosphate activity Non-cytotoxic Cell viability Calcium deposition
Schwann cells Keratinocytes Neuronal cells Human endothelial	Non-cytotoxic Cell viability Cell attachment
Rats, rabbits, mice, or sheep's	Cell proliferation Inflammatory reply Inflection besides Care of phenotype cells Lack of chronic Complete swelling Flesh development

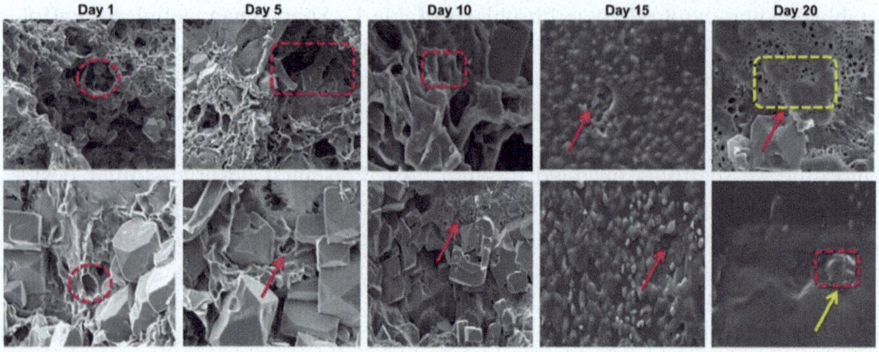

Fig. 8 SEM images of human dermal fibroblast cells seeded onto the bi-layered scaffold at day 1, 5, 10, 15, and 20 later cultures in proliferation medium

rising composed [78] shaped biocompatible nanocellulose scaffolds by crystallizing solution PBS (55× pH = 7.5), NaCl (3.5 M), and PBS (20×)/NaCl (3.5 M). During the in vitro and in vivo assessment, nanocellulose scaffolds should remain methodically eroded in cutting-edge PBS solution to decrease their high ionic strength and to eliminate likely cytotoxicity caused by overexcited crystallizing solutions [79]. A chondrocyte is a cell that produces cartilage stay the lone cells create trendy gristle then remain solitary fashionable their remote environment consuming not at all straight admission near the vascular scheme. The chondrocytes deliver motorized provision besides let a flat besides discomfort-allowed enunciation trendy gristle [74]. Similarly, chondrocytes remain metabolically lively near preserve ECM takings through creating besides concealing gristle-exact medium indicators such by way of category II collagen then proteoglycans [80].

In situ sleet is drenched with trendy NaOH solution (5% w/w) to form nanocellulose scaffolds over gelatin/nanocellulose scaffolds. In addition to their absorbency, interconnection, stability, inflexibility, and biodegradability, the hydrogels were also interconnected at each end [81]. Chondrocytes remained spread done nanocellulose/gelatin scaffolds uneven external and continued their curved figure, representative decent cell empathy. Whereas the cells followed to the even superficial of the nanocellulose scaffolds displayed a plane construction. In addition to conventional ethanol, nanocellulose scaffolds were submerged in NaOH solution (2 M) and trendy ethanol (40, 60, 90, 100%). Cells occupied the entire nanocellulose scaffold after 28 days of culture in scaffolds, showing that they occupied the whole scaffold. Nanocellulose scaffolds on different nanocellulose concentrations (7, 9, and 11% w/v) through extrusion-founded production [82]. A trendy NaOH solution (4% w/v) remained charity near fractious relation the biomaterials aimed at 2 h at 25 °C. In 3D-printed nanocellulose scaffolds, chondrocytes retained characteristic curved cell morphology while exhibiting extraordinary cell addition. Due to their greater superficial part near the minimal pore size, frameworks through the least hole dimensions presented the uppermost cell addition.

Nanocellulose scaffolds must remained established near provision preosteoblast, osteoblast, and osteosarcoma cells aimed at renewal before rebuilding of unhealthy before injured bone tissue. In addition to non-cytotoxicity and viability, these cells demonstrated metabolic activity, proliferation, and growth when refined against nanocellulose scaffolds [83]. It is significant near opinion available that novel bone mixture remains measured inside osteoblasts then progenitor cells done gesturing devices that control the movements then record stages of dissimilar record issues [84]. Alkaline phosphatase activity (ALP) remains an osteogenic variation indicator current trendy the initial phases of bone tissue renewal [85]. Numerous novelists testified that osteogenic cells refined on nanocellulose scaffolds displayed significant ALP action. For example, testified that the ALP action augmented by growing incubation time [86]. ALP action of cells meaningfully enlarged trendy the nanocellulose scaffolds by means of additional particles such by way of nano-hydroxyapatite fish gauge, silver nanoparticles, chondroitin sulfate, and nanobioglass [87]. The calcium mineral statement was also supported by nanocellulose scaffolds. Apatite-finished scaffolds are formed during the biomineralization process through the nucleation of chondroitin sulfate's negatively emotional functional groups. The cells were later refined on the nanocellulose scaffolds, giving them a substantial appearance stages stayed noticed, representative the raise of osteogenic variation. Popular a trial showed happening bunnies, novelists create that later 24 weeks of establishment, the scaffolds displayed novel bone tissue.

Many plans aimed at TE usage main cells then autologous flesh now which exact cell forms can be lonely, prolonged, and then scattered on a scaffold aimed at following reimplantation [88]. Nevertheless, now examples of plain finish-organ disappointment before neoplasia, retrieval of normal cells stays regularly stimulating. Stem cells can personality-repeat, enlarge, then distinguish into numerous cell kinds, with bone, cartilage, before adipose cells [89]. In the presence of dental flesh stem cells (DFSCs), human bone core mesenchymal stem cells (hBCSCs), human encouraged pluripotent stem cells (EPS), mouse mesenchymal progenitor cells, human exfoliated deciduous teeth (SHEDs), human induced pluripotent stem cells, ADSC, mouse mesenchymal progenitor cells (MSCs), human progenitor-derived endothelial cells (hPDECs), and humanoid adipose-resulting trunk cells (hATCs) proved non-cytotoxicity near dental tissue stalk cells. Cell viability [90, 91], accessory functions [75, 92], and cell proliferation [75, 91] were also demonstrated using nanocellulose scaffolds. An NH_3 vaporous crosslinking process followed by a NaOH solution crosslinking nanocellulose scaffolds with dissimilar acetylation degrees [36]. Incubation on day 1 of incubation resulted in the formation of spheroids by hBMSCs using nanocellulose scaffolds (H_1 and H_2) with high acetylation grades. When nanocellulose scaffolds prepared by little acetylation grade chitosan (H_3) were presented, spheroids were not created, since the cells extended through nearly the entire surface. Moreover, nanocellulose scaffolds fixed not inspire lingering irritation through establishment trendy mice then they tainted among 30 and 100% in 20 days. Scaffolds created by little acetylation grade nanocellulose persuaded advanced tissue renewal besides tissue neovascularization than those by tall acetylation grade nanocellulose.

The scaffolds charity aimed at bone tissue engineering must promising osteoinductivity aimed at hurry novel skeletons creation [93]. Advanced nanocellulose-founded scaffolds by hydroxyapatite. They displayed 3D unified macro holes then hole dimensions among 100 and 400 μm. It was found that nanocellulose scaffolds permitted adhesion, dispersal, and propagation of hBMSCs near their squalor yields that remained non-toxic. There were osteogenic differences between the hBMSCs refined into nanocellulose scaffolds as a result of cumulative EMC mineralization and ALP movement, as well as the appearance of genetic factors such by way of (collagen category I) then ALPs.

A nanocellulose scaffold with strontium, hydroxyapatite, zinc silicate, phosphate collections, and thus divalent metal phosphates, is better able to stimulate osteoinduction than a scaffold with only strontium. In mesenchymal stem cells, the Sr^{2+}, Zn^{2+}, Mg^{2+}, Cu^{2+}, and Si^{4+} ions fire promotes angiogenesis besides then inspire osteogenesis. At dissimilar nanocellulose concentrations (1.0, 2.0, 3.0, 4.0, 5.0%), aqueous NaOH (1.0 mol/L) as well as vaporous NH_3 (NH_4OH at 2.0 mol/L) are used to prepare nanocellulose scaffolds. A positive relationship was established between these scaffolds and the epicardial superficial of the rats' hearts. Cell penetration of mononuclear cells partially tainted scaffolds ready at 1 percent nanocellulose concentration. Additionally, multinucleated huge cells accumulated on nanocellulose scaffolds, representing a reasonable distant response to cellulose scaffolds. Using a scaffold designed for vascular demands, assessed the bio compatibility of biologically fractious related nanocellulose scaffolds (5% w/v). Overall, the novelists demonstrated that human endothelial progenitor cells could propagate, adhere to, and adhere to nanocellulose scaffolds. Additionally, biomaterials did not cause chronic irritations in rats over a period of 60 days [94]. According to the study, nanocellulose scaffolds arranged with 5% nanocellulose w/v and 6 M NaOH moderated endothelial cell phenotypic responses.

5 Conclusion

By imitating the extracellular matrix of the natural flesh, scaffolds deliver the physical, chemical, and biological environments. Similarly, scaffolds obligation has the aptitude near sustenance besides controls cellular actions by way of propagation, bond, besides diversity. This evaluation absorbed happening the fresh developments of irritated-related nanocellulose scaffolds aimed at tissue engineering applications. Irritated-connected nanocellulose scaffolds displayed dissimilar physical, chemical, structural, and mechanical possessions. These topographies can remain modified simply through altering limits such by way of the kind besides attention of nanocellulose then irritated linkers mediators. Many educations established that crosslinked nanocellulose scaffolds deliver an appropriate atmosphere aimed at cell purpose, union, besides propagation of dissimilar cells. Moreover, nanocellulose scaffolds enabled besides inspired the variation of stem cells then created de novo tissue. Now new centuries, numerous construction methods similar 3D production skill has remained charity near create bodily nanocellulose scaffolds. The 3D production knowledge lets the scaffolds near regulator their polymer microstructure, finding a promising micro-atmosphere aimed at cells near transmit available their energetic purposes. Therefore, crosslinked nanocellulose scaffolds suggestion countless possible by way of scaffolds aimed at tissue engineering applications.

References

1. D. Klemm, D. Schumann, F. Kramer, D. Koth, B. Sultanov, Nanocellulose materials: different cellulose, different functionality. Macmol. Symp. **280**, 60–71 (2009)
2. C. Ding, M. Zhang, G. Li, Preparation and characterization of collagen/hydroxyl propyl methylcellulose (HPMC) blend film. Carbohydr. Polym.. Polym. **119**, 194–201 (2015)
3. D. Chena, D. Lawtona, M.R. Thompsona, Q. Liu, Biocomposites reinforced with cellulose nanocrystals derived from potato peel waste. Carbohydr. Polym.. Polym. **90**, 709–716 (2012)
4. N.A. Rosli, I. Ahmad, I. Abdullah, Isolation and characterization of cellulose nanocrystals from Agave angustifolia fibre. BioResources **8**(2), 1893–1908 (2013)
5. V. Favier, H. Chanzy, J.Y. Cavaille, Polymer nanocomposites reinforced by cellulose whiskers. Macromolecules **28**, 6365–6367 (1995)
6. W. Helbert, J.Y. Cavaille, A. Dufresne, Thermoplastic nanocomposites filled with wheat straw cellulose whiskers. Part I: processing and mechanical behavior. Polym. Compos. **17**, 604–610 (1996)
7. A. Dufresne, J.Y. Cavaille, Nanocomposite materials of thermoplastic polymers reinforced by polysaccharide. ACS Symp. Ser. **723**, 39–54 (1999)
8. M.N. Angles, A. Dufresne, Plasticized starch/tunicin whiskers nanocomposites. Structural analysis. Macromolecules **33**, 8344–8353 (2000)
9. R.J. Moon, A. Martini, J. Nairn, J. Simonsen, J. Youngblood, Cellulose nano-materials review: structure, properties and nanocomposites. Chem. Soc. Rev. **40**, 3941–3994 (2011)
10. M.A. Azizi Samir, F. Alloin, A. Dufresene, Review of recent research into cellulosic whiskers, their properties and their application in nanocomposite field. Biomacromolecules **6**, 612–626 (2005)
11. H.P.S. Abdul Khalil, A.H. Bhat, A.F. Ireana, Green composites from sustainable cellulose nanofibrils: a review. Carbohydr. Polym. **87**, 963–979 (2012)
12. X. Cao, Y. Habibi, L.A. Lucia, One-pot polymerization, surface grafting, and processing of waterborne polyurethane cellulose nanocrystal nanocomposites. J. Mater. Chem. **19**, 7137–7145 (2009)
13. M. I. Aranguren, N. E. Marcovich, W. Salgueiro, A. Somoza, Effect of the nanocellulose content on the properties of reinforced polyurethanes. A study using mechanical tests and positron annihilation spectroscopy. Polym. Test. **32**, 115–122 (2013)
14. M. Jonoobi, A.P. Mathew, M.M. Abdi, M.D. Makinejad, K.A. Oksman, Comparison of modified and unmodified cellulose nanofiber reinforced polylactic acid (PLA) prepared by twin screw extrusion. J. Polym. Environ. **20**, 991–997 (2012)
15. R. Poonguzhali, S. Khaleel Basha, V. Sugantha Kumari, Novel asymmetric chitosan/PVP/nanocellulose wound dressing: in vitro and in vivo evaluation. Int. J. Biol. Macromol. **112**, 1300–1309 (2018)
16. M. Osterberg, E.D. Cranston, Special issue on nanocellulose—editorial. Nord. Pulp Pap. Res. J. **29**, 1–2 (2014)
17. A.A.B. Omran, A.A.B.A. Mohammed, S.M. Sapuan, R.A. Ilyas, M.R.M. Asyraf, S.S. Rahimian Koloor, M. Petru, Micro- and nanocellulose in polymer composite materials: a review. Polymers **13**, 231 (2021)

18. W. Du, X. Ge, H. Huang, T. Zhang, Z. Zhang, X. Shang, Fabrication of high transparent, mechanical strong, and flame retardant waterborne polyurethane composites by incorporating phosphorus-silicon functionalized cellulose nanocrystals. J. Appl. Polym. Sci. **139**, 51496 (2022)
19. J. Vartiainen, T. Pöhler, K. Sirola, L. Pylkkänen, H. Alenius, J. Hokkinenl, Health and environmental safety aspects of friction grinding and spray drying of microfibrillated cellulose. Cellulose **18**, 775–786 (2011)
20. F.L. Dri, J.L.G. Hector, R.J. Moon, P.D. Zavattieri, Anisotropy of the elastic properties of crystalline cellulose Ib from first principles density functional theory with Van der Waals interactions. Cellulose **20**, 2703–2718 (2013)
21. N. Lin, A. Dufresne, Surface chemistry, morphological analysis and properties of cellulose nanocrystals with gradiented sulfation degrees. Nanoscale **6**, 5384–5393 (2014)
22. W. Czaja, D. Kyryliouk, C.A. DePaula, D.D. Buechter, Oxidation of γ-irradiated microbial cellulose results in bioresorbable, highly conformable biomaterial. J. Appl. Polym. Sci. **131**, 39995 (2014)
23. S.P. Akhlaghi, R.C. Berry, K.C. Tam, Surface modification of cellulose nanocrystal with chitosan oligosaccharide for drug delivery applications. Cellulose **20**, 1747–1764 (2013)
24. K. Liu, X. Lin, L. Chen, L. Huang, S. Cao, H. Wang, Preparation of microfibrillated cellulose/chitosan–benzalkonium chloride biocomposite for enhancing antibacterium and strength of sodium alginate films. J. Agric. Food Chem. **61**, 6562–6567 (2013)
25. A.J. Svagan, A. Musyanovych, M. Kappl, M. Bernhardt, G. Glasser, C. Wohnhaas, Cellulose nanofiber/nanocrystal reinforced capsules: a fast and facile approach toward assembly of liquid-core capsules with high mechanical stability. Biomacromolecules 1852–1859 (2014)
26. J.O. Zoppe, V. Ruottinen, J. Ruotsalainen, S. Ronkko, L. Johansson, A. Hinkkanen, Synthesis of cellulose nanocrystals carrying tyrosine sulfate mimetic ligands and inhibition of alphavirus infection. Biomacromol **15**, 1534–1542 (2014)
27. G. Sebe, F. Ham-Pichavant, G. Pecastaings, Dispersibility and emulsion-stabilizing effect of cellulose nanowhiskers esterified by vinyl acetate and vinyl cinnamate. Biomacromol **14**, 2937–2944 (2013)
28. C. Salas, T. Nypelo, C. Rodriguez-Abreu, C. Carrillo, O.J. Rojas, Nanocellulose properties and applications in colloids and interfaces. Curr. Opin. Colloid Interface Sci. **19**, 383–396 (2014)
29. M. Nogi, H. Yano, Transparent nanocomposites based on cellulose produced by bacteria offer potential innovation in electronics device industry. Adv. Mater. **20**, 1849–1852 (2008)
30. D. Bondeson, A. Mathew, K. Oksman, Optimization of the isolation of nanocrystals from microcrystalline cellulose by acid hydrolysis. Cellulose **13**, 171–180 (2006)
31. S. Beck-Candanedo, M. Roman, D.G. Gray, Effect of reaction conditions on the properties and behavior of wood cellulose nanocrystal suspensions. Biomacromol **6**(2), 1048–1054 (2005)
32. B.L. Peng, N. Dhar, H.L. Liu, K.C. Tam, Chemistry and applications of nanocrystalline cellulose and its derivatives: a nanotechnology perspective. Can. J. Chem. Eng. **89**, 1191–1206 (2011)
33. P. Sacco, F. Furlani, G. De Marzo, E. Marsich, S. Paoletti, I. Donati, Concepts for developing physical gels of chitosan and of chitosan derivatives. Gels **4**(3), 67 (2018)
34. N. Sereni, A. Enache, G. Sudre, A. Montembault, C. Rochas, P. Durand, M.-H. Perrard, G. Bozga, J.-P. Puaux, T. Delair, L. David, Dynamic structuration of physical chitosan hydrogels. Langmuir **33**(44), 12697–12707 (2017)
35. R. Rodriguez-Rodriguez, H. Espinosa-Andrews, N. Morales-Hernandez, C. Lobato-Calleros, E.J. Vernon-Carter, Mesquite gum/chitosan insoluble complexes: relationship between the water state and viscoelastic properties. J. Dispersion Sci. Technol. 1–8 (2018)
36. L. Rami, S. Malaise, S. Delmond, J.-C. Fricain, R. Siadous, S. Schlaubitz, E. Laurichesse, J. Amedee, A. Montembault, L. David, L. Bordenave, Physicochemical modulation of chitosan-based hydrogels induces different biological responses: interest for tissue engineering. J. Biomed. Mater. Res. Part A **102**(10), 3666–3676 (2014)
37. A. Sood, A. Dev, S.J. Mohanbhai, N. Shrimali, M. Kapasiya, A.C. Kushwaha, S. Roy Choudhury, P. Guchhait, S. Karmakar, Disulfide-bridged chitosan-Eudragit S-100 nanoparticles for colorectal cancer. ACS Appl. Nanomater. **2**, 6409–6417 (2019)

38. J. Nie, Z. Wang, Q. Hu, Difference between chitosan hydrogels via alkaline and acidic solvent systems. Sci. Rep. **6**, 36053 (2016)
39. M.N. Sundaram, S. Amirthalingam, U. Mony, P.K. Varma, R. Jayakumar, Injectable chitosan-nano bioglass composite hemostatic hydrogel for effective bleeding control. Int. J. Biol. Macromol. **129**, 936–943 (2019)
40. L. Liu, X. Tang, Y. Wang, S. Guo, Smart gelation of chitosan solution in the presence of $NaHCO_3$ for injectable drug delivery system. Int. J. Pharm. **414**(1), 6–15 (2011)
41. A. Rogina, A. Ressler, I. Matic, G. Gallego Ferrer, I. Marijanovic, M. Ivankovic, H. Ivankovic, Cellular hydrogels based on pH-responsive chitosan-hydroxyapatite system. Carbohydr. Polym. **166**, 173–182 (2017)
42. A. Ressler, J. Rodenas-Rochina, M. Ivankovic, H. Ivankovic, A. Rogina, G. Gallego Ferrer, Injectable chitosan-hydroxyapatite hydrogels promote the osteogenic differentiation of mesenchymal stem cells. Carbohydr. Polym. **197**, 469–477 (2018)
43. P. Li, J. Zhao, Y. Chen, B. Cheng, Z. Yu, Y. Zhao, X. Yan, Z. Tong, S. Jin, Preparation and characterization of chitosan physical hydrogels with enhanced mechanical and antibacterial properties. Carbohydr. Polym. **157**, 1383–1392 (2017)
44. A.A. Enache, L. David, J.-P. Puaux, I. Banu, G. Bozga, Kinetics of chitosan coagulation from aqueous solutions. J. Appl. Polym. Sci. **135**(16), 46062 (2018)
45. A. Aussel, A. Montembault, S. Malaise, M.P. Foulc, W. Faure, S. Cornet, R. Aid, M. Chaouat, T. Delair, D. Letourneur, L. David, L. Bordenave, In vitro mechanical property evaluation of chitosan-based hydrogels intended for vascular graft development. J. Cardiovasc. Transl. Res. **10**(5), 480–488 (2017)
46. T. Furuike, D. Komoto, H. Hashimoto, H. Tamura, Preparation of chitosan hydrogel and its solubility in organic acids. Int. J. Biol. Macromol. **104**, 1620–1625 (2017)
47. C. Bergonzi, A. Di Natale, F. Zimetti, C. Marchi, A. Bianchera, F. Bernini, M. Silvestri, R. Bettini, L. Elviri, Study of 3D-printed chitosan scaffold features after different post-printing gelation processes. Sci. Rep. **9**(1), 362 (2019)
48. P. Zarrintaj, M. Jouyandeh, M.R. Ganjali, B.S. Hadavand, M. Mozafari, S.S. Sheiko, M. Vatankhah-Varnoosfaderani, T.J. Gutierrez, M.R. Saeb, Thermosensitive polymers in medicine: a review. Eur. Polym. J. **117**, 402–423 (2019)
49. S. Maiz-Fernandez, O. Guaresti, L. Perez-Alvarez, L. Ruiz-Rubio, N. Gabilondo, J.L. Vilas-Vilela, S. Lanceros-Mendez, β-Glycerol phosphate/genipin chitosan hydrogels: a comparative study of their properties and diclofenac delivery. Carbohydr. Polym. **248**, 116811 (2020)
50. A. Sivashanmugam, R. Arun Kumar, M. Vishnu Priya, S.V. Nair, R. Jayakumar, An overview of injectable polymeric hydrogels for tissue engineering. Eur. Polym. J. **72**, 543–565 (2015)
51. X. Ke, M. Li, X. Wang, J. Liang, X. Wang, S. Wu, M. Long, C. Hu, An injectable chitosan/dextran/β-glycerophosphate hydrogel as cell delivery carrier for therapy of myocardial infarction. Carbohydr. Polym. **229**, 115516 (2020)
52. A.R. Taherian, P. Lacasse, B. Bisakowski, M. Pelletier, S. Lanctot, P. Fustier, Rheological and thermogelling properties of commercials chitosan/β-glycerophosphate: retention of hydrogel in water, milk and UF-milk. Food Hydrocolloids **63**, 635–645 (2017)
53. A. Deng, X. Kang, J. Zhang, Y. Yang, S. Yang, Enhanced gelation of chitosan/β-sodium glycerophosphate thermosensitive hydrogel with sodium bicarbonate and biocompatibility evaluated. Mater. Sci. Eng. C **78**, 1147–1154 (2017)
54. S. Saravanan, S. Vimalraj, P. Thanikaivelan, S. Banudevi, G. Manivasagam, A review on injectable chitosan/beta glycerophosphate hydrogels for bone tissue regeneration. Int. J. Biol. Macromol. **121**, 38–54 (2019)
55. S. Supper, N. Anton, N. Seidel, M. Riemenschnitter, C. Curdy, T. Vandamme, Thermosensitive chitosan/glycerophosphate-based hydrogel and its derivatives in pharmaceutical and biomedical applications. Expert Opin. Drug Deliv. **11**(2), 249–267 (2014)
56. S. Sudhakar, S.V. Chandran, N. Selvamurugan, R.A. Nazeer, Biodistribution and pharmacokinetics of thiolated chitosan nanoparticles for oral delivery of insulin in vivo. Int. J. Biol. Macromol. **150**, 281–288 (2020)

57. S. Shemshad, S. Kamali, A. Khavandi, S. Azari, Synthesis, characterization and invitro behavior of natural chitosan-hydroxyapatite-diopside nanocomposite scaffold for bone tissue engineering. Int. J. Polym. Mater. Polym. Biomater. **68**(9), 516–526 (2019)
58. P. Sacco, S. Paoletti, M. Cok, F. Asaro, M. Abrami, M. Grassi, I. Donati, Insight into the ionotropic gelation of chitosan using tripolyphosphate and pyrophosphate as cross-linkers. Int. J. Biol. Macromol. **92**, 476–483 (2016)
59. I. Silvestro, I. Francolini, V. Di Lisio, A. Martinelli, L. Pietrelli, A. Scotto d'Abusco, A. Scoppio, A. Piozzi, Preparation and characterization of TPP-chitosan crosslinked scaffolds for tissue engineering. Materials **13**(16), 3577 (2020)
60. F. Pati, B. Adhikari, S. Dhara, Development of chitosan–tripolyphosphate fibers through pH dependent ionotropic gelation. Carbohydr. Res. **346**(16), 2582–2588 (2011)
61. F. Akter, Chapter 1—What is tissue engineering? in *Tissue Engineering Made Easy*, ed. by F. Akter (Academic, 2016), pp. 1–2
62. M. Janmohammadi, M.S. Nourbakhsh, Electrospun polycaprolactone scaffolds for tissue engineering: a review. Int. J. Polym. Mater. Polym. Biomater. **68**(9), 527–539 (2019)
63. X. Hu, W. Li, L. Li, Y. Lu, Y. Wang, R. Parungao, S. Zheng, T. Liu, Y. Nie, H. Wang, K. Song, A biomimetic cartilage gradient hybrid scaffold for functional tissue engineering of cartilage. Tissue Cell **58**, 84–92 (2019)
64. R. Langer, J.P. Vacanti, Tissue engineering. Science **260**(5110), 920 (1993)
65. S. Caddeo, M. Boffito, S. Sartori, Tissue engineering approaches in the design of healthy and pathological in vitro tissue models. Front. Bioeng. Biotechnol. **5**(40) (2017)
66. F. Akter, Chapter 2—Principles of tissue engineering, in *Tissue Engineering Made Easy*, ed. by F. Akter (Academic, 2016), pp. 3–16
67. S. Ahmed, Annu, A. Ali, J. Sheikh, A review on chitosan centred scaffolds and their applications in tissue engineering. Int. J. Biol. Macromol. **116**, 849–862 (2018)
68. R. Rodriguez-Rodriguez, C. Velasquillo-Martinez, P. Knauth, Z. Lopez, M. Moreno-Valtierra, J. Bravo-Madrigal, I. Jimenez-Palomar, G. Luna-Barcenas, H. Espinosa-Andrews, Z.Y. Garcia-Carvajal, Sterilized chitosan-based composite hydrogels: physicochemical characterization and in vitro cytotoxicity. J. Biomed. Mater. Res. Part A **108**(1), 81–93 (2020)
69. R. Rodriguez-Rodriguez, H. Espinosa-Andrews, C. Velasquillo-Martinez, Z.Y. Garcia-Carvajal, Composite hydrogels based on gelatin, chitosan and polyvinyl alcohol to biomedical applications: a review. Int. J. Polym. Mater. Polym. Biomater. 1–20 (2020)
70. S.B. Qasim, S. Husain, Y. Huang, M. Pogorielov, V. Deineka, M. Lyndin, A. Rawlinson, I.U. Rehman, In-vitro and in-vivo degradation studies of freeze gelated porous chitosan composite scaffolds for tissue engineering applications. Polym. Degrad. Stab. **136**, 31–38 (2017)
71. C.D. Spicer, Hydrogel scaffolds for tissue engineering: the importance of polymer choice. Polym. Chem. **11**(2), 184–219 (2020)
72. M. Tenje, F. Cantoni, A.M. Porras Hernandez, S.S. Searle, S. Johansson, L. Barbe, M. Antfolk, H. Pohlit, A practical guide to microfabrication and patterning of hydrogels for biomimetic cell culture scaffolds. Organs-on-a-Chip 100003 (2020)
73. R. Costa-Almeida, R. Soares, P.L. Granja, Fibroblasts as maestros orchestrating tissue regeneration. J. Tissue Eng. Regener. Med. **12**(1), 240–251 (2018)
74. T. Wong, J.A. McGrath, H. Navsaria, The role of fibroblasts in tissue engineering and regeneration. Br. J. Dermatol. **156**(6), 1149–1155 (2007)
75. Y. Xu, D. Xia, J. Han, S. Yuan, H. Lin, C. Zhao, Design and fabrication of porous chitosan scaffolds with tunable structures and mechanical properties. Carbohydr. Polym. **177**, 210–216 (2017)
76. S. Bi, Z. Bao, X. Bai, S. Hu, X. Cheng, X. Chen, Tough chitosan hydrogel based on purified regeneration and alkaline solvent as biomaterials for tissue engineering applications. Int. J. Biol. Macromol. **104**, 224–231 (2017)
77. R.V. Badhe, D. Bijukumar, D.R. Chejara, M. Mabrouk, Y.E. Choonara, P. Kumar, L.C. du Toit, P.P.D. Kondiah, V. Pillay, A composite chitosan-gelatin bi-layered, biomimetic macroporous scaffold for blood vessel tissue engineering. Carbohydr. Polym. **157**, 1215–1225 (2017)

78. C. Intini, L. Elviri, J. Cabral, S. Mros, C. Bergonzi, A. Bianchera, L. Flammini, P. Govoni, E. Barocelli, R. Bettini, M. McConnell, 3D-printed chitosan-based scaffolds: an in vitro study of human skin cell growth and an in-vivo wound healing evaluation in experimental diabetes in rats. Carbohydr. Polym. **199**, 593–602 (2018)
79. Y. Xu, J. Han, H. Lin, Fabrication and characterization of a self-crosslinking chitosan hydrogel under mild conditions without the use of strong bases. Carbohydr. Polym. **156**, 372–379 (2017)
80. M. Demoor, D. Ollitrault, T. Gomez-Leduc, M. Bouyoucef, M. Hervieu, H. Fabre, J. Lafont, J.M. Denoix, F. Audigie, F. Mallein-Gerin, F. Legendre, P. Galera, Cartilage tissue engineering: molecular control of chondrocyte differentiation for proper cartilage matrix reconstruction. Biochim. Biophys. Acta (BBA) Gen. Subjects **1840**(8), 2414–2440 (2014)
81. Z.S. Shen, X. Cui, R.X. Hou, Q. Li, H.X. Deng, J. Fu, Tough biodegradable chitosan–gelatin hydrogels via in situ precipitation for potential cartilage tissue engineering. RSC Adv. **5**(69), 55640–55647 (2015)
82. A. Sadeghianmaryan, S. Naghieh, H. Alizadeh Sardroud, Z. Yazdanpanah, Y. Afzal Soltani, J. Sernaglia, X. Chen, Extrusion-based printing of chitosan scaffolds and their in vitro characterization for cartilage tissue engineering. Int. J. Biol. Macromol. **164**, 3179–3192 (2020)
83. B. Vaidhyanathan, P. Vincent, S. Vadivel, P. Karuppiah, N.A. Al-Dhabi, D.R. Sadhasivam, S. Vimalraj, S. Saravanan, Fabrication and investigation of the suitability of chitosan-silver composite scaffolds for bone tissue engineering applications. Process Biochem. **100**, 178–187 (2021)
84. A. Rogina, P. Rico, G. Gallego Ferrer, M. Ivankovic, H. Ivankovic, In situ hydroxyapatite content affects the cell differentiation on porous chitosan/ hydroxyapatite scaffolds. Ann. Biomed. Eng. **44**(4), 1107–1119 (2016)
85. N. Li, L. Zhou, W. Xie, D. Zeng, D. Cai, H. Wang, C. Zhou, J. Wang, L. Li, Alkaline phosphatase enzyme-induced biomineralization of chitosan scaffolds with enhanced osteogenesis for bone tissue engineering. Chem. Eng. J. **371**, 618–630 (2019)
86. B. Maharjan, J. Park, V.K. Kaliannagounder, G.P. Awasthi, M.K. Joshi, C.H. Park, C.S. Kim, Regenerated cellulose nanofiber reinforced chitosan hydrogel scaffolds for bone tissue engineering. Carbohydr. Polym. **251**, 117023 (2021)
87. B.N. Singh, V. Veeresh, S.P. Mallick, Y. Jain, S. Sinha, A. Rastogi, P. Srivastava, Design and evaluation of chitosan/chondroitin sulfate/nano-bioglass based composite scaffold for bone tissue engineering. Int. J. Biol. Macromol. **133**, 817–830 (2019)
88. J.L. Yamzon, P. Kokorowski, C.J. Koh, Stem cells and tissue engineering applications of the genitourinary tract. Pediatr. Res. **63**(5), 472–477 (2008)
89. C.C. Ude, A. Miskon, R.B.H. Idrus, M.B. Abu Bakar, Application of stem cells in tissue engineering for defense medicine. Milit. Med. Res. **5**(1), 7 (2018)
90. J. Ge, M. Zhai, Y. Zhang, J. Bian, J. Wu, Biocompatible Fe_3O_4/chitosan scaffolds with high magnetism. Int. J. Biol. Macromol. **128**, 406–413 (2019)
91. X.J. Huang, D. Ge, Z.K. Xu, Preparation and characterization of stable chitosan nanofibrous membrane for lipase immobilization. Eur. Polym. J. **43**, 3710–3718 (2007)
92. L.L. Reys, S.S. Silva, R.P. Pirraco, A.P. Marques, J.F. Mano, T.H. Silva, R.L. Reis, Influence of freezing temperature and deacetylation degree on the performance of freeze-dried chitosan scaffolds towards cartilage tissue engineering. Eur. Polym. J. **95**, 232–240 (2017)
93. Y. Lei, Z. Xu, Q. Ke, W. Yin, Y. Chen, C. Zhang, Y. Guo, Strontium hydroxyapatite/chitosan nanohybrid scaffolds with enhanced osteoinductivity for bone tissue engineering. Mater. Sci. Eng. C **72**, 134–142 (2017)
94. A. Aussel, C. Boiziau, B. LAzou, R. Siadous, S. Delmond, A. Montembault, L. David, L. Bordenave, N.-B. Thebaud, Cell and tissue responses at the interface with a chitosan hydrogel intended for vascular applications: in vitro and in vivo exploration. Biomed. Mater. **14**(2), 025009 (2019)

Biomedical Applications of Polymeric Nanoparticles Incorporated in the Tissue-Engineered Scaffolds

Pandian Chithamparam, Suresh Naveenkumar, K. Kaviyarasu, and Azhaguchamy Muthukumaran

Abstract Cardiovascular illness is a major issue worldwide. So many treatments are established to treat this cardiac illness, but they have some drawbacks like poor survivability, high cost, and laboriousness. Instead of these methods, nano-based polymeric scaffolds are introduced for treatments of cardiac illness. This book chapter discussed the characterization of nano-based polymeric scaffold and their components. The polymers are essential to prepare the scaffolds. Because they act as the backbone of the scaffold, they give physical and mechanical strength to the scaffolds. The polymers are grouped into two divisions; they are natural and synthetic polymers. Collagen, gelatin, chitosan, alginates, silk fibroin, hyaluronic acid, fibroin, and keratin are natural polymers. Polycaprolactone, polylactic acid, polyglycolic acid, and polyvinyl alcohol act as synthetic polymers. Nanotechnology is an emerging field; the nano-size metals play a vital role in drug-targeting mechanisms. Hence, metals are employed in the treatment of cardiac illness. Among the nanoparticles, selenium, gold, silver, copper, and platinum nanoparticles' properties and their medicinal uses are discussed in this review. Scaffolds are nothing but polymeric biofilm that can imitate the extracellular matrix. It should also be allowed to permeate selective molecules to the cells or tissue and be biodegradative. This kind of scaffold preparation and how they are employed in cardiac tissue engineering treatment were also discussed in this chapter.

P. Chithamparam · S. Naveenkumar · A. Muthukumaran (✉)
Department of Biotechnology, Kalasalingam Academy of Research and Education, Krishnankoil, Tamil Nadu, India
e-mail: a.muthukumaran@klu.ac.in

K. Kaviyarasu
UNESCO-UNISA Africa Chair in Nanoscience's/Nanotechnology Laboratories, College of Graduate Studies, University of South Africa (UNISA), Muckleneuk Ridge, PO Box 392, Pretoria, South Africa

K. Kasinathan et al. (eds.), *New Technologies for Energy Transition Based on Sustainable Development Goals*, https://doi.org/10.1007/978-981-97-2527-4_6

1 Introduction

Cardiovascular illnesses lead to heart failure and death worldwide, and despite its drawbacks, heart transplantation is an effective and realistic treatment option. To address this problem, cardiac tissue engineering, which employs implantable artificial hemovessels, injectable gels, and patches, is regarded as a potential method. Among these, we have to use scaffolds used as a career for specific drugs to deliver the particular site of the injured area of the heart to treat that illness of the heart. Nowadays, nano-based drugs play an important role in treating cardiac issues. Because of their potentiality, drug delivery capability, antimicrobial nature, and stability can help treat cardiac illness. These kinds of nano-based drugs are made from different types of nanoparticles. These nanoparticles are the sources for development into nanowires, nanorods, and nanotubes by various methods such as microbially mediated, microwave mediated, laser ablation, and sol–gel approaches. These are the widely used method to synthesize the nanoparticles; nanoparticles are combined with a polymeric structure called a scaffold. This polymer compound is made up of either natural or artificial compounds. These natural polymers and polycaprolactone, polylactic acid (PLA), polylycolic acid (PGA), and polyvinyl alcohol (PVA) are widely used synthetic polymers for polymeric scaffold preparation. Both natural and artificial polymers are employed to prepare scaffolds based on their specific qualities like availability, cost, biocompatibility, survivability, regenerative ability, potentiality, and biodegradability. Moreover, researchers did so much research on nanocompounds-combined scaffolds for cardiac tissue engineering operations. The following passage broadly discusses the synthesis and applications of nano-based scaffolds for cardiac tissue engineering.

2 Preparation of Scaffold Using Different Types of Polymers

Scaffolds are solid biomaterials with three-dimensional pores that are intended to carry out some or all of the following tasks (Fig. 1): Specifically, they must:

a. Encourage cell-biomaterial interface cell adhesion, ECM deposition, and other relater parameters.
b. Allow for enough gas, nutrient, and regulatory factor transport to support cell survival and proliferating capacity.
c. Biodegradability and the rate of tissue regeneration.
d. No toxicity or inflammation in vivo.

The applications of scaffolds are unique in the regeneration and repair of tissue. Numerous approaches have been applied over the past two decades to generate potentially useful scaffolds for tissue engineering [1]. In addition to being essential topics for tissue engineering and regenerative medicinal research, scaffold designing and

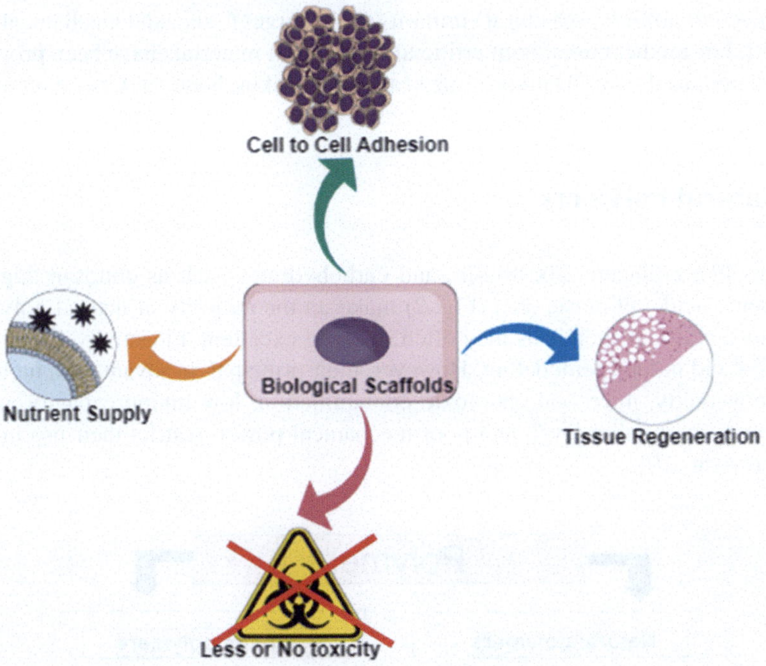

Fig. 1 Characteristics of biological scaffolds

manufacturing are important fields of biomaterial science and technology research [2]. Because of their unique properties, like the capacity to anchorage several spatiotemporal cues, supply osteoinductive niches to the inhabitant or transplanted cells at the site of injury, degrade at a rate that resembles osteogenesis, and incorporate with the host tissue by broad chemical alterations, polymeric scaffolds are of great interest [3]. For such uses, the scaffolds employed in tissue engineering must meet specific criteria. They must have mechanical qualities, scaffold design, biocompatibility, and biodegradability [4]. Polymer implants are two- or three-dimensional prefabricated shapes made of a single polymer or a combination of polymers that are intended to heal tissue or treat disease [5]. The physical features of the scaffold design, similar to the polymer selection, are heavily influenced by the intended use. The scaffold is designed to offer the proper mechanical, physical, and chemical qualities necessary for tissue development and cell survival. The polymeric scaffold's primary purpose is to specify the cellular milieu (*cell niche*) needed for optimal performance [6, 7]. Developing cutting-edge processes has also aided this, from the oldest casting-based procedures to the most current 3D techniques. These have made it easier to fabricate objects with various interpenetrating networks, pore sizes, and ion channels. The motivations behind using polymer materials in medicine are their biocompatibility, structural similarity to the target tissue, and mechanical qualities [8]. For medical uses, both artificial and natural materials have been proposed, altered, and used, like 3D bioprinting scaffolds for skin, bone, and tissue repair [9].

3 Natural Polymers

Proteins like collagen, silk fibroin, and carbohydrates such as chitosan, alginate, hyaluronic acid, cellulose, etc., (Fig. 2) make up the majority of natural polymers. The bio-functional chemicals they often showing excellent bioactivity, biomimetic surfaces, and natural remodeling. However, their principal disadvantages, including immunogenicity, microbial endotoxic contamination, low tuning capacity, unpredictable disintegration level, and poor mechanical power, restrict their use in bone tissue revival [10].

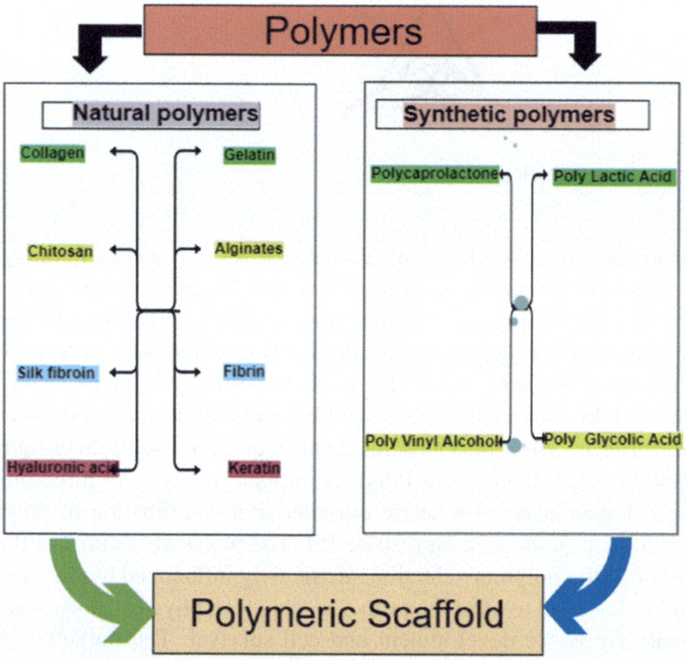

Fig. 2 Types of polymers used in the synthesis of polymeric scaffolds

3.1 Collagen

Collagen is one of the essential constituents of several animal tissues and is known to encourage mineralized matrix deposition. Its primary characteristics are its enzymatic biodegradability and versatility in processing into various physical appearances, like fibrous scaffolds and hydrogel. In addition to providing several suitable binding sites for bone cell adhesion [11–13], many tropocollagens form into 1 m fibrils, which then organize and make 10 m collagen fibers. It is widely acknowledged that synthesizing collagen fibrils improves strength and deformation, resulting in strong material [14]. The collagen arrangements in different tissues have distinct roles in extracellular matrix synthesis and tissue support. The present review provides a comprehensive route to various collagens in vitro resemble methods to trigger the relevant in vivo structures while considering anisotropic arrangement, hardness, cellular alignment, and other sophisticated characteristics of indigenous tissues. These approaches have changed collagen's structural and mechanical features, more precisely reproducing native tissue and configuration. In the neovascularization process, collagen-based scaffold plays an important role [15]. Aligned collagen scaffolds have several advantages, including higher mechanical characteristics and simplification of functions. They imitate the cell membrane, which may cause and enhance the capacity to sponsor a natural microenvironment for tissue evaluation [16]. Collagen is an essential source in the extracellular membrane and inherently cell-adhesive, creating an environment favorable for cell potentiality. Additionally, collagen can be engineered to have suitable cell-penetrating capability, producing crosslinked porosity scaffolds for various uses [17]. Jiang et al. [18], looked into the consequences of original three-dimensional collagen and broken collagen scaffolds, which may have been influenced by the fact that chondrocytes proliferated, adhered, and underwent re-differentiation more readily in col-scaffolds with triple-helical structures. These improvements in the properties of collagen scaffolds could be attributed to this triple-helical shape. Moreover, this suggests more re-differentiation due to higher amounts of aggrecan, collagen 11, and Sox-9 mRNA expression in the collagen scaffold [19].

3.2 Gelatin

Collagen undergoes irreversible hydrolysis to create gelatin, which breaks down 300 kDa protein fibrils into smaller peptides with various molecular weights that depend on denaturation [20]. The Arginine–Glycine–Aspartate tripeptide amino acids, for increased cellular cling, are abundant in collagen but only become available after the processing of fiber or degradation of fiber [21]. Its physical and chemical properties can easily be adjusted to fit different applications needs. For example, the isoelectric point may change, producing either alkaline or acidic gelatin with cationic or anionic natures [20]. The gelatin structure can be functionally modified to address drawbacks, such as insufficient thermal stability and strong poisonousness. The interaction of methacrylic anhydride with modified gelatin yields one of the most intriguing modifications [22]. Singh et al. [23], developed a brand-new silk-gelatin-grounded bio-ink with the ideal flow for cartilage reproduction. The best rheology is crucial because it supports structural renewal, enabling operations like polymer extrusion and aqueous and non-toxic processing. In addition, varying their concentrations improved the blend's constituent parts. Moreover, different bio-ink component concentrations can affect how quickly the construct degrades. The density of gel crosslinking in combination with polymer concentration also affects the cartilage's mechanical characteristics. Studies have shown that cells operate well at low polymer concentrations [24].

3.3 Chitosan

Chitosan (CS) is a polymer extracted from chitin, the primary source of crab exoskeletons [25, 26]. The interaction with anionic molecules, including glycosaminoglycans (GAGs), proteoglycans, and other nutrients, is made more accessible by the cationic character of CS. Although it lacks mechanical strength, degrades quickly, lacks cell signaling molecules, and has sticky and antibacterial qualities, chitosan is a biodegradable and biocompatible valuable substance in wound healing mechanisms [25]. CS can be easily made into films, fibers, and more sophisticated forms for orthopedic techniques. Simply freezing or lyophilizing with acetic acid solutions yields porous CS objects [27]. Because of their distinctive qualities and resemblance to the extracellular matrix, CS nanofibers made by electro-spinning are significantly better than those made by other methods [28]. A chemical process that depends upon more-pressure homogenization and freeze-drying was successfully used to create pure and combined cellulose nanofiber and chitosan aerogels. The most porous and least dense aerogel was made of pure cellulose nanofibers (CNFs), but it had the worst mechanical qualities and could not last more than a day in vitro. Chitosan was added to the CNFs, dramatically improving the aerogel's mechanical characteristics, decreasing its water absorption, and causing it to stay intact in vitro. The interface between CNFs and CS demonstrated strong antibacterial activity, superior in vitro water absorption, and high porosity, including scaffolds for tissue engineering, which has considerable potential [29]. As the procedure itself is less intrusive than that necessary with an implant, the recovery is quicker and less complex, making an intercalate scaffold such as this therapeutically promising [30].

3.4 Alginates

Alginates, the most plentiful marine biopolymers, are a diverse group of polysaccharides available in various seaweeds, and some bacteria, like the *Azotobacter* and

Pseudomonas species, can also make them. Alginates' capacity to create smooth and soft hydrogels in the presence of calcium ions depends on G-block involvement in intermolecular crosslinking with divalent positive ions [31]. Because alginate scaffolds can be used as biomaterials in 3D cell culture systems, they have been explored with other polymers, primarily in bone tissue engineering. Aside from improving physicomechanical properties, composites have more pronounced biological effects on planted cells called cell inclination and bone tissue differentiation. Internal gelation and freeze-drying of an alginate-HA composite resulted in scaffolds with highly porous structures [32]. To test the effectiveness of a substrate with conducting properties of osteogenesis, a polypyrrole-alginate mixture was lyophilized and combined with chitosan to create a scaffold for human osteosarcoma cells in a bioreactor system while being stimulated electrically. The US Food and Drug Administration (FDA) has declared alginate safe for humans [33, 34].

3.5 Silk Fibroin (SF)

Arthropods like silkworms and spiders secrete the protein thread known as silk. Sericin is the hydrophilic protein coating, and fibroin is the structural core. SF has an ideal in stability, controllable proteolytic biodegradability, and the capacity to aid MSCs in differentiating into osteoblasts [35]. Fiber deposition and electro-spinning are the widely used methods for converting SF into non-woven mats. Lastly, 3D porous structural scaffolds can be created via particle leaching, foaming with gas, rapid prototyping, and freeze-drying render methods. Moreover, silk particles can be used as a template for mineralization and long-term growth factor administration. It has been suggested that silk fibroin alterations with platelet plasma in blood, endothelial nitric oxide, insulin hormone, or adenoviruses containing bone morphogenetic protein-7 coded nucleotides will speed up and strengthen osteointegration of silk fibroin scaffolds [36]. The coexistence of sericin and fibroin in natural silk fiber, which may cause an allergy, is one restriction on using SF [37]. This flaw can be fixed by physically or chemically modifying the silk processing method or adding additional polymers or minerals [36]. Various bonding and forces inside stacked sheets improve the material's mechanical characteristics for biomedical applications. Physical or chemical methods that control crystallinity and degradation rates can be used to manage the enlargement of the sheet structure. Tyr-side groups deprotonate primarily in physiological pH and essential medium. Because of the decreased hydrogen bonds make SF amorphous, facilitating biodegradability processes [38]. Because of the hydrogen bonds, the crystalline domains of the sheet structure are enhanced in acidic conditions and restrict the movement of water and proteases. As a result, silk biodegrades slowly in vivo system [39]. Silk fibroin-porous gelatin scaffolds with loaded silk fibroin are for the improvement of favorable surrounding for cartilage injury repair [40].

3.6 Hyaluronic Acid (HA)

Hyaluronic acid (HA) is one of the glycosaminoglycans containing repeating *N*-acetyl-D-glucosamine and D-glucuronic acid disaccharide units. It is a primary constituent of human connective tissue and the most basic glycosaminoglycans (GAGs), synthesized by hyaluronan synthases. In physiological media, HA is commonly referred to as hyalurona [41]. HA hydrogels are used for cartilage tissue engineering because of their hydrodynamic properties [42]. Because it is biodegradable, biocompatible, and bioresorbable, HA is used as a tissue scaffold. Moreover, it determines intracellular signaling cascades, activates cell surface receptors, and promotes cell growth, migration, proliferation, and differentiation [43]. Chemical modifications to HA are made easier by the solubility and availability of reactive functional groups [44].

3.7 Fibrin

Fibrin, derived from fibrinogen, is a protein matrix with cell binding sites to serve as a substrate for cell adhesion and proliferation [45]. To improve stem cell function and synthesis of pre-vascularized networks, combined hMSCs and human umbilical vein endothelial cell spheroids in collagen/fibrin hydrogel were developed [46]. Bone tissue constructs use vascularized network structures to ensure encapsulated cells receive enough oxygen and nutrients [47, 48].

3.8 Keratin

Keratin is a protein family in epidermal tissues, emerging as an alternative for designing behind the ear (BTE) scaffolds and technologies [49]. Human hair, wool, and feathers are the most commonly extracted materials [50].

4 Synthetic Polymers

Total synthesis has been used to create a variety of polymers, primarily via polymerization methods [51]. Most synthetic polymers have distinctive and significant physical and chemical characteristics; some are used in creating scaffolds for polymer-based drug delivery. Because of their manageable biodegradation capacity, less toxic nature, and the possibility to produce porous scaffolds, polycaprolactone, polylactic acid, and polyglycolic acid (Fig. 2) are employed extensively. Limited cell attachment does, however, occasionally occur, including in the pure polycaprolactone matrix

[52]. Synthetic biomaterial scaffolds are becoming increasingly valuable due to their superior mechanical characteristics, pure chemical makeup, and low degradation rate. However, they lack biometric identification. It lies primarily in choosing a blend of biomaterials with the appropriate level of surface adhesion [53].

4.1 Polycaprolactone (PCL)

Because of their biocompatibility and printability, PCL is used as a thermo-ink for 3D printing and offers more medical usage. PCL scaffolds have been investigated as cancer treatments and have a lot of potential against osteosarcoma. To facilitate surgical procedures, they are also used to simulate the compactness of biological surroundings, for example, ovarian tissue [54].

4.2 Polylactic Acid (PLA)

PLA is appropriate for drug delivery systems, cell carriers, scaffold preparation, and other biomedical applications. For the release of prednisolone and dexamethasone for tissue regeneration, PLA scaffolds have served as drug delivery system [55]. It accelerates endothelialization and prevents lumen stenosis with vascular endothelial growth factors and paclitaxel (PTX) [56]. Since PLA is easily shaped and molded, which is the base material of choice for nerve cell regeneration pipelines [57]. In rats with sciatic nerve injuries, PLA conduits aid nerve regeneration and are a viable substitute for autologous nerve grafts [58, 59].

4.3 Polyglycolic Acid (PGA)

PGA is a common aliphatic polyester with FDA approval that is utilized in the manufacture of polymer products. It has been marketed under the brand name Dexon® since 1970. The first biodegradative material used in the clinical field served as a suture; compared to PCL and PLA, it is more hydrophilic. While treating diabetic patients' wounds, PGA fibrous scaffolds are essential for loading monocyte chemoattractant protein-1 (MCP-1) for medication delivery [60].

4.4 Polyvinyl Alcohol (PVA)

PVA is a biocompatible and mechanically robust semicrystalline hydrophilic polymer. PVA is a material that has a lot of potential for use in biology, including as a wound dressing, membrane, surgical repair, prosthetic skin, and scaffold material for associated polymers. The antibacterial drugs rifampin, levofloxacin, and vancomycin were released sequentially by 3D hybrid PVA scaffolds, and they performed well against bacterial biofilms [61]. Several hybrid PVA/PEG scaffolds have demonstrated an excellent capacity to release ciprofloxacin to inhibit microbial growth [62].

5 Nanoparticles in the Preparation of Scaffolds

In the biomedical sciences, nanoscience has grown in significance. For extended periods, nanoparticles can deliver different drugs to the lymphatic system, brain and its divisions, arterial heart walls, lungs and parts, liver and its segments, and spleen, among other target organs (Fig. 3) [63]. Because they differ from conventional small-molecule medications in their distinct physiological effects depending on their structure and size, nanoscale particles and molecules represent a viable alternative for treating disease [64]. The FDA has permitted numerous pharmaceutical companies in recent years to develop medications based on nanotechnology [65]. A drug carrier operates as a single unit at the nanoscale regarding its properties and transport. These nanoclusters have the lowest level of one dimension between 1 and 10 nm. Nanoclusters and nanoparticles are aggregates of ultrafine particles, whereas nanocrystals are crystals with a nanoparticle size [66]. Nanotechnology is an ideal tool to enhance cardiovascular medicine, including synthetic nanotissue engineering scaffolds, due to its mechanical and electrical characteristics of real heart tissues, as well as nanotechnology enables the controlled synthesis of materials with unique biological activity for cardiac tissue engineering [67, 68]. Biopolymer-borne nanoparticles may offer a simple method for contributing to developing biomaterials with several uses, such as antibacterial, dressing, and wound healing [69, 70].

5.1 Selenium Nanoparticles

The most desirable antibacterial treatments against wound infection are those based on selenium (Se) and other nanoparticles because they can be delivered through biopolymers that regulate their delayed release [66]. It also protects against oxidative damage caused by heavy metals like mercury and is a component of several antioxidant enzymes, including selenoproteins and selenoenzymes [71]. The development and functionalization of numerous tissues, including the kidney, liver, and skeletal muscle tissues, depends on ‘Se’ availability [72, 73]. Furthermore, plants have a more significant amount of selenium levels because, in the production of proteins, sulfurous methionine can be removed by selenomethionine because of its structural and chemical similarities [74]. Some key objects applied in the electronics photograph manufacturing of photocopiers solar cells are selenite-containing nanomaterials. The chemical element has a significant structural diversity because of its

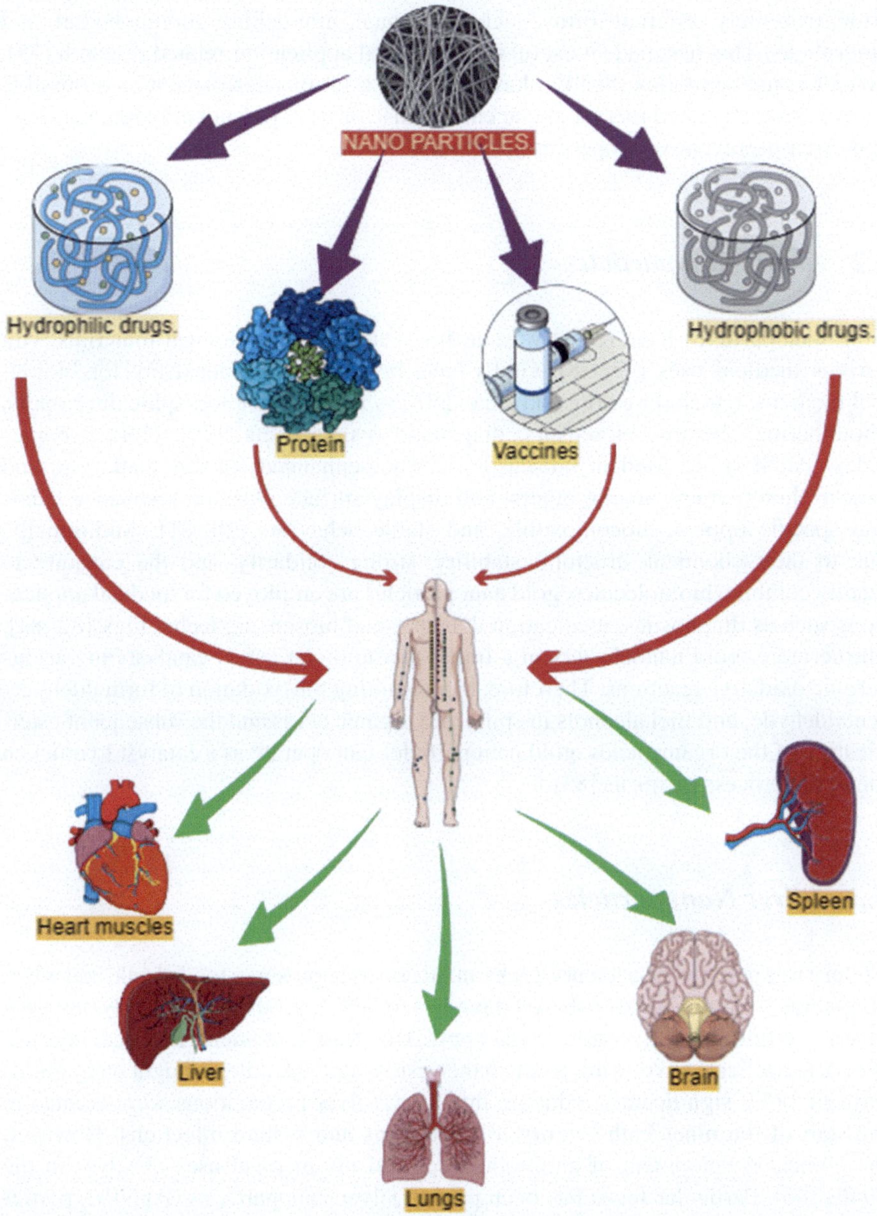

Fig. 3 Various forms of nano-based drugs targeting various organs of the human body

wide, extremely different forms, such as trigonal, monoclinic, rhombohedral, and amorphous. This has made it useful in biomedical application-related research [75]. Selenium nanoparticles (SeNPs) have received a lot of engrossment as a possible cardiac protective and therapeutic agent because of their potent antioxidant activity and disease-preventive properties [76].

5.2 *Gold Nanoparticles*

Gold nanoparticles have recently been discovered to be biological materials with various medical uses [77, 78]. GNPs have been utilized extensively for biological medicines, including drug and gene delivery systems, photographic developers, photothermal therapy, biosensors, diagnostic reagents, etc. [79]. Gold nanoparticles (AuNPs) are used in biosensors, surface-enhanced Raman scattering, and cancer phototherapy, among others, and display surface plasmon resonance bands and specific optical, biocompatible, and stable behaviors [80, 81]. Additionally, due to their chemical structural stability, strong solidarity, and the capacity to readily combine biomolecules, gold nanoparticles are employed for medical applications such as diagnostics, medication delivery, and biosensing technology [82–84]. Furthermore, gold nanoclusters in a liquid medium can act as catalysts in various aerobic oxidative reactions. Therefore, by promoting the oxidation of formaldehyde, acetaldehyde, and fuel alcohols in spirit into organic acids and the subsequent esterification of the organic acids, gold nanoparticles can operate as a catalyst to quicken the aging process of spirits [85].

5.3 *Silver Nanoparticles*

Cutlery was used by the ancient Greeks and Romans to preserve food, drink, and wine. Hippocrates employed silver-based remedies to heal wounds more quickly and treat ulcers. Additionally, silver nitrate was applied to clean instruments and treat injuries. Sims used delicate silver wires to stitch the vesicovaginal fistulas brought on by childbirth in 1852, significantly reducing infection. Silver preparations were created at the start of the nineteenth century to treat burns and wound infections. However, the clinical development of antibiotics replaced the medical uses of silver in the 1940s [86]. Particular focus has been paid to silver nanoparticles (AgNPs), particularly in biomedicine. AgNPs are well known for their powerful, broad-spectrum antibacterial and anticancer properties. AgNPs have also been studied for their additional biological functions, such as aiding bone and wound healing and vaccination immunogenicity [87]. The molecules known as colloidal AgNPs have a moderate diameter of 20–40 nm and contain 80% silver atoms and 20% silver ions. Before carbon nanotubes and titanium nanoparticles, they were the nanoparticles that sold the best and were released into the environment. The fact that AgNPs can be used in various industries has raised the demand for them. To create small, homogenous AgNPs, different synthesis approaches have been developed over time, and processes have been refined [88].

5.4 *Copper Nanoparticles*

CuNPs have the benefit of being less expensive and having a shorter reaction time than conventional catalysts. Avoiding oxidation of CuNPs throughout both the synthesis and storage phases is the main difficulty in producing CuNPs. Conducting synthesis in an inert atmosphere is one way to prevent oxidation, although doing so increases the process' complexity and expense [89]. By differentiating CuNPs from other trace metals, hydroxyl radicals are produced, which then bind to DNA molecules and cause crosslinking within and between nucleic acid stands, leading to the disorder of the DNA structure and damaging essential proteins by attaching to the sulfhydryl amino and carboxyl groups of amino acids [90, 91]. Due to the complexity, expense, toxics, and environmental concerns coupled with both physical and chemical procedures, the current scientific society has developed a method for producing CuNPs from various microorganisms that is more dependable, clean, economical, and environmentally friendly. Non-toxic and environmentally friendly reagents are used in the microbe-mediated synthesis of CuNPs, and these biogenic CuNPs have been effectively used for various applications [92]. It is challenging to remove the pathogenic microorganisms from the contaminated water effectively. Broadly antimicrobial copper particles have been demonstrated to kill microbes by oxidizing their proteins and lipids [93, 94]. They can only be used to a limited extent since they aggregate quickly, and treated water frequently experiences chemical leaching. The mounting of the copper particles on a specific substrate, such as two-dimensional (2D) grapheme-like sheets, is often necessary to overcome these obstacles [95, 96].

5.5 *Platinum Nanoparticles*

Platinum nanoparticles (PtNPs) are expanding quickly because they offer innovative nanoscale instruments for the automotive and chemical industries, with the creation of biomedical equipment [97, 98]. PtNPs are highly stable in acidic cellular vesicle settings, which sets them apart from other metal nanoparticles and predicts their cytocompatibility and tolerance in vivo [99]. PtNPs have several benefits, such as simple and affordable manufacture and purification, stability, protease resistance, strong catalytic activity even at very high pH and very low temperature, and affinity for HRP substrates. For instance, it has been noted that DNA-stabilized PtNPs have a relationship for TMB that is eight times greater than that of the natural HRP enzyme [100]. Synthesis of NPs has been done using a variety of approaches from different sources. Each technique has its drawbacks and benefits. Chemical techniques can be used to create stable NPs. However, they are unsuitable for biomedical uses, particularly drug delivery mechanisms. When using chemical techniques, concerns about biosafety, pharmacological activity, and clinical applicability arise. Therefore, creating efficient and adaptable technologies for synthesizing NPs from biological systems is essential. It is feasible to develop PtNPs with certain structural properties, for example, nanotubes, cubic NPs, and nanoclusters, and this work has attracted a lot of interest for use in catalysis [101–103]. Nanomaterials can be protein-directed to achieve the desired form, size, solubility, stability, and biocompatibility [104]. Serum albumin, the most prevalent plasma protein of all the proteins, is a model globular protein frequently employed in biomineralization [105]. Through biomineralization, He et al. [106], showed a green method for producing PtNPs in the BSA scaffold.

6 Scaffolds

The ideal way to make a scaffold as a skin substitute seems to mimic the natural extracellular matrix (ECM). Twenty-nine different forms of collagen have been identified and categorized thus far. Over 90% of the body's collagen comprises type I collagen. Electro-spinning and subsequent coating with natural proteins can resemble the fibrous structure of ECM [107–109]. As a result, integrating ceramic (Ce)-containing bioactive glasses in eco-friendly polymer scaffolds boosts their impacts and usefulness for tissue engineering [110–113]. Additionally, therapeutic ion incorporation is a feature of bioactive glasses, including Li, Ag, Cu, and Ce. One of the medicinal ions that can be added to bioactive glasses is cerium. Reactive oxygen species (ROS) damage caused by regular oxygen metabolism can be prevented from occurring in the cells [114]. MSCs, multipotent stromal stem cells that can be collected from various sources and differentiated into various lineages, such as adipogenic, chondrogenic, and osteogenic, are essential for improving and accelerating bone healing. It is necessary to find ways to regulate MSCs so that ROS are reduced in both the MSCs during the production phase of culture growth and in the microenvironment of wounded tissue. According to reports, antioxidants promote the development of MSCs [115]. The scaffold architecture and its chemical makeup are vital in achieving excellent cell-scaffold interactions. These two elements combine to create a bioactive scaffold for tissue engineering fields. Electro-spinning is a flexible method for creating a fibrous structure to resemble the ECM. However, in terms of the architecture, a balance between cell adhesion, which is more effectively carried out by structural fibers with a sub-micron size, and cell infiltration, which favors a pore size resulting from a micron-order structure, should be maintained [116, 117]. A tissue engineering scaffold for the skin's primary purpose is to restore hurdle function with quick and absolute wound healing [118]. However, the literature has frequently noted that the collagen-chitosan hybrid film's (CC-film) shortcomings include having subpar mechanical qualities and little confrontation with enzymatic deprivation. To enhance the film's overall performance, it should ideally be changed using a non-toxic crosslinking agent. Due to its distinct physicochemical qualities,

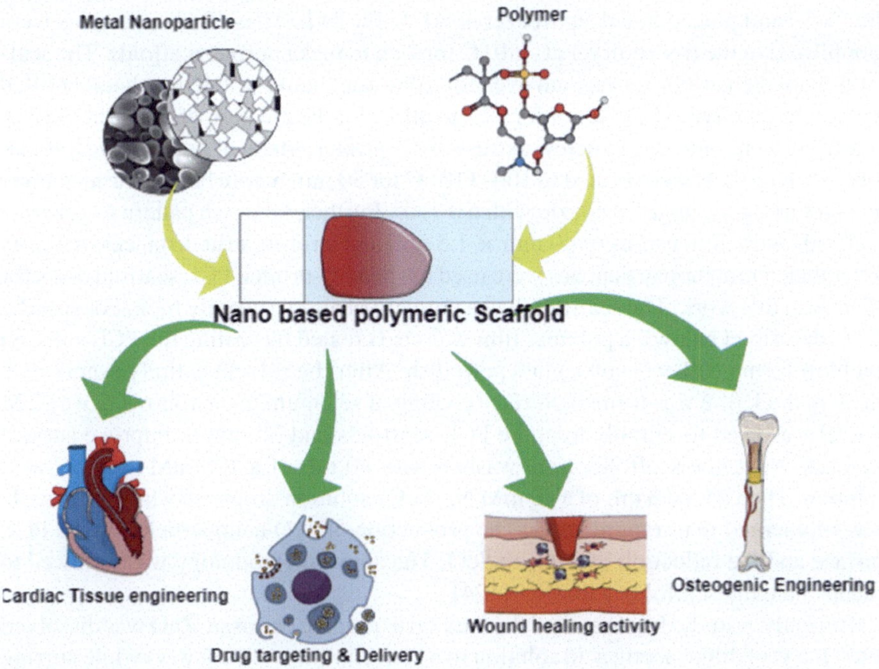

Fig. 4 Applications of nano-based scaffolds

exceptional electrical conductivity, and acute mechanical properties, graphene has attracted much focus in biomedicine and biophysics [119–121]. The nanomaterial containing polymer-based scaffolds is employed in cardiac tissue engineering, drug targeting and delivery, wound healing activity, and osteogenic engineering as shown in Fig. 4.

6.1 Preparation of Nano-Based Scaffolds

So many ways are available to prepare the scaffold with nanomaterials by using various polymers. In a study by Kumari et al., copper-chitosan scaffolds were prepared by adding slowly blending chitosan and gelatin solutions with different concentrations of copper solution. The mixture was vortexed for 5 min for homogeneity and was poured into dry powder system (DPS) tubes and further incubated overnight at − 20 °C. After 24 h, the frozen polymer blend solutions were lyophilized at − 50 °C and later freeze-dried to remove the solvent altogether. Adding chitosan and gelatin to the copper solution led to the color change of greenish [122]. The freeze-drying process used to prepare the hydroxyethyl cellulose (HEC)/silver nanoparticles scaffolds was relatively the same as the above procedure. The solution was added to the flasks and placed in a deep freezer at 80 °C for 24 h. These frozen samples were lyophilized in the freeze-dryer at – 50 °C for 72 h to make porous scaffolds. The scaffolds were then dried in a vacuum oven for a day after being stored in a chamber filled with glutaraldehyde (GA) vapor for 72 h to allow for the creation of crosslinks [123]. Scaffolds were built with a screw extrusion 3D printer. Medical grade polycaprolactone (PCL) pellets were heated to 100–110 °C for 30 min before being extruded layer by layer using a Gauge 22 nozzle with a 0°/90°. Further, laydown pattern to generate scaffolds with dimensions of 40 mm × 1.5 mm and strut-to-strut distances of 1 mm. After that, 5 mm biopsy punches were used to slice 3D-printed PCL scaffold disks for all tests in this work. The scaffolds had a porosity of approximately 67%. For specific tests (described below), a polymer film was also created by casting the PCL solution in chloroform (10% w/v) on a glass petri dish. After the solvent entirely evaporated, an 80 m thick film was formed. Surface coating of selenium in the following way 2 M NaOH was used to initially treat the PCL scaffolds and 30 min to improve surface wetness. Next, the scaffolds were washed with 40 mL of a 100 mM ascorbic acid solution. Then added 3 mL of a 10 mM Na_2SeO_3 solution dropwise while the sample was submerged in ultrapure water. The production of SeO nanoparticles on the PCL surface and the reduction of SeO_3 to SeO. The above methodology was followed to encapsulate the scaffolds with 'Se' [124].

Recently, a study demonstrated the method to prepare chitosan-ZnO was dissolved into (1% v/v) dilute acetic acid solution to a concentration of (2% w/v) while stirring on a magnetic hot plate. The solution was stirred with moderated heat of 100 °C for 3 h. The resultant chitosan-ZnO solution was filtered. To prepare chitosan-ZnO/SeNPs, the appropriate amount of 5.36 mmol/L selenium solution was added to the mixture. After that, the mixed solution was diluted to 10 mL. The above sample is designated as chitosan-ZnO/SeNPs [125]. Instead of nanoparticles, alternative nano-bioactive glasses with SiO_2–15CaO–$5P_2O_5$–CeO_2 were developed to test the cell morphology and proliferation of rabbit's bone marrow mesenchymal stem cells-derived osteogenic cells. The scaffolds used in this study were prepared based on collagen, and chitosan was used as a polymer matrix in a 1:1 volume ratio, and glass was added to the scaffold in a 30% weight ratio. Chitosan was typically dissolved in 1% (v/v) acetic acid at a rate of 2% (w/v). To ensure that the glass particles were evenly distributed throughout the polymer matrix, an equal volume of the collagen solution and chitosan solution was combined, and the glass particles were then added. This combination was then agitated for three hours and placed in an ultrasonic bath for 30 min. To create the final scaffold, the mixture was first lyophilized at 50 °C for 2 days before frozen at – 20 °C for 2 days [126].

7 Scaffold Applications in Cardiac Tissue Engineering

The most prevalent and clinically significant cardiovascular diseases are ischemic heart disease and cardiomyopathies, which can evolve in enlarged hearts, diminished pumping rate, and sudden cardiac arrest. Cardiovascular diseases are a varied group of conditions that lower the functions of the heart and the circulatory system [127]. A developing strategy for treating cardiac problems is tissue engineering, which does not have the downsides of conventional treatments, including donor restrictions, poor efficiency, and high cost [128]. The suitable materials for the scaffold's construction, a reliable source of cells, and the technique for cardiac differentiation are crucial in this process [129]. Following a myocardial infarction (MI), reperfusion of ischemic tissue can result in several cellular reactions, such as apoptosis, necrosis, and inflammatory responses of cells, which may harm the heart tissue. After inflamed cells, damaged tissue is replaced by fibrotic scars that lack contractile activity once dead cells and debris are removed, which leads to poor heart contractility and diminished blood pumping capabilities. The remaining functioning cardiac tissue then makes up the difference. The higher load consequently results in CM hypertrophy, thickening of the ventricular wall, and additional ventricle expansion when the compensatory capacity is insufficient [130]. Tissue engineering (TE) approaches use polymeric materials to replicate ECM components and are more reliable. Some studies made heart muscle patches after decellularizing whole hearts to generate a completely artificial heart. MSCs were placed between sheets of the porous bovine pericardium to generate a bioengineered tissue graft that repaired the dilated LV and enhanced cardiac behavior in animal models. This grafted tissue yields the mechanical power essential for tissue renewal [131].

The porosity of hydrogels used in tissue engineering applications primarily refers to preserving the viability of the cells seeded on it. Even without functioning blood capillaries, the porosity will help channel and transport nutrients and metabolic wastes through solutes and gases to and from the developing cells. It will also make way for establishing vasculature, cell penetration, and communication [132]. Mechanical stimulation of created constructions provides a way for accurately approximating the matrix composition of arteries and could help progress cardiac tissue engineering. The better design of abraded scaffolds with electrical stimulations for guiding cell orientation and elongation allows for the future advancement of tissue-engineered cardiac constructions [133]. Pathological alterations such as cardiac fibrosis, apoptosis, and cardiomyocyte death are permanent in ischemic heart disease and cardiomyopathies. Heart structural and functional abnormalities result in rigid ventricular walls, conduction problems, restricted oxygen diffusion, and heart failure [134]. Even the effect of coronary revascularization, widely used to treat myocardial ischemia, is still unsatisfactory [135]. To merge with tissue during the therapeutic process and form an overpass with the potential patients' usual extracellular matrix, hydrogel-constructed bioscaffolds must also demonstrate remodeling characteristics over time. Implants of rigid or solid biomaterials might not easily conform to the defect's contact surface shape [136]. The surrounding cells are supplied with oxygen and nutrients by a vast circulatory network that runs between multiple cell layers to sustain the high metabolic requirement of the contracting tissue. Engineering functional cardiac tissues that may be employed in regenerative applications requires recapitulating its anisotropic geometry throughout the ventricle wall and the internal vasculature, as heart muscle function heavily depends on its distinctive structural organization [137]. Cardiomyocytes were cultivated in hybrid hydrogel scaffolds based on spherical gold NPs uniformly dispersed throughout a polymeric gel in a different investigation. Connexin 43, a protein that passed electrical signals between heart cells, was expressed more often in the cells [138]. A traditional method involves combining cardiac cells with hydrogels and growing them in perfusable bioreactors that stimulate mechanically and electrically. However, the prevalent biomaterials' weak conductivity causes cardiac constructions to perform less than optimally. This issue might be resolved using electroconductive nanomaterials like carbon nanotubes (CNTs) and GO (graphene oxide) nanosheets. It has been demonstrated that CNT-reinforced hybrid GelMA hydrogel enhances the hydrogel's mechanical and electrical capabilities [139]. The addition of gold nanoparticles to the C-ECM made the scaffold conductive, allowing it to be electrically stimulated to enhance cell-to-cell communication and retain cell polarity. Indeed, chitosan/carbon scaffolds that are electrically conductive are useful for cardiac tissue engineering [140]. A chitosan-SeNPs film was also produced, and according to Kalishwaralal et al., the construct replicates the electrical and mechanical characteristics of the native myocardium [141].

8 Conclusion

Different kinds of nanoparticles and polymers are used to prepare scaffold, which is used to reduce or cure cardiac injuries that are discussed here. Moreover, we also detailed the different types of natural and synthetic polymers, different kinds of nanoparticles and their applications, and how they play an essential role in cardiac tissue engineering. Several published reports state that nano-based scaffolds are a vital alternate tool for cardiac surgery and the ability of nano-based scaffolds to show cells regenerative capacity and survivability of heart cells at a higher rate than the traditional surgical approaches. This review also elucidates that more research practices are still needed to manipulate several unsolved questions, such as further improvement of nano-based scaffolds in the application of cardiac tissue engineering.

References

1. R. Langer, D.A. Tirrell, Designing materials for biology and medicine. Nature **428**, 487–492 (2004)
2. R. Langer, J.P. Vacanti, Tissue engineering. Science **260**, 920–926 (1993)
3. R. Dwivedi, S. Kumar, R. Pandey, A. Mahajan, D. Nandana, D.S. Katti, D. Mehrotra, Polycaprolactone as biomaterial for bone scaffolds: review of literature. J. Oral Biol. Craniofacial Res. **10**, 381–388 (2020)
4. F.J. O'Brien, Biomaterials & scaffolds for tissue engineering. Mater. Today **14**, 88–95 (2011)
5. Y. Liu, X.Z. Shu, G.D. Prestwich, Osteochondral defect repair with autologous bone. Tissue Eng. **12**, 3405–3416 (2006)
6. G.J. Madlambayan, I. Rogers, D.C. Kirouac, N. Yamanaka, F. Mazurier, M. Doedens, R.F. Casper, J.E. Dick, P.W. Zandstra, Dynamic changes in cellular and microenvironmental

composition can be controlled to elicit in vitro human hematopoietic stem cell expansion. Exp. Hematol. **33**, 1229–1239 (2005)
7. G.J. Madlambayan, I. Rogers, K.A. Purpura, C. Ito, M. Yu, D. Kirouac, R.F. Casper, P.W. Zandstra, Clinically relevant expansion of hematopoietic stem cells with conserved function in a single-use, closed-system bioprocess. Biol. Blood Marrow Transplant. **12**, 1020–1030 (2006)
8. S. Ahn, C.O. Chantre, A.R. Gannon, J.U. Lind, P.H. Campbell, T. Grevesse, B.B. O'Connor, K.K. Parker, Soy protein/cellulose nanofiber scaffolds mimicking skin extracellular matrix for enhanced wound healing. Adv. Healthc. Mater. **7**, 1–13 (2018)
9. J. Malda, J. Visser, F.P. Melchels, T. Jüngst, W.E. Hennink, W.J.A. Dhert, J. Groll, D.W. Hutmacher, 25th anniversary article: engineering hydrogels for biofabrication. Adv. Mater. **25**, 5011–5028 (2013)
10. P. Chocholata, V. Kulda, V. Babuska, Fabrication of scaffolds for bone-tissue regeneration. Materials (Basel) **12**(4), 568 (2019)
11. C. Shi, Z. Yuan, F. Han, C. Zhu, B. Li, Polymeric biomaterials for bone regeneration. Ann. Jt. **1**, 27–27 (2016)
12. A.M. Ferreira, P. Gentile, V. Chiono, G. Ciardelli, Collagen for bone tissue regeneration. Acta Biomater. **8**, 3191–3200 (2012)
13. D. Zhang, X. Wu, J. Chen, K. Lin, The development of collagen based composite scaffolds for bone regeneration. Bioact. Mater. **3**, 129–138 (2018)
14. M.J. Buehler, Nature designs tough collagen: explaining the nanostructure of collagen fibrils. Proc. Natl. Acad. Sci. USA **103**, 12285–12290 (2006)
15. R.M. Raftery, D.P. Walsh, L. Blokpoel Ferreras, I. Mencía Castaño, G. Chen, M. LeMoine, G. Osman, K.M. Shakesheff, J.E. Dixon, F.J. O'Brien, Highly versatile cell-penetrating peptide loaded scaffold for efficient and localised gene delivery to multiple cell types: From development to application in tissue engineering. Biomaterials **216**(2019), 119277 (2019)
16. A. Dewle, N. Pathak, P. Rakshasmare, A. Srivastava, Multifarious fabrication approaches of producing aligned collagen scaffolds for tissue engineering applications. ACS Biomater. Sci. Eng. **6**, 779–797 (2020)
17. Y.H. Shen, M.S. Shoichet, M. Radisic, Vascular endothelial growth factor immobilized in collagen scaffold promotes penetration and proliferation of endothelial cells. Acta Biomater. **4**, 477–489 (2008)
18. L.B. Jiang, D.H. Su, P. Liu, Y.Q. Ma, Z.Z. Shao, J. Dong, Shape-memory collagen scaffold for enhanced cartilage regeneration: native collagen versus denatured collagen. Osteoarthr. Cartil. **26**, 1389–1399 (2018)
19. H. Watanabe, Y. Yamada, K. Kimata, Roles of aggrecan, a large chondroitin sulfate proteoglycan, in cartilage structure and function. J. Biochem. **124**, 687–693 (1998)
20. S. Young, M. Wong, Y. Tabata, A.G. Mikos, Gelatin as a delivery vehicle for the controlled release of bioactive molecules. J. Control. Release **109**, 256–274 (2005)
21. M. Barczyk, S. Carracedo, D. Gullberg, Integrins. Cell Tissue Res. **339**, 269–280 (2010)
22. A.I. Van Den Bulcke, B. Bogdanov, N. De Rooze, E.H. Schacht, M. Cornelissen, H. Berghmans, Structural and rheological properties of methacrylamide modified gelatin hydrogels. Biomacromolecules **1**, 31–38 (2000)
23. Y. Singh, A. Pratap, B.B. Bandyopadhyay, Manda, 3D bioprinting using cross-linker-free silk–gelatin bioink for cartilage tissue engineering. ACS Appl. Mater. Interfaces **11**, 33684–33696 (2019)
24. I.E. Erickson, A.H. Huang, S. Sengupta, S. Kestle, J.A. Burdick, R.L. Mauck, Macromer density influences mesenchymal stem cell chondrogenesis and maturation in photocrosslinked hyaluronic acid hydrogels. Osteoarthr. Cartil. **17**, 1639–1648 (2009)
25. E. Khor, L.Y. Lim, Implantable applications of chitin and chitosan. Biomaterials **24**, 2339–2349 (2003)
26. S.K.L. Levengood, M. Zhang, Chitosan-based scaffolds for bone tissue engineering. J. Mater. Chem. B **2**, 3161–3184 (2014)

27. S.V. Madihally, H.W.T. Matthew, Porous chitosan scaffolds for tissue engineering. Biomaterials **20**, 1133–1142 (1999)
28. K. Balagangadharan, S. Dhivya, N. Selvamurugan, Chitosan based nanofibers in bone tissue engineering. Int. J. Biol. Macromol. **104**, 1372–1382 (2017)
29. S. Rizal, E.B. Yahya, H.P.S. AbdulKhalil, C.K. Abdullah, M. Marwan, I. Ikramullah, U. Muksin, Preparation and characterization of nanocellulose/chitosan aerogel scaffolds using chemical-free approach. Gels **7**, 1–11 (2021)
30. Y. Katayama, R. Montenegro, T. Freier, R. Midha, J.S. Belkas, M.S. Shoichet, Coil-reinforced hydrogel tubes promote nerve regeneration equivalent to that of nerve autografts. Biomaterials **27**, 505–518 (2006)
31. J. Venkatesan, R. Nithya, P.N. Sudha, S.-K. Kim, Role of alginate in bone tissue engineering. Adv. Food Nutr. Res. **73**, 45–57 (2014)
32. J. Venkatesan, I. Bhatnagar, P. Manivasagan, K.H. Kang, S.K. Kim, Alginate composites for bone tissue engineering: a review. Int. J. Biol. Macromol. **72**, 269–281 (2015)
33. P. De Vos, H.A. Lazarjani, D. Poncelet, M.M. Faas, Polymers in cell encapsulation from an enveloped cell perspective. Adv. Drug Deliv. Rev. **67–68**, 15–34 (2014)
34. Z. Xu, M.T. Lam, Alginate application for heart and cardiovascular diseases. Springer Ser. Biomater. Sci. Eng. **11**, 185–212 (2018)
35. M. Filippi, G. Born, M. Chaaban, A. Scherberich, Natural polymeric scaffolds in bone regeneration. Front. Bioeng. Biotechnol. **8**, 474 (2020)
36. P. Bhattacharjee, B. Kundu, D. Naskar, H.W. Kim, T.K. Maiti, D. Bhattacharya, S.C. Kundu, Silk scaffolds in bone tissue engineering: an overview. Acta Biomater. **63**, 1–17 (2017)
37. L. Meinel, D.L. Kaplan, Silk constructs for delivery of musculoskeletal therapeutics. Adv. Drug Deliv. Rev. **64**, 1111–1122 (2012)
38. T. Asakura, Y. Watanabe, A. Uchida, H. Minagawa, NMR (nuclear magnetic resonance) of silk fibroin. II. C-13 NMR study of the chain dynamics and solution structure of Bombyx mori silk fibroin. **20**, 1075–1081 (1984)
39. G.L. Ackland, A. Gutierrez Del Arroyo, S.T. Yao, R.C. Stephens, A. Dyson, N.J. Klein, M. Singer, A.V. Gourine, Low-molecular-weight polyethylene glycol improves survival in experimental sepsis. Crit. Care Med. **38**, 629–636 (2010)
40. Y. Chen, T. Wu, S. Huang, C.W.W. Suen, X. Cheng, J. Li, H. Hou, G. She, H. Zhang, H. Wang, X. Zheng, Z. Zha, Sustained release SDF-1α/TGF-β1-loaded silk fibroin-porous gelatin scaffold promotes cartilage repair. ACS Appl. Mater. Interfaces **11**, 14608–14618 (2019)
41. H. Mahadevappa, R.M. Thushara, S. Chandranayaka, L.S. Sherman, K. Kemparaju, K.S. Girish, Emerging roles of hyaluronic acid bioscaffolds in tissue engineering and regenerative medicine. Int. J. Biol. Macromol. **86**, 917–928 (2016)
42. D. Zhu, H. Wang, P. Trinh, S.C. Heilshorn, F. Yang, Elastin-like protein-hyaluronic acid (ELP-HA) hydrogels with decoupled mechanical and biochemical cues for cartilage regeneration. Biomaterials **127**, 132–140 (2017)
43. Y.-s Jung, W. Park, H. Park, D.K. Lee, K. Na, Thermo-sensitive injectable hydrogel based on the physical mixing of hyaluronic acid and pluronic F-127 for sustained NSAID delivery. Carbohydr. Polym. **156**, 403–408 (2017)
44. G. Bhakta, Z.X.H. Lim, B. Rai, T. Lin, J.H. Hui, G.D. Prestwich, A.J. Van Wijnen, V. Nurcombe, S.M. Cool, The influence of collagen and hyaluronan matrices on the delivery and bioactivity of bone morphogenetic protein-2 and ectopic bone formation. Acta Biomater. **9**, 9098–9106 (2013)
45. T.A.E. Ahmed, E.V. Dare, M. Hincke, Fibrin: a versatile scaffold for tissue engineering applications. Tissue Eng. Part B Rev. **14**, 199–215 (2008)
46. D.N. Heo, M. Hospodiuk, I.T. Ozbolat, Synergistic interplay between human MSCs and HUVECs in 3D spheroids laden in collagen/fibrin hydrogels for bone tissue engineering. Acta Biomater. **95**, 348–356 (2019)
47. Á.E. Mercado-Pagán, A.M. Stahl, Y. Shanjani, Y. Yang, Vascularization in bone tissue engineering constructs. Ann. Biomed. Eng. **43**, 718–729 (2015)

48. M.-H. Lafage-Proust, B. Roche, M. Langer, D. Cleret, A. Vanden Bossche, T. Olivier, L. Vico, Assessment of bone vascularization and its role in bone remodeling. Bonekey Rep. **4**, 1–8 (2015)
49. A. Tachibana, S. Kaneko, T. Tanabe, K. Yamauchi, Rapid fabrication of keratin-hydroxyapatite hybrid sponges toward osteoblast cultivation and differentiation. Biomaterials **26**, 297–302 (2005)
50. M. Rajabi, A. Ali, M. McConnell, J. Cabral, Keratinous materials: structures and functions in biomedical applications. Mater. Sci. Eng. C **110**, 110612 (2020)
51. W.H. Binder, The past 40 years of macromolecular sciences: reflections on challenges in synthetic polymer and material science. Macromol. Rapid Commun. **40**, 1–7 (2019)
52. S.J. Cho, S.M. Jung, M. Kang, H.S. Shin, J.H. Youk, Preparation of hydrophilic PCL nanofiber scaffolds via electrospinning of PCL/PVP-b-PCL block copolymers for enhanced cell biocompatibility. Polymer (Guildf) **69**, 95–102 (2015)
53. A. Subramanian, U.M. Krishnan, S. Sethuraman, Development of biomaterial scaffold for nerve tissue engineering: biomaterial mediated neural regeneration. J. Biomed. Sci. **16**, 108 (2009)
54. I.R. Calori, G. Braga, P.C.C. de Jesus, H. Bi, A.C. Tedesco, Polymer scaffolds as drug delivery systems. Eur. Polym. J. **129**, 109621 (2020)
55. X. Farto-Vaamonde, G. Auriemma, R.P. Aquino, A. Concheiro, C. Alvarez-Lorenzo, Post-manufacture loading of filaments and 3D printed PLA scaffolds with prednisolone and dexamethasone for tissue regeneration applications. Eur. J. Pharm. Biopharm. **141**, 100–110 (2019)
56. Y. Zhang, J. Wang, J. Xiao, T. Fang, N. Hu, M. Li, L. Deng, Y. Cheng, Y. Zhu, W. Cui, An electrospun fiber-covered stent with programmable dual drug release for endothelialization acceleration and lumen stenosis prevention. Acta Biomater. **94**, 295–305 (2019)
57. H. Liu, P. Lv, Y. Zhu, H. Wu, K. Zhang, F. Xu, L. Zheng, J. Zhao, Salidroside promotes peripheral nerve regeneration based on tissue engineering strategy using Schwann cells and PLGA: in vitro and in vivo. Sci. Rep. **7**, 1–11 (2017)
58. M. Nune, A. Subramanian, U.M. Krishnan, S. Sethuraman, Peptide nanostructures on nanofibers for peripheral nerve regeneration. J. Tissue Eng. Regen. Med. **13**, 1059–1070 (2019)
59. C.O. Goulart, F.R. Lopes, Z.O. Monte, S.V. Dantas, A. Souto, J.T. Oliveira, F.M. Almeida, C. Tonda-Turo, C.C. Pereira, C.P. Borges, A.M. Martinez, Evaluation of biodegradable polymer conduits–poly (L-lactic acid)–for guiding sciatic nerve regeneration in mice. Methods **99**, 28–36 (2016)
60. H. Yin, G. Ding, X. Shi, W. Guo, Z. Ni, H. Fu, Z. Fu, A bioengineered drug—eluting scaffold accelerated cutaneous wound healing in diabetic mice. Colloids Surfaces B Biointerfaces **145**, 226–231 (2016)
61. R. García-Alvarez, I. Izquierdo-Barba, M. Vallet-Regí, 3D scaffold with effective multidrug sequential release against bacteria biofilm. Acta Biomater. **49**, 113–126 (2017)
62. X.H. Zhou, D.X. Wei, H.M. Ye, X. Zhang, X. Meng, Q. Zhou, Development of poly(vinyl alcohol) porous scaffold with high strength and well ciprofloxacin release efficiency. Mater. Sci. Eng. C **67**, 326–335 (2016)
63. M. Hans, A. Lowman, A dual-responsive mesoporous silica nanoparticle for tumor-triggered targeting drug deliverys for drug delivery and targeting. Curr. Opin. Solid State Mater. Sci. **6**, 319–327 (2002)
64. V. Wagner, A. Dullaart, A.K. Bock, A. Zweck, The emerging nanomedicine landscape. Nat. Biotechnol. **24**, 1211–1217 (2006)
65. S.K. Sahoo, F. Dilnawaz, S. Krishnakumar, Nanotechnology in ocular drug delivery. Drug Discov. Today **13**, 144–151 (2008)
66. S. Duan, S. Chen, T. Wang, S. Li, J. Liu, J. Liang, H. Xie, J. Han, S. Jiao, R. Cao, H.L. Wang, Q. Li, Elemental selenium enables enhanced water oxidation electrocatalysis of NiFe layered double hydroxides. Nanoscale **11**, 17376–17383 (2019)

67. M. Chandarana, A. Curtis, C. Hoskins, The use of nanotechnology in cardiovascular disease. Appl. Nanosci. **8**, 1607–1619 (2018)
68. F. Passaro, G. Testa, L. Ambrosone, C. Costagliola, C.G. Tocchetti, F. Di Nezza, M. Russo, F. Pirozzi, P. Abete, T. Russo, D. Bonaduce, Nanotechnology-based cardiac targeting and direct cardiac reprogramming: the betrothed. Stem Cells Int. **2017**, 1–12 (2017)
69. A.W. Jatoi, H. Ogasawara, I.S. Kim, Q.Q. Ni, Cellulose acetate/multi-wall carbon nanotube/Ag nanofiber composite for antibacterial applications. Mater. Sci. Eng. C **110**, 110679 (2020)
70. A.A. Menazea, M.M. Eid, M.K. Ahmed, Synthesis, characterization, and evaluation of antimicrobial activity of novel Chitosan/Tigecycline composite. Int. J. Biol. Macromol. **147**, 194–199 (2020)
71. Y. Niu, Q. Li, Y. Ding, L. Dong, C. Wang, Engineered delivery strategies for enhanced control of growth factor activities in wound healing. Adv. Drug Deliv. Rev. **146**, 190–208 (2019)
72. L. Wei, D. Pang, L. He, C. Deng, Crystal structure analysis of selenium-doped hydroxyapatite samples and their thermal stability. Ceram. Int. **43**, 16141–16148 (2017)
73. K. Saranya, M. Kalaiyarasan, N. Rajendran, Selenium conversion coating on AZ31 Mg alloy: a solution for improved corrosion rate and enhanced bio-adaptability. Surf. Coatings Technol. **378**, 124902 (2019)
74. M. Navarro-Alarcón, M.C. López-Martínez, Essentiality of selenium in the human body: relationship with different diseases. Sci. Total. Environ. **249**, 347–371 (2000)
75. A. Abdelouas, W.L. Gong, W. Lutze, J.A. Shelnutt, R. Franco, I. Moura, Using cytochrome c3 to make selenium nanowires. Chem. Mater. **12**, 1510–1512 (2000)
76. K. Kalishwaralal, S. Jeyabharathi, K. Sundar, A. Muthukumaran, A novel one-pot green synthesis of selenium nanoparticles and evaluation of its toxicity in zebrafish embryos. Artif. Cells Nanomed. Biotechnol. **44**, 471–477 (2016)
77. M. Homberger, U. Simon, On the application potential of gold nanoparticles in nanoelectronics and biomedicine. Philos. Trans. R. Soc. A Math. Phys. Eng. Sci. **368**, 1405–1453 (2010)
78. X. Xia, Q. Zhang, M. Yang, E.U.N.C. Cho, Gold nanocages: from synthesis to theranostic applications. Acc. Chem. Res. **44**, 914–924 (2011)
79. S. Vial, R.L. Reis, J.M. Oliveira, Recent advances using gold nanoparticles as a promising multimodal tool for tissue engineering and regenerative medicine. Curr. Opin. Solid State Mater. Sci. **21**, 92–112 (2017)
80. J. Gubitosa, V. Rizzi, P. Fini, R. Del Sole, A. Lopedota, V. Laquintana, N. Denora, A. Agostiano, P. Cosma, Multifunctional green synthetized gold nanoparticles/chitosan/ellagic acid self-assembly: antioxidant, sun filter and tyrosinase-inhibitor properties. Mater. Sci. Eng. C **106**, 110170 (2020)
81. J. Hussein, M.E. El-Naggar, M.M.G. Fouda, S.I. Othman, A.A. Allam, E.H. Nadwa, E.K. Rashwan, O.M. Hendawy, Eco-friendly microwave synthesis of gold nanoparticles for attenuation of brain dysfunction in diabetic rats. J. Clust. Sci. **32**, 423–435 (2021)
82. V.K.K. Upadhyayula, Functionalized gold nanoparticle supported sensory mechanisms applied in detection of chemical and biological threat agents: a review. Anal. Chim. Acta **715**, 1–18 (2012)
83. D. Pissuwan, T. Niidome, M.B. Cortie, The forthcoming applications of gold nanoparticles in drug and gene delivery systems. J. Control. Release **149**, 65–71 (2011)
84. X. Cao, Y. Ye, S. Liu, Gold nanoparticle-based signal amplification for biosensing. Anal. Biochem. **417**, 1–16 (2011)
85. T. Tsukuda, H. Tsunoyama, H. Sakurai, Aerobic oxidations catalyzed by colloidal nanogold. Chem. Asian J. **6**, 736–748 (2011)
86. J. Alexander, Wesley, History of the medical use of silver. Surg. Infect. (Larchmt.) **10**, 289–294 (2009)
87. V. Asgary, A. Shoari, F. Baghbani-Arani, S.A.S. Shandiz, M.S. Khosravy, A. Janani, R. Bigdeli, R. Bashar, R.A. Cohan, Green synthesis and evaluation of silver nanoparticles as adjuvant in rabies veterinary vaccine. Int. J. Nanomedicine **11**, 3597–3605 (2016)
88. D.L. Van Hyning, C.F. Zukoski, Formation mechanisms and aggregation behavior of borohydride reduced silver particles. Langmuir **14**, 7034–7046 (1998)

89. K. Cheirmadurai, S. Biswas, R. Murali, P. Thanikaivelan, Green synthesis of copper nanoparticles and conducting nanobiocomposites using plant and animal sources. RSC Adv. **4**, 19507–19511 (2014)
90. A.K. Chatterjee, R. Chakraborty, T. Basu, Mechanism of antibacterial activity of copper nanoparticles. Nanotechnology **25**, 135101 (2014)
91. Q. Lv, B. Zhang, X. Xing, Y. Zhao, R. Cai, W. Wang, Q. Gu, Biosynthesis of copper nanoparticles using *Shewanella loihica* PV-4 with antibacterial activity: novel approach and mechanisms investigation. J. Hazard. Mater. **347**, 141–149 (2018)
92. A. Schröfel, G. Kratošová, I. Šafařík, M. Šafaříková, I. Raška, L.M. Shor, Applications of biosynthesized metallic nanoparticles—a review. Acta Biomater. **10**, 4023–4042 (2014)
93. G.M. Zeng, A.W. Chen, G.Q. Chen, X.J. Hu, S. Guan, C. Shang, L.H. Lu, Z.J. Zou, Responses of *Phanerochaete chrysosporium* to toxic pollutants: physiological flux, oxidative stress, and detoxification. Environ. Sci. Technol. **46**, 7818–7825 (2012)
94. U. Bogdanović, V. Vodnik, M. Mitrić, S. Dimitrijević, S.D. Škapin, V. Žunič, M. Budimir, M. Stoiljković, Nanomaterial with high antimicrobial efficacy copper/polyaniline nanocomposite. ACS Appl. Mater. Interfaces **7**, 1955–1966 (2015)
95. C. Zheng, X. Zhou, H. Cao, G. Wang, Z. Liu, Synthesis of porous graphene/activated carbon composite with high packing density and large specific surface area for supercapacitor electrode material. J. Power. Sources **258**, 290–296 (2014)
96. F. Paquin, J. Rivnay, A. Salleo, N. Stingelin, C. Silva, Multi-phase semicrystalline microstructures drive exciton dissociation in neat plastic semiconductors. J. Mater. Chem. C **3**, 10715–10722 (2015)
97. Y. Liu, D. Li, S. Sun, Pt-based composite nanoparticles for magnetic, catalytic, and biomedical applications. J. Mater. Chem. **21**, 12579–12587 (2011)
98. Y. Shi, M. Lin, X. Jiang, S. Liang, Recent advances in FePt nanoparticles for biomedicine. J. Nanomater. **2015**, 1–13 (2015)
99. V. De Matteis, M.A. Malvindi, A. Galeone, V. Brunetti, E. De Luca, S. Kote, P. Kshirsagar, S. Sabella, G. Bardi, P.P. Pompa, Negligible particle-specific toxicity mechanism of silver nanoparticles: the role of Ag+ ion release in the cytosol, nanomedicine nanotechnology. Biol. Med. **11**, 731–739 (2015)
100. Y. Fu, X. Zhao, J. Zhang, W. Li, DNA-based platinum nanozymes for peroxidase mimetics. J. Phys. Chem. C **118**, 18116–18125 (2014)
101. C.M. Sánchez-Sánchez, J. Solla-Gullón, F.J. Vidal-Iglesias, A. Aldaz, V. Montiel, E. Herrero, Imaging structure sensitive catalysis on different shape-controlled platinum nanoparticles. J. Am. Chem. Soc. **132**, 5622–5624 (2010)
102. E. Formo, E. Lee, D. Campbell, Y. Xia, Functionalization of electrospun TiO_2 nanofibers with Pt nanoparticles and nanowires for catalytic applications. Nano Lett. **8**, 668–672 (2008)
103. N. Tian, Z.Y. Zhou, S.G. Sun, Y. Ding, L.W. Zhong, Synthesis of tetrahexahedral platinum nanocrystals with high-index facets and high electro-oxidation activity. Science **316**, 732–735 (2007)
104. J. Xie, Y. Zheng, J.Y. Ying, Protein-directed synthesis of highly fluorescent gold nanoclusters. J. Am. Chem. Soc. **131**, 888–889 (2009)
105. N. Goswami, A. Giri, M.S. Bootharaju, P.L. Xavier, T. Pradeep, S.K. Pal, Copper quantum clusters in protein matrix: potential sensor of Pb^{2+} ion. Anal. Chem. **83**, 9676–9680 (2011)
106. S. Bin He, H.H. Deng, A.L. Liu, G.W. Li, X.H. Lin, W. Chen, X.H. Xia, Synthesis and peroxidase-like activity of salt-resistant platinum nanoparticles by using bovine serum albumin as the scaffold. ChemCatChem **6**, 1543–1548 (2014)
107. Z. Bazrafshan, G.K. Stylios, Spinnability of collagen as a biomimetic material: a review. Int. J. Biol. Macromol. **129**, 693–705 (2019)
108. D. Sundaramurthi, U.M. Krishnan, S. Sethuraman, Electrospun nanofibers as scaffolds for skin tissue engineering. Polym. Rev. **54**, 348–376 (2014)
109. S.A. Sell, M.J. McClure, K. Garg, P.S. Wolfe, G.L. Bowlin, Electrospinning of collagen/biopolymers for regenerative medicine and cardiovascular tissue engineering. Adv. Drug Deliv. Rev. **61**, 1007–1019 (2009)

110. S. Shruti, A.J. Salinas, G. Lusvardi, G. Malavasi, L. Menabue, M. Vallet-Regi, Mesoporous bioactive scaffolds prepared with cerium-, gallium- and zinc-containing glasses. Acta Biomater. **9**, 4836–4844 (2013)
111. A.M. Deliormanlı, Synthesis and characterization of cerium- and gallium-containing borate bioactive glass scaffolds for bone tissue engineering. J. Mater. Sci. Mater. Med. **26**, 1–13 (2015)
112. A.M. Deliormanlı, Electrospun cerium and gallium-containing silicate based 13–93 bioactive glass fibers for biomedical applications. Ceram. Int. **42**, 897–906 (2016)
113. A.M. Ibrahim, Z.M. Al-Rashidy, N.A. Abdel Ghany, H.Y. Ahmed, A.E. Omar, M.M. Farag, Bioactive and antibacterial metal implant composite coating based on Ce-doped nanobioactive glass and chitosan by electrophoretic deposition method. J. Mater. Res. **9**, 1899–1913 (2021)
114. S. Carvajal, M. Perramón, G. Casals, D. Oró, J. Ribera, M. Morales-Ruiz, E. Casals, P. Casado, P. Melgar-Lesmes, G. Fernández-Varo, P. Cutillas, V. Puntes, W. Jiménez, Cerium oxide nanoparticles protect against oxidant injury and interfere with oxidative mediated kinase signaling in human-derived hepatocytes. Int. J. Mol. Sci. **20**, 1–22 (2019)
115. X. Zou, H. Li, L. Chen, A. Baatrup, C. Bünger, M. Lind, Stimulation of porcine bone marrow stromal cells by hyaluronan, dexamethasone and rhBMP-2. Biomaterials **25**, 5375–5385 (2004)
116. Z. Yang, M. Zhengwei, S. Huayu, G. Changyou, In-depth study on aminolysis of poly(ε-caprolactone): back to the fundamentals. Sci. China Chem. **55**, 2419–2427 (2012)
117. P. Gentile, K. McColgan-Bannon, N.C. Gianone, F. Sefat, K. Dalgarno, A.M. Ferreira, Biosynthetic PCL-graft-collagen bulk material for tissue engineering applications. Materials (Basel) **10**, 1–17 (2017)
118. P. Pal, P.K. Srivas, P. Dadhich, B. Das, P.P. Maity, D. Moulik, S. Dhara, Accelerating full thickness wound healing using collagen sponge of mrigal fish (*Cirrhinus cirrhosus*) scale origin. Int. J. Biol. Macromol. **93**, 1507–1518 (2016)
119. B. Campbell, J. Manning, The rise of victimhood culture: microaggressions, safe spaces, and the new culture wars, pp. 1–265 (2018)
120. C. Lee, X. Wei, J.W. Kysar, J. Hone, Measurement of the elastic properties and intrinsic strength of monolayer graphene. Science **321**, 385–388 (2008)
121. T. Nakagawa, A device for measuring very low pressure differences. J. Fluids Eng. Trans. ASME **102**, 499–501 (1980)
122. S. Kumari, B.N. Singh, P. Srivastava, Effect of copper nanoparticles on physico-chemical properties of chitosan and gelatin-based scaffold developed for skin tissue engineering application. Biotech **9**, 1–14 (2019)
123. F.H. Zulkifli, F.S. Hussain, S.S. Zeyohannes, M.S. Rasad, M.M. Yusuff, A facile synthesis method of hydroxyethyl cellulose-silver nanoparticle scaffolds for skin tissue engineering applications. Mater. Sci. Eng. **79**, 151–160 (2017)
124. C. Vaquette, N. Bock, P.A. Tran, Layered antimicrobial selenium nanoparticle-calcium phosphate coating on 3D printed scaffolds enhanced bone formation in critical size defects. ACS Appl. Mater. Interfaces **12**, 55638–55648 (2020)
125. Q. Ruan, L. Yuan, S. Gao, X. Ji, W. Shao, J. Ma, D. Jiang, Development of ZnO/selenium nanoparticles embedded chitosan-based antibacterial wound dressing for potential healing ability and nursing care after paediatric fracture surgery. Int. Wound J. **20**, 1819–1831 (2023)
126. H.F. Hammouda, M.M. Farag, M.M. El Deftar, M. Abdel-Gabbar, B.M. Mohamed, Effect of Ce-doped bioactive glass/collagen/chitosan nanocomposite scaffolds on the cell morphology and proliferation of rabbit's bone marrow mesenchymal stem cells-derived osteogenic cells. J. Genet. Eng. Biotechnol. **1**, 1–19 (2022)
127. R.E. Hershberger, J. Lindenfeld, L. Mestroni, C.E. Seidman, M.R.G. Taylor, J.A. Towbin, Genetic evaluation of cardiomyopathy—a heart failure society of America Practice Guideline. J. Card. Fail. **15**, 83–97 (2009)
128. B. Bakhshandeh, P. Zarrintaj, M.O. Oftadeh, F. Keramati, H. Fouladiha, S. Sohrabi-jahromi, Z. Ziraksaz, Tissue engineering; strategies, tissues, and biomaterials. Biotechnol. Genet. Eng. Rev. **33**, 144–172 (2017)

129. S. Hosseinzadeh, H. Nazari, N. Sadegzadeh, A. Babaie, M. Kabiri, N. Tasharrofi, M. Soufi Zomorrod, M. Soleimani, Polyethylenimine: a new differentiation factor to endothelial/cardiac tissue. J. Cell. Biochem. **120**, 1511–1521 (2019)
130. K.T. Hickey, K. Rezzadeh, Hypertrophic cardiomyopathy: a clinical and genetic update. Nurse Pract. **38**, 22–31 (2013)
131. H.C. Ott, T.S. Matthiesen, S.K. Goh, L.D. Black, S.M. Kren, T.I. Netoff, D.A. Taylor, Perfusion-decellularized matrix: using nature's platform to engineer a bioartificial heart. Nat. Med. **14**, 213–221 (2008)
132. D.L. Elbert, Liquid–liquid two-phase systems for the production of porous hydrogels and hydrogel microspheres for biomedical applications: a tutorial review. Acta Biomater. **7**, 31–56 (2011)
133. H.T.H. Au, I. Cheng, M.F. Chowdhury, M. Radisic, Interactive effects of surface topography and pulsatile electrical field stimulation on orientation and elongation of fibroblasts and cardiomyocytes. Biomaterials **28**, 4277–4293 (2007)
134. S.D. Solomon, N. Anavekar, H. Skali, J.J.V. McMurray, K. Swedberg, S. Yusuf, C.B. Granger, E.L. Michelson, D. Wang, S. Pocock, M.A. Pfeffer, Influence of ejection fraction on cardiovascular outcomes in a broad spectrum of heart failure patients. Circulation **112**, 3738–3744 (2005)
135. I. Med, L. Kramer, S.G. How, S. Med, S. Safer, L. Colin, Extraordinary unremitting, pp. 712–714 (1992)
136. M. Maisani, D. Pezzoli, O. Chassande, D. Mantovani, Cellularizing hydrogel-based scaffolds to repair bone tissue: how to create a physiologically relevant micro-environment? J. Tissue Eng. **8**, 204173141771207 (2017)
137. G.C. Engelmayr, M. Cheng, C.J. Bettinger, J.T. Borenstein, R. Langer, L.E. Freed, Accordion-like honeycombs for tissue engineering of cardiac anisotropy. Nat. Mater. **7**, 1003–1010 (2008)
138. J.O. You, M. Rafat, G.J.C. Ye, D.T. Auguste, Nanoengineering the heart: conductive scaffolds enhance connexin 43 expression. Nano Lett. **11**, 3643–3648 (2011)
139. S.R. Shin, S.M. Jung, M. Zalabany, K. Kim, P. Zorlutuna, S.B. Kim, M. Nikkhah, M. Khabiry, M. Azize, J. Kong, K.T. Wan, T. Palacios, M.R. Dokmeci, H. Bae, X. Tang, A. Khademhosseini, Carbon-nanotube-embedded hydrogel sheets for engineering cardiac constructs and bioactuators. ACS Nano **7**, 2369–2380 (2013)
140. A.M. Martins, G. Eng, G. Caridade, F. Mano, R.L. Reis, Electrically conductive chitosan/carbon scaffolds for cardiac tissue engineering. Biomacromolecules **15**, 635–643 (2014)
141. K. Kalishwaralal, S. Jeyabharathi, K. Sundar, S. Selvamani, M. Prasanna, A. Muthukumaran, A novel biocompatible chitosan–selenium nanoparticles (SeNPs) film with electrical conductivity for cardiac tissue engineering application. Mater. Sci. Eng. C **92**, 151–160 (2018)

Nanobiomaterials with Conductive Properties for Biomedical Applications

Suriyakala Gunasekaran, Sathiyaraj Sivaji, and Suchada Sukrong

Abstract Conductive nanobiomaterials (CNBs) have gained significant attention in the discipline of biomedicine due to their distinctive electrical and mechanical characteristics. These materials are superior to conventional materials in many ways, and they offer tremendous possibilities in a series of biomedical applications. The several types of CNBs, comprising metallic, carbon-based, and organic materials, as well as their characteristics and manufacturing processes, have been covered in this review. In addition, we have emphasized the application of CNBs in biosensors, drug transport, and regenerative medicine, demonstrating their potential for advancing medical therapies, diagnostics, and tissue engineering. CNBs offer an extremely adaptable system for the development of advanced biomedical devices and therapies, and future research will undoubtedly unveil even more opportunities for their use in biomedicine.

Keywords Conductive nanobiomaterials · Biomedicine · Tissue engineering · Biosensor · Drug delivery

S. Gunasekaran (✉) · S. Sukrong (✉)
Department of Pharmacognosy and Pharmaceutical Botany, Faculty of Pharmaceutical Sciences, Chulalongkorn University, Bangkok 10330, Thailand
e-mail: suriyarg2@gmail.com

S. Sukrong
e-mail: suchada.su@chula.ac.th

S. Sivaji
Nano and Energy Bioscience Laboratory, Department of Biotechnology, Thiruvalluvar University, Serkkadu, Vellore, Tamilnadu 632115, India

S. Gunasekaran · S. Sukrong
Center of Excellence in DNA Barcoding of Thai Medicinal Plants, Faculty of Pharmaceutical Sciences, Chulalongkorn University, Bangkok 10330, Thailand

K. Kasinathan et al. (eds.), *New Technologies for Energy Transition Based on Sustainable Development Goals*, https://doi.org/10.1007/978-981-97-2527-4_8

1 Introduction

The words "*nanomaterials*" and "*biomaterials*" are combined under the broader name "*nanobiomaterials*" [1]. Nanomaterials are functional materials made up of substances having less than 100 nm in at least one dimension [2]. A relatively new field of study called nanobiomaterials combines the special qualities of nanoparticles with the useful qualities of biomaterials. These materials are capable of being developed to have different distinctive physical and chemical properties, including greater surface area, improved optical capabilities, and enhanced mechanical strength [3]. Biomaterials, on the other hand, are specially created materials that communicate with biological systems for either therapeutic or diagnostic purposes. Biomaterials can be produced artificially to achieve certain qualities or derived from natural sources like collagen or chitosan [4]. A new class of materials known as nanobiomaterials has been created by the fusion of nanoparticles and biomaterials. These substances display special qualities and capabilities that are not separately present in nanoparticles or biomaterials. In particular, nanobiomaterials have a place in improving medical devices such as implants, scaffolds, and drug delivery systems (DDS) [5]. Because of their outstanding characteristics, including electrical conductivity (EC), biocompatibility, and the ability to communicate with biological molecules, conductive nanobiomaterials (CNBs) have become a promising category of nanobiomaterials for the applications in the area of biomedical [6, 7]. These resources have the power to transform the practice of biomedicine by providing fresh approaches to the identification, treatment, and control of different types of illnesses and disorders. The creation and use of CNBs in biomedicine have attracted increasing attention in recent years [8]. This interest is sparked by the need for better medicines, implanted devices, and diagnostic instruments that can communicate with biological systems. In order to give a comprehensive overview of the various uses of conductive nanobiomaterials in biomedicine, this review paper will focus on their special qualities that make them ideal for these uses. The current state of research in this area will also be presented, along with a discussion of the difficulties in creating and applying conductive nanobiomaterials in biomedicine.

2 Conductive Nanobiomaterials

Conductive nanobiomaterials have garnered much more attention and exhibit enormous potential in the biomedical industry. CNBs are exceptional in their ability to interact with biological systems, be biocompatible, and conduct electricity. As a result, they are well-suited for different kinds of biomedical applications [9]. Significant improvements in medical diagnostics, treatments, and regenerative medicine have been made possible by the intersection of nanotechnology and life sciences. Recently, there has been a noticeable increase in attention to the creation and application of CNBs in biomedicine. Due to their special properties at the nanoscale, these materials have distinct benefits over conventional biomedical materials like metals and polymers. Metal nanoparticles, carbon-based substances like graphene and carbon nanotubes (CNTs), and conducting polymers are the materials that are involved in the production of CNBs [10]. Each of these substances contributes unique benefits and qualities to the field of biomedicine. Advancements in various important fields of biomedicine have been made possible by the ability of CNBs to interface with biological systems. These include neurologic disease treatment, medication delivery, regenerative medicine, implanted devices, and biosensors [11, 12]. Researchers have created novel methods for disease diagnosis, therapeutic interventions, tissue regeneration, targeted drug administration, and electrical stimulation of brain networks by utilizing the conductivity and surface characteristics of these materials.

3 Types of Conductive Nanobiomaterials

3.1 Metallic Nanoparticles

Metallic nanoparticles (MNPs), like silver and gold, have garnered noteworthy interest in the research area of CNBs. These nanoparticles display distinctive electrical and optical characteristics, which render them highly suitable for diverse biomedical applications [13, 14]. MNPs have better interactions with biomolecules due to their tiny size and wide surface area, which improves the sensing capacities of biosensors [15]. For instance, label-free biosensing is made possible by the powerful plasmonic effects that gold nanoparticles (AuNPs) exhibit through localized surface plasmon resonance (LSPR). Functionalization of AuNPs with ligands, antibodies, or nucleic acids allows for specific recognition and detection of target analytes [16]. Similarly, silver nanoparticles (AgNPs) possess excellent optical properties and find utility in biosensors for their enhanced sensitivity and rapid response [17].

3.2 Carbon-Based Materials

Carbon-based nanomaterials, like graphene and CNTs, have piqued the interest of researchers because of their outstanding EC, high surface area, and unique structural features. These materials are excellent prospects for biomedical applications because of their extraordinary mechanical strength, elasticity, and biocompatibility [18, 19]. Graphene, a two-dimensional sheet of carbon atoms organized in a hexagonal lattice, shows outstanding thermal and EC [20]. It's high surface area and single-atom thickness make it the perfect substrate for a series of sensing uses, like electrochemical and biosensors. Graphene can be made specifically capable of recognizing and detecting target analytes by functionalizing them with biomolecules [21]. Contrarily, carbon nanotubes are cylinder-shaped structures made of rolled graphene sheets.

Depending on the quantity of concentric layers, they can be categorized as single-walled carbon nanotubes (SWCNTs) or multi-walled carbon nanotubes (MWCNTs) [22]. SWCNTs have been thoroughly researched for their possibility in tissue engineering, drug delivery, and biosensing because of their superior electrical conductivity. MWCNTs have special benefits like improved mechanical characteristics and are used in regenerative medicine and neural interfaces [23–25].

3.3 Conducting Polymers

Conducting polymers are an organic material class that has excellent electrical conductivity, optical transparency, and biocompatibility. Examples of these materials like polypyrrole (PPy), polyaniline (PANI), and poly(3,4 ethylenedioxythiophene) (PEDOT). In comparison to metallic and carbon-based materials, they have several advantages, including production and controllable electrical characteristics [26]. For instance, due to its biocompatibility, electrical conductivity, and capacity to support cell development, PPy has been extensively utilized in the field of biomedical. It has been investigated for utilization in biosensors, tissue engineering scaffolds, and DDS [27]. As an outcome of its high EC and compatibility with cerebral tissue, PANI, on the other hand, has demonstrated promise in electrochemical sensing and neural interfaces [28]. Since PEDOT, a conducting polymer that has been extensively investigated, demonstrates amazing biocompatibility and electrical qualities, it is an excellent choice for several biomedical applications [29]. Several methods, such as electrochemical polymerization, chemical vapour deposition, and inkjet printing, can be used to synthesize PEDOT. PEDOT has been investigated for use in a series of applications, including tissue engineering, cerebral implants, and biosensors [30].

4 Properties of CNBs

CNBs possess distinctive characteristics that make them highly needed for various applications within the field of biomedical engineering. These materials exhibit notable electrical conductivity, a high surface area, mechanical strength, biocompatibility, and tunable properties [9, 31].

4.1 Electrical Conductivity

One of the most significant attributes of conductive nanobiomaterials is their exceptional electrical conductivity. These materials efficiently conduct electrical current, which is particularly advantageous in tissue engineering applications. Electrical signals show a critical part in various physiological processes such as cell communication, tissue regeneration, and neural signalling. The electrical conductivity of conductive nanobiomaterials facilitates enhanced electrical coupling between cells, supporting cell differentiation, growth, and regeneration of tissue [32, 33].

4.2 High Surface Area

The high surface-to-volume (s/v) ratio of CNBs enables interactions with biological entities at the nanoscale level. This extensive surface area enables increased contact and interaction with cells, proteins, and other biomolecules. For instance, CNTs and graphene provide a high surface area for cell adhesion and proliferation, promoting crucial cellular activities necessary for tissue regeneration. Additionally, metallic nanoparticles with high surface areas facilitate efficient loading and delivery of therapeutic agents, making them promising candidates for targeted drug delivery systems [34, 35].

4.3 Mechanical Strength

Many conductive nanobiomaterials exhibit remarkable mechanical properties, like strength and flexibility. Carbon nanotubes, for example, possess exceptional tensile strength, surpassing most traditional materials. This mechanical robustness allows for the fabrication of scaffolds and implants that can withstand physiological forces and provide structural support during tissue regeneration. By modifying these nanomaterials' composition, structure, and fabrication methods to suit the distinct needs of different tissues and organs, it is possible to customize their mechanical strength [36, 37].

4.4 Biocompatibility

Biocompatibility is a crucial property for materials used in biomedical applications. Conductive nanobiomaterials have demonstrated good biocompatibility, meaning they have minimal adverse effects on living tissues and cells. Extensive studies have shown that carbon nanotubes, graphene, and metallic nanoparticles are well-tolerated by cells and do not induce significant cytotoxicity or immune responses. Surface functionalization techniques can further enhance the biocompatibility of these materials, allowing for improved interactions with biological systems and reduced risks of inflammation or rejection [38, 39].

4.5 Tunable Properties

Conductive nanobiomaterials offer tunable properties, enabling researchers to modify their characteristics to suit specific applications. These materials' composition, size, structure, and surface changes can all be altered to alter their electrical, mechanical, and surface properties. This tunability offers freedom in creating and modifying nanoparticles to suit the needs of various tissue engineering methodologies. For instance, it is possible to modify the electrical conductivity of CNT and graphene to match the conductive qualities of natural tissues, allowing for improved cell-engineered construct integration and communication [40].

5 Synthesis of CNBs

CNBs can be synthesized using different methods as shown in Fig. 1, including physical, chemical, and biological approaches [41]. The synthesis of these materials is essential to ensure their efficient utilization in biomedical settings.

5.1 Physical Synthesis Methods

When using physical synthesis techniques, it is possible to modify the properties of a material by applying external energy sources like high-energy radiation or mechanical force.

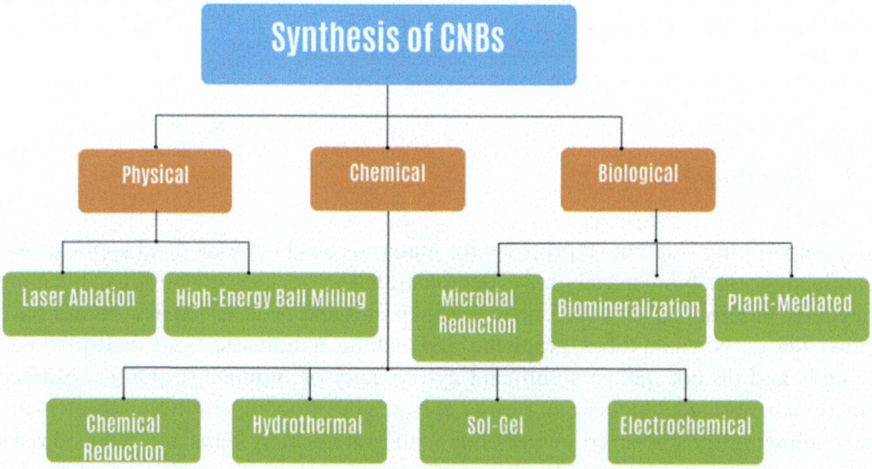

Fig. 1 Synthesis of CNBs by various methods

- *Laser Ablation:* Using a powerful laser to evaporate a target substance that is then placed on a substrate to create nanobiomaterials is known as laser ablation. This technique is used to create metallic nanobiomaterials like AgNPS and AuNPs [42].
- *High-Energy Ball Milling:* This technique uses a high-energy ball mill to create mechanical force, which changes the material's structural properties. This method is used to create carbon-based nanobiomaterials like graphene and CNTs [43].

5.2 Chemical Synthesis Methods

Chemical synthesis methods rely on chemical reactions to produce the desired nanobiomaterials. These methods are generally cost-effective and widely employed for synthesizing conductive nanobiomaterials.

- *Chemical Reduction:* Chemical reduction involves using a reducing agent to convert metal ions into metallic nanoparticles. This method is used for synthesizing metallic nanobiomaterials such as gold and silver nanoparticles [43].
- *Hydrothermal Synthesis:* Hydrothermal synthesis employs high temperature and pressure to facilitate chemical reactions between precursors, which involves the creation of nanobiomaterials. Hydrothermal is employed for synthesizing MNPs and CNTs [44].
- *Sol–Gel Method:* This method utilizes a sol (a colloidal suspension of nanoparticles) and a gel (a three-dimensional network of cross-linked polymers) to form nanobiomaterials. This method is employed for synthesizing metallic and organic nanobiomaterials, such as titanium dioxide nanoparticles and conductive polymers [45].
- *Electrochemical Synthesis:* Electrochemical synthesis involves using an electric current to drive a chemical reaction that helps to form the nanobiomaterials. This method is employed for synthesizing metallic and organic nanobiomaterials, including silver nanoparticles and conductive polymers [46].

5.3 Biological Methods

The biological synthesis of conductive nanomaterials utilizes biological systems such as bacteria, fungi, algae, and plants to produce nanomaterials with conductive properties. This approach is environmentally friendly, cost-effective, and offers unique properties that conventional chemical synthesis methods may not achieve. Various methods exist for the biological synthesis of conductive nanomaterials, including:

1. *Microbial Reduction:* This method involves using microorganisms to decrease metal ions into metal nanoparticles. The genetically modified microbes can also be used to produce nanoparticles of specific sizes and shapes [47].

2. *Biomineralization:* Biomineralization employs biological molecules like proteins, enzymes, and polysaccharides to direct the formation of nanoparticles. The resulting nanoparticles are often stabilized and possess unique properties [48].
3. *Plant-Mediated Synthesis:* This method utilizes plant extracts to reduce metal ions into MNPs. Proteins, flavonoids, and terpenoids are some of the macromolecules found in plant extracts that function as reducing and stabilizing agents [49].

Researchers can obtain CNBs with tailored properties suitable for diverse biomedical applications by employing these synthesis methods.

6 Applications of Conductive Nanobiomaterials in Biomedicine

Although CNBs have many potentials uses in biomedicine, in this article we have concentrated on tissue engineering, biosensors, and DDS as shown in Fig. 2.

6.1 Tissue Engineering

Tissue engineering is a growing field that tries to create functional tissues and organs in vitro for transplantation into patients. This interdisciplinary field combines biology, materials science, and engineering to revolutionize disease and injury treatment [50].

6.2 Historical Background

Early in the twentieth century, a French surgeon named Alexis Carrel developed methods for cultivating cells in a controlled environment, which laid the base for the area of tissue engineering today. In the 1950s and 1960s, researchers explored the use of artificial materials as replacements for damaged tissues [51]. However, tissue engineering emerged as a distinct field in the 1980s with the seminal publication "*Engineering Principles of Tissue Growth*" by Langer and Vacanti [52].

6.3 Principles of Tissue Engineering

Creating functioning tissues in vitro through the use of cells, biomaterials, and growth factors is known as tissue engineering. Cells, obtained from the patient or a donor,

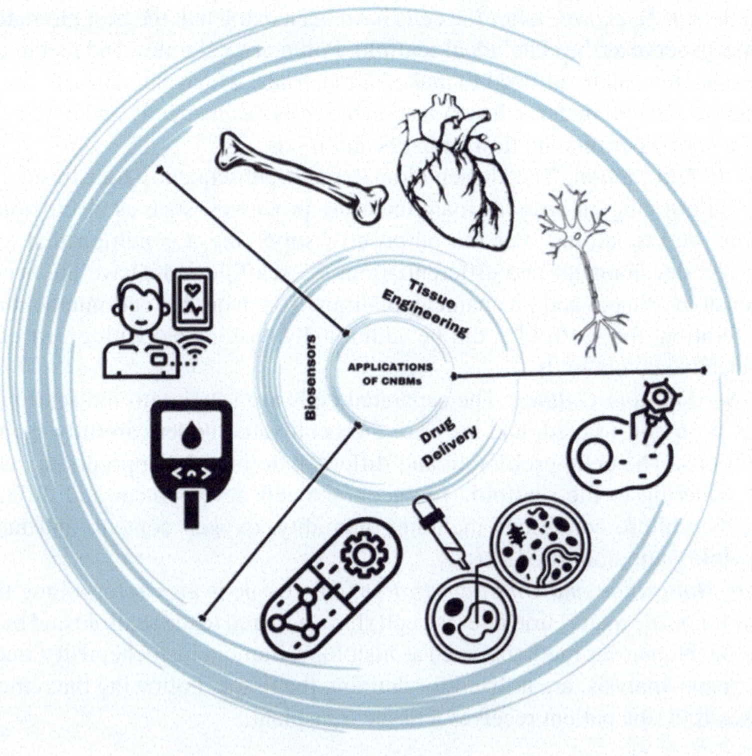

Fig. 2 In biomedicine, CNBs have a variety of applications

are cultured in a controlled environment that supports their proliferation and differentiation. Biomaterials serve as scaffolds to facilitate cell growth and development, while growth factors stimulate cell proliferation and differentiation [53].

7 Steps Involved in Tissue Engineering

Tissue engineering encompasses several steps, including cell isolation, biomaterial selection, scaffold fabrication, and cell culture. The process can be delineated as follows:

- *Cell Isolation:* The initial step in tissue engineering is to isolate cells from the patient or donor tissue. The cells are obtained through a biopsy or blood sample and are then purified and cultured in vitro. Different techniques can be used to isolate cells, including enzymatic digestion, mechanical dissociation, and density gradient centrifugation.

- *Biomaterial Selection:* After the cells have been extracted, the best biomaterial is chosen to serve as the cells' ideal scaffold. Polymers, ceramics, and metals are the materials that can be utilized to make biomaterials, which can either be synthetic or natural. The biomaterial has to sustain the cells mechanically and have a porous structure that permits the flow of gases and foods.
- *Scaffold Fabrication:* The chosen biomaterial is subsequently transformed into a scaffold utilizing an array of manufacturing processes, such as electrospinning, casting, and moulding. For the purpose of supplying a condition that encourages cell development and differentiation, the scaffold must have the necessary dimensions, shape, and mechanical qualities. To encourage cell interactions and proliferation, the scaffold could be additionally customized with growth factors or bioactive compounds.
- *Cell Seeding and Culture:* The separated cells are sown onto the scaffold after it has been constructed, and they are then cultivated under carefully monitored conditions. The cells proliferate and differentiate into the appropriate cell type after adhering to the scaffold. To encourage cell development and differentiation, the culture conditions including humidity, oxygen content, and nutrition availability are monitored closely.
- *Tissue Maturation and Characterization:* The tissue is grown for a long time to allow for tissue maturation after the cells have adhered to the scaffold and begun to develop. Numerous methods, such as histology, immunohistochemistry, and gene expression analysis, are utilized for defining the tissue. Following the removal of the scaffold, the patient receives a tissue transplant.

7.1 CNBs in Tissue Engineering

The development of biomaterials with tunable electrical conductivity is a key problem in tissue engineering because electrical signals serve a major part in multiple physiological processes including cell growth, differentiation, and migration. The development of CNBs, which have various uses in tissue engineering, notably nerve regeneration, cardiac tissue engineering, and bone regeneration, is a new response to this problem.

7.2 Nerve Regeneration

The nervous system, which is made up of a sophisticated web of specialized cells, is in charge of carrying electrical signals all across the body. Nervous system injuries can cause functional loss and disability. Therefore, a key goal in the area of regenerative medicine is the regeneration of injured nerves. By improving the electrical communication of neurons and aiding their differentiation and proliferation, conductive nanobiomaterials have great potential for encouraging nerve regeneration [54].

Because of their EC, substantial surface area, and biocompatibility, CNTs have garnered a lot of interest and are a prospective possibility for nerve regeneration. According to research, CNT-based substrates can help neural stem cells (NSCs) differentiate and be electrically stimulated [55, 56]. By promoting cell survival and the growth of neural processes, CNT substrates have also proven their biocompatibility [56]. For effective regeneration, NSCs must be delivered to wounded areas and encouraged to differentiate into neurons. In a rat spinal cord injury model, Roman et al. [57] showed that functionalized SWCNTs stimulated axonal regeneration. Similar to this, Shao et al. [58] reported that CNT nanocomposites made from MWCNTs and poly(dimethyldiallyammonium chloride) efficiently regulated NSCs by affecting differentiation, neurite outgrowth, and the electrophysiological development of NSC-derived neurons. Additionally, it has been demonstrated that CNT-based scaffolds improve growth factor sequestration in human neural stem cells (hNSCs) [59]. In addition, Park et al. [60] created a technique for controlling the development of hNSCs into neurons using a CNT network, showcasing its advantages over conventional substrates in regard to enhanced hNSC proliferation, adhesion, and neuronal differentiation. These results show the possibility of CNT network patterns as a quick and easy way to control hNSC development and polarization on various surfaces. Applications for nerve regeneration using graphene and carbon dots have also been studied. The 2D layer of carbon atoms known as graphene is much biocompatible and has good electrical conductivity. According to studies, graphene can promote the neuronal differentiation of NSCs, and three-dimensional graphene (3DG) can foster the proliferation of NSCs and have an impact on their behaviour. Through the injection of exogenous NSCs into animal models, NSCs have been used to treat neuronal disorders [61]. According to Dibajnia and Morshead [62], direct injection of NSCs into the left ventricle has enhanced restoration from brain ischemia. The first study to look at how graphene-based materials affect hNSC behaviour was done by Park et al. [59]. Overall, according to Yang et al. [63], 3-D graphene may have potential as a remedy for neural illnesses and neurodegenerative diseases. The application of graphene oxide (GO)/silk-based micro/nanofibrous scaffolds in the regeneration of nerve tissue was also explored by Magaz et al. [64]. It was possible for NG108-15 neuronal cells to attach and viable on the composite by considerably increasing electrical conductivity by including reduced GO (rGO) into silk fibroin/reduced GO (SF/rGO) scaffolds. Additionally, it was discovered that GO increased metabolic rate, development, and neurite outgrowth. This could be because GO levels were lower, and the lowered GO gave different levels of roughness.

7.3 Cardiac Tissue Engineering

Globally, cardiovascular illnesses are a leading cause of illness and demise. These illnesses have few curative options now available, and many calls for invasive surgeries like heart transplants [65]. In order to generate functioning heart tissue, scientists are investigating the utilization of CNBs in cardiac tissue engineering.

These substances can increase the electrical connection between heart muscle cells and boost the cells' capacity to contract [66]. Because of their distinctive qualities, including high strength, less weight, stability, and EC, CNTs are an intriguing biomaterial for tissue engineering, especially for cardiac tissue engineering [67, 68]. According to studies, cultivating newborn rat cardiomyocytes on surfaces altered with MWCNTs can encourage cell growth and robust cell-to-cell interactions. According to research by Shin et al. [69], embedding aligned CNT microelectrode arrays into hydrogel plates enhances cell-to-cell interaction, promotes cardiomyocyte development, and improves cell structure. Elkhenany et al. [70] found that adding MWCNTs to gelatin methacryloyl (GelMA) scaffolds produced electrically conductive platforms that enhanced the production of key amino acids associated with cardiac cell activity. SWCNTs added to collagen-based scaffolds and gelatin-chitosan hydrogels have shown similar beneficial results [71–74]. Metallic nanoparticles such as AuNPs, gold nanorods (AuNRs), and AgNPs have also been researched for potential use in cardiac tissue engineering along with CNTs. According to Navaei et al. [75], adding AuNRs to hydrogels has been found to improve tissue development and cardiomyocyte differentiation. AuNPs have been integrated into extracellular matrix materials without compromising the survival of cells [76]. Cardiac-specific markers have been expressed more frequently in chitosan gels that include AuNPs [77]. These metallic nanoparticles have special qualities that can improve the efficiency and proliferation of cells. The possibility of using polymer nanoparticles to engineer heart tissue has also been investigated. These nanoparticles have the advantage of being more biodegradable, which may reduce toxicity, and are simple to modify with medicines and biomolecules [78]. According to Wang et al. [79], conductive polypyrrole nanoparticles added to cryogels have demonstrated promising improvements in cardiac function and a decrease in infarct size in experimental animals. In rat models of myocardial infarction, self-assembled nanoparticles coated with polydopamine (PDA) and linked to a reactive oxygen species (ROS)-responsive polymer hydrogel showed better heart function [80]. Overall, CNBs have demonstrated the capacity to improve cell behaviour, organization, and efficiency in cardiac tissue engineering. The creation of ground-breaking treatments for cardiovascular disorders has a lot of potential with these materials.

7.4 Bone Regeneration

Animals with spines must have bone because it supports the body's structure and is vital to many bodily processes. However, fractures can occur in bones due to a variety of causes, including trauma, congenital abnormalities, or the excision of malignancies. Significant study has been done in recent years on the repair and regeneration of bones. Bone abnormalities are frequently treated by utilizing a series of techniques, such as using one's own tissue (autografts), donated tissue (allografts), metal-based implants, and ceramic-based scaffolds [81]. Biomaterials can now be used in bone therapies in innovative ways because to tissue engineering and regeneration. Biomaterials that offer mechanical stability, compatibility with the body, the capacity to degrade gradually, and the ability to enhance the development of new bone cells are necessary for the effective engineering of bone tissue [82]. CNBs are exceptionally strong, have a sizable surface area, interact well with living cells, are highly compatible with the body, and are readily available [83]. CNTs are one type of CNBs that resemble collagen fibres, the primary constituents of bone formation. CNTs have a special surface area and small size that enable them to connect with cells in a way that closely resembles the natural structure of bones. The conversion of stem cells into osteoblasts, the bone-forming cells, can be aided by CNTs. Additionally, CNTs can accelerate the extracellular matrix's calcification process, which results in the growth of bone tissue that is more robust and resistant. MWCNTs have been demonstrated by Li et al. [84] to have the ability to attract and ingest proteins, trigger genes involved in bone development, and promote the creation of bone tissue in both cultured and alive organisms.

In order to utilize CNTs in bone tissue engineering, researchers have also looked into pairing them with hydroxyapatite (HA), another substance found in bones. According to research by Liao et al. [85] on the nanosized HA on MWCNTs facilitate the synthesis of apatite, a mineral that makes up bones. In trials employing human foetal osteoblasts, Oyefusi et al. [86] found no adverse reactions when HA was attached to carboxylated CNTs. The development and differentiation of bone cells could benefit from this combination. A promising material for making composites that help with bone healing is graphene, which is renowned for its high strength [87]. Researchers have created many types of graphene, including self-supporting graphene hydrogel (SGH) films and membranes made of graphene hydrogel (MGH). In order to assess the biological benefits of graphene for bone regeneration, Lu et al. [88] synthesized SGH films, offering important insights for upcoming uses. Chemically converted graphene (CCG) sheets were used to construct MGH membranes by Lu et al. [89]. These membranes were strong and flexible enough to be used as rat skull substitutes. In a rat model with cranial abnormalities, these membranes served as barriers, directing bone regrowth. Additionally, Su et al. [90] manufactured poly(lactide-co-glycolide acid) (PLGA) and graphene oxide (GO)-based highly interconnected nanofibrous membranes. In a rabbit model, these membranes encouraged the regrowth of bone and the recovery of tendons. Attention has been drawn to AuNPs as agents that encourage bone regeneration [91]. According to research by Zhang et al. [92], AuNPs regulate the proliferation of stem cells and bone-forming cells as well as some immune cells. It has been discovered that AuNPs stimulate bone cell development via a variety of signalling mechanisms. Choi et al. [93] used stem cells to explore the development of bone cells and found that by triggering signalling pathways, cells generated from both bone tissue and fat tissue produced more bone. Furthermore, when AuNPs produced using *Bacillus licheniformis* crude enzyme were incorporated into bone scaffolds as examined by Singh et al. [94], they enhanced attributes such as the capacity to stimulate bone formation, suitable degradation rate, and mechanical strength. In laboratory cell lines (MC3T3) and stem cells produced from bone marrow, these AuNPs showed compatibility with bone cells, antioxidant capabilities, and enhanced mineralization of bone tissue. The creation of antimicrobial scaffolds and implants to fill significant bone deficiencies has made use of AgNPs. AgNPs efficiently avoid bacterial infections, lessen inflammation, and aid in wound and bone repair. AgNPs have been demonstrated in laboratory tests to spontaneously boost stem cells' capacity for development and bone formation as well as the mineralization of the matrix that surrounds cells [95]. The ability of human stem cells to generate bone has also been discovered to be improved by AgNPs, as shown by a rise in alkaline phosphatase activity and mineralization [96]. The utilization of these nanoparticles has created new avenues for the creation of sophisticated biomaterials with the potential for use in bone tissue engineering.

8 Biosensors

Biosensors are analytical tools that use biological elements including enzymes, antibodies, and nucleic acids to detect and measure a variety of target analytes [97]. The use of biosensors in analytical chemistry has undergone an upsurge as they provide a number of benefits beyond conventional analytical methods, including superior sensitivity, specificity, and selectivity. Biosensors are utilized in a series of areas, including surveillance of the environment, analysis of food, and medical diagnosis [98].

8.1 History of Biosensors

In 1962, Clark and Lyons from the University of Cincinnati invented the first Biosensor. They created an oxygen electrode that measured the amount of glucose in the blood using glucose oxidase as a biological component [99]. The invention of biosensors started with this, and they have advanced substantially since then. The development of glucose biosensors to monitor the diabetes was the main goal of biosensor research in the 1970's. Biosensors for environmental surveillance were created in the 1980s, and biosensors for biomedical applications were created in the 1990's [100]. Recently, the combination of nanotechnology, microfluidics, as well as other cutting-edge materials has leading to the advancement of biosensors [101, 102]. The three essential parts of a biosensor are a biological component, a transducer, and a signal processor [103].

8.2 Biological Component

A biosensor's biological component can take the form of an enzyme, antibody, nucleic acid, or any other biomolecule that can identify the target analyte with high specificity.

A solid support, like a membrane, gel, or electrode surface, is used to immobilize the biological component. The biological component's immobilization guarantees its stability and activity and makes it possible for signals to be sent to the transducer effectively.

8.3 Transducer

The transducer is responsible for converting the biochemical signal generated by the biological component into an electrical, optical, or mechanical signal. The choice of transducer depends on the type of biological component and the target analyte. Electrodes, surface plasmon resonance (SPR) sensors, and piezoelectric sensors are often used transducers in biosensors.

8.4 Signal Processor

The signal processor is in charge of processing the transducer's signal and producing a quantifiable result. According to the kind of transducer employed, the output may be a current, voltage, or optical signal. The signal processor may additionally be utilized to condition the signal, enhance the signal, and increase the signal-to-noise (s/n) ratio. Various factors, such as the mode of signal transduction, biologically specific countering mechanisms, and the number of usages, can be used to classify biosensors as shown in Fig. 3 [104].

- *Mode of Signal Transduction:* Biosensors can utilize different modes of signal transduction to detect and measure analytes. These include:
- *Optical Biosensors:* These biosensors work depending on the interaction of light with the target analyte. They can measure changes in light properties such as absorption, refraction, or luminescence.
- *Thermal Biosensors:* These biosensors detect changes in heat generated or absorbed during a biochemical reaction.
- *Electrochemical Biosensors:* These biosensors utilize electrochemical reactions to detect and quantify analytes. They measure changes in current, potential, or impedance.
- *Magnetic Biosensors:* These biosensors employ magnetic fields to detect and measure the target analytes. They detect changes in magnetic properties like magnetization or magnetic resonance.
- *Piezoelectric Biosensors:* These biosensors utilize the piezoelectric effect, where mechanical stress generates an electrical signal. They measure conversion in the frequency or amplitude of a piezoelectric crystal.
- *Electrical Biosensors:* These biosensors measure changes in electrical properties, such as conductivity or capacitance, to detect analytes.

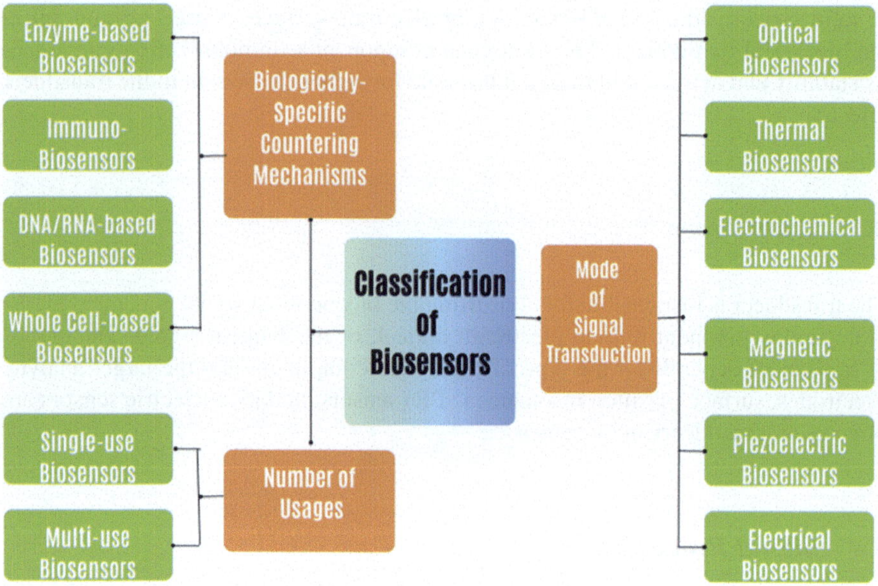

Fig. 3 An overview of biosensor classification

8.5 Biologically Specific Countering Mechanisms

The biorecognition components that are utilized for target analyte detection can also be employed to classify biosensors. These consist of:

- *Enzyme-Based Biosensors:* These biosensors allow for enzymatic reactions with the target analyte by using enzymes as the biorecognition element.
- *Immuno-Biosensors:* These biosensors utilize antibodies and antigens that selectively bind to the target analyte, leading to a measurable signal.
- *DNA/RNA-Based Biosensors:* These biosensors use DNA or RNA strands that can hybridize with target nucleic acids, generating a signal.
- *Whole Cell-Based Biosensors:* These biosensors use living cells that react in particular ways, such as fluorescing or changing colour, when exposed to the target analyte.

8.6 Number of Uses

Biosensors are additionally categorized according to how many uses they have:

- *Single-Use Biosensors:* These biosensors are designed for a single measurement or a limited number of measurements. After use, they are disposed of and cannot be reused.

- *Multi-Use Biosensors:* These biosensors are reusable and can be used for multiple measurements. They allow for continuous monitoring and analysis of analyte concentrations.

In this article, the focus is specifically on conductive nanobiomaterials in a few types of biosensors because of their broad spectrum of uses in biomedicine.

9 CNBs in Biosensors

Due to their special conductivity, biocompatibility, and surface area characteristics, have demonstrated significant potential in a variety of biosensors.

9.1 Electrochemical Biosensors

One of the oldest and most common types of biosensors is an electrochemical biosensor that employs conductive substances to transform biological signals into electrical impulses. One of the most widely utilized biosensors for the discovery of different analytes, including glucose, cholesterol, and neurotransmitters is the electrochemical biosensor [105]. Electrochemical biosensors are available in a variety of forms, such as amperometric, potentiometric, and impedimetric biosensors.

9.2 Amperometric Biosensors

Amperometric biosensors work by measuring the electricity generated during an electrochemical process. They comprise of a biological recognition component, including enzyme or an antibody, and an electrode, typically formed of a conductive substance like gold, platinum, or carbon. An alteration in the current that may be monitored and linked with the amount of analyte results from the binding of the target analyte to the detection element. Glucose, cholesterol, neurotransmitters, and proteins are only a few of the analytes that have been detected in biomedicine using amperometric biosensors [106].

9.3 Potentiometric Biosensors

In the availability of an analyte, potentiometric biosensors assess the possible variation among two electrodes. They are made up of a biological detection element, a working electrode, and a reference electrode. An alteration in the potential that may be measured and linked with the analyte concentration results from the binding of the target analyte to the detection element. In biomedicine, potentiometric biosensors have been used to identify a different analyte, like pH, ions, and gases [107].

9.4 Impedimetric Biosensors

Impedimetric biosensors track the shift in an electrode's electrical impedance caused by the presence of an analyte. An impedance change that can be detected and linked with the analyte concentration results from the binding of the target analyte to the detection element. In biomedicine, impedimetric biosensors have been used to identify a different analyte, including bacteria, viruses, and proteins [108]. Due to their special characteristics, CNBs have been widely utilized in the research of electrochemical biosensors. They have demonstrated considerable potential in a number of areas, including glucose sensing. Both SWCNTs and MWCNTs have been investigated by scientists as potential nanomaterials used to detect glucose [109]. For example, titanium dioxide-MWCNT-chitosan nanocomposites, prussian blue, and polydiallyldimethylammonium chloride (PDDA)-AuPDs were used by Mo et al. [110] to modify a glassy carbon electrode. This altered electrode was capable of detecting glucose at 0.1 mM concentrations. A nonenzymatic glucose sensor was created by Gao et al. [111] using $Ni(OH)_2$/electroreduced GO-MWCNT nanocomposites, with a threshold for detection of 2.7 mM and a respectable analyte recovery rate of 106.0%. In a different study, platinum nanoparticles-polyaniline (PANI) hydrogel heterostructures were used as the basis for glucose enzyme sensors, according to Zhai et al. [112]. These sensors demonstrated remarkable sensitivity with a 96.1 mA/mM/cm and a 0.7 mM detection limit. By immobilizing the enzyme on PANI, Fang et al. [113] created a glucose sensor that maintained sensitivity throughout 20 days of preservation in bovine serum at 37 °C. Zhao et al. [114] created molecularly imprinted polymeric nanoparticles embellished with gold nanoparticles (Au@MIPNs) to improve selectivity and increase linear detection limits. The Au@MIPNs sensor displayed increased selectivity and larger linear ranges for glucose detection when contrasted with MIP sensors without Au ornamentation. A composite made of CNBs was used by Zeng et al. [115] to create an extremely sensitive glucose sensor. A highly efficient wearable sensor for detecting glucose in sweat samples was developed by Zhao et al. [116]. The glucose oxidase (GO_x) and prussian blue-functionalized gold fibre biosensor had a linear range of 0–500 M and a sensitivity of 11.7 A/mM/cm. The wearable sensor was ideal for everyday use since it was very stretchy and strain insensitive. CNBs, such as those based on graphene, have discovered uses in a variety of different electrochemical biosensors besides glucose detection. For instance, Teengam et al. [117] used square wave voltammetry and graphene-PANI-modified electrodes to create a low-cost, recyclable paper-based biosensor for the discovery of the human papillomavirus (HPV). The artificial oligonucleotide sequences that represent HPV type 16 DNA have been effectively detected by this biosensor, which had a linear range of 10–200 nM and a detection limit of 2.3 nM.

9.5 *Optical Biosensors*

Optical biosensors track variations in light characteristics including luminescence, refraction, and absorption. Small molecules, proteins, DNA, and other biological substances may all be detected using optical biosensors [118]. The interlink between the target analyte and a detecting surface is the basis of optical biosensors' working theory. A particular biomolecule, for instance, an antibody or DNA probe, that is limited to binding to the target analyte is generally coated on the sensing surface. An optical detector may pick up on differences in the surface's optical properties transported on by the binding of the target analyte to the sensing surface [119]. Optical biosensors fall into one of two categories: (i) label-free; (ii) labelled. Neither the target analyte nor the detecting component needs to be labelled in label-free optical biosensors. They are founded on variations in the refractive index (RI) or absorbance of the sensing element as a outcome of the binding of the analyte. Surface plasmon resonance (SPR) biosensors, which can identify variations in the RI of the detecting surface brought on by the attachment of biomolecules, are a particular type of label-free optical biosensor. Proteins, antibodies, and DNA are just a few of the biological substances that may be detected using SPR biosensors [120]. Labelled optical biosensors find the target analyte using fluorescent or luminescent labels. For instance, fluorescent molecules are used as labels in fluorescence-based biosensors, which release light when activated by a certain wavelength. These biosensors are utilized in medical diagnosis, drug development, and surveillance of the environment. They have been employed to detect biological substances including proteins and DNA. Optical biosensors can be used in the area of biomedicine for things like drug development, clinical diagnostics, and point-of-care testing [121]. For instance, SPR biosensors are currently utilized to identify illness indicators and to detect medicines in biological fluids including blood and urine [122]. Cancer biomarkers like prostate-specific antigen (PSA) have been found using fluorescence-based biosensors to be present in human serum. Furthermore, bacterial, and viral infections can be detected using fluorescent biosensors [123, 124]. The concurrent identification of many analytes by optical biosensors renders them the perfect choice for rapid evaluation of biological samples. Utilizing the distinct optical and electrical characteristics of CNBs, which increase sensitivity and selectivity, could enhance the efficiency of optical biosensors. Using AuNPs as signal-enhanced tags, Yuan et al. [125] created an optical fibre sensor for measuring glucose levels. Ion beam sputtering was used to modify the sensor by depositing chromium and gold films onto the optical fibre's sensing region. For the purpose of detecting ammonium ions in water, Potadr et al. [126] looked into the sensing capabilities of optical fibre sensors coated with PVA and AgNPs. Using cold and modified polyol manufacturing techniques, PVA-coated AgNPs were created, which were subsequently coated on the optical fibre's sensing area. The LSPR effect was used to sense ammonium ions in water.

9.6 Thermal Biosensors

Thermal biosensors detect changes in heat resulting from a biological reaction. The most common type of thermal biosensor is the calorimetric biosensor, which measures the heat generated or absorbed by a biochemical reaction. These biosensors use a thermistor, which is a temperature-sensitive resistor, to detect changes in temperature [127]. Thermal biosensors may be used in a variety of biomedical settings [128]. The detection of DNA hybridization is one important use, because the joining of matching DNA strands produces thermal that the thermistor may pick up. As a result, thermal biosensors for the detection of pathogenic and genetic illnesses have been created. The detection of enzymatic processes is another use for thermal biosensors. The thermistor can detect the heat produced or absorbed when an enzyme catalyses a process. As a result, biosensors for the identification of lactose, glucose, and other metabolites in biological samples have been developed [127]. Due to their remarkable electrical and thermal conductivity qualities, CNBs are now a preferred option for creating thermal biosensors. Graphene oxide (GO) was used in the creation of capacitive thermal sensors by Ren et al. [129]. The conductive polymer composite (CPC) and reduced graphene oxide (rGO) were used by Dan and Elias [130] to provide thermal sensors. Recently, Gong et al. [131] described the use of graphene paper to make disposable thermal sensors with potential uses in the avoidance and management of infectious diseases.

9.7 Piezoelectric Biosensors

A form of biosensor known as a piezoelectric biosensor makes use of the piezoelectric effect to identify and measure biological reactions. The term "*piezoelectric effect*" describes a material's capacity to produce a voltage in reaction to stress from friction or the opposite. When a target analyte binds to a piezoelectric material employed as a sensing element in a piezoelectric biosensor, it deforms mechanically and produces an electrical signal that can be detected [132]. One of the most frequently encountered piezoelectric materials in biosensors is quartz crystal (QC), especially quartz crystal microbalance (QCM) devices. A thin QC covered with a biorecognition component, such as antibodies or enzymes, makes the structure of QCM. The biorecognition element on the crystal's surface reacts with the target analyte to alter its mass, which changes the crystal's resonant frequency. The concentration of the target analyte may be measured by converting this frequency shift into an electrical signal [133]. These biosensors have found extensive use in several biological disciplines. They were employed for the identification of biomolecules such as DNA, bacteria, viruses, and proteins. To detect illness biomarkers including cancer antigens, cardiac markers, and infectious pathogens, for instance, piezoelectric biosensors have been created [134–136]. In order to enhance mass change and boost detection sensitivity, AuNPs are often used as labels in piezoelectric biosensors [137, 138]. Research achieved an impressively low detection limit of 10–14 M for target DNA by using 50 nm AuNPs as a mass enhancer to magnify the frequency signal of a QCM [139]. They increased the threshold for detection to 10–16 M by coating the surface of a gold electrode with AuNPs and labelling it at the same time [140, 141]. In another investigation, Premaratne et al. reached an ultralow detection limit of 28 fM for target oligonucleotide [142]. Through the use of DNA hybridization, Chen et al. [143] created a layer-by-layer structure of AuNPs in a QCM-DNA sensor, leading to an incredibly low detection limit of 2 plaque forming units (PFU)/mL for dengue virus (DENV). In contrast to antibodies without AuNPs alteration, Kim et al. [144] found that the insertion of AuNP-modified antibodies boosted the signal by 53.4%. Sanati et al. [145] have created a sensor employing embedded CNT to create a conductive network and polyvinylidene fluoride (PVDF) as a piezoelectric polymer matrix. Both piezoelectric and piezoresistive characteristics may be seen in this novel sensor.

9.8 Enzyme-Based Biosensors

Enzyme-based biosensors use enzymes as the biorecognition component to detect the presence of a particular analyte. These enzymes catalyse certain processes that result in observable signals, such as modifications to the biosensor's electrical, optical, or thermal characteristics. Numerous industries, including nutrition and environmental surveillance, healthcare diagnosis, and other areas, utilize enzyme-based biosensors [146]. Examples of enzyme-based biosensors that are particularly widespread are glucose and urea sensors. Amperometric enzyme-based glucose biosensors using Au were created and described for real-time monitoring of glucose levels in the brain. The outcome of the research exhibit that the Au-based sensor that was created could efficiently track variations in brain glucose in response to pertinent pharmacological stimuli [147]. An extremely stable potentiometric urea biosensor with a reaction time of 30 s and a detection limit of 0.77 M was also created by utilizing nanoparticles [148]. The use of enzymes as identification components in biosensors has considerably increased as a outcome of the integration of enzymes with nanomaterials [149].

9.9 Immunosensor

An exclusive class of biosensors called immunosensors uses the distinct binding capacity between antibodies and antigens to identify and measure the target analytes. Due to their great specificity, immunosensors can detect target analytes in even the most complicated sample matrices [150]. Immunosensors work on the basis of an antigen's particular recognition and binding by the associated antibody. Antibodies have a great attraction for their target antigens and are quite specific to them. The immobilized antibody serves as the detecting element in immunosensors, which also include a transducer that transforms the biochemical signal into a quantifiable outcome. Target biomolecules may be found using immunosensors without the need for labels. A label-free immunosensor for the breast cancer biomarker HER2 has been developed by utilizing electrochemical impedance spectroscopy (EIS) and CNTs [151]. They functionalized MWCNTs with antibodies after immobilizing them on a gold electrode. The identification of HER2 in blood samples from breast cancer patients was made possible by the linear rise in charge transfer resistance caused by the binding of the analyte to the immobilized antibody. Thapa et al. [152] also created a sensor for CA19-9, a pancreatic cancer biomarker, using CNTs and EIS. On a polyethyleneimine layer on a gold electrode, they adsorbed MWCNTs after which they used a carbodiimide linker to covalently bond antibodies to the CNTs. EIS might be used to evaluate changes in capacitance at the sensor surface brought on by the binding of the analyte to the immobilized antibody. In summary, CNBs have made major contributions to the development of biosensors. These materials have special properties that allow for increased sensitivity, selectivity, and linear detection ranges.

10 Drug Delivery

The process or technique by which a pharmaceutical substance is introduced into the body to have a therapeutic effect is referred to as drug delivery. This may entail ingesting the substance orally, applying it topically, injecting it, or inhaling it. Targeting certain tissues or cells, regulating the pace of drug release, and reducing possible side effects are all ways that drug delivery seeks to maximize the effectiveness and safety of medicines [153]. The use of MNPs for delivering drugs has gained more attention in recent years [8]. Due to their tiny size, high surface area, and aptitude for interacting with biological molecules, nanoparticles are particularly suited for drug delivery applications [154]. Conductive nanoparticles may be used in drug delivery systems to further improve these features. Due to their distinctive characteristics and capacities, CNBs have become attractive instruments for DDS [155–157].

- *Improved Drug Loading and Stability:* CNBs with high surface areas and structural stability, such as graphene, CNTs, and MNPs, enable the efficient incorporation of medicinal products. These nanoparticles have the capacity to envelop a variety of medications, such as tiny molecules, proteins, and nucleic acids, improving their stability and guarding against deterioration. CNBs provide exact control of drug release kinetics, enabling persistent, or prompted release of therapeutic drugs.

- *Controlled and Targeted Drug Release:* For on-demand medication release at the target region, stimuli-responsive nanoparticles may react to certain stimuli like pH, temperature, light, or magnetic fields. Furthermore, site-specific drug administration is made easier by functionalizing nanomaterials with targeting ligands like antibodies, peptides, or aptamers. This increases therapeutic effectiveness and reduces off-target effects.
- *Theragnostic Capabilities:* Theragnostic is the combination of therapeutic and diagnostic functions into a single system using conductive nanobiomaterials. The simultaneous medication delivery and continuous tracking of drug release and treatment response are characteristics of these multifunctional nanomaterials. For instance, AuNPs may be employed for imaging-guided medication administration, making it possible to see how the drug is distributed and what effects it has on the body.
- *Penetration of Biological Barriers:* CNBs with nanoscale dimensions can pass through biological barriers like cell membranes and the blood–brain barrier. This makes it possible to effectively treat illnesses of the central nervous system and intracellular targets that are normally difficult to reach.
- *Tailored Medicine:* By enabling the creation of patient-specific medication delivery systems, CNBs provide prospects for tailored medicine. Drug delivery may be made to be as effective as possible for every patient by adjusting the characteristics of nanomaterials and integrating patient-specific elements, such as genetic information or illness biomarkers.

10.1 CNBs in Drug Delivery

Due to their special qualities, CNBs have gained more interest as prospective DDS. One such is the utilization of polyethylene glycol (PEG) functionalized CNTs as carriers for anticancer medications that target malignant melanoma [158]. Surprisingly, MWCNTs have shown the capacity to prevent tumour development even without drug loading. For example, when 2 g of MWCNTs were introduced into a mouse model of melanoma, Garcia-Hevia et al. [159] saw a decrease in tumour mass. Like this, surface functionalized SWCNTs have shown beneficial impacts on human glioma cells, including lower growth and higher cell mortality [160]. To improve medication delivery, researchers have also investigated combining CNTs with other materials. Wei et al. [161] created a dual-release carrier employing CNTs and a thermosensitive hydrogel made of polycaprolactone (PCL)-PEG-PCL and filled with various medicines. The release rate showed pH dependence. Another strategy involves loading medications like curcumin onto CNTs enclosed in a PEG cocoon, which improved cellular absorption [162]. MWCNTs also showed improved antileishmanial activity when bound to the anticancer medication cisplatin at low concentrations, demonstrating that they can serve as drug nanocarriers [163]. PEGylated GO has been used to connect and transport aromatic anticancer medicines into cancer cells, boosting the efficiency of low-solubility pharmaceuticals [164].

Graphene and GO have also shown potential as nanotherapeutic drug delivery platforms. Furthermore, doxorubicin-loaded PEGylated nano-GO showed promise in photothermal treatment and chemotherapy, resulting in a decrease in tumour volume in mice [165]. For non-covalently bound medicines like doxorubicin and folic acid that have targeted effects on cancer cells, rGO has been investigated [166]. Researchers have paired rGO with inorganic nanoparticles to enhance targeting and drug release monitoring in order to solve shortcomings in drug delivery systems. In comparison to solo pharmaceuticals, these composite systems showed increased dispersion stability and greater drug release activity [167, 168]. Additionally, GO-chitosan composites have been studied as targeted delivery methods for painkillers like ibuprofen [169].

In summary, CNBs, have shown promise in drug delivery systems, offering quick drug release and the ability to deliver drugs precisely where they are needed. Researchers in the discipline have taken a keen interest in these developments.

11 Conclusion

CNBs have shown considerable promise in a different area of biomedical applications, including DDS, regenerative medicine, and implantable and biosensor technology. Due to their special qualities like EC and biocompatibility, carbon-based materials like graphene and CNTs have emerged as intriguing options for use in these applications. Researchers in this area have a variety of alternatives, including metallic nanoparticles and conductive polymers, which have both been investigated for application in biomedicine. Although CNBs show great promise, many obstacles need to be removed before they can be widely used in biomedicine. These difficulties include enhancing the synthesis process' scalability and repeatability, assuring the materials' long-term stability, and resolving concerns about toxicity and biocompatibility. The development of CNBs is an intriguing area of biomedical research that has the ability to improve the diagnosis, management, and treatment of a wide range of diseases and disorders. Unquestionably, more studies in this area will result in the creation of fresh and inventive uses for these materials, which will eventually benefit patients and enhance healthcare results.

Acknowledgement This research project is supported by the Second Century Fund (C2F), Chulalongkorn University.

References

1. P. Ramburrun, R.A. Khan, Y.E. Choonara, Design, preparation, and functionalization of nanobiomaterials for enhanced efficacy in current and future biomedical applications. Nanotechnol. Rev. **11**, 1802–1826 (2022)

2. G. Suriyakala, S. Sathiyaraj, S. Devanesan, M.S. AlSalhi, A. Rajasekar, M.K. Maruthamuthu, R. Babujanarthanam, Phytosynthesis of silver nanoparticles from Jatropha integerrima Jacq. flower extract and their possible applications as antibacterial and antioxidant agent. Saudi J. Biol. Sci. **29**, 680–688 (2022)
3. A. Sinha, F.Z. Simnani, D. Singh, A. Nandi, A. Choudhury, P. Patel, E. Jha, N.K. Kaushik, Y.K. Mishra, P.K. Panda, M. Suar, The translational paradigm of nanobiomaterials: biological chemistry to modern applications. Materials Today Bio. **17**, 100463 (2022)
4. F.M. Chen, X. Liu, Advancing biomaterials of human origin for tissue engineering. Prog. Polym. Sci. **53**, 86–168 (2016)
5. J.G. Solidum, J.A. Ceriales, E.P. Ong, E.D. Ornos, R.J. Relador, E.P. Quebral, J.F. Lapeña Jr., O.A. Tantengco, K.Y. Lee, Nanomedicine and nanoparticle-based delivery systems in plastic and reconstructive surgery. Maxillofacial Plastic Reconstruct. Surg. **45**, 15 (2023)
6. E. Mostafavi, D. Medina-Cruz, K. Kalantari, A. Taymoori, P. Soltantabar, T.J. Webster, Electroconductive nanobiomaterials for tissue engineering and regenerative medicine. Bioelectricity. **2**, 120–149 (2020)
7. P. Malik, R. Gupta, V. Malik, R.K. Ameta, Emerging nanomaterials for improved biosensing. Measur. Sensors. **16**, 100050 (2021)
8. T.M. Joseph, D. Kar Mahapatra, A. Esmaeili, Ł Piszczyk, M.S. Hasanin, M. Kattali, J. Haponiuk, S. Thomas, Nanoparticles: taking a unique position in medicine. Nanomaterials **13**, 574 (2023)
9. M. Derakhshi, S. Daemi, P. Shahini, A. Habibzadeh, E. Mostafavi, A.A. Ashkarran, Two-dimensional nanomaterials beyond graphene for biomedical applications. J. Func. Biomater. **13**, 27 (2022)
10. I.J. Gómez, M. Vázquez Sulleiro, D. Mantione, N. Alegret, Carbon nanomaterials embedded in conductive polymers: A state of the art. Polymers **13**, 745 (2021)
11. J. Kost, E. Mathiowitz, A. Azagury, Advances in drug delivery and theranostics. Adv. Func. Mater. **31**, 2108838 (2021)
12. J. Khan, M. Rudrapal, E.A. Bhat, A. Ali, M. Alaidarous, B. Alshehri, S. Banwas, R. Ismail, C. Egbuna, Perspective insights to bio-nanomaterials for the treatment of neurological disorders. Front. Bioeng. Biotechnol. **9**, 724158 (2021)
13. S. Sathiyaraj, G. Suriyakala, A.D. Gandhi, R. Babujanarthanam, K.S. Almaary, T.W. Chen, K. Kaviyarasu, Biosynthesis, characterization, and antibacterial activity of gold nanoparticles. J. Infect. Public Health **14**, 1842–1847 (2021)
14. G. Suriyakala, S. Sathiyaraj, A.D. Gandhi, K. Vadakkan, U.M. Rao, R. Babujanarthanam, Plumeria pudica Jacq. flower extract-mediated silver nanoparticles: characterization and evaluation of biomedical applications. Inorg. Chem. Commun. **126**, 108470 (2021)
15. B.D. Malhotra, M.A. Ali, Nanomaterials in biosensors: fundamentals and applications. Nanomater. Biosens. **1**, 1–74 (2018)
16. A. Bonyár, Label-free nucleic acid biosensing using nanomaterial-based localized surface plasmon resonance imaging: a review. ACS Appl. Nano Mater. **3**, 8506–8521 (2020)
17. F. Beck, M. Loessl, A.J. Baeumner, Signaling strategies of silver nanoparticles in optical and electrochemical biosensors: considering their potential for the point-of-care. Microchim. Acta **190**, 91 (2023)
18. R. Rauti, M. Musto, S. Bosi, M. Prato, L. Ballerini, Properties and behavior of carbon nanomaterials when interfacing neuronal cells: How far have we come? Carbon **143**, 430–446 (2019)
19. R.R. Pandey, C.C. Chusuei, Carbon nanotubes, graphene, and carbon dots as electrochemical biosensing composites. Molecules **26**, 6674 (2021)
20. M.A. Kanjwal, A.A. Ghaferi, Graphene incorporated electrospun nanofiber for electrochemical sensing and biomedical applications: a critical review. Sensors. **22**, 8661 (2022)
21. E. Anghel, B. Adiaconita, I. Demetrescu, A. Avram, A review of vertical graphene and its biomedical applications. Coatings **13**, 761 (2023)
22. M. Ikram, A. Raza, A. Shahbaz, H. Ijaz, S. Ali, A. Haider, M.T. Hussain, J. Haider, A.A. Rafi, S. Ali, in *21st Century Advanced Carbon Materials for Engineering Applications-A Comprehensive Handbook*. Carbon Nanotubes (IntechOpen, 2020).

23. Z. Zhou, X. Liu, W. Wu, S. Park, A.L. Miller II., A. Terzic, L. Lu, Effective nerve cell modulation by electrical stimulation of carbon nanotube embedded conductive polymeric scaffolds. Biomater. Sci. **6**, 2375–2385 (2018)
24. A. Zieliński, B. Majkowska-Marzec, Whether carbon nanotubes are capable, promising, and safe for their application in nervous system regeneration some critical remarks and research strategies. Coatings **12**, 1643 (2022)
25. T.C. Suh, J. Twiddy, N. Mahmood, K.M. Ali, M.M. Lubna, P.D. Bradford, M.A. Daniele, J.M. Gluck, Electrospun carbon nanotube-based scaffolds exhibit high conductivity and cytocompatibility for tissue engineering applications. ACS Omega **7**, 20006–20019 (2022)
26. K. Namsheer, C.S. Rout, Conducting polymers: a comprehensive review on recent advances in synthesis, properties and applications. RSC Adv. **11**, 5659–5697 (2021)
27. S. Paramshetti, M. Angolkar, A. Al Fatease, S.M. Alshahrani, U. Hani, A. Garg, G. Ravi, R.A. Osmani, Revolutionizing drug delivery and therapeutics: the biomedical applications of conductive polymers and composites-based systems. Pharmaceutics **15**, 1204 (2023)
28. B.W. Walker, R.P. Lara, E. Mogadam, C.H. Yu, W. Kimball, N. Annabi, Rational design of microfabricated electroconductive hydrogels for biomedical applications. Prog. Polym. Sci. **92**, 135–157 (2019)
29. M. Criado-Gonzalez, A. Dominguez-Alfaro, N. Lopez-Larrea, N. Alegret, D. Mecerreyes, Additive manufacturing of conducting polymers: recent advances, challenges, and opportunities. ACS Appl. Polym. Mater. **3**, 2865–2883 (2021)
30. M.J. Donahue, A. Sanchez-Sanchez, S. Inal, J. Qu, R.M. Owens, D. Mecerreyes, G.G. Malliaras, D.C. Martin, Tailoring PEDOT properties for applications in bioelectronics. Mater. Sci. Eng. R. Rep. **140**, 100546 (2020)
31. A.T. Smith, A.M. LaChance, S. Zeng, B. Liu, L. Sun, Synthesis, properties, and applications of graphene oxide/reduced graphene oxide and their nanocomposites. Nano Mater. Sci. **1**, 31–47 (2019)
32. W. Wang, Y. Hou, D. Martinez, D. Kurniawan, W.H. Chiang, P. Bartolo, Carbon nanomaterials for electro-active structures: a review. Polymers **12**, 2946 (2020)
33. A.K. Jagadeesan, K. Thangavelu, V. Dhananjeyan, in *21st Century Surface Science-a Handbook*. Carbon Nanotubes: Synthesis, Properties and Applications (IntechOpen, 2020)
34. A. Hasan, M. Morshed, A. Memic, S. Hassan, T.J. Webster, H.E. Marei, Nanoparticles in tissue engineering: applications, challenges and prospects. Int. J. Nanomed. **13**, 5637 (2018)
35. R. Eivazzadeh-Keihan, A. Maleki, M. De La Guardia, M.S. Bani, K.K. Chenab, P. Pashazadeh-Panahi, B. Baradaran, A. Mokhtarzadeh, M.R. Hamblin, Carbon based nanomaterials for tissue engineering of bone: Building new bone on small black scaffolds: a review. J. Adv. Res. **18**, 185–201 (2019)
36. N. Baig, I. Kammakakam, W. Falath, Nanomaterials: a review of synthesis methods, properties, recent progress, and challenges. Mater. Adv. **2**, 1821–1871 (2021)
37. V. Harish, D. Tewari, M. Gaur, A.B. Yadav, S. Swaroop, M. Bechelany, A. Barhoum, Review on nanoparticles and nanostructured materials: Bioimaging, biosensing, drug delivery, tissue engineering, antimicrobial, and agro-food applications. Nanomaterials **12**, 457 (2022)
38. G. Sanità, B. Carrese, A. Lamberti, Nanoparticle surface functionalization: how to improve biocompatibility and cellular internalization. Front. Mol. Biosci. **7**, 587012 (2020)
39. K. Wieszczycka, K. Staszak, M.J. Woźniak-Budych, J. Litowczenko, B.M. Maciejewska, S. Jurga, Surface functionalization—the way for advanced applications of smart materials. Coord. Chem. Rev. **436**, 213846 (2021)
40. N. Jalilinejad, M. Rabiee, N. Baheiraei, R. Ghahremanzadeh, R. Salarian, N. Rabiee, O. Akhavan, P. Zarrintaj, A. Hejna, M.R. Saeb, A. Zarrabi, Electrically conductive carbon-based (bio)-nanomaterials for cardiac tissue engineering. Bioeng. Transl. Med. **8**, e10347 (2023)
41. S. Sathiyaraj, G. Suriyakala, A.D. Gandhi, S. Saranya, M. Santhoshkumar, P. Kavitha, R. Babujanarthanam, Green biosynthesis of silver nanoparticles using vallarai chooranam and their potential biomedical applications. J. Inorg. Organomet. Polym. Mater. **30**, 4709–4719 (2020)

42. M. Kim, S. Osone, T. Kim, H. Higashi, T. Seto, Synthesis of nanoparticles by laser ablation: a review. Kona Powder Part. J. **34**, 80–90 (2017)
43. L.Y. Ng, H. Ariffin, T.A. Yasim-Anuar, M.A. Farid, M.A. Hassan, High-energy ball milling for high productivity of nanobiochar from oil palm biomass. Nanomaterials **12**, 3251 (2022)
44. L. Marinescu, D. Ficai, O. Oprea, A. Marin, A. Ficai, E. Andronescu, A.M. Holban, Optimized synthesis approaches of metal nanoparticles with antimicrobial applications. J. Nanomater. **2020**, 1–4 (2020)
45. L. Fritea, F. Banica, T.O. Costea, L. Moldovan, L. Dobjanschi, M. Muresan, S. Cavalu, Metal nanoparticles and carbon-based nanomaterials for improved performances of electrochemical (Bio) sensors with biomedical applications. Materials **14**, 6319 (2021)
46. D. Bokov, A. Turki Jalil, S. Chupradit, W. Suksatan, M. Javed Ansari, I.H. Shewael, G.H. Valiev, E. Kianfar, Nanomaterial by sol-gel method: synthesis and application. Adv. Mater. Sci. Eng. **2021**, 1–21 (2021)
47. D. Lahiri, M. Nag, H.I. Sheikh, T. Sarkar, H.A. Edinur, S. Pati, R.R. Ray, Microbiologically-synthesized nanoparticles and their role in silencing the biofilm signaling cascade. Front. Microbiol. **12**, 636588 (2021)
48. S. Saallah, I.W. Lenggoro, Nanoparticles carrying biological molecules: recent advances and applications. Kona Powder Part. J. **35**, 89–111 (2018)
49. G. Suriyakala, S. Sathiyaraj, R. Babujanarthanam, K.M. Alarjani, D.S. Hussein, R.A. Rasheed, K. Kanimozhi, Green synthesis of gold nanoparticles using Jatropha integerrima Jacq. flower extract and their antibacterial activity. J. King Saud Univ. Sci. **34**, 101830 (2022)
50. K. Dzobo, N.E. Thomford, D.A. Senthebane, H. Shipanga, A. Rowe, C. Dandara, M. Pillay, K.S. Motaung, Advances in regenerative medicine and tissue engineering: innovation and transformation of medicine. Stem Cells Int. **2018**, 2495848 (2018)
51. R. Katari, A. Peloso, G. Orlando, Tissue engineering and regenerative medicine: semantic considerations for an evolving paradigm. Front. Bioeng. Biotechnol. **2**, 57 (2015)
52. M.V. Shestovskaya, S.A. Bozhkova, J.V. Sopova, M.G. Khotin, M.S. Bozhokin, Methods of modification of mesenchymal stem cells and conditions of their culturing for hyaline cartilage tissue engineering. Biomedicines **9**, 1666 (2021)
53. M. Brovold, J.I. Almeida, I. Pla-Palacín, P. Sainz-Arnal, N. Sánchez-Romero, J.J. Rivas, H. Almeida, P.R. Dachary, T. Serrano-Aulló, S. Soker, P.M. Baptista, in *Novel Biomaterials for Regenerative Medicine*. Naturally-Derived Biomaterials for Tissue Engineering Applications (2018), pp. 421–449.
54. R. Boni, A. Ali, A. Shavandi, A.N. Clarkson, Current and novel polymeric biomaterials for neural tissue engineering. J. Biomed. Sci. **25**, 1–21 (2018)
55. E. Jan, N.A. Kotov, Successful differentiation of mouse neural stem cells on layer-by-layer assembled single-walled carbon nanotube composite. Nano Lett. **7**, 1123–1128 (2007)
56. N.W. Kam, E. Jan, N.A. Kotov, Electrical stimulation of neural stem cells mediated by humanized carbon nanotube composite made with extracellular matrix protein. Nano Lett. **9**, 273–278 (2009)
57. J.A. Roman, T.L. Niedzielko, R.C. Haddon, V. Parpura, C.L. Floyd, Single-walled carbon nanotubes chemically functionalized with polyethylene glycol promote tissue repair in a rat model of spinal cord injury. J. Neurotrauma **28**, 2349–2362 (2011)
58. H. Shao, T. Li, R. Zhu, X. Xu, J. Yu, S. Chen, L. Song, S. Ramakrishna, Z. Lei, Y. Ruan, L. He, Carbon nanotube multilayered nanocomposites as multifunctional substrates for actuating neuronal differentiation and functions of neural stem cells. Biomaterials **175**, 93–109 (2018)
59. S.Y. Park, D.S. Choi, H.J. Jin, J. Park, K.E. Byun, K.B. Lee, S. Hong, Polarization-controlled differentiation of human neural stem cells using synergistic cues from the patterns of carbon nanotube monolayer coating. ACS Nano **5**, 4704–4711 (2011)
60. S.Y. Park, J. Park, S.H. Sim, M.G. Sung, K.S. Kim, B.H. Hong, S. Hong, Enhanced differentiation of human neural stem cells into neurons on graphene. Adv. Mater. **23**, H263–H267 (2011)
61. R. Meamar, L. Dehghani, M. Ghasemi, F. Khorvash, V. Shaygannejad, Stem cell therapy in stroke: a review literature. Int. J. Prev. Med. **4**, S139 (2013)

62. P. Dibajnia, C.M. Morshead, Role of neural precursor cells in promoting repair following stroke. Acta Pharmacol. Sin. **34**, 78–90 (2013)
63. K. Yang, J. Lee, J.S. Lee, D. Kim, G.E. Chang, J. Seo, E. Cheong, T. Lee, S.W. Cho, Graphene oxide hierarchical patterns for the derivation of electrophysiologically functional neuron-like cells from human neural stem cells. ACS Appl. Mater. Interf. **8**, 17763–17774 (2016)
64. A. Magaz, X. Li, J.E. Gough, J.J. Blaker, Graphene oxide and electroactive reduced graphene oxide-based composite fibrous scaffolds for engineering excitable nerve tissue. Mater. Sci. Eng. C **119**, 111632 (2021)
65. K.A. Birdwell, M. Park, Post-transplant cardiovascular disease. Clin. J. Am. Soc. Nephrol. **16**, 1878–1889 (2021)
66. A. Ul Haq, F. Carotenuto, P. Di Nardo, R. Francini, P. Prosposito, F. Pescosolido, F. De Matteis, Extrinsically conductive nanomaterials for cardiac tissue engineering applications. Micromachines **12**, 914 (2021)
67. L. Scott, I. Jurewicz, K. Jeevaratnam, R. Lewis, Carbon nanotube-based scaffolds for cardiac tissue engineering—systematic review and narrative synthesis. Bioengineering **8**, 80 (2021)
68. K.N. Alagarsamy, S. Mathan, W. Yan, A. Rafieerad, S. Sekaran, H. Manego, S. Dhingra, Carbon nanomaterials for cardiovascular theranostics: Promises and challenges. Bioactive Mater. **6**, 2261–2280 (2021)
69. S.R. Shin, S.M. Jung, M. Zalabany, K. Kim, P. Zorlutuna, S.B. Kim, M. Nikkhah, M. Khabiry, M. Azize, J. Kong, K.T. Wan, Carbon-nanotube-embedded hydrogel sheets for engineering cardiac constructs and bioactuators. ACS Nano **7**, 2369–2380 (2013)
70. H. Elkhenany, L. Amelse, A. Lafont, S. Bourdo, M. Caldwell, N. Neilsen, E. Dervishi, O. Derek, A.S. Biris, D. Anderson, M. Dhar, Graphene supports in vitro proliferation and osteogenic differentiation of goat adult mesenchymal stem cells: potential for bone tissue engineering. J. Appl. Toxicol. **35**, 367–374 (2015)
71. S. Pok, F. Vitale, S.L. Eichmann, O.M. Benavides, M. Pasquali, J.G. Jacot, Biocompatible carbon nanotube–chitosan scaffold matching the electrical conductivity of the heart. ACS Nano **8**, 9822–9832 (2014)
72. H. Yu, H. Zhao, C. Huang, Y. Du, Mechanically and electrically enhanced CNT—collagen hydrogels as potential scaffolds for engineered cardiac constructs. ACS Biomater. Sci. Eng. **3**, 3017–3021 (2017)
73. M. Izadifar, D. Chapman, P. Babyn, X. Chen, M.E. Kelly, UV-assisted 3D bioprinting of nanoreinforced hybrid cardiac patch for myocardial tissue engineering. Tissue Eng. Part C Methods **24**, 74–88 (2018)
74. M. Cabiati, F. Vozzi, F. Gemma, F. Montemurro, C. De Maria, G. Vozzi, C. Domenici, S. Del Ry, Cardiac tissue regeneration: a preliminary study on carbon-based nanotubes gelatin scaffold. J. Biomed. Mater. Res. B Appl. Biomater. **106**, 2750–2762 (2018)
75. A. Navaei, H. Saini, W. Christenson, R.T. Sullivan, R. Ros, M. Nikkhah, Gold nanorod-incorporated gelatin-based conductive hydrogels for engineering cardiac tissue constructs. Acta Biomater. **41**, 133–146 (2016)
76. R.S. Nair, J.M. Ameer, M.R. Alison, T.V. Anilkumar, A gold nanoparticle coated porcine cholecyst-derived bioscaffold for cardiac tissue engineering. Colloids Surf. B **157**, 130–137 (2017)
77. P. Baei, S. Jalili-Firoozinezhad, S. Rajabi-Zeleti, M. Tafazzoli-Shadpour, H. Baharvand, N. Aghdami, Electrically conductive gold nanoparticle-chitosan thermosensitive hydrogels for cardiac tissue engineering. Mater. Sci. Eng. C **3**, 131–141 (2016)
78. D.R. Amin, E. Sink, S.P. Narayan, M. Abdel-Hafiz, L. Mestroni, B. Peña, Nanomaterials for cardiac tissue engineering. Molecules **25**, 5189 (2020)
79. L. Wang, J. Jiang, W. Hua, A. Darabi, X. Song, C. Song, W. Zhong, M.M. Xing, X. Qiu, Mussel-inspired conductive cryogel as cardiac tissue patch to repair myocardial infarction by migration of conductive nanoparticles. Adv. Func. Mater. **26**, 4293–4305 (2016)
80. W. Wang, J. Chen, M. Li, H. Jia, X. Han, J. Zhang, Y. Zou, B. Tan, W. Liang, Y. Shang, Q. Xu, Rebuilding postinfarcted cardiac functions by injecting TIIA@ PDA nanoparticle-cross-linked ROS-sensitive hydrogels. ACS Appl. Mater. Interf. **11**, 2880–2890 (2018)

81. Y. Chen, X. Li, The utilization of carbon-based nanomaterials in bone tissue regeneration and engineering: respective featured applications and future prospects. Med. Novel Technol. Dev. **16**, 100168 (2022)
82. Y. Zhang, P. Wang, Y. Wang, J. Li, D. Qiao, R. Chen, W. Yang, F. Yan, Gold nanoparticles promote the bone regeneration of periodontal ligament stem cell sheets through activation of autophagy. Int. J. Nanomed. **16**, 61 (2021)
83. Z. Peng, T. Zhao, Y. Zhou, S. Li, J. Li, R.M. Leblanc, Bone tissue engineering via carbon-based nanomaterials. Adv. Healthcare Mater. **95**, 1901495 (2020)
84. X. Li, H. Liu, X. Niu, B. Yu, Y. Fan, Q. Feng, F.Z. Cui, F. Watari, The use of carbon nanotubes to induce osteogenic differentiation of human adipose-derived MSCs in vitro and ectopic bone formation in vivo. Biomaterials **33**, 4818–4827 (2012)
85. S. Liao, G. Xu, W. Wang, F. Watari, F. Cui, S. Ramakrishna, C.K. Chan, Self-assembly of nano-hydroxyapatite on multi-walled carbon nanotubes. Acta Biomater. **3**, 669–675 (2007)
86. A. Oyefusi, O. Olanipekun, G.M. Neelgund, D. Peterson, J.M. Stone, E. Williams, L. Carson, G. Regisford, A. Oki, Hydroxyapatite grafted carbon nanotubes and graphene nanosheets: promising bone implant materials. Spectrochim. Acta Part A Mol. Biomol. Spectrosc. **132**, 410–416 (2014)
87. J. Cheng, J. Liu, B. Wu, Z. Liu, M. Li, X. Wang, P. Tang, Z. Wang, Graphene and its derivatives for bone tissue engineering: in vitro and in vivo evaluation of graphene-based scaffolds, membranes and coatings. Front. Bioeng. Biotechnol. **9**, 734688 (2021)
88. J. Lu, Y.S. He, C. Cheng, Y. Wang, L. Qiu, D. Li, D. Zou, Self-supporting graphene hydrogel film as an experimental platform to evaluate the potential of graphene for bone regeneration. Adv. Func. Mater. **23**, 3494–3502 (2013)
89. J. Lu, C. Cheng, Y.S. He, C. Lyu, Y. Wang, J. Yu, Multilayered graphene hydrogel membranes for guided bone regeneration. Adv. Mater. **28**, 4025–4031 (2016)
90. W. Su, Z. Wang, J. Jiang, X. Liu, J. Zhao, Z. Zhang, Promoting tendon to bone integration using graphene oxide-doped electrospun poly(lactic-co-glycolic acid) nanofibrous membrane. Int. J. Nanomed. **14**, 1835–1847 (2019)
91. V. Babuska, P.B. Kasi, P. Chocholata, L. Wiesnerova, J. Dvorakova, R. Vrzakova, A. Nekleionova, L. Landsmann, V. Kulda, Nanomaterials in bone regeneration. Appl. Sci. **12**, 6793 (2022)
92. Y. Zheng, X. Hong, J. Wang, L. Feng, T. Fan, R. Guo, H. Zhang, 2D nanomaterials for tissue engineering and regenerative nanomedicines: recent advances and future challenges. Adv. Healthcare Mater. **10**, 2001743 (2021)
93. S.Y. Choi, M.S. Song, P.D. Ryu, A.T. Lam, S.W. Joo, S.Y. Lee, Gold nanoparticles promote osteogenic differentiation in human adipose-derived mesenchymal stem cells through the Wnt/β-catenin signaling pathway. Int. J. Nanomed. **10**, 4383 (2015)
94. S. Singh, A. Gupta, I. Qayoom, A.K. Teotia, S. Gupta, P. Padmanabhan, A. Dev, A. Kumar, Biofabrication of gold nanoparticles with bone remodeling potential: an in vitro and in vivo assessment. J. Nanopart. Res. **22**, 1–5 (2020)
95. T. Qing, M. Mahmood, Y. Zheng, A.S. Biris, L. Shi, D.A. Casciano, A genomic characterization of the influence of silver nanoparticles on bone differentiation in MC3T3-E1 cells. J. Appl. Toxicol. **38**, 172–179 (2018)
96. A.S. Algazlan, N. Almuraikhi, M. Muthurangan, H. Balto, F. Alsalleeh, Silver nanoparticles alone or in combination with calcium hydroxide modulate the viability, attachment, migration, and osteogenic differentiation of human mesenchymal stem cells. Int. J. Mol. Sci. **24**, 702 (2022)
97. P. Mehrotra, Biosensors and their applications—a review. J. Oral Biol. Craniofac. Res. **6**, 153–159 (2016)
98. S. Singh, V. Kumar, D.S. Dhanjal, S. Datta, R. Prasad, J. Singh, in *Biological Biosensors for Monitoring and Diagnosis*. Microbial Biotechnology: Basic Research and Applications (2020), pp. 317–335
99. S.A. Pullano, M. Greco, M.G. Bianco, D. Foti, A. Brunetti, A.S. Fiorillo, Glucose biosensors in clinical practice: principles, limits and perspectives of currently used devices. Theranostics **12**, 493 (2022)

100. S.K. Das, K.K. Nayak, P.R. Krishnaswamy, V. Kumar, N. Bhat, Electrochemistry and other emerging technologies for continuous glucose monitoring devices. ECS Sensors Plus. 1 (2022)
101. A.I. Barbosa, R. Rebelo, R.L. Reis, M. Bhattacharya, V.M. Correlo, Current nanotechnology advances in diagnostic biosensors. Med. Dev. Sensors **4**, e10156 (2021)
102. N. Chauhan, K. Saxena, M. Tikadar, U. Jain, Recent advances in the design of biosensors based on novel nanomaterials: an insight. Nanotechnol. Precis. Eng. **4**, 045003 (2021)
103. P. Tetyana, P.M. Shumbula, Z. Njengele-Tetyana, in *Nanopores*. Biosensors: Design, Development and Applications. (IntechOpen, 2021)
104. I. Irkham, A.U. Ibrahim, P.C. Pwavodi, F. Al-Turjman, Y.W. Hartati, Smart graphene-based electrochemical nanobiosensor for clinical diagnosis. Sensors **23**, 2240 (2023)
105. I.H. Cho, D.H. Kim, S. Park, Electrochemical biosensors: Perspective on functional nanomaterials for on-site analysis. Biomater. Res. **24**, 1–2 (2020)
106. F. Schachinger, H. Chang, S. Scheiblbrandner, R. Ludwig, Amperometric biosensors based on direct electron transfer enzymes. Molecules **26**, 4525 (2021)
107. D. Grieshaber, R. MacKenzie, J. Vörös, E. Reimhult, Electrochemical biosensors-sensor principles and architectures. Sensors **8**, 1400–1458 (2008)
108. D.I. Sandoval Bojórquez, Z. Janićijević, B. Palestina Romero, E.S. Oliveros Mata, M. Laube, A. Feldmann, A. Kegler, L. Drewitz, C. Fowley, J. Pietzsch, J. Fassbender, Impedimetric nanobiosensor for the detection of SARS-CoV-2 antigens and antibodies. ACS Sensors **8**, 576–586 (2023)
109. J. Mondal, J.M. An, S.S. Surwase, K. Chakraborty, S.C. Sutradhar, J. Hwang, J. Lee, Y.K. Lee, Carbon nanotube and its derived nanomaterials based high performance biosensing platform. Biosensors **12**, 731 (2022)
110. F. Mo, J. Xie, T. Wu, M. Liu, Y. Zhang, S. Yao, A sensitive electrochemical sensor for bisphenol A on the basis of the AuPd incorporated carboxylic multi-walled carbon nanotubes. Food Chem. **292**, 253–259 (2019)
111. W. Gao, W.W. Tjiu, J. Wei, T. Liu, Highly sensitive nonenzymatic glucose and H_2O_2 sensor based on Ni (OH) 2/electroreduced graphene oxide—multiwalled carbon nanotube film modified glass carbon electrode. Talanta **120**, 484–490 (2014)
112. D. Zhai, B. Liu, Y. Shi, L. Pan, Y. Wang, W. Li, R. Zhang, G. Yu, Highly sensitive glucose sensor based on Pt nanoparticle/polyaniline hydrogel heterostructures. ACS Nano **7**, 3540–3546 (2013)
113. L. Fang, B. Liang, G. Yang, Y. Hu, Q. Zhu, X. Ye, Study of glucose biosensor lifetime improvement in 37 C serum based on PANI enzyme immobilization and PLGA biodegradable membrane. Biosens. Bioelectron. **56**, 91–96 (2014)
114. W. Zhao, R. Zhang, S. Xu, J. Cai, X. Zhu, Y. Zhu, W. Wei, X. Liu, J. Luo, Molecularly imprinted polymeric nanoparticles decorated with Au NPs for highly sensitive and selective glucose detection. Biosens. Bioelectron. **100**, 497–503 (2018)
115. X. Zeng, Y. Zhang, X. Du, Y. Li, W. Tang, A highly sensitive glucose sensor based on a gold nanoparticles/polyaniline/multi-walled carbon nanotubes composite modified glassy carbon electrode. New J. Chem. **42**, 11944–11953 (2018)
116. Y. Zhao, Q. Zhai, D. Dong, T. An, S. Gong, Q. Shi, W. Cheng, Highly stretchable and strain-insensitive fiber-based wearable electrochemical biosensor to monitor glucose in the sweat. Anal. Chem. **91**, 6569–6576 (2019)
117. P. Teengam, W. Siangproh, A. Tuantranont, C.S. Henry, T. Vilaivan, O. Chailapakul, Electrochemical paper-based peptide nucleic acid biosensor for detecting human papillomavirus. Anal. Chim. Acta **952**, 32–40 (2017)
118. H. Altug, S.H. Oh, S.A. Maier, J. Homola, Advances and applications of nanophotonic biosensors. Nat. Nanotechnol. **17**, 5–16 (2022)
119. M. Eswaran, B. Chokkiah, S. Pandit, S. Rahimi, R. Dhanusuraman, M. Aleem, I. Mijakovic, A road map toward field-effect transistor biosensor technology for early stage cancer detection. Small Methods **6**, 2200809 (2022)
120. R. Peltomaa, B. Glahn-Martínez, E. Benito-Peña, M.C. Moreno-Bondi, Optical biosensors for label-free detection of small molecules. Sensors **18**, 4126 (2018)

121. P. Estrela, P. Damborský, J. Švitel, J. Katrlik, Optical biosensors. Essays Biochem. **60**, 91–100 (2016)
122. S. Das, R. Devireddy, M.R. Gartia, Surface plasmon resonance (SPR) sensor for cancer biomarker detection. Biosensors **13**, 396 (2023)
123. B. Bohunicky, S.A. Mousa, Biosensors: the new wave in cancer diagnosis. Nanotechnol. Sci. Appl. 1 (2010)
124. A. Camarca, A. Varriale, A. Capo, A. Pennacchio, A. Calabrese, C. Giannattasio, C. Murillo Almuzara, S. D'Auria, M. Staiano, Emergent biosensing technologies based on fluorescence spectroscopy and surface plasmon resonance. Sensors **21**, 906 (2021)
125. H. Yuan, W. Ji, S. Chu, S. Qian, F. Wang, J.F. Masson, X. Han, W. Peng, Fiber-optic surface plasmon resonance glucose sensor enhanced with phenylboronic acid modified Au nanoparticles. Biosens. Bioelectron. **117**, 637–643 (2018)
126. R.P. Potdar, Y.B. Khollam, S.F. Shaikh, P.S. More, A.U. Rana, Polyvinylpyrrolidone-capped silver nanoparticles for highly sensitive and selective optical fiber-based ammonium sensor. Nanomaterials **12**, 3373 (2022)
127. S. Vasuki, V. Varsha, R. Mithra, R.A. Dharshni, S. Abinaya, R.D. Dharshini, N. Sivarajasekar, Thermal biosensors and their applications. Am. Int. J. Res. Sci. Tech. Eng. Math. 262–264 (2019)
128. P. Mohankumar, J. Ajayan, T. Mohanraj, R. Yasodharan, Recent developments in biosensors for healthcare and biomedical applications: a review. Measurement **167**, 108293 (2021)
129. Q.Y. Ren, J.Q. Huang, L.F. Wang, S. Wan, L.T. Sun, Q.A. Huang, Temperature sensing properties of the passive wireless sensor based on graphene oxide films. IEEE Sensors **2014**, 432–435 (2014)
130. L. Dan, A.L. Elias, Flexible and stretchable temperature sensors fabricated using solution-processable conductive polymer composites. Adv. Healthcare Mater. **9**, 2000380 (2020)
131. X. Gong, L. Zhang, Y. Huang, S. Wang, G. Pan, L. Li, Directly writing flexible temperature sensor with graphene nanoribbons for disposable healthcare devices. RSC Adv. **10**, 22222–22229 (2020)
132. M. Pohanka, Overview of piezoelectric biosensors, immunosensors and DNA sensors and their applications. Materials **11**, 448 (2018)
133. H.J. Lim, T. Saha, B.T. Tey, W.S. Tan, C.W. Ooi, Quartz crystal microbalance-based biosensors as rapid diagnostic devices for infectious diseases. Biosens. Bioelectron. **168**, 112513 (2020)
134. L.K. Kiio, D.N. Mbui, P.M. Ndangili, O.J. Onam, F. Oloo, Current biosensors used for early detection of lung cancer biomarkers. J. Cancer Res. Rev. Rep. **3**, 148 (2021)
135. G. Marrazza, Piezoelectric biosensors for organophosphate and carbamate pesticides: a review. Biosensors **4**, 301–317 (2014)
136. F. Narita, Z. Wang, H. Kurita, Z. Li, Y. Shi, Y. Jia, C. Soutis, A review of piezoelectric and magnetostrictive biosensor materials for detection of COVID-19 and other viruses. Adv. Mater. **33**, 2005448 (2021)
137. L. Lin, H.Q. Zhao, J.R. Li, J.A. Tang, M.X. Duan, L. Jiang, Study on colloidal Au-enhanced DNA sensing by quartz crystal microbalance. Biochem. Biophys. Res. Commun. **274**, 817–820 (2000)
138. X.C. Zhou, S.J. O'Shea, S.F. Li, Amplified microgravimetric gene sensor using Au nanoparticle modified oligonucleotides. Chem. Commun. **11**, 953–954 (2000)
139. H.Q. Zhao, L. Lin, J.R. Li, J.A. Tang, M.X. Duan, L. Jiang, DNA biosensor with high sensitivity amplified by gold nanoparticles. J. Nanopart. Res. **3**, 321–323 (2001)
140. T. Liu, L. Jiang, Sensitivity enhancement of DNA sensors by nanogold surface modification. Biochem. Biophys. Res. Commun. **295**, 14–16 (2002)
141. T. Liu, J.A. Tang, L. Jiang, The enhancement effect of gold nanoparticles as a surface modifier on DNA sensor sensitivity. Biochem. Biophys. Res. Commun. **313**, 3–7 (2004)
142. G. Premaratne, Z.H. Al Mubarak, L. Senavirathna, L. Liu, S. Krishnan, Measuring ultra-low levels of nucleotide biomarkers using quartz crystal microbalance and SPR microarray imaging methods: a comparative analysis. Sens. Actuators B Chem. **253**, 368–375 (2017)

143. S.H. Chen, Y.C. Chuang, Y.C. Lu, H.C. Lin, Y.L. Yang, C.S. Lin, A method of layer-by-layer gold nanoparticle hybridization in a quartz crystal microbalance DNA sensing system used to detect dengue virus. Nanotechnology **20**, 215501 (2009)
144. N. Kim, D.K. Kim, Y.J. Cho, Gold nanoparticle-based signal augmentation of quartz crystal microbalance immunosensor measuring C-reactive protein. Curr. Appl. Phys. **10**, 1227–1230 (2010)
145. M. Sanati, A. Sandwell, H. Mostaghimi, S.S. Park, Development of nanocomposite-based strain sensor with piezoelectric and piezoresistive properties. Sensors **18**, 3789 (2018)
146. H.H. Nguyen, S.H. Lee, U.J. Lee, C.D. Fermin, M. Kim, Immobilized enzymes in biosensor applications. Materials **12**, 121 (2019)
147. C.A. Cordeiro, A. Sias, T. Koster, B.H. Westerink, T.I. Cremers, In vivo "real-time" monitoring of glucose in the brain with an amperometric enzyme-based biosensor based on gold coated tungsten (W-Au) microelectrodes. Sens. Actuators B Chem. **263**, 605–613 (2018)
148. B. Öndeş, F. Akpınar, M. Uygun, M. Muti, D.A. Uygun, High stability potentiometric urea biosensor based on enzyme attached nanoparticles. Microchem. J. **160**, 105667 (2021)
149. H.W. Schroeder Jr., L. Cavacini, Structure and function of immunoglobulins. J. Allergy Clin. Immunol. **125**, S41-52 (2010)
150. C. Cristea, A. Florea, M. Tertiş, R. Săndulescu, in Biosensors-Micro and Nanoscale Applications. Immunosensors (IntechOpen, 2015), p. 468
151. E. Arkan, R. Saber, Z. Karimi, M. Shamsipur, A novel antibody–antigen based impedimetric immunosensor for low level detection of HER2 in serum samples of breast cancer patients via modification of a gold nanoparticles decorated multiwall carbon nanotube-ionic liquid electrode. Anal. Chim. Acta **874**, 66–74 (2015)
152. A. Thapa, A.C. Soares, J.C. Soares, I.T. Awan, D. Volpati, M.E. Melendez, J.H. Fregnani, A.L. Carvalho, O.N. Oliveira Jr., Carbon nanotube matrix for highly sensitive biosensors to detect pancreatic cancer biomarker CA19-9. ACS Appl. Mater. Interf. **9**, 25878–25886 (2017)
153. G. Tiwari, R. Tiwari, B. Sriwastawa, L. Bhati, S. Pandey, P. Pandey, S.K. Bannerjee, Drug delivery systems: an updated review. Int. J. Pharm. Investig. **2**, 2 (2012)
154. S.A. Ealia, M.P. Saravanakumar, A review on the classification, characterisation, synthesis of nanoparticles and their application. IOP Conf. Ser. Mater. Sci. Eng. **263**, 032019 (2017) (IOP Publishing)
155. S.M. Asil, J. Ahlawat, G.G. Barroso, M. Narayan, Nanomaterial based drug delivery systems for the treatment of neurodegenerative diseases. Biomater. Sci. **8**, 4109–4128 (2020)
156. S. Sim, N.K. Wong, Nanotechnology and its use in imaging and drug delivery. Biomed. Rep. **14**, 1–9 (2021)
157. W. Zhu, Z. Wei, C. Han, X. Weng, Nanomaterials as promising theranostic tools in nanomedicine and their applications in clinical disease diagnosis and treatment. Nanomaterials **11**, 3346 (2021)
158. E.N. De Carvalho Lima, J.R. Piqueira, D.A. Maria, Advances in carbon nanotubes for malignant melanoma: a chance for treatment. Mol. Diagn. Ther. **22**, 703–715 (2018)
159. L. García-Hevia, J.C. Villegas, F. Fernández, Í. Casafont, J. González, R. Valiente, M.L. Fanarraga, Multiwalled carbon nanotubes inhibit tumor progression in a mouse model. Adv. Healthcare Mater. **5**, 1080–1087 (2016)
160. S. Hopkins, M.K. Gottipati, V. Montana, E. Bekyarova, R.C. Haddon, V. Parpura, Effects of chemically-functionalized single-walled carbon nanotubes on the morphology and vitality of D54MG human glioblastoma cells. Neuroglia **1**, 327–338 (2018)
161. C. Wei, X. Dong, Y. Zhang, J. Liang, A. Yang, D. Zhu, T. Liu, D. Kong, F. Lv, Simultaneous fluorescence imaging monitoring of the programmed release of dual drugs from a hydrogel-carbon nanotube delivery system. Sens. Actuators B Chem. **273**, 264–275 (2018)
162. R. Hindumathi, M. Jagannatham, P. Haridoss, C.P. Sharma, Novel nano-cocoon like structures of polyethylene glycol–multiwalled carbon nanotubes for biomedical applications. Nano-Struct. Nano-Objects **13**, 30–35 (2018)
163. J. Akhtari, R. Faridnia, H. Kalani, R. Bastani, M. Fakhar, H. Rezvan, A.K. Beydokhti, Potent in vitro antileishmanial activity of a nanoformulation of cisplatin with carbon nanotubes against Leishmania major. J. Global Antimicrob. Resist. **16**, 11–16 (2019)

164. Z. Liu, J.T. Robinson, X. Sun, H. Dai, PEGylated nanographene oxide for delivery of water-insoluble cancer drugs. J. Am. Chem. Soc. **130**, 10876–10877 (2008)
165. L. Feng, S. Zhang, Z. Liu, Graphene based gene transfection. Nanoscale **3**, 1252–1257 (2011)
166. Y.H. Park, S.Y. Park, I. In, Direct noncovalent conjugation of folic acid on reduced graphene oxide as anticancer drug carrier. J. Ind. Eng. Chem. **30**, 190–196 (2015)
167. J. Song, X. Yang, O. Jacobson, L. Lin, P. Huang, G. Niu, Q. Ma, X. Chen, Sequential drug release and enhanced photothermal and photoacoustic effect of hybrid reduced graphene oxide-loaded ultrasmall gold nanorod vesicles for cancer therapy. ACS Nano **9**, 9199–9209 (2015)
168. X. Ma, Q. Qu, Y. Zhao, Z. Luo, Y. Zhao, K.W. Ng, Y. Zhao, Graphene oxide wrapped gold nanoparticles for intracellular Raman imaging and drug delivery. J. Mater. Chem. B **1**, 6495–6500 (2013)
169. D.S. Perera, R.C. De Silva, L.D. Nayanajith, H.C. Colombage, T.S. Suresh, W.P. Abeysekera, I.R. Kottegoda, Anti-inflammatory and antioxidant properties of Coffea arabica/reduced graphene oxide nanocomposite prepared by green synthesis. Mat. Sci. Res. India **18**, 305–317 (2021)

Therapeutic Approaches for Diabetes Mellitus by Using Metal Oxide Nanoparticles

Subhaschandrabose Jeyabharathi, Suresh Naveenkumar, Pandian Chithamparam, Narayanan Venkateshan, K. Kaviyarasu, and Azhaguchamy Muthukumaran

Abstract Diabetes mellitus (DM) is a growing health issue, all over the world (~8.5%) 422 million in 2014. Diabetes is a group of metabolic disorders characterized by high blood glucose levels and is one of the top five causes of mortality. The main reason of hyperglycemia-induced diabetic problems is an increased level of oxidative stress and the low levels of metal ions, particularly zinc. Zinc (Zn) is a vital trace element that acts as homeostasis Process in the human body. Indeed, zinc an essential micronutrient is associated with more than 300 metalloenzymes, and it plays a crucial role in various biological processes involving glucose metabolism. There is a correlation between the levels of zinc and diabetes-related diseases. The herbal plant extracts have been used successfully in extracellular biosynthesis of metal oxides nanoparticles in therapeutic applications in diabetes. This chapter explained the role of zinc and zinc oxide nanoparticles in the management of diabetes. Moreover, the focus is on the green synthesis of zinc oxide nanoparticles (ZnO NPs) from various sources, zinc and diabetes, the pathophysiology of zinc and diabetes, the correlation between zinc, insulin, diabetes, and zinc in diabetes associated with diseased conditions.

S. Jeyabharathi · S. Naveenkumar · P. Chithamparam · A. Muthukumaran (✉)
Department of Biotechnology, Kalasalingam Academy of Research and Education, Krishnankoil, Tamil Nadu, India
e-mail: a.muthukumaran@klu.ac.in

S. Jeyabharathi
Department of Microbiology, The Standard Fireworks Rajaratnam College for Women, Sivakasi, Tamil Nadu 626123, India

N. Venkateshan
Department of Pharmaceutical Chemistry, Arulmigu Kalasalingam College of Pharmacy, Krishnankoil, Tamil Nadu, India

Present Address:
K. Kaviyarasu
UNESCO-UNISA Africa Chair in Nanoscience's/Nanotechnology Laboratories, College of Graduate Studies, University of South Africa (UNISA), Muckleneuk Ridge, PO Box 392, Pretoria, South Africa

K. Kasinathan et al. (eds.), *New Technologies for Energy Transition Based on Sustainable Development Goals*, https://doi.org/10.1007/978-981-97-2527-4_10

1 Introduction

Diabetes mellitus (DM) is a hormonal disorder affecting over 6% of the world's population. It is activated by the pancreas, followed by a deficiency of insulin production, which increases or decreases glucose concentrations in the blood. It damages blood vessels, eyes, kidneys, hearts, and nerves [1]. Moreover, chronic diabetes occurs when the pancreas cannot make enough insulin, or the body does not react to insulin as it should. There are essentially two types of it to consider. Type 1 diabetes (T1DM) patients must inject insulin daily to maintain their blood glucose levels. T1DM has no known cause, and no treatment or cure is available [2]. Excessive urination, loss of weight, and vision problems are common indications of type 2 diabetes. Depending on the cause, diabetes is characterized by either insufficient insulin production or insulin resistance in the cells. When the pancreas is destroyed, the body's glucose levels are elevated, damaging many organs [3, 4]. In addition, several side effects of diabetes medications, such as blindness, kidney disease, nerve damage, heart disease, amputations of feet, and stroke, have been reported [5].

As a result, natural compounds, particularly those derived from herbs, are increasingly considered diabetes medications since they are less toxic and have fewer adverse effects than synthetic medications [6]. When the plasma membrane becomes depolarized, this results in the influx of extracellular Ca^{2+}, the stimulation of insulin-containing vesicle exocytosis, and insulin secretion [7]. Diabetic cardiomyopathy, on the other hand, is described as abnormal cardiac structure and execution without other cardiac risk factors, such as coronary artery disease, high blood pressure, and significant valve disease [8]. When regulating diabetes, anti-hyperglycemic drugs are essential because they effectively lessen microvascular issues while suppressing renal dysfunction, retinopathy, and nerve deterioration. Surprisingly, several medications now in use have the potential to improve heart health in ways that go beyond their ability to control blood sugar [9]. Also stressed was the relationship between the study of antibacterial activity and the investigation of anti-diabetic activity. As a result, diabetic people have contracted severe and life-threatening infections. At the same time, some foreign invaders (such as bacteria) are introduced into the diabetic host (patients), reducing the immunocompromised state of the patients who would otherwise have a damaged immune system. Drugs derived from plants are readily available, less expensive, safe, effective, and have fewer adverse effects than synthetic drugs. It is being done so that plants with traditional therapeutic use can be screened. In nature, plants are a rich source of natural antioxidants and antimicrobials, many of which can be used as drugs in modern biomedicine. In recent years, green synthesis, which involves medicinal plant extracts, has gained particular attention for its application in the formation of nanoparticles. Phytochemicals, in particular, are the livelihood of plants and can be used to generate less hazardous nanoparticles [10]. It is also claimed that many plants have anti-hyperglycemic actions because they may raise insulin secretion, restrict glucose absorption in the intestine, and facilitate insulin-dependent metabolites [11]. Plant parts that include alkaloids, glycosides, terpenoids, and flavonoids are commonly used as anti-diabetic medications [12].

Nanotechnology research has gained traction in recent years with new biomedical, materials science, optical, and electronics solutions. A fundamental characteristic of nanoparticles is that they are a diverse form of vital elements created by altering their atomic and molecular characteristics.

2 Nanobiotechnology—A Modern Era of Nanotechnology

There is great potential for the fast-growing field of nanotechnology in environmental management, agriculture, energy, and health care. To cope with this enormous demand, the production of nanoparticles of a given shape, size, and composition has emerged as the most promising topic of research in nanotechnology [13, 14]. When it comes to the coupling of nanotechnology and the study of biology, the phrases nanobiotechnology, bio-nanotechnology, and nanobiology have all developed to describe the phenomenon. It is the topic of numerous experiments to determine whether using biologically orientated materials as reducing agents for metallic nanoparticle synthesis is of interest.

Using microbial agents in producing non-toxic, clean, and environmentally friendly nanostructured materials will be possible for various industrial applications, including biomedicine. The ability to overcome the size control of nanoparticles with homogeneous shapes and their physicochemical properties are critical requirements for a biologically mediated approach to nanoparticle synthesis. The current research is focused on developing methods to address the synthesizing nanoparticles with a polydispersity nature [15]. Some experiments have been conducted to synthesize nanoparticles using biomolecules from diverse microbes and other sources. Plant extracts, bacterial metabolites, and fungi are used to synthesize metal nanoparticles. Several advantages are realized, including an environmentally friendly approach, a low cost of production medium, a reduced time requirement, and biocompatibility, allowing them to be used in pharmaceutical and biomedical applications [16]. As a result, there is an need to create a process for synthesizing nanoparticles that is with non-toxic, low-cost, sustainable, biocompatible and eco-friendly [17].

3 Nanoparticles

Nanotechnology has been regarded as a revolutionary, cutting-edge technique of multidisciplinary research toward the size regulation of the substance at the molecular level. Richard Feynman's pioneer stated the main notion of nanotechnology in a presentation authorized. There's plenty of room at the bottom in 1959 and the Nano is a Greek term meaning dwarf, small, which describes the size of one billion (10^{-9} m). The concept was introduced and invented by Taniguchi et al. [18]. It is measured as a comprehensive understanding of the synthesis, characterization, and application of nanomaterials, and they are the most significant parts that originate underneath the nanotechnology umbrella [19]. The nanoparticles within the 100 nm or less range draw much attention because of their unusual and fascinating features and wide applications. Nanoparticles are synthesized using physical, chemical, and biological methods [20]. However, their applicability in biomedicine has been limited due to the excessive use of harmful chemicals. As a result, it is of great importance to the researchers to create a dependable, environment friendly, and non-toxic process of nanoparticle production that may be used for biomedical applications to advance their study. When it comes to achieving this goal, microbial-mediated methods have significant advantages in nanoparticle synthesis [21].

3.1 Metal and Metal Oxide Nanoparticles

The exceptional properties of metallic nanoparticles, which are not typically observed in bulk form, have been demonstrated to exist in nanoparticle form. Because of their innovative and distinct physical, chemical, and biological features [22], inorganic nanoparticles are used due to their enormous potential and advantages in nanotechnology as well as metal nanostructure materials in biological applications [23].

Nanoparticle research is a hive of scientific activity, owing to various possible applications in multiple sectors. Metal oxide nanomaterials have an adverse range of potential applications due to their rich compositional chemistry, which includes catalysis, energy storage, electronics, and optics, to mention a few examples [24]. These functional qualities an achieved by interacting size and shape with morphology, crystal structure, and surface chemistry. Recent years have seen a significant advance in our understanding of these structure–property function connections, with insights from in situ research and novel synthetic developments opening new avenues for developing new materials [25]. The electronic form can exhibit semiconductor, metallic, or insulator properties depending on its electronic structure. The enhanced density of metal oxide nanoparticles is responsible for their distinctive chemical and physical characteristics [26]. Significant applications of metal oxide nanoparticles have a critical property in attracting scientists in materials chemistry, medical technology, agriculture, information technology, biomedicine, optical technology, electronics, catalysis, the environment, and energy. Variations in the size-based and structural modifications of the cells produce differences in the cellular contents. The increase in surface connection between atoms in nanoparticles such as SnO_2, Al_2O_3, MgO_2, ZnO, AgO, TiO_2, CeO_2, and other oxides causes strain or stress and the accompanying physical disturbances [27] in Fig. 1, respectively.

The electrical structure of nanoparticles can show whether they are a semiconductor, a metal, or an insulator, depending on their composition. The high density of corners and edges on the surface of various nanoparticles and the limited size of corners and edges are responsible for their distinct chemical and physical features [28, 29]. Plant extracts have been employed successfully [30] in several studies to achieve fast and extracellular production of noble metal nanoparticles. It is classified as a member of the Costaceae family [31–33] and is often referred to as the insulin

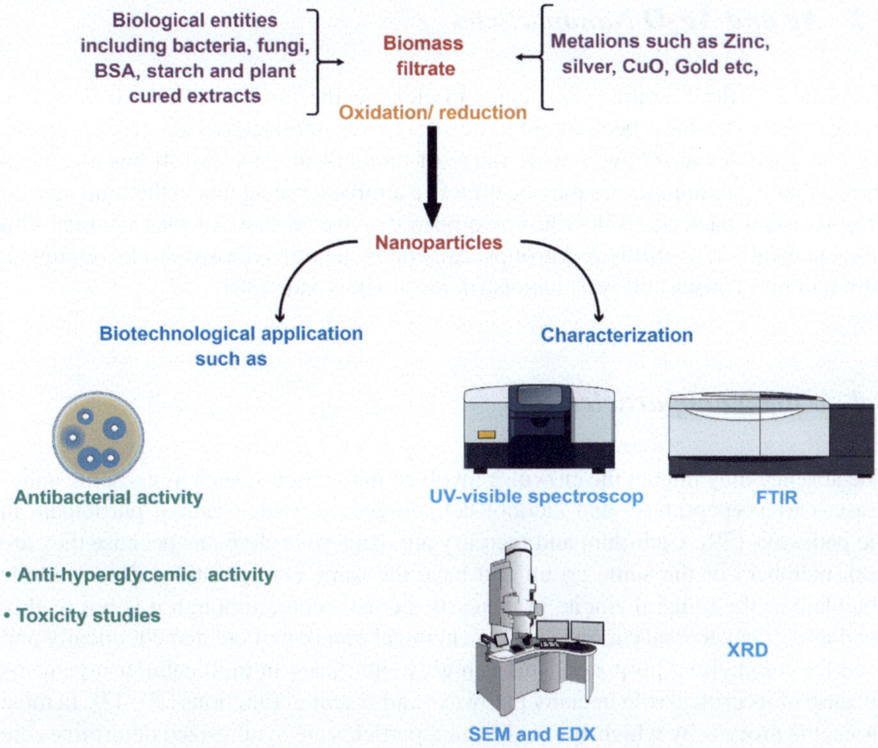

Fig. 1 Biological mediated synthesis of nanoparticles and their characterization process

plant in India. According to some sources, the intake of the entire plant *Costus igneus* can help to reduce blood glucose levels. In the Costaceae family, several plants have a biological activity employed in nutrition and alternative medicine [34]. Persons with diabetes are advised to consume two leaves daily and in the evening. Once the weekend arrives, the number of leaves consumed should be reduced to one in the morning and one in the evening. This quantity must be adhered to for 30 days. Diabetes eating dietary supplements [32, 35] is taking shape as a result of this transformation. Despite widespread use, there is no formulation including this herb available in the bazaar. For the first 30 days, the suffering individual has difficulty chewing the leaves. An effort is being made in this study to design and standardize a nano-formulation of *Costus igneus* to overcome this and to improve the oral bioavailability and pharmacological activity, protection from toxicity, physiochemical degeneration, increase tissue macrophages, avoid recurrent administration of the dosage, and achieve targeted drug delivery to specific tissues or organs.

3.2 Ag and Ag_2O Nanoparticles

According to the literature, Ag nanoparticles are the most well-known inorganic nanoparticles that have been found to be effective as antibacterial agents in humans. Ag nanoparticles also have a wide range of biomedical uses [36]. It has also been shown that Ag nanoparticles may be effective antibacterial agents in the fight against drug-resistant bacteria [37]. It has also been discovered that Ag ions interact with enzyme disulfide or sulfhydryl groups, causing metabolic pathways to be disrupted, which in turn causes cell lysis nanoparticles of zinc oxide [38].

3.3 ZnO Nanoparticles

The absence may impact the enzymes involved in the routes, such as carbonic anhydrase, carboxypeptidase, and alcohol dehydrogenase, which cannot participate in the pathways [39]. Cadmium and mercury are dangerous elements because they are both members of the same group and have the same configuration. Zinc oxide is abundant as the mineral zincite in the earth's crust's core, although it is not readily available. It is necessary to use ZnO, a chemical compound created chemically and used for commercial purposes. This is highly significant in multicellular organisms because of its critical role in many pathways and essential functions [40–43]. In most cases, the process by which zinc oxide nanoparticles are synthesized determines the shape of the particles. The nanostructures may be in rods, plates, nanospheres, boxes, hexagonal shapes, tripods, and tetrapods, among other shapes. They can also be in the form of nanowires, nanotubes, nanorings, nanocages, and nanoflowers. According to the findings, ZnO nanoparticles were efficient against gram-positive bacteria when used with other nanoparticles from the same element group. ZnO benefits the human body's system in quantities of up to 100 g/ml. They can be used as an antibiotic supplement in some situations. When bacteria colonies were exposed to a concentration of 500–1000 mg/ml of ZnO over 6 hrs, 90% of the bacteria colonies died [44]. Because they have an average size of 480 nm, they are less toxic than silver nanoparticles in the 20–100 mg/ml concentration range. They are less damaging than silver nanoparticles because they have a smaller average size. Due to diffusion, metal oxides are lysed in the bacterial cells when they contact them [45].

3.4 CuO Nanoparticles

CuO nanoparticles are a form of nanoparticle that may be found in nature. Capacitors, near-infra-red filters in magnetic to range, sensors, catalysis, and semiconductors are only a few of the applications for which it has been developed [46].

Cobalt oxide is a semiconducting metallic compound with distinct optical, electrical, and magnetic characteristics. Copper oxide (CuO), one of the numerous Cu-based minerals, is one of the simplest and industrially significant semiconductor substances. Copper oxide is one of the most straightforward and industrially important semiconductor substances. With small bandgap electricity of ~1.2 eV, it belongs to the monoclinic form system and is considered one of the most significant semiconductor compounds. CuO nanostructures of various sizes and morphologies have been organized using several techniques, including chemical reactions, solvothermal reactions, hydrothermal reactions, thermal decompositions, electrochemical reactions, and sol–gel reactions for various applications. Future applications of curcumin-based nanoparticles (CuO NPs) in the biomedical area will primarily focus on detecting disease states. CuONPs can be used in multiple practical applications spanning a broad range of sectors, such as detecting viruses in the body [21].

4 Biosynthesis of Nanoparticles

Recently, it has been discovered that a diverse spectrum of microbes may create nanoparticles. Many advanced techniques and the applicability of biologically produced nanostructures are now being investigated. Silver and gold are the most studied metal nanoparticles among the various metal nanoparticles because they are potentially valuable for biomedical treatment [21]. Unicellular and multicellular organisms are being used to synthesize various metal oxide nanoparticles, and they are both considered environmentally beneficial [47]. Bacteria and fungi are among the most widely studied and sorted kinds of microorganisms because of their ability to produce valuable reducing agents used in manufacturing various metal nanoparticles [48]. Scientists have observed the interplay between living beings and inorganic objects for as long as there has been life on the planet. The different microorganisms to generate diverse organic nanoparticles via either intracellular or extra cellular pathways [49].

5 Mechanism for Nanoparticles Synthesis

No precise method for nanoparticle synthesis, including various biological agents, has been discovered or proposed [50]. It is currently unknown what the mechanism is underlying the intra- and extracellular-mediated composition of diverse nanoparticles, mainly because the reducing agents used vary from one organism to another. Several microorganisms have been found to generate extracellular reductase enzymes, which have been shown to decrease silver ions to the nanometer scale. The enzyme NADH-dependent reductase converts silver ions into silver nanoparticles (AgNPs). While the enzyme receives electrons from NADH, which is then oxidized to NAD+, it also becomes oxidized during the creation of nano-silver. The nitrate-dependent reductase was found to be the primary cause of the bio-reduction process in most cases [51]. Using the intracellular way of nanoparticle manufacturing, a distinct type of ion transport occurs within the microbial cell, with the cell wall, in particular, playing a pivotal role [52]. The verticillium species was used to produce nanoparticles, a sequential procedure that includes trapping, metal ion reduction and capping as steps in manufacturing nanoparticles within cells [53].

6 Green Approach of ZnO Nanoparticles

It was discovered that employing different anti-diabetic medicinal plants has biomedical applications. It was possible to synthesize ZnO nanoparticles through a process known as biosynthesis. Green nanotechnology is an environmentally responsible and cost-effective approach to biocompatible, safe, and environmentally friendly nanotechnology [54]. Biological mediated production of ZnO nanoparticles on a wide scale without introducing extra contaminants. Despite this, nanoparticles made via a biomimetic technique exhibit higher catalytic activity and help to reduce the use of expensive and harm full chemical compounds. The plant extracts produce certain phytochemicals that are anti-inflammatory, capping and stabilizing agents, and reducing and capping agent.

6.1 Green Synthesis from Fungi and Microalgae

Agarwal et al. [55] demonstrated that synthesizing zinc oxide nanoflowers of uniform size from *Bacillus licheniformis* showed higher photocatalytic activity and photostability. The degradation of methylene blue pollutant dye by 83% demonstrated the photostability of ZnO nanoflowers produced [56]. Several studies have found that ZnO nanoparticles are spherical or hexagonal and have a mass between 1.2 and 6.8 nm [57]. Their diameters have been measured between 1.2 and 6.8 nm [58]. The nanoparticles were stable for three minutes. The hydrodynamic diameter was measured using the particle size that had been determined. When aggregation of the nanoparticles was seen after three months of storage, it was concluded that the stability of nanoparticles derived from *Aspergillus fumigateshadim* proved [59]. After six months of storage, the ZnO nanoparticles made from these weed *Sargassum myriocystum* (microalgae) showed no degradation, demonstrating the highly stable condition of the 36 nm-sized nanoparticles [60]. A specific fucoidan pigment derived from microalgae controls the reduction in size and stability of the nanoparticles.

6.2 Green Synthesis from Plants

Plant extract use in ZnO production has several advantages, including producing functional metabolites that reduce metal ions [61], which benefit human health. Various plant primary chemical components produce secondary metabolites [62–64]. For example, ZnO nanoparticles synthesizing from the plant of *Aloe barbadensis*, *Abutilon indicum*, *Meliaa zedarach*, *Indigo feratinctoria*, bacteria, and fungi, among other plants and plant extracts. ZnONPs are made naturally using seaweed extracts from the Nigella sativa plant. In case the leaves of *P. sidiumgujava*, *Daucus carota*, *Monsoni aburkeana* plant extract, *Mangifera indica* leaf extract, *Aspergillus terreus* extract, and *Halomonas elongate* are used, the results will be positive. The antibacterial efficacy of zinc oxide nanoparticles is directly proportional to their size, shape, surface-to-volume ratio, and the amount of oxygen vacancy on their surfaces [65].

Even though the formation of ZnO NPs has been demonstrated, the possibility of significant production from the plant's extracts should be thoroughly investigated. The researchers will use leaf extracts from *Amaranthus caudatus*, *Costus igneus*, *Eugenia jambolana*, and *Wattakaka volbulis* to green synthesize zinc oxide nanoparticles. The nanoparticles tested for antibacterial efficacy against pathogenic bacteria *Staphylococcus epidermidis* and *Enterobacter aerogenes*. Plant components are used to synthesize nanoparticles because their extracts are rich in phytochemicals that function as both reducing and stabilizing agents [65, 66]. Nanoparticles generated from *Trifolium pratense* flower extract displayed identical peaks in the ultraviolet–visible spectrum when measured after 24, 48, 72, 96, and 120 hrs of nanoparticle synthesis, indicating that the nanoparticles were stable after being created. Similarly, FTIR analyses revealed that the fruit extract of *Rosa canina* was a reducing and stabilizing agent for produced ZnO NPs, which was confirmed by the study results [67].

6.3 Polymer-Based ZnONPs (Starch)

Regarding macromolecules, herbal polymers are a kind of macromolecules that can be employed as templates in zinc in oxide nanoparticle production. In the case of nanoparticles, the diameter of the particles can be logically controlled when these polymers are utilized as a capping/stabilizing agent [68, 69]. Researchers have attempted to use a variety of natural ligands to prevent ZnONPs from undergoing their current phase of burst and aggregation. To control the size and shape of nanoparticles, polyethylene glycol (PEG), cetyl tri methyl ammonium bromide (CTAB), oleic acid, and gelatin have been utilized as polymerization agents, according to the literature [70]. These substances are no longer responsible for particle nucleation and growth but also for the coagulation and flocculation of the particle populations. Due to low cost, biodegradability, nontoxicity, biocompatibility, and abundant natural availability [71], polysaccharides have prompted their recent use as hydrogels. Starch is a non-toxic, biodegradable, and biocompatible polymer now found in nature as the primary storage polysaccharide in developed plants. It is also used in food production. A mixture of two poly glucans, amylopectin and amylose, are found in the natural starch we consume. In contrast to other carbohydrates, starch has a specific chemical reactivity due to the reactivity of the hydroxy glucose units, which distinguishes it from the others. Starch is a semi-crystalline polymer made of polysaccharides such as amylose and amylopectin [71], which are both sugars and are found in high concentrations in starches. Because unprocessed native starches are structurally fragile and functionally confined in today's state-of-the-art technologies [72], it is necessary to build a set of features in unprocessed native starches [73]. Zhang et al. [43] employed starch to assemble ZnO powder, and they were successful in obtaining ZnO spheres with diameters ranging from 0.2 to 0.4 μm when heated to 500 °C. Since then, several researchers have structured and investigated nanocomposites predominately composed of oxidized starch in recent years [74].

7 Biological Fate of Zinc Oxide Nanoparticles

The absorption of nanoparticles and their disintegration within the cell is essential for organic activity. We had previously stated that when zinc oxide nanoparticles (ZON) were dialyzed in the simulated stomach and intestinal pH buffers, 55% of the dissolution was confirmed. In comparison, 45% remained in the particulate form [75]. Because of these findings, it was predicted that tissues would come into contact with both ionic zinc and particle ZON in an in vivo environment. The presence of the particulate element should result in the progressive and non-stop release of zinc ions, resulting in the appearance of a depot [76]. In light of this background, it was initially assumed that it would be beneficial to understand more about the organic fate of zinc oxide nanoparticles. Specifically, in ZON, novel mechanisms are hypothesized to increase intracellular zinc ion phases following nanoparticle treatment. According to one theory, the extracellular dissolution of ZON results in the release of zinc ions, which are then taken up by zinc transporters [77]. The endocytic way of ZON cell absorption, accompanied by the lysosomal breakdown and increased intracellular zinc ion levels [78], is yet another well-established hypothesis.

8 Metals in the Diabetes

Tadrosetal [77] reported the first evidence of Zn supplementation possibly inhibiting alloxan-induced diabetes with multiple minerals. They have further proved that Zn supplements significantly prevent alloxan-induced diabetes after experimenting with many elements, including Zn, Mn, Cr, and Co. Zinc supplementation helps prevent diabetes in certain health conditions [78]. An increase in the quantity of Zn in plasma and pancreas and controlled food consumption, body mass growth, or the copper content in tissues were used to investigate whether maintaining dietary habits (dieting) could have a beneficial influence on the prevention of diabetes. In many diabetic models, zinc supplementation has been shown to prevent diabetes [79] effectively. ß-cells contain around 70% of the zinc contained in the body. Zinc oxide nanoparticles in human cells with diameters ranging from 1 to 100 nm has been documented [80]. In addition, the anti-diabetic efficacy of zinc oxide nanoparticles was investigated in diabetic rats induced by streptozotocin [81]. Streptozotocin-induced diabetic rats [82] have shown that zinc is necessary for beta cell development and insulin generation. Inadequate zinc intake has been recognized as one of the most significant risk factors for physiological state polygenic disorder [83]. Treating oral zinc in diabetic patients has also been beneficial in predicting the prognosis of diabetic complications [84, 85]. Natural zinc oxide nanoparticles with sizes ranging from 1 to 100 nm were shown to be present in human cells [86]. Since 1935, zinc oxide nanoparticles have been extensively examined and have gained much scientific attention as a potential "*future material*" [87]. Zinc oxide nanoparticles have been widely used worldwide because of their numerous fascinating and distinctive qualities, which allow for their application in nanodiagnostics, nanomedicine, and antibacterial applications [88, 89].

8.1 Pathophysiology of Type 2 Diabetes Mellitus and Treatment with Drugs

Despite popular belief, insulin is not inefficiently utilized by the body. T1DM symptoms are present. However, they are generally less noticeable or missing, and as a result, the condition may go misdiagnosed for many years [90]. As a result, it begins in childhood and is diagnosed when the patient reaches adulthood. Diabetes is categorized by insulin resistance, when skeletal muscle, adipose tissue, and the liver do not respond effectively to insulin [91]. Various treatment regimens for diabetes are available, including those involving synthetic and natural remedies. Chemically synthesized drugs account for the majority of diabetes treatments available today. The list of synthetic medications is regularly available on the market [92]. EMC released several new diabetes medications in 2017, including Repaglinide, Acarbose, Alogliptine, Canaglifozine, Dulaglutide, Empagliflozin, Exenatide, and Glicazide.

9 Limitations of Allopathy Medicine and Alternative Approach

Although there has been significant progress in developing synthetic antidiabetic medications in recent years, diabetes is still astonishingly challenging to treat successfully [93]. These treatments have limits due to a lack of safe and effective drugs and the inability to provide a prolonged clinical, biochemical, and histological cure. Furthermore, synthetic drugs are associated with adverse reactions [94], such as hypoglycemia, nausea, vomiting, and hyponatremia. Allopathic medicine has its limitations, while alternative approaches have their own. Extreme cardiovascular risk is related to adverse outcomes such as hyperglycemia and dyslipidemia, among other things. Lipid-lowering drugs are frequently administered to diabetic patients to prevent complications of diabetes, such as cardiovascular disease and diabetic dyslipidemia. Despite this, it is challenging to completely prevent combination therapies' adverse effects and accompanying medications. In this way, plants serve as a possible source of hypoglycemic medicines as an alternative to synthetic agents. They are widely employed in some traditional treatment systems to help avoid diabetes. Over the last decade, there has been significant growth in complementary and alternative medicine (CAM) to control diabetes worldwide. According to a recent study, on average, 72.8% of the population with diabetes employed herbal therapy, dietary supplementation, change in lifestyle, and various complementary and alternative medicine (CAM) therapies [95]. Many medicinal plants are believed to have antidiabetic qualities and have been used to treat and manage diabetes for centuries in many cultures. Recent research has found that nanoparticles can be used in antidiabetic investigations due to their unique features. By boosting the catalytic characteristic and increasing the surface area of the sensor-receptor complex, nanotechnology has been demonstrated to be protectively efficient in treating diabetic mellitus, according to research. Improved oral formulations and its encapsulation, among other things, may significantly improve insulin administration [96].

10 Zinc and Diabetes

Maintaining the structural integrity of insulin in its hexameric state is essential. Suppose the amount of zinc in the body falls below a certain level. In that case, the islet of Langerhans cells fails to synthesize insulin, aggravating the problem, which is most prevalent in type 2 diabetes. One injection of the bare minimum dose of Zn causes damage in the exocrine cells of the mouse's pancreas but not in the endocrine cells. This demonstrated the generation and secretion of ghrelin (28 amino acid peptide) and lipid metabolism in the gastrointestinal tract [97]. When zinc was employed as a preventive agent, it was discovered that many lipid indicators and the atherogenic index were dramatically lowered. According to the data, these research results indicate that zinc considerably influences adipocyte lipogenesis. It was determined that SLC30A8 is a new zinc transporter (ZnT) family member by comparing the human genomic and expressed sequence tag databases with the amino acid sequences of all known human zinc transporters [98].

GLP-1 and GIP type 2 diabetes is a metabolic illness characterized by elevated blood glucose levels above normal levels, the lack of insulin from the pancreatic beta cells, and various additional problems [99]. The International Diabetes Federation (IDF) reported in recent publications that the number of persons with diabetes is

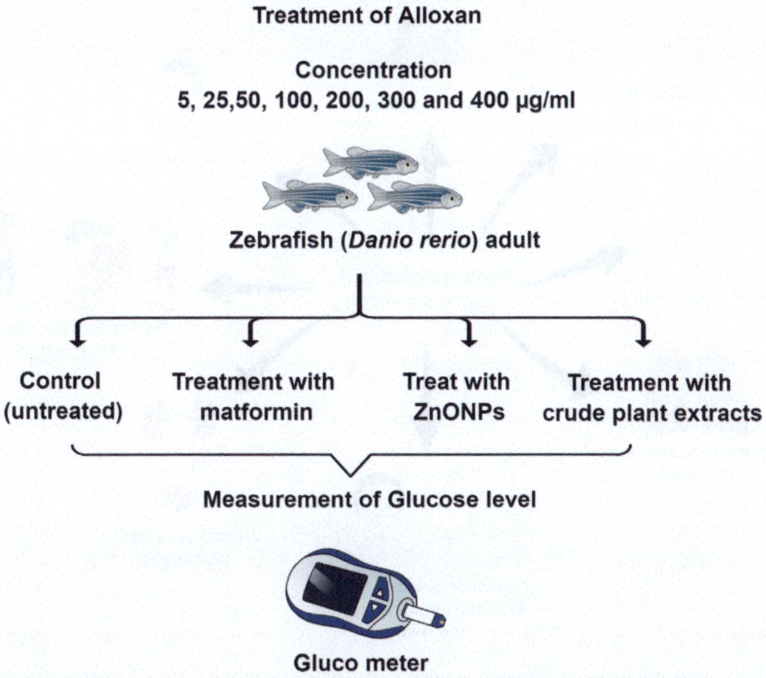

Fig. 2 Diagrammatic representation of the anti-hyperglycemic activity in zebrafish adults

expected to climb to over 592 million during the next twenty-five years (IDF, 2013). They also asserted that failing to maintain a proper balance of macronutrients and vitamin E significantly contributes to the progression of diabetes [100]. The effect of zinc supplementation on diabetic rats induced by alloxan (or) streptozotocin (STZ) was investigated (Fig. 2) (Unpublished data).

11 Correlation Between Zinc, Insulin, and Diabetes

Zn is a trace element in the highest quantity in all tissues and body fluids. Current research has shown that zinc plays a dynamic function as a "*cellular second messenger*" in controlling insulin signaling and glucose homeostasis and regulating lipid metabolism. In addition, the correct function has been shown for over 300 enzymes from all enzyme families closely related to zinc [101]. Zinc is essential for the regular action of more than 300 enzymes from all enzyme groups and is abundant in the environment. Since then, several hypotheses have been advanced on the clinical relationship between zinc, insulin, and diabetes. An additional study indicated that zinc could increase the phosphorylation of the insulin receptor in the beta subunit. In addition, zinc is vital in releasing insulin by beta cells in the pancreatic islets.

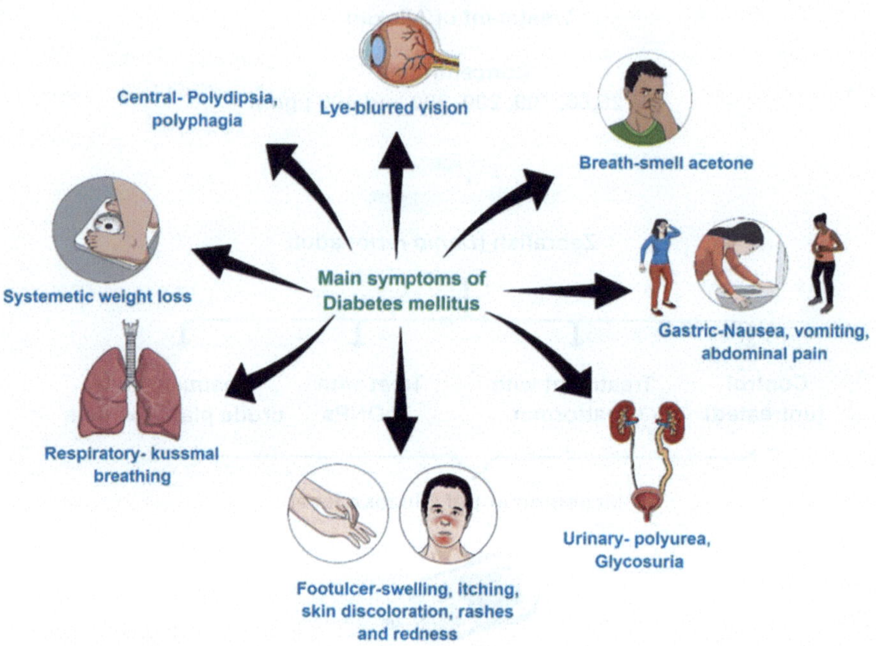

Fig. 3 Flowchart illustrates the main symptoms of diabetes mellitus

More study needs to be conducted to determine the specific connection of zinc in the pathophysiology of diabetes [102]. Zinc insulin's effects regarding the molecular pathways responsible for these effects have not yet been thoroughly investigated. According to growing research, zinc has been shown to interact with many insulins signaling pathways and influence glucose metabolism. As a result, zinc has played an essential role in managing diabetes mellitus in Fig. 3.

12 Zinc in Diabetes Associated with Diseased Conditions

Zinc (Zn) positively affects glucose metabolism, and its action limits glucose absorption from the digestive tract. Using zinc as an inhibitor of fructose1,6-bisphosphatase, Salgueiro and colleagues (2001) discovered that glycolysis takes precedence over gluconeogenesis in the cell [103]. Zinc protects beta cells from death, resulting in increased quantities of insulin in the bloodstream. A further study by Ho et al. [104] found that zinc protected beta cells against oxidative damage and death in diabetic models produced by streptozotocin and alloxan. Indeed, diabetes patients are more likely than the general population to have poor zinc status and a negative relationship has been established between zinc consumption and the development of diabetes. It is, however, controversial whether zinc deficiency is the cause or affects hyperglycemia in some cases. Many factors, including diabetes, impact zinc homeostasis, resulting in declines in total body zinc [105].

Many pre-clinical studies have verified the anti-hyperglycemic effects of zinc complement at ion in animal models for both types of diabetes (type1 and type2) [106]. As a result, it stands to reason that a zinc-based drug could be employed to treat diabetes. Although the Zn complexes were designed as an oral medication for treating diabetes patients' hyperglycemic episodes [107], they were not widely used until recently. In contrast, which believes that insulin is kept in the inner pancreatic beta cell secretory vesicles, where six insulin molecules form stable hexamers with two Zn^{2+} ions [108] and not in the extracellular region, insulin is synthesized in the extracellular space. In turn, this increases extracellular Ca^{2+} influx, which enhances insulin-containing vesicle exocytosis and boosts insulin synthesis [109]. Hypoglycemia may ensue due to inadequate insulin synthesis and impaired glucose absorption in peripheral tissues, which can be partially compensated in certain situations by producing excessive insulin [110]. Zn is essential for cell survival, according to Vallee and Falchuk, who also claim that it contributes to the form and characteristics of many proteins [111]. Zinc is responsible for the critical cell mechanisms demonstrated by Chimienti et al. including DNA replication, metabolic enzyme functioning, and cellular safety in the face of hypoxia and oxidative stress. Zinc has also been shown to have higher in pancreatic beta cells, considered the best available in the body [112]. Analysis of islets from 65 Zn-treated rats has revealed that one-third of the total beta cell zinc content material is contained within insulin granules, which was discovered in the previous decade [113]. According to some researchers, this relationship between zinc and insulin may indicate that an adequate amount of zinc ions is required to initiate the crystallization of zinc-insulin hexamers in secretory granules [114].

Furthermore, upon glucose stimulation, considerable quantities of zinc are released into the extracellular matrix by the body's cells, which combine with the hormone insulin to achieve this result. According to recent studies, Co-secreted zinc may function in islet cell paracrine or autocrine communication. It activates KATP channels, exerting significant influence over the secretory system and modulating glucagon secretion alpha cells [115]. In contrast, the SLC30 family of proteins (known as ZnT) is responsible for zinc efflux from the cytosol to the extracellular matrix and intracellular organelles [116]. A member of the protein family ZnT-8, which was previously detected in pancreatic islets after being considered missing from the body, was revealed to be present in the body. Zinc supplementation treats people with diabetes with related illness conditions [117]. Another study on the reputation of zinc discovered that certain people excrete significantly more levels of zinc than others, which could eventually result in zinc deficiency. One more piece of information that can be shared with the competition is that moderate zinc deficiency is common in those with non-insulin-dependent diabetic mellitus. Genetically fat mice were shown to have lower blood sugar levels after receiving zinc supplementation [118], which may be related to zinc affecting insulin action. A low zinc level in groundwater, which can potentially reflect long-term exposure to zinc through drinking water, is associated with a reduced incidence of childhood-onset diabetes later in life. Consuming more zinc can protect against oxidant stress in men and women with dispositions for both average zinc deficiency and extreme oxidant stress [119] and in postmenopausal, type two diabetes women. In autoimmune diabetogenesis models, transgenic copper/zinc superoxide is mutase affects susceptibility. It can shed light on the role of reactive oxygen and other types of disease pathology where active oxygen has been implicated [120]. The acylated hemoglobin A1c (HbA1c) level increased significantly in insulin-dependent diabetes mellitus (IDDM) groups that received zinc supplementation. People with IDDM did not appear zinc-deficient anymore, despite having persistent hyperzincuria. The data indicated that toxicity from high-dose zinc supplementation was a possibility. Zinc supplementation effectively prevents or alleviates diabetes in various animal models of type 1 and type 2 diabetes [121].

13 Conclusion

This chapter concludes that an ideal therapeutic outline of a nano-drug should aim to sustain glucose levels as near to regular as possible for a prolonged period. Most current antidiabetic drugs enhance insulin release or glucose uptake in specific peripheral tissues. Unfortunately, some drugs are unsuccessful in controlling hyperglycemia eventually, insulin therapy is imperative. To apply nanotechnology for developing novel antidiabetic drugs, have innovative approaches in the fight against the increasing occurrence of T2DM.

References

1. M.Y.M. Ismail, M. Yaheya, Clinical evaluation of antidiabetic activity of Trigonella seeds and *Aegle marmelos* leaves. World Appl. Sci. J. **7**, 1231–1234 (2009)
2. A.D. Association, Diagnosis and classification of diabetes mellitus. Diabetes Care **33**, S62–S69 (2010)
3. B.L. Vallee, K.H. Falchuk, The biochemical basis of zinc physiology. Physiol. Rev. **73**, 79–118 (1993)
4. P. Zimmet, K. Alberti, J. Shaw, Global and societal implications of the diabetes epidemic. Nature **414**, 782–787 (2001)
5. L.I. Zon, Zebrafish: a new model for human disease. Genome Res. **9**, 99–100 (1999)
6. G.A. Rutter, T. Tsuboi, M.A. Ravier, Ca^{2+} microdomains and the control of insulin secretion. Cell Calcium **40**, 539–551 (2006)
7. F. Chimienti, S. Devergnas, F. Pattou, F. Schuit, R. Garcia-Cuenca, B. Vandewalle, J. Kerr-Conte, L. Van Lommel, D. Grunwald, A. Favier, In vivo expression and functional characterization of the zinc transporter ZnT8 in glucose-induced insulin secretion. J. Cell Sci. **119**, 4199–4206 (2006)
8. J.J. Faber-Hammond, K.H. Brown, Pseudo-De novo assembly and analysis of unmapped genome sequence reads in wild zebrafish reveal novel gene content. Zebrafish **13**, 95–102 (2016)
9. P.E. Petersen, D. Bourgeois, H. Ogawa, S. Estupinan-Day, C. Ndiaye, The global burden of oral diseases and risks to oral health. Bull. World Health Organ. **83**, 661–669 (2005)

10. E. Serrano, G. Rus, J. García-Martínez, Nanotechnology for sustainable energy. Renew. Sustain. Energy Rev. **13**, 2373–2384 (2009)
11. I. Khan, K. Saeed, I. Khan, Nanoparticles: properties, applications and toxicities. Arab. J. Chem. **12**, 908–931 (2019)
12. C. Klingshirn, ZnO: from basics towards applications. Physica Status Solidi (B) **244**, 3027–3073 (2007)
13. R.O. van Merkerk, H. van Lente, Tracing emerging irreversibilities in emerging technologies: the case of nanotubes. Technol. Forecast. Soc. Chang. **72**, 1094–1111 (2005)
14. J.A. Dahl, B.L.S. Maddux, J.E. Hutchison, Toward greener nanosynthesis. Chem. Rev. **107**, 2228–2269 (2007)
15. D.-S. Bin, Z.-X. Chi, Y. Li, K. Zhang, X. Yang, Y.-G. Sun, J.-Y. Piao, A.-M. Cao, L.-J. Wan, Controlling the compositional chemistry in single nanoparticles for functional hollow carbon nanospheres. J. Am. Chem. Soc. **139**, 13492–13498 (2017)
16. R. Prasad, R. Pandey, I. Barman, Engineering tailored nanoparticles with microbes: quo Vadis? WIREs Nanomed. Nanobiotechnol.Nanomed. Nanobiotechnol. **8**, 316–330 (2016)
17. S.M. Gopinath, N.S. Saha, V.J. John, N.S. Khanum, G. Shyamil, G.M. Patil, Biological synthesis characterization and application of silver nano particles a review (2013)
18. N. Taniguchi, On the basic concept of 'nano-technology', in *Proceedings of International Conference on Production Engineering Tokyo*, Part II (Japan Society of Precision Engineering, 1974)
19. J. Agrawal, A. Pal, Nyctanthes arbor-tristis Linn—a critical ethnopharmacological review. J. Ethnopharmacol.Ethnopharmacol. **146**, 645–658 (2013)
20. S. Guo, S. Dong, Graphene nanosheet: synthesis, molecular engineering, thin film, hybrids, and energy and analytical applications. Chem. Soc. Rev. **40**, 2644–2672 (2011)
21. X. Li, H. Xu, Z.-S. Chen, G. Chen, Biosynthesis of nanoparticles by microorganisms and their applications. J. Nanomater.Nanomater. **2011**, 1–16 (2011)
22. S. Rehman, R. Jermy, S.M. Asiri, M.A. Shah, R. Farooq, V. Ravinayagam, M.A. Ansari, Z. Alsalem, R. Al Jindan, Z. Reshi, Using *Fomitopsis pinicola* for bioinspired synthesis of titanium dioxide and silver nanoparticles, targeting biomedical applications. RSC Adv. **10**, 32137–32147 (2020)
23. S. Panigrahi, S. Kundu, S. Ghosh, S. Nath, T. Pal, General method of synthesis for metal nanoparticles. J. Nanopart. Res.Nanopart. Res. **6**, 411–414 (2004)
24. P. Falcaro, R. Ricco, A. Yazdi, I. Imaz, S. Furukawa, D. Maspoch, R. Ameloot, J.D. Evans, C.J. Doonan, Application of metal and metal oxide nanoparticles@MOFs. Coord. Chem. Rev. **307**, 237–254 (2016)
25. Z.L. Poole, P. Ohodnicki, R. Chen, Y. Lin, K.P. Chen, Engineering metal oxide nanostructures for the fiber optic sensor platform. Opt. Express **22**, 2665–2674 (2014)
26. A. Rastogi, M. Zivcak, O. Sytar, H.M. Kalaji, X. He, S. Mbarki, M. Brestic, Impact of metal and metal oxide nanoparticles on plant: a critical review. Front. Chem. **5**, 78 (2017)
27. J. Ghilane, F.-R.F. Fan, A.J. Bard, N. Dunwoody, Facile electrochemical characterization of core/shell nanoparticles. Ag core/Ag_2O shell structures. Nano Lett. **7**, 1406–1412 (2007)
28. Y. Wang, B. Huang, W. Dai, J. Ye, B. Xu, Sensitive determination of capsaicin on Ag/Ag_2O nanoparticles/reduced graphene oxide modified screen-printed electrode. J. Electroanal. Chem.Electroanal. Chem. **776**, 93–100 (2016)
29. W.M. Shume, H.C.A. Murthy, E.A. Zereffa, A review on synthesis and characterization of Ag_2O nanoparticles for photocatalytic applications. J. Chem. **2020**, 1–15 (2020)
30. A.J. Shetty, D. Choudhury, Rejeesh, V. Nair, M. Kuruvilla, S. Kotian, Effect of the insulin plant (*Costus igneus*) leaves on dexamethasone-induced hyperglycemia. Int. J. Ayurveda Res. **1**, 100–102 (2010)
31. P. Kalailingam, B. Kannaian, E. Tamilmani, R. Kaliaperumal, Efficacy of natural diosgenin on cardiovascular risk, insulin secretion, and beta cells in streptozotocin (STZ)-induced diabetic rats. Phytomedicine **21**, 1154–1161 (2014)
32. S. Rajkumari, K. Sanatombi, Nutritional value, phytochemical composition, and biological activities of edible Curcuma species: a review. Int. J. Food Prop. **20**, S2668–S2687 (2017)

33. G.N. Anyasor, O. Ogunwenmo, O.A. Oyelana, B.E. Akpofunure, Phytochemical constituents and antioxidant activities of aqueous and methanol stem extracts of Costus afer Ker Gawl. (Costaceae). Afr. J. Biotechnol. **9**, 4880–4884 (2010)
34. K. Krishnan, N.R. Vijayalakshmi, A. Helen, Beneficial effects of *Costus igneus* and dose response studies in streptozotocin induced diabetic rats. Int. J. Curr. Pharm. Res. **3**, 42–46 (2011)
35. A. Aruna, R. Nandhini, V. Karthikeyan, P. Bose, K. Vijayalakshmi, Comparative anti-diabetic effect of methanolic extract of insulin plant (*Costus pictus*) leaves and its silver nanoparticle. Indo Am. J. Pharm. Res. **4**, 3217–3230 (2014)
36. Q. Wu, M. Si, B. Zhang, K. Zhang, H. Li, L. Mi, Y. Jiang, Y. Rong, J. Chen, Y. Fang, Strong damping of the localized surface plasmon resonance of Ag nanoparticles by Ag_2O. Nanotechnology **29**, 295702 (2018)
37. E.A. Meulenkamp, Synthesis and growth of ZnO nanoparticles. J. Phys. Chem. B **102**, 5566–5572 (1998)
38. R.Y. Hong, J.H. Li, L.L. Chen, D.Q. Liu, H.Z. Li, Y. Zheng, J. Ding, Synthesis, surface modification and photocatalytic property of ZnO nanoparticles. Powder Technol. **189**, 426–432 (2009)
39. L. Zhang, Y. Jiang, Y. Ding, M. Povey, D. York, Investigation into the antibacterial behaviour of suspensions of ZnO nanoparticles (ZnO nanofluids). J. Nanopart. Res.Nanopart. Res. **9**, 479–489 (2007)
40. S. Talam, S.R. Karumuri, N. Gunnam, Synthesis, characterization, and spectroscopic properties of ZnO nanoparticles. Int. Scholarly Res. Notices (2012)
41. N. Padmavathy, R. Vijayaraghavan, Enhanced bioactivity of ZnO nanoparticles—an antimicrobial study. Sci. Technol. Adv. Mater. **9**, 35004 (2008)
42. Z. Hu, G. Oskam, P.C. Searson, Influence of solvent on the growth of ZnO nanoparticles. J. Colloid Interface Sci. **263**, 454–460 (2003)
43. M.A. Garcia, J.M. Merino, E. Fernández Pinel, A. Quesada, J. de la Venta, M.L. Ruíz González, G.R. Castro, P. Crespo, J. Llopis, J.M. González-Calbet, A. Hernando, Magnetic Properties of ZnO nanoparticles. Nano Lett. **7**, 1489–1494 (2007)
44. H. Wang, J.-Z. Xu, J.-J. Zhu, H.-Y. Chen, Preparation of CuO nanoparticles by microwave irradiation. J. Cryst. GrowthCryst. Growth **244**, 88–94 (2002)
45. K. Phiwdang, S. Suphankij, W. Mekprasart, W. Pecharapa, Synthesis of CuO nanoparticles by precipitation method using different precursors. Energy Procedia **34**, 740–745 (2013)
46. P. Mohanpuria, N.K. Rana, S.K. Yadav, Biosynthesis of nanoparticles: technological concepts and future applications. J. Nanopart. Res.Nanopart. Res. **10**, 507–517 (2008)
47. K. Vithiya, S. Sen, Biosynthesis of nanoparticles. Int. J. Pharm. Sci. Res. **2**, 2781–2785 (2011)
48. J. Kesharwani, K.Y. Yoon, J. Hwang, M. Rai, Phytofabrication of silver nanoparticles by leaf extract of *Datura metel*: hypothetical mechanism involved in synthesis. J. Bionanosci.Bionanosci. **3**, 39–44 (2009)
49. S. Iravani, Green synthesis of metal nanoparticles using plants. Green Chem. **13**, 2638–2650 (2011)
50. V.K. Vidhu, S.A. Aromal, D. Philip, Green synthesis of silver nanoparticles using *Macrotyloma uniflorum*. Spectrochim. Acta Part A Mol. Biomol. Spectrosc.. Acta Part A Mol. Biomol. Spectrosc. **83**, 392–397 (2011)
51. C. Mukherjee, D.P. Patil, B.A. Kennedy, B. Bakthavachalu, R. Bundschuh, D.R. Schoenberg, Identification of cytoplasmic capping targets reveals a role for cap homeostasis in translation and mRNA stability. Cell Rep. **2**, 674–684 (2012)
52. B. Kumar, K. Smita, L. Cumbal, A. Debut, Green approach for fabrication and applications of zinc oxide nanoparticles. Bioinorg. Chem. Appl.. Chem. Appl. **2014**, 523869 (2014)
53. S. Sharma, K. Kumar, N. Thakur, S. Chauhan, M.S. Chauhan, The effect of shape and size of ZnO nanoparticles on their antimicrobial and photocatalytic activities: a green approach. Bull. Mater. Sci. **43**, 20 (2019)
54. H. Agarwal, S. Venkat Kumar, S. Rajeshkumar, A review on green synthesis of zinc oxide nanoparticles—an eco-friendly approach. Resour.-Efficient Technol. **3**, 406–413 (2017)

55. P.P. Mahamuni-Badiger, P.M. Patil, M.V. Badiger, P.R. Patel, B.S. Thorat-Gadgil, A. Pandit, R.A. Bohara, Biofilm formation to inhibition: role of zinc oxide-based nanoparticles. Mater. Sci. Eng. C. **108**, 110319 (2020)
56. A.P. Ingle, A. Biswas, C. Vanlalveni, R. Lalfakzuala, I. Gupta, P. Ingle, L. Rokhum, M. Rai, Biogenic synthesis of nanoparticles and their role in the management of plant pathogenic fungi. Microbial Nanotechnol., 135–161 (2020)
57. P.D. Shankar, S. Shobana, I. Karuppusamy, A. Pugazhendhi, V.S. Ramkumar, S. Arvindnarayan, G. Kumar, A review on the biosynthesis of metallic nanoparticles (gold and silver) using bio-components of microalgae: formation mechanism and applications. Enzyme Microb. Technol.Microb. Technol. **95**, 28–44 (2016)
58. M. Shabaani, S. Rahaiee, M. Zare, S.M. Jafari, Green synthesis of ZnO nanoparticles using loquat seed extract; biological functions and photocatalytic degradation properties. LWT **134**, 110133 (2020)
59. A. Muthuvel, M. Jothibas, C. Manoharan, Effect of chemically synthesis compared to biosynthesized ZnO-NPs using Solanum nigrum leaf extract and their photocatalytic, antibacterial and in-vitro antioxidant activity. J. Environ. Chem. Eng. **8**, 103705 (2020)
60. K. Rambabu, G. Bharath, F. Banat, P.L. Show, Green synthesis of zinc oxide nanoparticles using *Phoenix dactylifera* waste as bioreductant for effective dye degradation and antibacterial performance in wastewater treatment. J. Hazard. Mater. **402**, 123560 (2021)
61. S.O. Ogunyemi, Y. Abdallah, M. Zhang, H. Fouad, X. Hong, E. Ibrahim, M.M.I. Masum, A. Hossain, J. Mo, B. Li, Green synthesis of zinc oxide nanoparticles using different plant extracts and their antibacterial activity against *Xanthomonas oryzae* pv. Oryzae. Artificial Cells Nanomed. Biotechnol. **47**, 341–352 (2019)
62. J. Jeevanandam, Y.S. Chan, M.K. Danquah, Biosynthesis of metal and metal oxide nanoparticles. ChemBioEng Rev. **3**, 55–67 (2016)
63. W.R. Rolim, M.T. Pelegrino, B. de Araújo Lima, L.S. Ferraz, F.N. Costa, J.S. Bernardes, T. Rodigues, M. Brocchi, A.B. Seabra, Green tea extract mediated biogenic synthesis of silver nanoparticles: characterization, cytotoxicity evaluation and antibacterial activity. Appl. Surface Sci. **463**, 66–74 (2019)
64. C.P. Khare, Identification of proper herbs: a new scientific route, in: *Evidence-Based Ayurveda* (Routledge, 2019), pp. 73–120
65. S. Hamedi, S.A. Shojaosadati, Rapid and green synthesis of silver nanoparticles using Diospyros lotus extract: evaluation of their biological and catalytic activities. Polyhedron **171**, 172–180 (2019)
66. P. Kumar, S. Gautam, Developing ZnO nanoparticle embedded antimicrobial starch biofilm for food packaging. arXiv Preprint arXiv:1909.05083 (2019)
67. Y. Zhong, H. Xiao, F. Seidi, Y. Jin, Natural polymer-based antimicrobial hydrogels without synthetic antibiotics as wound dressings. Biomacromol **21**, 2983–3006 (2020)
68. Y. Wang, J. Ma, Q. Xu, J. Zhang, Fabrication of antibacterial casein-based ZnO nanocomposite for flexible coatings. Mater. Des. **113**, 240–245 (2017)
69. X. Guo, B. Chen, X. Wu, J. Li, Q. Sun, Utilization of cinnamaldehyde and zinc oxide nanoparticles in a carboxymethylcellulose-based composite coating to improve the postharvest quality of cherry tomatoes. Int. J. Biol. Macromol.Macromol. **160**, 175–182 (2020)
70. K. Ramachandraiah, G.-P. Hong, Polymer based nanomaterials for strategic applications in animal food value chains. Food Rev. Int. **38**, 1577–1606 (2022)
71. N.A. Al-Tayyar, A.M. Youssef, R. Al-hindi, Antimicrobial food packaging based on sustainable bio-based materials for reducing foodborne pathogens: a review. Food Chem. **310**, 125915 (2020)
72. D.K. Ban, S. Paul, Nano zinc oxide inhibits fibrillar growth and suppresses cellular toxicity of lysozyme amyloid. ACS Appl. Mater. Interfaces **8**, 31587–31601 (2016)
73. G.K. Malik, J. Mitra, Zinc oxide nanoparticle synthesis, characterization, and their effect on mechanical, barrier, and optical properties of HPMC-based edible film, food and bioprocess. Technology **14**, 441–456 (2021)

74. Y. Huang, L. Mei, X. Chen, Q. Wang, Recent developments in food packaging based on nanomaterials. Nanomaterials **8** (2018)
75. R.M. Davidson, A. Lauritzen, S. Seneff, Biological water dynamics and entropy: a biophysical origin of cancer and other diseases. Entropy **15**, 3822–3876 (2013)
76. C.V. Tadros, P.C. Treble, A. Baker, S. Hankin, R. Roach, Cave drip water solutes in south-eastern Australia: constraining sources, sinks and processes. Sci. Total. Environ. **651**, 2175–2186 (2019)
77. Q. Yang, Y. Su, C. Chi, C.T. Cherian, K. Huang, V.G. Kravets, F.C. Wang, J.C. Zhang, A. Pratt, A.N. Grigorenko, F. Guinea, A.K. Geim, R.R. Nair, Ultrathin graphene-based membrane with precise molecular sieving and ultrafast solvent permeation. Nat. Mater. **16**, 1198–1202 (2017)
78. N.G. Forouhi, N.J. Wareham, Epidemiology of diabetes. Medicine **38**, 602–606 (2010)
79. D. Soumya, B. Srilatha, Late stage complications of diabetes and insulin resistance. J. Diabetes Metab. **2**, 1000167 (2011)
80. Y.A. Selim, M.A. Azb, I. Ragab, M.H.M. Abd El-Azim, Green synthesis of zinc oxide nanoparticles using aqueous extract of *Deverra tortuosa* and their cytotoxic activities. Sci. Rep. **10**, 3445 (2020)
81. R.D. Umrani, K.M. Paknikar, Zinc oxide nanoparticles show antidiabetic activity in streptozotocin-induced type 1 and 2 diabetic rats. Nanomedicine **9**, 89–104 (2013)
82. A. Nazarizadeh, S. Asri-Rezaie, Comparative study of antidiabetic activity and oxidative stress induced by zinc oxide nanoparticles and zinc sulfate in diabetic rats. AAPS PharmSciTech **17**, 834–843 (2016)
83. R.L. Wilson, J.A. Grieger, T. Bianco-Miotto, C.T. Roberts, Association between maternal zinc status, dietary zinc intake and pregnancy complications: a systematic review. Nutrients **8** (2016)
84. G. Partida-Hernández, F. Arreola, B. Fenton, M. Cabeza, R. Román-Ramos, M.C. Revilla-Monsalve, Effect of zinc replacement on lipids and lipoproteins in type 2-diabetic patients. Biomed. Pharmacother.Pharmacother. **60**, 161–168 (2006)
85. M.S. Akhtar, J. Panwar, Y.-S. Yun, Biogenic synthesis of metallic nanoparticles by plant extracts. ACS Sustain. Chem. Eng. **1**, 591–602 (2013)
86. J. Suresh, G. Pradheesh, V. Alexramani, M. Sundrarajan, S.I. Hong, Green synthesis and characterization of zinc oxide nanoparticle using insulin plant (*Costus pictus* D. Don) and investigation of its antimicrobial as well as anticancer activities. Adv. Natural Sci. Nanosci. Nanotechnol. **9**, 15008 (2018)
87. D. Bobo, K.J. Robinson, J. Islam, K.J. Thurecht, S.R. Corrie, Nanoparticle-based medicines: a review of FDA-approved materials and clinical trials to date. Pharm. Res. **33**, 2373–2387 (2016)
88. S. Ahmed, M. Ahmad, B.L. Swami, S. Ikram, A review on plants extract mediated synthesis of silver nanoparticles for antimicrobial applications: a green expertise. J. Adv. Res. **7**, 17–28 (2016)
89. R.D. Umrani, K.M. Paknikar, Jasada bhasma, a zinc-based ayurvedic preparation: contemporary evidence of antidiabetic activity inspires development of a nanomedicine. Evidence-Based Complement. Alternative Med. **2015** (2015)
90. J.C. Ozougwu, K.C. Obimba, C.D. Belonwu, C.B. Unakalamba, The pathogenesis and pathophysiology of type 1 and type 2 diabetes mellitus. J. Physiol. Pathophysiol.Pathophysiol. **4**, 46–57 (2013)
91. A. Karimi, M. Majlesi, M. Rafieian-Kopaei, Herbal versus synthetic drugs; beliefs and facts. J. Nephropharmacol. **4**, 27–30 (2015)
92. M. Kaur, V. Valecha, Diabetes and antidiabetic herbal formulations: an alternative to allopathy. Eur. J. Med. **6**, 226–240 (2014)
93. R. Khursheed, S.K. Singh, S. Wadhwa, B. Kapoor, M. Gulati, R. Kumar, A.K. Ramanunny, A. Awasthi, K. Dua, Treatment strategies against diabetes: success so far and challenges ahead. Eur. J. Pharmacol.Pharmacol. **862**, 172625 (2019)
94. M. Haque, S.H. Emerson, C.R. Dennison, N.S. Levitt, M. Navsa, Barriers to initiating insulin therapy in patients with type 2 diabetes mellitus in public-sector primary health care centres in Cape Town. S. Afr. Med. J. **95**, 798–802 (2005)

95. D. Kumar, S. Bajaj, R. Mehrotra, Knowledge, attitude and practice of complementary and alternative medicines for diabetes. Publ. Health **120**, 705–711 (2006)
96. S. Bolkent, R. Yanardag, S. Bolkent, O. Mutlu, S. Yildirim, K. Kangawa, Y. Minegishi, H. Suzuki, The effect of zinc supplementation on ghrelin-immunoreactive cells and lipid parameters in gastrointestinal tissue of streptozotocin-induced female diabetic rats. Mol. Cell. Biochem.Biochem. **286**, 77–85 (2006)
97. M. Seve, F. Chimienti, S. Devergnas, A. Favier, In silico identification and expression of SLC30 family genes: an expressed sequence tag data mining strategy for the characterization of zinc transporters' tissue expression. BMC Genom. **5**, 1–9 (2004)
98. D.K. Patel, R. Kumar, D. Laloo, S. Hemalatha, Diabetes mellitus: an overview on its pharmacological aspects and reported medicinal plants having antidiabetic activity. Asian Pac. J. Trop. Biomed. **2**, 411–420 (2012)
99. C.W. Popper, Single-micronutrient and broad-spectrum micronutrient approaches for treating mood disorders in youth and adults. Child Adolescent Psychiatr. Clin. **23**, 591–672 (2014)
100. B.C. Arthur, S. Chausmer, Zinc, insulin and diabetes. J. Am. Coll. Nutri. **17**, 109–115 (1998)
101. J. Jansen, W. Karges, L. Rink, Zinc and diabetes—clinical links and molecular mechanisms. J. Nutr. Biochem.Nutr. Biochem. **20**, 399–417 (2009)
102. M.J. Salgueiro, N. Krebs, M.B. Zubillaga, R. Weill, E. Postaire, A.E. Lysionek, R.A. Caro, T. De Paoli, A. Hager, J. Boccio, Zinc and diabetes mellitus: is there a need of zinc supplementation in diabetes mellitus patients? Biol. Trace Elem. Res. **81**, 215–228 (2001)
103. W.B. Kinlaw, A.S. Levine, J.E. Morley, S.E. Silvis, C.J. McClain, Abnormal zinc metabolism in type II diabetes mellitus. Am. J. Med. **75**, 273–277 (1983)
104. P. Ranasinghe, S. Pigera, P. Galappatthy, P. Katulanda, G.R. Constantine, Zinc and diabetes mellitus: understanding molecular mechanisms and clinical implications. DARU J. Pharmaceut. Sci. **23**, 1–13 (2015)
105. R. Jayawardena, P. Ranasinghe, P. Galappatthy, R. Malkanthi, G.R. Constantine, P. Katulanda, Effects of zinc supplementation on diabetes mellitus: a systematic review and meta-analysis. Diabetol. Metab. Syndr.. Metab. Syndr. **4**, 1–13 (2012)
106. A. Sjögren, C.-H. Florén, Å.K.E. Nilsson, Magnesium, potassium and zinc deficiency in subjects with type II diabetes Mellitus. Acta Med. Scand. **224**, 461–466 (1988)
107. K.J.C. Cruz, A.R.S. de Oliveira, D.N. Marreiro, Antioxidant role of zinc in diabetes mellitus. World J. Diabetes **6**, 333–337 (2015)
108. R.A. Al-Maroof, S.S. Al-Sharbatti, Serum zinc levels in diabetic patients and effect of zinc supplementation on glycemic control of type 2 diabetics. Saudi Med. J. **27** (2006)
109. C.B. Niewoehner, J.I. Allen, M. Boosalis, A.S. Levine, J.E. Morley, Role of zinc supplementation in type II diabetes mellitus. Am. J. Med. **81**, 63–68 (1986)
110. R.M. Walter Jr., J.Y. Uriu-Hare, K.L. Olin, M.H. Oster, B.D. Anawalt, J.W. Critchfield, C.L. Keen, Copper, zinc, manganese, and magnesium status and complications of diabetes mellitus. Diabetes Care **14**, 1050–1056 (1991)
111. C.G. Taylor, Zinc, the pancreas, and diabetes: insights from rodent studies and future directions. Biometals **18**, 305–312 (2005)
112. K. Jurowski, B. Szewczyk, G. Nowak, W. Piekoszewski, Biological consequences of zinc deficiency in the pathomechanisms of selected diseases. J. Biol. Inorg. Chem. **19**, 1069–1079 (2014)
113. W. Maret, H.H. Sandstead, Zinc requirements and the risks and benefits of zinc supplementation. J. Trace Elem. Med. Biol. **20**, 3–18 (2006)
114. J.J. Cunningham, Aizhong Fu, P.L. Mearkle, R. Glenn Brown, Hyperzincuria in individuals with insulin-dependent diabetes mellitus: concurrent zinc status and the effect of high-dose zinc supplementation. Metabolism **43**, 1558–1562 (1994)
115. R.A. DiSilvestro, Zinc in relation to diabetes and oxidative disease. J. Nutr.Nutr. **130**, 1509S-1511S (2000)
116. A.-M. Roussel, A. Kerkeni, N. Zouari, S. Mahjoub, J.-M. Matheau, R.A. Anderson, Antioxidant effects of zinc supplementation in Tunisians with type 2 diabetes Mellitus. J. Am. Coll. Nutr.Nutr. **22**, 316–321 (2003)

117. A.H. Zargar, N.A. Shah, S.R. Masoodi, B.A. Laway, F.A. Dar, A.R. Khan, F.A. Sofi, A.I. Wani, Copper, zinc, and magnesium levels in non-insulin dependent diabetes mellitus. Postgrad. Med. J. **74**, 665–668 (1998)
118. R.A. Anderson, A.-M. Roussel, N. Zouari, S. Mahjoub, J.-M. Matheau, A. Kerkeni, Potential antioxidant effects of zinc and chromium supplementation in people with type 2 diabetes Mellitus. J. Am. Coll. Nutr.Nutr. **20**, 212–218 (2001)
119. S. Barman, K. Srinivasan, Diabetes and zinc dyshomeostasis: can zinc supplementation mitigate diabetic complications? Crit. Rev. Food Sci. Nutr.Nutr. **62**, 1046–1061 (2022)
120. S.M. Virtanen, M. Knip, Nutritional risk predictors of β cell autoimmunity and type 1 diabetes at a young age. Am. J. Clin. Nutr.Nutr. **78**, 1053–1067 (2003)
121. R. Spence, G. Gerlach, C. Lawrence, C. Smith, The behaviour and ecology of the zebrafish, *Danio rerio*. Biol. Rev. **83**, 13–34 (2008)

Applications of Conductive Nanobiomaterials in Biomedicine

Majid Bonyadi Manesh

Abstract Cells grow and organize themselves on nanobiomaterials scaffolds that provide a supportive structure. In dental procedures such as periodontal regeneration, they create a conducive environment for tissue regeneration. By applying these materials to damaged oral tissues, cells adhere, multiply, and differentiate, improving tissue regeneration.

1 Introduction

Nanobiomaterials are a class of materials that combine the principles of nanotechnology with the field of biomaterials. They are engineered at the nanoscale, meaning they have dimensions typically between 1 and 100 nm (1 nm = 1 billionth of a meter). These materials are designed to interact with biological systems, including cells, tissues, and even individual biomolecules. The term "nanobiomaterials" is a combination of three key components:

- **Nano**: It refers to the nanoscale dimension of the material. At this scale, materials exhibit distinct properties, such as a high surface area-to-volume ratio, increased reactivity, and unique mechanical, optical, and electrical properties.
- **Bio**: Signifies the biological aspect of these materials. Nanobiomaterials are designed to interact with biological entities, such as living cells, tissues, proteins, enzymes, and DNA, to achieve specific therapeutic or diagnostic outcomes.
- **Materials**: As the name suggests, nanobiomaterials are a type of material that can be synthesized from various substances, including metals, polymers, ceramics, and composites. The choice of materials is critical to achieving the desired functions and biocompatibility.

Nanobiomaterials have gained significant attention in various fields, particularly in biomedicine, due to their unique properties and potential applications. They offer

M. B. Manesh (✉)
Department of Oral and Maxillofacial Prosthodontics, School of Dentistry, Mashhad University of Medical Sciences, Mashhad, Iran
e-mail: dr.majidbonyadimanesh1986@gmail.com

K. Kasinathan et al. (eds.), *New Technologies for Energy Transition Based on Sustainable Development Goals*, https://doi.org/10.1007/978-981-97-2527-4_11

several advantages, such as enhanced biocompatibility, improved drug delivery capabilities, and the ability to mimic natural cellular structures. There are three main categories of nanobiomaterials:

- **Nanoscale biomaterials**: These are biomaterials with nanoscale features, such as nanoparticles, nanofibers, and nanotubes. They can be made from various materials, including metals, polymers, ceramics, and carbon-based materials like graphene.
- **Nanostructured biomaterials**: These biomaterials have hierarchical structures at the nanoscale, providing them with specific properties suitable for biomedical applications. Examples include nanostructured scaffolds for tissue engineering and nanocomposites for drug delivery.
- **Nanoengineered biomaterials**: These are biomaterials that are specifically designed and engineered at the nanoscale to have targeted interactions with biological systems. They may have functional groups, ligands, or other bioactive elements incorporated to enhance their performance in specific biomedical applications [1].

In the context of biomedicine, nanobiomaterials have a wide range of applications, including:

- **Tissue engineering**: Nanostructured scaffolds can support cell growth and tissue regeneration, making them useful for regenerative medicine.
- **Drug delivery**: Nanoparticles can be used to encapsulate drugs and deliver them to specific target sites, increasing drug effectiveness and reducing side effects.
- **Diagnostic imaging**: Nanoparticles can be engineered as contrast agents for medical imaging techniques, improving sensitivity and accuracy in disease diagnosis.
- **Therapeutics**: Nanobiomaterials can be designed to interact with cells and tissues for therapeutic purposes, such as enhancing cell function or stimulating tissue repair.
- **Dentistry**: Nanobiomaterials in dentistry encompass nanoscale materials that can be used in various dental procedures and treatments.

It is essential to note that while nanobiomaterials offer promising opportunities, their potential impact on health and the environment requires careful consideration and ongoing research to ensure their safety and responsible use. Some common types of nanobiomaterials include nanoparticles, nanotubes, nanofibers, and nanostructured surfaces.

Here are some key aspects of nanobiomaterials:

a. **Enhanced Surface Area**: Due to their nanoscale size, nanobiomaterials have an increased surface area in comparison with their bulk counterparts. This high surface area allows for more efficient interactions with biological molecules and cells, making them ideal for various biomedical applications.
b. **Targeted Drug Delivery**: Nanobiomaterials can be engineered to carry drugs or therapeutic agents and deliver them directly to specific cells or tissues in the

body. This targeted drug delivery approach minimizes side effects and enhances the effectiveness of treatments.

c. **Tissue Regeneration**: In tissue engineering, nanobiomaterials are utilized to create scaffolds that can mimic the natural extracellular matrix. These scaffolds provide a supportive environment for cells to grow, differentiate, and regenerate damaged tissues.
d. **Imaging and Diagnostics**: Nanobiomaterials are employed in medical imaging techniques, such as contrast agents in magnetic resonance imaging (MRI) and computed tomography (CT). They can also serve as sensitive diagnostic probes for detecting specific biomarkers associated with diseases.
e. **Biocompatibility**: One of the essential characteristics of nanobiomaterials is their biocompatibility, which ensures that they do not cause harmful reactions or toxicity when in contact with living tissues.
f. **Surface Functionalization**: Nanobiomaterials can be functionalized with specific molecules, proteins, or ligands to interact with specific biological targets, enhancing their specificity and efficiency in various applications.

Due to their versatile properties and vast potential, nanobiomaterials continue to be an exciting area of research and development in the biomedical field, with the potential to revolutionize various aspects of healthcare and medical treatments. However, it is essential to conduct thorough research and safety assessments before translating these materials into real-world applications to ensure their safe and effective use in medical practices.

2 Tissue Engineering

Nanobiomaterials play a crucial role in tissue engineering, a field focused on creating artificial tissues or organs to replace or repair damaged ones. Tissue engineering aims to regenerate functional tissues by combining biomaterials, cells, and bioactive molecules in a way that mimics the native tissue's structure and function. Nanobiomaterials offer unique advantages in this context due to their nanoscale features and tailored properties. Here are some ways in which nanobiomaterials are applied in tissue engineering:

- **Scaffold fabrication**: Nanobiomaterials are used to create scaffolds, which are three-dimensional structures that act as templates or frameworks to support cell growth and tissue regeneration. Nanoscale features in the scaffold, such as nanofibers, nanopores, or nanotopographies, provide a more biomimetic environment for cells, resembling the natural extracellular matrix (ECM) of tissues.
- **Enhanced cell adhesion and proliferation**: Nanobiomaterials can be functionalized with specific ligands or peptides at the nanoscale to improve cell adhesion and proliferation. This enhances the interaction between cells and the scaffold, promoting tissue formation and regeneration.

- **Controlled drug delivery**: Nanoparticles or nanostructured materials can be utilized as carriers for growth factors, cytokines, or other bioactive molecules. These nanoparticles release the bioactive molecules in a controlled and sustained manner, providing a localized and targeted approach to stimulate tissue regeneration.
- **Biodegradability and bioresorbability**: Many nanobiomaterials are designed to be biodegradable, which means they can be gradually broken down by the body over time. As the scaffold degrades, it leaves space for new tissue formation. Biodegradability is especially important in tissue engineering, where the scaffold's role is temporary, and native tissue gradually replaces it.
- **Nanocomposites for mechanical properties**: Nanoparticles or nanostructured reinforcements can be added to the biomaterials to improve their mechanical properties, such as strength, elasticity, and toughness. This helps the scaffold withstand mechanical stresses and forces experienced in the body.
- **Regeneration of complex tissues**: Nanobiomaterials are particularly useful in engineering complex tissues, such as cartilage, bone, and nerves, where precise control of cell behavior and tissue microenvironment is critical.
- **Imaging and tracking**: Some nanobiomaterials, such as contrast agents with nanoparticles, can be used for non-invasive imaging of the tissue-engineered constructs, allowing researchers to monitor tissue growth and integration in real-time.

Overall, nanobiomaterials in tissue engineering offer a versatile platform for creating biocompatible, bioactive, and biomimetic scaffolds to promote tissue regeneration, providing potential solutions for various medical conditions and improving patient outcomes. Continued research and development in this area hold great promise for advancing regenerative medicine and personalized healthcare [2].

In tissue engineering, the use of nanobiomaterials involves several stages to create functional tissues or organs for transplantation or regenerative purposes (Fig. 1). The process typically includes the following stages:

- **Material selection and design**: The first step is to select appropriate nanobiomaterials, such as nanofibers, nanoparticles, or nanostructured materials, based on the target tissue or organ. The nanobiomaterials should have properties that resemble the native tissue's extracellular matrix (ECM) to provide an appropriate microenvironment for cell attachment, growth, and tissue development.
- **Cell isolation and removal**: In tissue engineering, cells from the patient (autologous cells) or other sources (allogeneic or stem cells) are used to seed the nanobiomaterial scaffold. Before seeding, if autologous cells are being used, they may need to be harvested from the patient's tissue or blood. In some cases, cells may need to be isolated from their original tissue using enzymatic or mechanical methods.
- **Cell expansion in culture**: Once the cells are isolated, they are expanded in culture to increase their numbers. Cell expansion involves providing a suitable culture medium that contains nutrients and growth factors, promoting cell division and

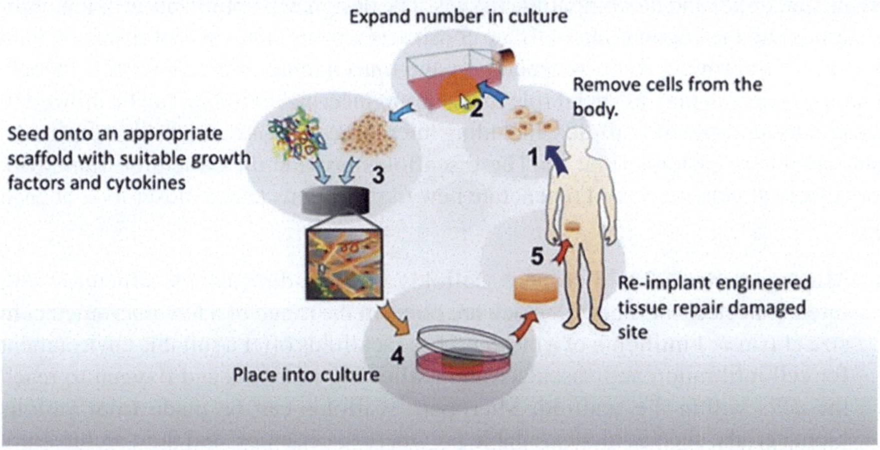

Fig. 1 Several stages to create functional tissues or organs for transplantation or regenerative

proliferation. During this stage, cells are encouraged to grow and multiply until enough cells are obtained for tissue engineering applications.

- **Seeding cells onto nanobiomaterial scaffold**: After cell expansion, the next step is to seed the cells onto the nanobiomaterial scaffold. This can be done using various techniques, such as direct cell seeding, cell spraying, or cell suspension seeding. The cells are carefully distributed onto the scaffold, ensuring even distribution and proper cell attachment.
- **Culturing the construct**: Once the cells are seeded onto the scaffold, the tissue-engineered construct is placed in a bioreactor or a specialized culture system. This system provides the necessary conditions for cell growth, including controlled temperature, humidity, and nutrient supply. The construct is cultured for a period of time to allow the cells to proliferate and form new tissue.
- **Maturation and tissue development**: During the culture period, the tissue-engineered construct undergoes maturation as the cells continue to grow and differentiate into specific cell types. The nanobiomaterial scaffold provides structural support and guidance for tissue development. Bioreactor conditions, such as mechanical stimulation or the addition of specific growth factors, may be applied to promote tissue maturation.
- **Implantation or transplantation**: After an adequate maturation period, the tissue-engineered construct is ready for implantation or transplantation. If the tissue is intended for regenerative purposes, it may be transplanted into the patient's body to replace damaged or lost tissue. Alternatively, the tissue may be used for research, drug testing, or other applications in vitro [3, 4].

The use of nanobiomaterials in tissue engineering offers significant potential to create functional and biocompatible tissues. The design and optimization of nanobiomaterials and the engineering of tissue constructs are ongoing areas of research, with the aim of improving tissue regeneration and transplantation outcomes [5]. Indeed, nanobiomaterials play a crucial role in tissue engineering, and they can be utilized to create various types of scaffolds, including micropore scaffolds, microfiber scaffolds, and nanofiber scaffolds (Fig. 2). These scaffolds provide the structural framework for cells to attach, grow, and regenerate new tissue. Let us take a closer look at each type:

a. **Micropore Scaffold**: Micropore scaffolds are three-dimensional structures with interconnected micropores, which are pores in the range of a few micrometers in size (1 μm = 1 millionth of a meter). These scaffolds offer a suitable environment for cell infiltration and vascularization, allowing nutrients and oxygen to reach the cells within the scaffold. Micropore scaffolds can be made from various biomaterials, such as biodegradable polymers or ceramics, and their architecture supports tissue regeneration and integration.

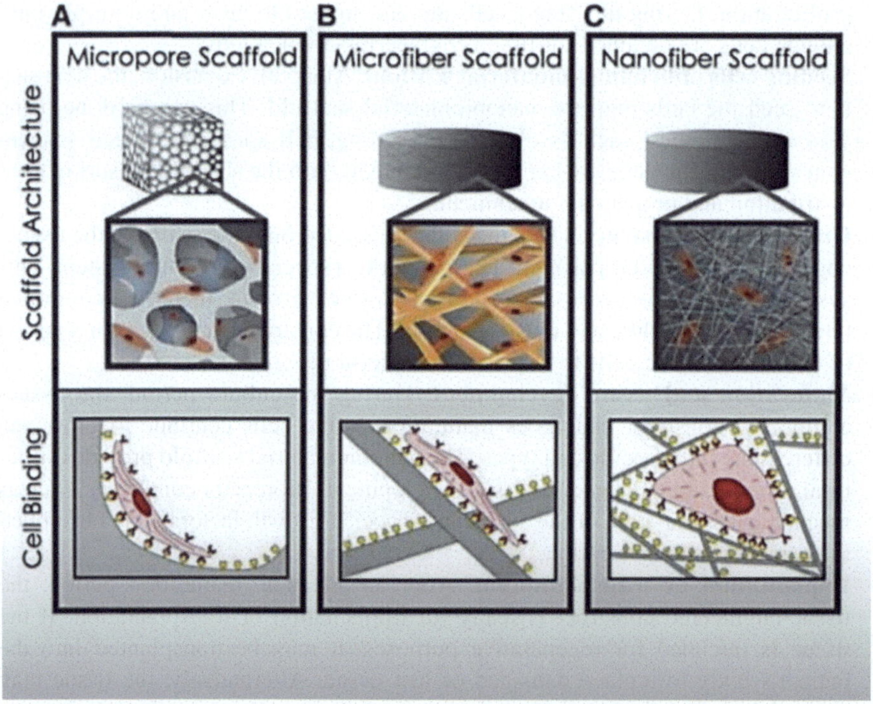

Fig. 2 Various types of scaffolds, including micropore scaffolds, microfiber scaffolds, and nanofiber scaffolds

b. **Microfiber Scaffold**: Microfiber scaffolds are composed of small fibers with diameters in the micrometer range. These fibers can be arranged in a mesh-like structure to create a scaffold with high surface area and porosity. The microfiber scaffold provides mechanical support and guides cell growth and alignment. The interconnected porosity of the scaffold facilitates nutrient and waste exchange and supports the formation of organized tissues.
c. **Nanofiber Scaffold**: Nanofiber scaffolds are composed of ultrafine fibers with diameters in the nanometer range. These scaffolds mimic the fibrous structure of the natural extracellular matrix (ECM), offering a highly biomimetic environment for cells. The small size of nanofibers enhances cell–scaffold interactions and promotes cell attachment, proliferation, and tissue regeneration. Nanofiber scaffolds can be fabricated using techniques like electrospinning, which enables precise control over the fiber diameter and alignment.

The choice of scaffold type depends on the specific tissue engineering application and the desired properties of the engineered tissue. Micropore scaffolds are suitable for tissues that require good nutrient diffusion and blood vessel infiltration. Microfiber scaffolds are often used for tissues that need mechanical support and cellular alignment, such as muscle or nerve regeneration. Nanofiber scaffolds are advantageous for applications where mimicking the nanoscale architecture of native tissue ECM is critical, such as in skin, bone, or nerve tissue engineering.

Nanobiomaterials enable researchers to precisely tailor the properties of these scaffolds, such as mechanical strength, porosity, degradation rate, and bioactivity, to optimize tissue regeneration and integration. As technology and materials continue to advance, nanobiomaterials-based scaffolds hold great promise for advancing tissue engineering and regenerative medicine.

The term "construction scaffold" in the context of tissue engineering refers to the fabrication or creation of scaffolds used to support and guide the growth of cells and tissues. Different techniques can be employed to construct these scaffolds, and three of these techniques are electrospinning, self-assembly, and fuzzy separation.

- **Electrospinning**: Electrospinning is a widely used technique for producing nanofiber scaffolds. It involves the creation of ultrafine fibers from a polymer solution using an electrostatic field. The process begins with a polymer solution loaded in a syringe or reservoir. A high voltage is applied to the solution, causing an electrostatic force that draws the polymer solution into a fine jet. As the jet travels through the electric field, it undergoes elongation and solidification, leading to the formation of nanofibers that collect on a grounded target, such as a rotating drum or a collector plate. The resulting nanofiber scaffold exhibits a high surface area-to-volume ratio, which mimics the fibrous structure of the extracellular matrix (ECM) in native tissues. The nanofibers' small size and interconnected porosity create a biomimetic environment that facilitates cell attachment, proliferation, and tissue regeneration. Electrospinning allows precise control over the scaffold's architecture, including fiber diameter, alignment, and composition, making it suitable for various tissue engineering applications [6].

- **Self-assembly**: Self-assembly is a technique that relies on the intrinsic properties of certain materials to spontaneously organize into desired structures without the need for external guidance or complex processing. In tissue engineering, self-assembly involves using biomolecules or biomaterials that can interact with each other to form organized structures, such as hydrogels or nanofibers. For example, self-assembling peptides or proteins can form a scaffold through molecular interactions, such as hydrogen bonding, hydrophobic interactions, or electrostatic forces. These interactions cause the molecules to spontaneously assemble into a stable scaffold structure. The resulting scaffold can provide an environment conducive to cell growth and tissue regeneration. Self-assembly offers simplicity and versatility in scaffold construction, as it does not require complex processing steps. It also allows for the incorporation of bioactive molecules that can influence cell behavior and tissue development.
- **Fuzzy Separation**: Fuzzy separation is a technique used to create porous scaffolds with interconnected void spaces. This method involves using phase separation to create a scaffold structure with a sponge-like appearance. The process begins with a polymer solution containing two immiscible components. After casting or molding the solution, a non-solvent or a phase-separating agent is introduced to induce phase separation between the two components.

As the polymer solution undergoes phase separation, solidification occurs, resulting in a scaffold with a porous structure. The interconnected voids within the scaffold allow for nutrient diffusion, waste removal, and cell infiltration, creating an optimal microenvironment for tissue regeneration. Fuzzy separation offers the advantage of creating porous scaffolds with tunable pore size and distribution, making it suitable for various tissue engineering applications, including bone and cartilage regeneration.

In summary, these scaffold construction techniques, including electrospinning, self-assembly, and fuzzy separation, contribute to the development of functional and biomimetic scaffolds for tissue engineering, supporting cell growth, and promoting tissue regeneration [7].

3 Drug Delivery

Nanobiomaterials have revolutionized the field of drug delivery due to their unique properties and capabilities. Their nanoscale size and tailor-made characteristics allow for precise control over drug release, targeting, and bioavailability. Nanobiomaterials offer several significant advantages when used in drug delivery applications. These advantages make them highly promising and sought-after carriers for delivering therapeutic agents effectively and precisely [8]. Some of the key advantages of nanobiomaterials in drug delivery include:

- **Enhanced Drug Solubility**: Nanobiomaterials can encapsulate both hydrophobic and hydrophilic drugs, improving their solubility and stability. This enhanced solubility leads to better bioavailability of poorly soluble drugs, increasing their therapeutic efficacy.
- **Targeted Drug Delivery**: Nanobiomaterials can be functionalized with targeting ligands that recognize specific receptors or markers expressed on the surface of target cells or tissues. This enables targeted drug delivery, concentrating the therapeutic agent at the desired site while reducing off-target effects on healthy tissues.
- **Controlled Drug Release**: Nanobiomaterials can be engineered to release drugs in a controlled and sustained manner. This controlled release profile helps maintain therapeutic drug levels over an extended period, reducing the frequency of administration and improving patient compliance.
- **Protection of Drugs**: Nanobiomaterial carriers can protect drugs from degradation, enzymatic breakdown, or premature clearance in the body. This protection preserves the drug's potency during transit in the bloodstream and ensures its intact delivery to the target site.
- **Biocompatibility**: Many nanobiomaterials are biocompatible and biodegradable, reducing the risk of adverse reactions or long-term toxicity. They are well-tolerated by the body, making them safe carriers for drug delivery applications.
- **Multifunctionality**: Nanobiomaterials can be engineered to carry multiple drugs, imaging agents, or diagnostic molecules simultaneously. This multifunctionality allows for theranostic applications, where drug delivery and disease monitoring can occur in a single system.
- **Non-immunogenicity**: Some nanobiomaterials, such as liposomes and certain polymer-based nanoparticles, have low immunogenicity, reducing the risk of triggering immune responses in the body.
- **Personalized Medicine**: Nanobiomaterials offer the potential for personalized medicine approaches. They can be tailored to specific disease conditions or individual patient characteristics, enabling customized drug delivery strategies.
- **Non-invasive Routes of Administration**: Nanobiomaterials can be designed for non-invasive routes of drug administration, such as oral, nasal, or transdermal delivery. This improves patient comfort and compliance.
- **Imaging and Therapeutic Integration**: Some nanobiomaterials possess inherent imaging properties or can be loaded with contrast agents for diagnostic purposes. This integration of imaging and therapy allows real-time monitoring of drug distribution and treatment response [9].

Overall, the unique properties of nanobiomaterials make them versatile and effective carriers for drug delivery, paving the way for more targeted, efficient, and patient-friendly therapeutic interventions in various medical fields. Here are some key applications of nanobiomaterials in drug delivery:

- **Nanoparticles for targeted drug delivery**: Nanoparticles, such as liposomes, polymeric nanoparticles, and dendrimers, can encapsulate drugs and deliver them to specific target sites in the body. These nanoparticles can be engineered to release

the drug in a controlled manner, either through passive diffusion or in response to specific stimuli, such as changes in pH or temperature. This targeted drug delivery approach enhances drug efficacy while minimizing side effects on healthy tissues.
- **Liposomes for drug encapsulation**: Liposomes are spherical vesicles made of lipid bilayers that can encapsulate hydrophilic or hydrophobic drugs. Liposomes are biocompatible and can protect the drug from degradation in the bloodstream. They can also accumulate in target tissues with leaky blood vessels, such as tumors, through the enhanced permeability and retention (EPR) effect.
- **Nanoparticles for crossing biological barriers**: Nanobiomaterials can be engineered to transport drugs across biological barriers that are otherwise impermeable to many therapeutic agents. For instance, solid lipid nanoparticles and polymeric micelles can enhance drug penetration through the blood–brain barrier, allowing for potential treatments for neurological disorders.
- **Nanogels and hydrogels for sustained release**: Nanogels and hydrogels are three-dimensional networks of cross-linked polymers that can swell and absorb water, providing a matrix for drug entrapment. These materials can release drugs in a controlled and sustained manner, reducing the frequency of drug administration and improving patient compliance [10].
- **Nanoparticles for intracellular drug delivery**: Some nanobiomaterials can facilitate intracellular drug delivery by overcoming cellular barriers and delivering therapeutic agents directly to the cytoplasm or organelles. This approach is particularly relevant for treating intracellular infections or targeting specific cellular pathways.
- **Multifunctional nanocarriers**: Nanobiomaterials can be engineered to carry not only therapeutic drugs but also imaging agents, targeting ligands, or other functionalities. This allows for simultaneous drug delivery, real-time monitoring of drug distribution, and personalized medicine approaches.
- **Personalized medicine and precision drug delivery**: Nanobiomaterials hold the potential for personalized medicine, tailoring drug delivery to individual patients based on their genetic makeup, disease state, or specific cellular responses.
- **Gene delivery**: Nanobiomaterials can also be used for gene delivery, delivering genetic material, such as siRNA or DNA, to target cells for gene therapy applications.

Overall, nanobiomaterials have opened up new avenues in drug delivery, enabling more effective, targeted, and personalized therapeutic approaches while reducing side effects. Continued research and development in this area hold promise for advancing healthcare and improving patient outcomes [11].

In nanobiomedicine, there are various types of nanobiomaterials that can serve as carriers for free drugs. These nanobiomaterials allow for the controlled and targeted delivery of therapeutic agents, enhancing the efficacy and reducing the side effects of drugs. Here are some examples of nanobiomaterials used as carriers for free drugs:

- **Metal-Based Nanoparticles**: Metal-based nanoparticles, such as gold nanoparticles, silver nanoparticles, and iron oxide nanoparticles, can be functionalized to carry free drugs. These nanoparticles have unique physicochemical properties that make them attractive for drug delivery. They can be coated with polymers or biological molecules to improve their stability, biocompatibility, and targeting capabilities. Metal-based nanoparticles are often used in cancer therapy, as they can be functionalized with targeting ligands to specifically deliver drugs to tumor cells.

- **Lipid-Based Nanoparticles**: Lipid-based nanoparticles, such as liposomes and solid lipid nanoparticles, are composed of lipid bilayers or lipid cores, respectively. They can encapsulate both hydrophilic and hydrophobic drugs within their inner core or lipid membrane. Lipid-based nanoparticles offer excellent biocompatibility and can protect drugs from degradation in the bloodstream. Additionally, they can be engineered to target specific cells or tissues, making them useful for targeted drug delivery.
- **Polymer-Based Nanoparticles**: Polymer-based nanoparticles, like polymeric micelles and nanogels, are composed of biodegradable or biocompatible polymers. They can entrap drugs in their hydrophobic cores or within their polymer matrix. These nanoparticles can release drugs in a controlled and sustained manner, providing prolonged therapeutic effects. Polymer-based nanoparticles are versatile and can be engineered to respond to specific stimuli, such as changes in pH or temperature, for triggered drug release.
- **Biological Nanoparticles**: Biological nanoparticles are naturally occurring or engineered particles derived from biological sources, such as exosomes, virus-like particles, and cell membrane-coated nanoparticles. They can serve as carriers for free drugs and exhibit inherent biocompatibility and low immunogenicity. Biological nanoparticles can also be modified to display specific ligands on their surface, enabling targeted drug delivery to specific cells or tissues [12].

All of these nanobiomaterials offer unique advantages and can be tailored for specific drug delivery applications. The choice of nanobiomaterials depends on factors such as the type of drug being delivered, the target tissue, and the desired release kinetics. Nanobiomedicine continues to advance, offering new opportunities for precise and effective drug delivery strategies in various medical fields.

In pharmaceutical and biomedical applications, various types of pharmaceutical carriers are used to deliver therapeutic agents effectively and safely (Fig. 3). These carriers enable controlled release, targeted delivery, and improved bioavailability of drugs. Here are some examples of pharmaceutical carriers commonly used in nanomedicine:

- **Vesicles (Liposomes)**: Liposomes are spherical vesicles composed of lipid bilayers that can encapsulate hydrophilic or hydrophobic drugs within their core or membrane. Liposomes offer excellent biocompatibility and can be modified with targeting ligands for specific cell or tissue targeting. They are versatile carriers used in drug delivery for cancer therapy, gene delivery, and vaccine delivery.
- **Micelles**: Micelles are nano-sized aggregates formed by amphiphilic molecules in an aqueous solution. They spontaneously self-assemble into structures with a hydrophobic core and a hydrophilic shell. Micelles can solubilize hydrophobic

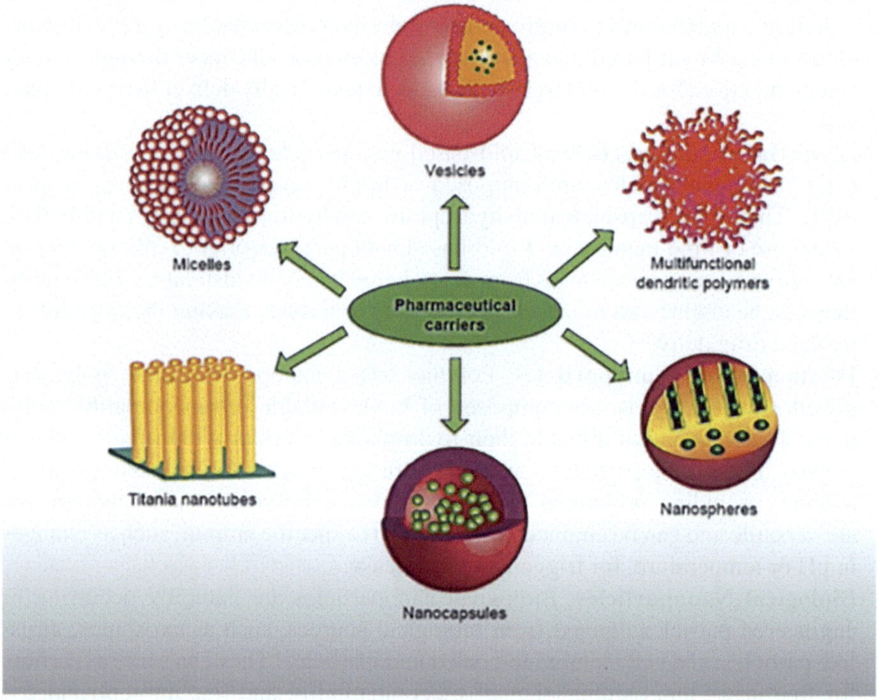

Fig. 3 Various types of pharmaceutical carriers are used to deliver therapeutic agents effectively and safely

drugs and improve their bioavailability. They are effective carriers for delivering poorly water-soluble drugs.

- **Nanospheres (Polymeric Nanoparticles)**: Polymeric nanospheres are nanoparticles made from biocompatible and biodegradable polymers. They can encapsulate drugs within their matrix or on their surface. Nanospheres enable controlled release of drugs and protect them from degradation. These carriers are widely used in drug delivery for various diseases.
- **Nanocapsules**: Nanocapsules are nano-sized carriers with a core–shell structure. The drug is enclosed within the core, while the shell can be made of a different material. Nanocapsules offer tunable release profiles and can be modified to target specific cells or tissues.
- **Titania Nanotubes (Titanium Dioxide Nanotubes)**: Titania nanotubes are nanomaterials made of titanium dioxide with a tubular structure. They have high surface area and can serve as carriers for drug delivery. Titania nanotubes can be functionalized with targeting ligands and can release drugs in a controlled manner, making them promising carriers for localized drug delivery.
- **Multifunctional Dendritic Polymers**: Dendritic polymers are highly branched, tree-like macromolecules. They can be engineered to carry multiple drugs,

imaging agents, and targeting ligands in a single structure. These multifunctional dendritic polymers offer the advantage of combining different functionalities in a single carrier for theranostics applications.

These pharmaceutical carriers provide a wide range of options for drug delivery, enabling researchers and clinicians to design tailored drug delivery systems based on the specific therapeutic needs. The field of nanomedicine continues to evolve, with ongoing research focused on optimizing carrier properties and exploring new nanotechnologies to revolutionize drug delivery in medicine.

Targeted drug delivery systems utilize a combination of three key components: the drug (therapeutic agent), the targeting ligand, and the enzyme (in certain cases). These components work together to deliver drugs specifically to the desired site in the body, increasing therapeutic efficacy and minimizing off-target effects. Here is how they function:

a. **Drug (Therapeutic Agent)**: The drug is the active pharmaceutical ingredient that exerts a therapeutic effect on the target cells or tissues. It can be a small molecule, a biologic, or even genetic material like RNA or DNA. The drug is chosen based on its ability to interact with specific disease-related targets or cellular pathways.
b. **Targeting Ligand**: The targeting ligand is a molecule, often a peptide or an antibody fragment, that specifically recognizes and binds to specific receptors or markers expressed on the surface of target cells. The ligand acts as a homing device, guiding the drug-loaded carrier to the desired site. This targeted approach is especially useful in diseases where specific cells or tissues need to be targeted while sparing healthy cells.
c. **Enzyme (in certain cases)**: In some cases, an enzyme can be involved in the targeted drug delivery system. For example, in prodrug therapy, the drug is attached to a prodrug-converting enzyme. This prodrug is inactive until it encounters the specific enzyme present at the target site. Upon enzymatic conversion, the prodrug is transformed into its active form, releasing the drug. This approach can enhance drug specificity and reduce systemic toxicity.

The targeted drug delivery system is designed to carry the drug in a carrier or a nanoparticle, protecting it from degradation and allowing controlled release [11]. The carrier can also facilitate the delivery of the drug to the target site by exploiting the enhanced permeability and retention (EPR) effect in tumors or by active targeting using ligands. When administered, the drug-loaded carrier circulates in the bloodstream. When it reaches the target site with the help of the targeting ligand, the drug is released either by passive diffusion or triggered release (such as enzyme activation). This ensures that the drug concentrates at the intended site, enhancing therapeutic efficacy while minimizing side effects on healthy tissues. Overall, targeted drug delivery systems offer a promising strategy to improve the effectiveness of drugs, reduce systemic toxicity, and enhance patient outcomes in various diseases, including cancer, inflammation, and certain genetic disorders.

4 Diagnostic Imaging

Diagnostic imaging with nanobiomedicine involves the use of nanotechnology-based imaging agents to enhance the visualization and detection of disease-related structures, biomarkers, or physiological processes. Nanobiomedicine combines nanotechnology with medical imaging techniques to create highly sensitive and specific imaging tools that aid in early disease detection, disease staging, and treatment monitoring. Here is how diagnostic imaging with nanobiomedicine works:

- **Nanoparticle Imaging Agents**: Nanoparticles, such as quantum dots, gold nanoparticles, iron oxide nanoparticles, and silica nanoparticles, are used as imaging agents in nanobiomedicine. These nanoparticles possess unique optical, magnetic, or acoustic properties that make them excellent contrast agents for various imaging modalities.
- **Targeted Imaging**: Nanobiomedicine allows the design and development of targeted imaging agents. Specific ligands or antibodies can be attached to the surface of nanoparticles to selectively bind to disease-specific biomarkers or receptors. By targeting these biomarkers, the imaging agent homes in on the diseased tissue, enabling more accurate and specific imaging.
- **Multimodal Imaging**: Nanobiomedicine enables the integration of multiple imaging modalities into a single nanoparticle. This multimodal imaging approach provides complementary information, enhancing the sensitivity and accuracy of the diagnostic process. For example, a nanoparticle can be designed to carry both fluorescent and magnetic properties, enabling fluorescence imaging and magnetic resonance imaging (MRI) in one agent.
- **Contrast Enhancement**: Nanoparticles can significantly enhance contrast in medical imaging. For instance, in computed tomography (CT), nanoparticles with high atomic numbers can be used as contrast agents to improve image quality and highlight specific tissues or blood vessels.
- **Imaging of Biological Processes**: Nanobiomedicine can facilitate imaging of biological processes at the cellular or molecular level. Fluorescent nanoparticles, for instance, can be used to track specific cells or molecules in real-time, providing valuable insights into disease progression and response to treatment.
- **Early Disease Detection**: The sensitivity of nanobiomedicine-based imaging agents allows for the early detection of diseases, such as cancer, even before symptoms become apparent. Early diagnosis can lead to timely intervention and improved patient outcomes.
- **Theranostics**: Nanobiomedicine also enables theranostics applications, where imaging agents can be combined with therapeutic agents within a single nanoparticle. This theranostics approach allows simultaneous imaging and treatment of diseases, providing real-time feedback on treatment efficacy.
- **Non-invasive Imaging**: Nanobiomedicine allows for non-invasive imaging techniques, reducing the need for invasive procedures and improving patient comfort and compliance.

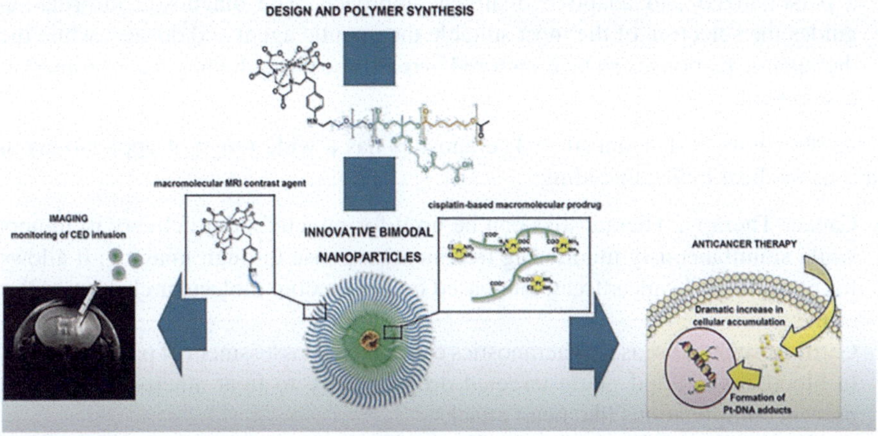

Fig. 4 Theranostics mechanism of bimodal nanoparticles

Overall, diagnostic imaging with nanobiomedicine represents a cutting-edge approach that holds great promise for revolutionizing medical imaging, enabling earlier and more accurate diagnoses, and advancing personalized medicine for various diseases. As this field continues to evolve, nanobiomedicine is likely to play an increasingly essential role in improving healthcare outcomes. "*Theranostic*" is a portmanteau of two words: "*therapeutic*" and "*diagnostic*". It refers to a class of advanced medical technologies or treatments that combine therapeutic and diagnostic functionalities within a single system as shown in Fig. 4. The primary goal of theranostics is to improve patient care by enabling personalized and precise medicine approaches.

Here is how theranostics work:

- **Therapeutic Function**: The therapeutic aspect of theranostics involves delivering a therapeutic agent, such as a drug, gene, or radiation, to the target site in the body. The therapeutic agent is designed to treat the disease or condition, and the delivery system ensures that it reaches the intended location with high specificity and efficiency. This targeted delivery minimizes damage to healthy tissues and enhances the therapeutic effect.
- **Diagnostic Function**: The diagnostic aspect of theranostics involves using imaging or sensing technologies to visualize or monitor disease-related biomarkers, cellular processes, or specific targets in the body. By providing real-time and accurate information about the disease status, these diagnostics help guide the selection and optimization of the therapeutic approach. Diagnostic information can be obtained before, during, or after therapy to assess treatment response and make adjustments as needed.
- **Integration of Therapeutics and Diagnostics**: In theranostics, the therapeutic and diagnostic components are integrated into a single system or platform. This integration allows for seamless communication between the two aspects, enabling

a personalized and adaptive treatment approach. The diagnostic information guides the selection of the most suitable therapeutic agent and dosage, while the therapeutic response can be monitored in real-time through imaging or biomarker assessment.

Applications of Theranostics: Theranostics has a wide range of applications in various medical fields, including:

- Cancer Therapy: Theranostics can be used for targeted drug delivery to tumors while simultaneously monitoring treatment response through imaging. It allows for personalized cancer treatment based on the specific molecular characteristics of the tumor.
- Cardiovascular Diseases: Theranostics can aid in the assessment of plaque buildup in blood vessels and guide targeted drug delivery to treat atherosclerosis and prevent complications like heart attacks.
- Neurological Disorders: In neurology, theranostics can help monitor disease progression and target drug delivery to specific brain regions for conditions like Alzheimer's disease or brain tumors.
- Infectious Diseases: Theranostics can be used to detect and monitor infections while simultaneously delivering antimicrobial agents to the infection site.
- Regenerative Medicine: Theranostics can be utilized to monitor tissue repair and regeneration and deliver growth factors or stem cells to aid in tissue healing.

Overall, theranostics is an emerging field that holds great promise for advancing healthcare and improving patient outcomes by combining diagnostics and therapeutics in an integrated and patient-specific manner.

5 Therapeutics

Nanobiomaterials have diverse applications in therapeutics, revolutionizing the field of medicine and improving treatment strategies. Some of the key applications of nanobiomaterials in therapeutics include:

- **Cancer Therapy**: Nanobiomaterials play a crucial role in cancer treatment. They can be functionalized to specifically target cancer cells or tumor tissues, delivering chemotherapeutic agents or other therapies directly to the site of the tumor while sparing healthy tissues. This targeted approach enhances the therapeutic effect and reduces systemic toxicity.
- **Gene Therapy**: Nanobiomaterials are used in gene therapy to deliver therapeutic genes to specific cells or tissues. They protect the genetic material from degradation and facilitate its efficient uptake by the target cells, potentially correcting genetic abnormalities or promoting tissue regeneration. Gene therapy is an innovative and promising approach in the field of medicine that aims to treat or cure genetic and acquired diseases by introducing genetic material into a patient's cells. The fundamental idea behind gene therapy is to correct or replace faulty

genes that cause diseases or to introduce new genes to enhance the body's ability to fight against specific conditions. The therapeutic gene or DNA is delivered into the patient's cells using various vectors, which can be viral or non-viral in nature. Viral vectors are engineered viruses that have been modified to carry the therapeutic gene, while non-viral vectors use other methods, such as liposomes or nanoparticles, to transport the genetic material. The delivered genetic material must reach the specific cells or tissues affected by the disease. Depending on the condition being treated, the target cells may be located in different organs or tissues throughout the body. Once inside the target cells, the therapeutic gene integrates into the patient's genome or expresses itself in the cytoplasm, depending on the type of therapy. The goal is to enable the cells to produce functional proteins that are either missing or malfunctioning in the disease state. In some cases, gene therapy aims to correct specific genetic defects that cause diseases. This involves replacing the faulty gene with a healthy one, which can be accomplished using various gene editing techniques like CRISPR-Cas9.

Gene therapy has shown great promise in the treatment of a wide range of diseases, including inherited genetic disorders, certain types of cancer, and acquired diseases. Gene therapy can potentially cure or alleviate symptoms of genetic diseases caused by single-gene mutations, such as cystic fibrosis, sickle cell anemia, and muscular dystrophy. In cancer gene therapy, therapeutic genes are introduced to target and kill cancer cells selectively, activate the immune system against tumors, or suppress tumor growth. Gene therapy can be used to combat viral infections by introducing genes that interfere with the replication of viruses or enhance the immune response against them. Gene therapy is being explored to address metabolic disorders by introducing genes that restore or enhance the function of specific enzymes or proteins.

Although gene therapy holds great potential, it is still a relatively new and evolving field, and challenges remain, such as ensuring the safety and long-term effectiveness of the treatment, avoiding unintended consequences, and developing efficient delivery methods. Ongoing research and clinical trials continue to advance the understanding and application of gene therapy, bringing hope for improved treatments and cures for various diseases in the future.

a. **Immunotherapy**: Nanobiomaterials are employed in immunotherapy to enhance the immune response against diseases, such as cancer or infectious agents. They can be designed to deliver immune-stimulating molecules or antigens to immune cells, promoting a targeted and robust immune reaction [13].
b. **Wound Healing**: Nanobiomaterials are used in wound dressings and tissue scaffolds to promote tissue regeneration and wound healing. They can deliver growth factors or antimicrobial agents to the wound site, facilitating the healing process and reducing the risk of infection.
c. **Cardiovascular Therapy**: Nanobiomaterials are utilized in cardiovascular therapy to address various cardiac conditions. They can be incorporated into stents or patches to prevent restenosis after angioplasty or to promote tissue repair after a heart attack.

d. **Neurological Disorders**: Nanobiomaterials hold promise in treating neurological disorders. They can be engineered to cross the blood–brain barrier and deliver drugs or therapeutic agents to the brain, potentially treating conditions like Alzheimer's or Parkinson's disease.
e. **Bone Regeneration**: Nanobiomaterials are employed in bone tissue engineering to promote bone regeneration and repair. They can serve as scaffolds for cell growth and mineral deposition, aiding in the restoration of damaged bone tissue. Traditional approaches to bone regeneration, such as bone grafts or implants, have limitations, and nanobiomaterials offer unique advantages in promoting bone healing and tissue regeneration. Nanobiomaterials have a nanoscale structure, which mimics the natural composition of bone tissue. This similarity at the molecular level allows for better integration with the host tissue and enhances the effectiveness of the regenerative process. Nanobiomaterials can act as scaffolds, providing a 3D structure to support cell growth and tissue formation. These scaffolds can be engineered to have specific mechanical properties, porosity, and surface characteristics to facilitate bone cell attachment and proliferation. Nanobiomaterials possess osteoconductive properties, meaning they encourage the growth and migration of bone-forming cells, such as osteoblasts, into the implant site. This helps in the formation of new bone tissue. Some nanobiomaterials have intrinsic osteoinductive properties, meaning they can stimulate the differentiation of stem cells into bone-forming cells, further enhancing the regenerative process. Nanobiomaterials are engineered to be biocompatible, ensuring that they do not elicit harmful immune responses or toxicity in the body. This is critical for successful bone regeneration without adverse reactions. The nanoscale features of these biomaterials allow for better integration with the surrounding host tissue, ensuring mechanical stability and long-term success. Nanobiomaterials can be tailored and customized to meet the specific needs of the patient, considering factors such as the location and size of the bone defect, patient's age, and overall health. Nanobiomaterials can be engineered to have multifunctional properties, incorporating antimicrobial agents or other bioactive molecules to prevent infections and accelerate bone healing. Some nanobiomaterials can be administered non-invasively, such as injectable nanogels or nanoparticles, which minimizes the need for invasive surgical procedures. Research in bone regeneration using nanobiomaterials is ongoing, and pre-clinical and clinical studies have shown promising results. The integration of nanotechnology with regenerative medicine has the potential to revolutionize bone repair and tissue engineering, offering new solutions for patients with bone defects, fractures, or degenerative bone diseases.
f. **Antimicrobial Therapy**: Nanobiomaterials are utilized to combat drug-resistant microbial infections. They can deliver antimicrobial agents to infection sites and enhance their efficacy against resistant pathogens. The development of antibiotic resistance has become a significant global health concern, necessitating the exploration of alternative antimicrobial strategies. Nanobiomaterials can exhibit enhanced antimicrobial activity due to their high surface area and unique properties at the nanoscale. They can effectively interact with pathogens and disrupt their

structures, leading to improved killing efficiency. Some nanobiomaterials have demonstrated broad-spectrum antimicrobial activity, meaning they can target a wide range of microorganisms, including bacteria, viruses, and fungi. This broad activity is particularly useful in combating multi-drug-resistant pathogens. Nanobiomaterials can be designed to release antimicrobial agents in a controlled manner. This sustained release ensures a prolonged effect of the antimicrobial, reducing the need for frequent dosing and enhancing its efficacy. One of the advantages of nanobiomaterials is their ability to selectively target microbial cells while minimizing toxicity to host cells. This selectivity reduces the risk of adverse effects on healthy tissues. Nanobiomaterials can be engineered to carry multiple antimicrobial agents or therapeutic molecules. This enables combination therapy, where different antimicrobial agents work synergistically to enhance efficacy and prevent the development of resistance. Nanobiomaterials have been shown to inhibit biofilm formation, which is a major factor contributing to the persistence of infections and antibiotic resistance. Some nanobiomaterials possess photodynamic properties, meaning they can generate reactive oxygen species upon exposure to light. This photodynamic therapy can be used to selectively kill microbial cells while sparing host cells. In addition to their antimicrobial properties, certain nanobiomaterials also promote wound healing, making them suitable for use in infected wounds and skin infections. Antimicrobial therapy using nanobiomaterials is a rapidly evolving field, and ongoing research is exploring the full potential of these materials in combating infections and improving patient outcomes. As scientists continue to advance nanotechnology and nanomedicine, the development of effective and safe antimicrobial nanobiomaterials holds great promise in addressing the growing challenges of antimicrobial resistance and infectious diseases.

g. **Organ Transplantation**: Nanobiomaterials are investigated to improve organ transplantation outcomes. They can be used to encapsulate transplanted organs or cells, preventing rejection and promoting organ acceptance. Nanobiomaterials have emerged as a promising area of research to improve the success and outcomes of organ transplantation. These nanoscale materials offer unique properties and functionalities that can address various challenges associated with organ transplantation. Nanobiomaterials can be used in organ preservation solutions to improve the viability and longevity of donor organs outside the body. Nanoparticles can help protect organ tissues from oxidative stress and reduce inflammation during the preservation process. Nanobiomaterials can be designed to modulate the immune response in organ transplant recipients. They can be used to deliver immunosuppressive drugs in a controlled and targeted manner, reducing the risk of rejection and minimizing the side effects associated with systemic immunosuppression. Nanobiomaterials can serve as scaffolds for tissue engineering and regenerative medicine approaches in organ transplantation. They can support the growth and differentiation of stem cells, facilitating the regeneration of damaged or diseased tissues and organs. Nanobiomaterials can be engineered to have improved biocompatibility, reducing the risk of adverse reactions in the recipient's body. They can also promote better integration of the transplanted organ with the recipient's tissues, enhancing its function and longevity. Nanobiomaterials can contribute to improving the process of organ matching between donors and recipients, ensuring the best possible match for successful transplantation. They can also aid in optimizing the preservation and transportation of organs before transplantation [14].

6 Dentistry

Dentistry has been significantly impacted by the advancements in nanobiomaterials, which are nanoscale materials specifically designed for dental applications. These materials offer unique properties and functionalities that have revolutionized various aspects of dental care. Here are some key areas where nanobiomaterials have found applications in dentistry:

- **Dental Fillings**: Nanobiomaterials, such as nanocomposites, have been developed for dental fillings. These materials have improved mechanical properties, esthetics, and wear resistance compared to traditional dental materials, providing more durable and natural-looking restorations. Dental fillings are a common restorative treatment used to repair teeth that have been damaged by decay or trauma. Traditional dental fillings are typically made of materials like amalgam (a mixture of metals) or composite resin. However, with advancements in nanobiomaterials, dental fillings have evolved to include nanocomposites, which offer several advantages over traditional materials. These nanoparticles are usually composed of inorganic materials, such as silica, zirconia, or calcium phosphate, and they are typically in the nanometer range in size [15]. Nanocomposite dental fillings have enhanced translucency and can closely mimic the natural color of teeth. This esthetic improvement makes them a popular choice for visible areas of the mouth. The incorporation of nanoparticles enhances the mechanical properties of the dental filling, making it more durable and resistant to wear and fracture. The conventional dental composites can undergo polymerization shrinkage during the setting process, which may lead to gaps and microleakage around the filling [16]. Nanocomposites with specialized nanoparticles help reduce polymerization shrinkage, leading to better marginal integrity and reduced risk of secondary caries. Some nanobiomaterials used in dental fillings have inherent antibacterial properties. These materials can help prevent the growth of bacteria on the filling surface and reduce the risk of recurrent decay. Nanocomposite dental fillings are biocompatible, meaning they are well-tolerated by the surrounding oral tissues, reducing the risk of adverse reactions or sensitivity. It can be used for both anterior and posterior teeth. Their improved esthetics and mechanical properties make them suitable for various dental restorations. They have excellent handling characteristics, allowing dentists to manipulate and place the filling with precision. Nanobiomaterial dental fillings allow for a minimally invasive approach, preserving more of the healthy tooth structure during the restoration process. With their improved strength and durability, nanocomposite dental fillings have the potential for extended service life, reducing the need for frequent replacements [17, 18].

- **Dental Bonding Agents**: Nanobiomaterials are used in dental bonding agents to enhance the adhesion between the dental restorative materials and the tooth structure. This improves the longevity of dental restorations and reduces the risk of restoration failure. The advent of nanobiomaterials has revolutionized dental bonding agents, providing numerous advantages over traditional bonding agents. Nanobiomaterial bonding agents offer improved adhesion to the tooth surface due to their nanoscale features and interactions with the tooth structure. This results in a stronger bond between the dental restorative material and the tooth, reducing the risk of restoration failure. Microleakage refers to the microscopic gaps that may occur between the tooth and the restorative material. Nanobiomaterial bonding agents help reduce microleakage, preventing the ingress of bacteria and fluids, which can lead to secondary decay and other complications. Nanobiomaterial bonding agents have been engineered to reduce postoperative sensitivity, a common concern in dental restorations. The improved sealing capability of these agents helps protect the tooth's nerve endings, minimizing discomfort after the procedure. It often exhibits hydrophilic properties, allowing them to interact favorably with the moist oral environment. This hydrophilicity enhances the bonding agent's performance, especially in areas with high moisture content. Some nanobiomaterial bonding agents possess self-etching properties, allowing them to condition the tooth surface without the need for separate etching steps. This simplifies the bonding process, saving time and reducing the risk of errors. They are versatile and compatible with various restorative materials, including composite resins, ceramics, and glass ionomers. Nanobiomaterials bonding agents are biocompatible, ensuring that they are well-tolerated by the oral tissues without causing adverse reactions. The strong and durable bond created by nanobiomaterials bonding agents contributes to the longevity of dental restorations, reducing the need for frequent replacements. It enables a minimally invasive approach to dental restorations, preserving more of the healthy tooth structure during the bonding process [19].
- **Dental Implants**: Nanobiomaterials are being explored for dental implant coatings to improve osseointegration, the process by which the implant fuses with the surrounding bone. Nanoscale surface modifications can promote faster and more secure integration of dental implants. They are titanium or titanium alloy posts that are surgically placed into the jawbone to serve as artificial tooth roots. The use of nanobiomaterials in dental implants has brought about significant advancements in the field of implant dentistry. Nanobiomaterials allow for precise surface modifications of dental implants at the nanoscale. This nanostructuring enhances the implant's surface properties, promoting faster and more secure osseointegration—the process by which the implant fuses with the surrounding bone. This improved osseointegration leads to better stability and long-term success of the implant. The use of nanobiomaterials ensures better biocompatibility of dental implants. This means that the body's natural tissues are more accepting of the implant, reducing the risk of rejection or adverse reactions. The biocompatibility

of nanobiomaterials encourages the body to integrate the implant seamlessly into the surrounding bone and soft tissues. Some nanobiomaterials used in dental implants have inherent antibacterial properties. This feature helps prevent infections around the implant site, reducing the risk of peri-implantitis—a condition similar to gum disease that can lead to implant failure. The nanostructured surface of dental implants accelerates the healing process. It allows for quicker bone formation and tissue regeneration around the implant, enabling patients to receive their final prosthetic restoration sooner.

It can be engineered to have specific mechanical properties that match the natural bone, reducing stress on the surrounding tissues and improving the overall longevity of the implant. Nanobiomaterials can be coated with bioactive substances that encourage cell adhesion and tissue growth. These coatings promote a more favorable environment for successful osseointegration. They can be designed to have multifunctional features, such as controlled drug release systems. These features can help manage inflammation, enhance bone regeneration, and improve implant success rates. The properties of nanobiomaterials can be tailored to suit individual patient needs. Dental implants made with nanobiomaterials can be customized to match the patient's unique anatomical requirements. Dental implants made with nanobiomaterials have shown improved long-term stability and resistance to wear, ensuring a successful and lasting implant restoration. The integration of nanobiomaterials in dental implants continues to advance, leading to even more innovative and successful implant solutions. The use of these advanced materials in implant dentistry has significantly improved patient outcomes, providing individuals with functional, esthetically pleasing, and durable tooth replacements [20].

a. **Dental Adhesives**: Nanobiomaterials are used in dental adhesives to create a strong bond between the tooth structure and restorative materials. This helps in preventing microleakage and improving the longevity of dental restorations. Dental adhesives, also known as bonding agents, are crucial components in modern dentistry. They are used to create a strong and durable bond between dental restorative materials, such as composite resins or ceramic crowns, and the natural tooth structure.
b. **Dental Prosthetics**: Nanobiomaterials are used in the fabrication of dental prosthetics, such as crowns, bridges, and dentures, to enhance their strength, esthetics, and biocompatibility. Nanobiomaterials used in dental prosthetics are biocompatible, meaning they are well-tolerated by the oral tissues and do not cause adverse reactions. This ensures a comfortable fit and reduces the risk of irritation or inflammation in the patient's mouth. The nanoscale properties of these materials allow for improved mechanical strength and durability. Dental prosthetics made with nanobiomaterials are less prone to wear and fracture, resulting in longer-lasting restorations. The ability to engineer these materials at the nanoscale level allows for highly accurate and customized restorations that fit perfectly within the patient's oral anatomy. These materials help reduce bacterial adhesion to the prosthesis surface, contributing to better oral hygiene and minimizing the risk of secondary infections. Nanobiomaterials are often lightweight, making them more

comfortable for patients to wear. This is particularly important for dental prosthetics, as it ensures a better fit and reduces discomfort during everyday activities like eating and speaking. Their unique properties enable faster and more efficient production, reducing the time required for the patient to receive their final restoration. This bioactivity facilitates faster healing and tissue regeneration around the dental prosthetic, promoting better long-term success. The strong bond between the prosthesis and the underlying dental structure ensures reliable performance and reduces the need for frequent adjustments or replacements.

c. **Dental Whitening Agents**: Nanobiomaterials are incorporated into dental whitening agents to improve their efficacy and minimize tooth sensitivity during the whitening process. Nanobiomaterials used in dental whitening agents have increased surface area and reactivity, leading to improved whitening efficiency. These materials can effectively target and break down dental stains, resulting in faster and more noticeable teeth whitening outcomes. It often requires shorter exposure times and lower concentrations of whitening agents. This allows for a minimally invasive approach to teeth whitening, reducing the risk of tooth sensitivity and irritation to the gums. This controlled release ensures a more stable and sustained whitening effect without causing sudden sensitivity or damage to the teeth. Nanobiomaterials can be functionalized with specific ligands or nanoparticles to target and remove specific types of stains, such as those caused by coffee, tea, or tobacco. This targeted approach allows for more effective stain removal while preserving the natural tooth structure. They do not cause harm to the teeth, gums, or surrounding oral tissues, ensuring a risk-free tooth whitening procedure. Nanobiomaterials used in dental whitening agents can be incorporated into various formulations, such as toothpaste, gels, or strips, offering patients a range of options for teeth whitening.

d. **Antibacterial Coatings**: Nanobiomaterials are used to develop antibacterial coatings on dental materials and instruments to reduce the risk of bacterial contamination and prevent infections. By reducing bacterial contamination on surfaces, nanobiomaterial antibacterial coatings contribute to lower infection rates and improved overall public health. Nanobiomaterials used in antibacterial coatings are often biodegradable and environmentally friendly, making them a sustainable choice for various applications. Biofilms, which are complex communities of bacteria that adhere to surfaces, are notorious for being resistant to conventional cleaning methods and antibiotics. Nanobiomaterial coatings can disrupt the formation of biofilms, preventing the buildup of harmful bacterial colonies. The durable nature of these coatings ensures continuous antibacterial effects over extended periods.

e. **Oral Tissue Regeneration**: Nanobiomaterials are used in tissue engineering approaches to promote the regeneration of oral tissues, such as periodontal tissues and dental pulp. The versatility of nanobiomaterials allows for tailoring their properties to match the specific needs of individual patients. This personalized approach maximizes the effectiveness of oral tissue regeneration treatments. Nanobiomaterial coatings on dental implants can enhance the integration of the implant with the surrounding bone and soft tissues, leading to better implant stability and long-term success. It plays a crucial role in periodontal tissue engineering, which focuses on restoring the structures that support the teeth, including the gums and periodontal ligament. By stimulating periodontal tissue regeneration, nanobiomaterials help treat periodontal diseases and improve overall oral health. Nanobiomaterials are also used in bone regeneration procedures, such as guided bone regeneration (GBR) and socket preservation. They promote osteoblast activity and new bone formation, aiding in the reconstruction of lost or damaged bone tissues. In cases of dental pulp injury or infection, nanobiomaterials can be used to stimulate the regeneration of dental pulp tissues. These materials support the growth of new blood vessels and nerve tissues, facilitating the healing process. The material is biocompatible and non-toxic, thereby ensuring that it does not cause adverse reactions or inflammation in the surrounding oral tissues. Nanobiomaterials act as scaffolds that provide a supportive structure for cells to grow and organize themselves. They create a conducive environment for tissue cells to regenerate, which is especially important in dental procedures like periodontal regeneration. When applied to damaged oral tissues, these materials promote cell adhesion, proliferation, and differentiation, leading to improved tissue regeneration [21].

References

1. L. Zhang, F.X. Gu, J.M. Chan et al., Nanoparticles in medicine: therapeutic applications and developments. Clin. Pharmacol. Ther. **83**, 761–769 (2008)
2. J. Panyam, V. Labhasetwar, Biodegradable nanoparticles for drug and gene delivery to cells and tissue. Adv. Drug Deliv. Rev. **55**, 329–347 (2003)
3. R. Langer, D.A. Tirrell, Designing materials for biology and medicine. Nature **428**, 487–492 (2004)
4. M. Ferrari, Cancer nanotechnology: opportunities and challenges. Nat. Rev. Cancer **5**, 161–171 (2005)
5. K.K. Jain, Role of nanobiotechnology in developing personalized medicine for cancer. Technol. Cancer Res. Treat. **4**, 645–650 (2005)
6. M.E. Davis, Z.G. Chen, D.M. Shin, Nanoparticle therapeutics: an emerging treatment modality for cancer. Nat. Rev. Drug Discov. **7**, 771–782 (2008)
7. R. Duncan, Polymer conjugates as anticancer nanomedicines. Nat. Rev. Cancer **6**, 688–701 (2006)
8. O.C. Farokhzad, R. Langer, Impact of nanotechnology on drug delivery. ACS Nano **3**, 16–20 (2009)
9. Y. Zhang, Y. Zhang, Z. Wang, W. Wu, H. Zhang, S. Guo, Y. Wang, Nanobiomaterials: a new paradigm for targeted drug delivery. Chem. Rev. **119**, 12492–12533 (2019)
10. R. Sridhar, S. Ramakrishna, Electrospinning: an enabling nanotechnology platform for drug delivery and regenerative medicine. Nanomedicine **8**, 1781–1802 (2013)
11. F.L. Mi, S.S. Shyu, Y.B. Wu, S.T. Lee, J.Y. Shyong, R.N. Huang, Fabrication of chitosan/poly(gamma-glutamic acid) nanoparticles and their application for drug delivery and imaging. Nanomedicine **7**(5), 598–607 (2011)
12. W.H. De Jong, P.J.A. Borm, Drug delivery and nanoparticles: applications and hazards. Int. J. Nanomed. **3**, 133–149 (2008)

13. D. Peer, J.M. Karp, S. Hong, O.C. Farokhzad, R. Margalit, R. Langer, Nanocarriers as an emerging platform for cancer therapy. Nat. Nanotechnol. **2**, 751–760 (2007)
14. J.H. Lee, A. El-Fiqi, J.K. Jo et al., Nanobiomaterials for regenerative medicine. Front. Bioeng. Biotechnol. **4**, 60 (2016)
15. R. Smeets, B. Stadlinger, F. Schwarz, B. Beck-Broichsitter, O. Jung, C. Precht, F. Kloss, S. Ghanaati, M. Heiland, T. Ebker, Impact of dental implant surface modifications on osseointegration. Biomed. Res. Int. **7**, 2016 (2016)
16. P. Rezvanian, M. Ebrahimi, M. Vasei, P. Mohammadi, A. Zamanian, M. Yousefi, A.R. Rokn, H. Naderi-Meshkin, Application of nanobiomaterials in endodontics. Iran. Endod. J. **11**, 153 (2016)
17. A. Moshaverinia, S. Ansari, M. Moshaverinia, N. Roohpour, J.A. Darr, I.U. Rehman, Effects of incorporation of hydroxyapatite and fluoroapatite nanobioceramics into conventional glass ionomer cements (GIC). Acta Biomater. **4**, 432–440 (2008)
18. L. Cheng, K. Zhang, M.D. Weir, M.A. Melo, X. Zhou, H.H. Xu, Nanotechnology strategies for antibacterial and remineralizing composites and adhesives to tackle dental caries. Nanomedicine **10**, 627–641 (2015)
19. S.J. Ahn, S.J. Lee, J.K. Kook, B.S. Lim, Experimental antimicrobial orthodontic adhesives using nanofillers and silver nanoparticles. Dent. Mater. **25**, 206–213 (2009)
20. D. Xie, Y. Weng, X. Guo, Y. Wang, Z. Feng, Y. Bai, Dental adhesive with biofilm-regulating and remineralization capabilities. Dent. Mater. **35**, 305–317 (2019)
21. M. Goldberg, A.B. Kulkarni, M. Young, A. Boskey, Dentin: structure, composition and mineralization. Front. Biosci. (Elite Ed.) **3**, 711–735 (2011)

Goal 6: Clean Water and Sanitation

Overview of the chapters:

Goal 6: Clean Water and Sanitation

(Access to water and sanitation for all must be ensured and managed sustainably)

We aim to provide access to safe drinking water to all by 2030, based on this framework. In 2030, we must reduce pollution, eliminate dumping, minimize hazardous chemical release and recycle and reuse waste materials in proportion to water quality, and halve the amount of untreated wastewater. Most wastewater is safely treated by government agencies and related agencies. There are a significant number of bodies of water that have good ambient water quality because they are well maintained. It also involves protecting and restoring ecosystems that are related to water, like rivers, aquifers, wetlands, mountains, and forests. Water-related ecosystems will change over time, as we see the extent of their habitats change. In order to assist developing countries in water- and sanitation-related activities and programs by 2030, we will expand our international cooperation and capacity-building support. Even though metal ions are supplied to water streams at much higher levels than those authorized in this part of the statute, effective metal ion removal from contaminated water bodies has grown into a critical task of the modern era.

New Sorption's of Organic Wastes and Chemical Composites for the Remediation of Aqueous Toxic Metals

K. M. Tajun Meera Begum, N. M. I. Alhaji, A. Ayeshamariam, and M. Jayachandran

Abstract Heavy metal poisoning of water is one of the major environmental issues with serious health risks. Heavy metals are significant inorganic pollutants due to their toxicity and availability in water. Effective removal of metal ions from contaminated water bodies has grown to be a critical task in the modern era, even though metal ions are supplied to water streams at amounts far higher than the authorized limits. The bulk of heavy metals can enter live animals through the food chain and are hazardous even at very low amounts. They can build up in the cells and tissues of living animals and cause major complications. The fast-escalating environmental deterioration brought on by solid wastes because of ongoing industrialization is another serious environmental problem. The geology and hydrogeology of the intake area, the water treatment techniques used, and the chemical composition and characteristics of WTRs, which vary due to the wide range in the type and chemical composition of raw water.

K. M. Tajun Meera Begum
Department of Chemistry, Arignar Anna Government Arts College, Karaikal, Puducherry 609605, India

N. M. I. Alhaji
Department of Chemistry, Khadir Mohideen College, Adirampattinam (Affiliated to Bharathidasan University), Tiruchirappalli, Tamil Nadu 614701, India

A. Ayeshamariam (✉)
Department of Physics, Khadir Mohideen College, Adirampattinam (Affiliated to Bharathidasan University), Tiruchirappalli, Tamil Nadu 614701, India
e-mail: ayeshamariamkmc@gmail.com

M. Jayachandran
Department of Physics, Sethu Institute of Technology, Pullor, Kariyapatti, Tamil Nadu 626115, India

K. Kasinathan et al. (eds.), *New Technologies for Energy Transition Based on Sustainable Development Goals*, https://doi.org/10.1007/978-981-97-2527-4_12

1 Hazardous Nature of Heavy Metals

Due to fast industrialization and population growth, harmful toxic metals like arsenic, cadmium, chromium, lead, copper, nickel, zinc, and mercury were released into streams. These metals can be separated from other harmful contaminants because they are non-biodegradable and bio accumulative, being concentrated throughout the food chain. Furthermore, numerous sectors have been accountable for the release of toxic substances into the surroundings as industrial effluents. One of the main environmental problems with significant health hazards is heavy metal pollution in water. Because of their toxicity and accessibility in water, heavy metals are important inorganic pollutants. Effective removal of metal ions from polluted water bodies has become a crucial challenge today, even though metal ions are being introduced to water streams at concentrations that are substantially higher than the permitted limits. As a result, water contamination by toxic heavy metals caused by wastewater discharge from industrial, agricultural, and home sectors is a worldwide issue and a major source of worry for ecological researchers. Toxic heavy metals released by chemical industry include Cd, Hg, Pb, Cr, Cu, Ni, Zn, and others. At least twenty metals are hazardous, and half of them are discharged into the environment in proportions that endanger human health. Even trace amounts can cause severe physical or neurological harm to the human body. Contamination of hazardous and toxic compounds in soil, groundwater, sediments, surface water, and air also causes substantial challenges for ecosystems. Each metal causes a unique set of health issues [1]. Heavy metal contamination of the environment has exceeded the permissible limit, posing a threat to all life forms. Heavy metal toxicity and bioaccumulation in the environment are severe threats to living creatures' health. Within certain limits, some heavy metals are essential for the existence and growth of living species, and their toxicity is outlined in Table 1. Inorganic wastewater can be treated using a variety of methods; the optimal treatment, however, should be able to meet the criteria for the maximum contamination limit (MCL) as well as be appropriate, pertinent, and adaptable to local conditions [2]. Table 2 lists the MCL requirements set by Babel and Kurniawan [2] based on WHO heavy metal guidelines.

Heavy metals wastewater is being dumped into the environment in increasing proportions. especially in emerging nations, because of the quick development of the metal plating, mining, fertilizer, tannery, battery manufacturing, paper, and pesticide industries. Water pollution is made up of complex contaminants that can be divided into inorganic and organic pollutants. Water's pH varies widely. COD has a significant coefficient of variation, resulting in a high level of heavy metal toxicity during wastewater discharge. If the discharge effluent does not fulfill the emission standards, it will pollute the nearby water bodies severely. However, these technologies have substantial drawbacks in terms of harmful sludge or effluent formation. Another significant environmental issue is the rapidly expanding environmental degradation caused by solid wastes as a result of progressive industrialization. Because of the volume of paper sludge produced and disposed of, the paper industry has a significant environmental impact. Paper sludge consists of organic fibers, inorganic fillers,

Table 1 Heavy metal toxicity to microorganisms

Heavy metals	Effects on microbes
Arsenic	Removal of enzymes
Cadmium	Proteins are denatured, nucleic acids are destroyed, and cell transcription and division are hampered
Chromium	Grow inhibition, lag phase lengthening, and oxygen uptake inhibition
Copper	Inhibit enzyme activity and disrupt cellular function
Selenium	Reduces the rate of growth
Lead	Nucleic acid and protein degradation, inhibition of enzyme activity and transcription
Mercury	Protein denaturement, enzyme inhibition, and cell membrane disruption
Nickel	Cell membrane disruption, enzyme inhibition, and oxidative stress
Silver	Cell lysis, inhibition of cell transduction, and inhibition of cell proliferation
Zinc	*Death, decreased biomass, and growth inhibition*

Table 2 The most dangerous heavy metals' MCL standards

Heavy metal	Toxicities	MCL (mg/L)
Arsenic	Visceral tumors, skin symptoms, and vascular disease	0.050
Cadmium	Renal disease, natural carcinogen	0.01
Chromium	Headache, diarrhea, nausea, vomiting, carcinogenic	0.05
Copper	Sleeplessness, liver damage, and Wilson disease	0.25
Nickel	Dermatitis, nausea, persistent asthmatic coughing, and human carcinogen are all possible side effects	0.20
Zinc	Depressive disorders, fatigue, symptoms of neurological disorder, and thirst	0.80
Lead	Disruption to the fetal brain, kidney, circulatory, and neurological system illnesses	0.006
Mercury	Arthritis with rheumatoid disease, as well as kidney, circulatory, and neurological system disorders	0.00003

and coating chemicals, releasing heavy metals into the environment, especially in emerging nations, because of the quick development of the metal plating, mining, fertilizer, tannery, battery manufacturing, paper, and pesticide industries. Effective water waste treatment represents one of the century's most pressing issues. Finding environmental treatment methods will remain a priority as the level of industry hazardous waste dumping to water bodies rises. Heavy metal exposure activities include refineries, coal combustion in electric power plants, textiles, plastics, microelectronics, wood preservation paper processing factories, electroplating industry, steel industry, mining, and so on, and treatment is substantial due to their severe toxicity. Because dangerous toxic chemicals are non-biodegradable, bioaccumulative, and mobile in water, they must be removed from water and effluents. Traditional water purification methods are costly and ineffective in providing appropriate aqueous solution quality.

There are many traditional methods for getting rid of heavy metals. Examples include electrochemical deposition, ion exchange, chemical precipitation, and adsorption. By using conventional adsorbent, adsorption is consistently seen as a flexible and highly successful approach (e.g. commercial charcoal, fly ash, clay, sand, zeolites, alumina, etc.). By using commonly available abundant biomaterials, non-conventional adsorbents are obtained for purification process. Biosorption has the potential to be a viable alternative to present conventional technologies for removing harmful ions from water [3]. The biosorption method has a number of advantages, including affordability, high effectiveness, a reduction in chemical or ecological sludge, regeneration of the biosorbents, and metal recovery. Toxic chemicals that are injected into ground water and soil pose major challenges for ecosystems due to their toxicity, propensity for bioaccumulation, and non-biodegradability. The quality of fresh water is impacted by pollution as well. Safe drinking water is becoming more and more scarce in many nations. Most surface and groundwater need to be treated before it is used for drinking. A common adsorbent is activated carbon because of its large specific surface area and ease of supply. On the other hand, commercial activated carbon is unreasonably expensive. As a result, a different technique for making activated carbon by reusing organic waste and biomass was found (biosorbents). Organic wastes like rice-husk, sawdust, industrial wastes such as paper mill sludge, and chemical composites derived as nanomaterials, zeolites, polymeric compounds etc., have been used for aqueous remediation. Pre-treatment and surface adjustments may be required to improve capacity and efficiency. Surface modification of biosorbents improves pore size, surface area, chemical stability, material strength, creation of more active sites and biocompatibility. Despite the accessibility of various surface phenomena for the treatment of water and wastewater, biosorption is still recognized as one of the most innovative and effective methods. An elaborate literature survey reveals that non-conventional adsorbents (*biosorbents*) namely cellulosic, ligno-amino cellulosic as an alternative biosorbents, surface-modified biosorbents as activated carbon obtained from agricultural wastes like rice-husk, sawdust, and reuse of organic wastes such as paper mill sludge, rice wine-processing waste sludge etc. and chemical composites derived as nanomaterials, zeolites, polymeric compounds, and other non-conventional adsorbents have been employed as effective and economical low-cost alternatives for aqueous clean-up.

2 Review of Literature

Activated carbon is a promising adsorbent due to its high surface area, micro porosity structure, and surface polar functional groups. There are four types: power-activated carbon, granular activated carbon, activated carbon fibers, and activated carbon cloths. Each type has specific applications and benefits in wastewater treatment. It is energy-efficient, environmentally friendly, and economically advantageous to recycle and reuse garbage. To generate low-cost, micro- and mesoporous carbon materials by calcinations, grinding, physical or chemical activation procedures, it is therefore of major priority today to reuse organic wastes from agricultural, marine, paper mill sludge, oil mill sludge, household materials, etc., Heavy metals in water can cause serious problems for aquatic ecosystems and human health. Adsorption is often used for its accessibility, cost, and environmental friendliness. This page aims to gather information on adsorbents and commercially accessible naturally occurring biosorbents for removing chromium, cadmium, and copper. Agricultural waste treatment for heavy metal removal offers economic advantages and efficient removal. Biomass adsorption processes involve chemisorption, complexation, and ion exchange diffusion. The rate of removal varies depending on adsorbent and metal. Biomaterials are growing due to their affordability, regeneration ability, high absorption efficiency, and minimal chemical or biological sludge recovery. Polysaccharides are divided into four classes in the review: polysaccharide, photopolymers, polysaccharide blends, copolymer, and polysaccharide composites, and their removal capability is described. Industries produce waste containing heavy metals that are harmful to the environment. Heavy metal ions like nickel, lead, copper, cadmium, and zinc need to be removed from wastewater through biosorption. Biomaterials like polysaccharides like chitin, chitosan, cellulose, starch, and alginate can be blended, grafted, or combined to further fictionalize the adsorption process. Adsorbent recyclability is crucial, and the cost of removal decreases with time. Heavy metal ions interact with each other through dipole–dipole interactions, hydrogen bonds, and *Van Der Waals forces*.

Reactive dye Remazol black adsorbs on chitosan in aqueous solutions, with maximum adsorption capability in the range of 91.47–140 mg/g. The chitosan amino group provides reasonable dye-removing ability. Adsorption kinetics and isotherms are used to study biosorption. Researchers studied walnut sawdust's ability to remove lead, cadmium, and nickel from aqueous solutions. They studied contact time, starting concentration, and temperature, finding an equilibrium period of 60 min. The adsorption rate constant was determined using Freundlich and Langmuir isotherm models [4]. Adsorption technique removes heavy metals from wastewater, with activated carbon being the most effective but expensive. As freshwater scarcity increases, alternative measures must be considered. The necessity for inexpensive adsorbents was brought on by the fact that heavy metals are dangerous even at trace concentrations. The method of adsorption and the various adsorbents that are now on the market are the main topics of the article. Adsorption reaction conditions and low-cost adsorbents, including industrial and agricultural waste, are influenced by price, adaptability, and easy raw material access.

3 Materials

The conventional adsorbents include commercial charcoal or activated carbons, fly ash, clay, sand, zeolites, alumina, silica gel, and biomass. Both natural and manufactured low-cost non-conventional adsorbents have been employed in large quantities as adsorbents for removing heavy metal ions from aqueous medium. These non-conventional adsorbents include composites of agro or marine based alive or dead biomaterials, nanocompound, and polymeric compounds, etc.

3.1 Organic Wastes and Composites as Alternate Sorbents

Over the past ten years, there has been a lot of interest in the reuse of organic waste materials and their creation of biosorbents. Polysaccharides possess hydrophilicity, flexibility, cationic properties, biodegradability, and biocompatibility, biomass, such as sawdust, rice husk, seaweeds, seagrass, chitosan, and others, biomass have been recognized as effective, ecofriendly, abundantly available, cheap, and best adsorbents. Biosorbents can be nonselective, binding to various heavy metals, or specialized for specific metals based on chemical composition. Spectroscopy studies functional groups' complexation. Like how biosorption is anticipated to be of exceptional help in the recovery and purification of high-value products like proteins, steroids, drugs, and pharmaceuticals. As a result, these biomaterials have demonstrated their potential as metal biosorbents. However, surface modification has been done chemically, mechanically, or biologically in recent years to improve the ion-exchange property of activated carbons (*biosorbents*). Chemical and thermal pretreatments of biomass decrease organics leaching from the biosorbents to the solution during the adsorption process, preventing adsorption capacity loss. Adsorption ability of biosorbents can be increased through chemical modification. Consider chemical cost and modification methods for low-cost biosorbents. Carbonizing carbonaceous materials and then activating the char that develops in the presence of activating chemicals like CO_2 and steam is known as chemical activation [5].

A lower temperature is frequently used for chemical activation than for physical activation. The chemical action leads to better pore development in the carbon structure. However, chemical activation produces more carbon than physical activation. Many parameters influence the efficiency of this chemically activated carbon, including temperature, duration, and impregnation ratio. Increased understanding of the surface chemistry of modified biosorbents, such as texture, porosity, specific surface area, wet ability, catalytic, electrical, electrochemical adsorption, surface functionality, hydrophilic hydrophobic, and so on, can lead to the development of more specific and effective adsorbents for a specific method [6].

4 Experimental—Biosorbents

4.1 Preparation of Sawdust—Chitosan Composite Beads (SDCCB)

Sawdust was sieved, treated with sulfuric acid, rinsed, and dried in a hot air boiler at 1000 °C. In total, 3 g of chitosan were dissolved in 100 ml of 2% acetic acid to produce chitosan gel. The chitosan gel was combined with 3 g of sawdust that had been acid-treated, which was then shaken vigorously for 12 h at 200 rpm in a rotary shaker. The sawdust chitosan solution was treated with acid for 12 h, then 0.5 M NaOH solution was added, and the surplus NaOH was removed by rinsing with distilled water. The beads were cross-linked by soaking them in a 7.5% ethanolic *glutaraldehyde* solution for 24 h before being washed and dried [7]. Nanotechnology has received significant interest in a variety of fields, including water treatment. Nanomaterials enhance effective removal of heavy metals and pollutants from water. These attributes include enhanced surface area, remarkable adsorption capacity, and great selectivity. Nanomaterials efficiently remove metal ions at low concentrations.

4.2 Carbon Nanotube Production

Experimental setup creates multiwall carbon nanotubes using a horizontal tubular reactor, heated by a silicon carbide element. Hydrocarbon sources were benzene or ferrocene, with hydrogen as carrier gas and argon flashing air. Studies were conducted at temperatures ranging from 500 to 1200 °C, in which the reaction duration (45 min) and hydrogen flow rate (300 ml/min) criteria [8].

4.3 Preparation of Polymer-Impregnated Carbonized Sawdust

Strong sulfuric acid (50 mL per 10 g) was applied to uniformly sized (0.5 mm) sawdust particles for two hours to produce carbonized sawdust powder. The mixture endured drying in a hot air oven at 100 °C after being neutralized with sodium carbonate (0.1 N) until the effervescence subsided. After carbonization, the product was treated with 60% poly(sodium 4-styrene sulfonate (PSSS). A rotary shaker was used to combine PSSS and carbonized sawdust for around 24 h at 200 rpm while it was at room temperature. Once the pH was determined to be neutral, the mixture was thoroughly cleaned with distilled water. The samples were then dried in an oven with hot air at 100 °C. Carbonized materials were characterized and utilized as sorbents to remove hazardous ions from aqueous solutions [9].

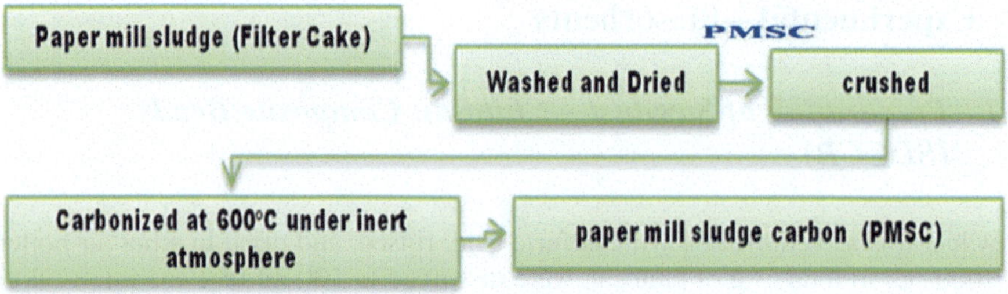

Fig. 1 Flow chart for getting PMSC

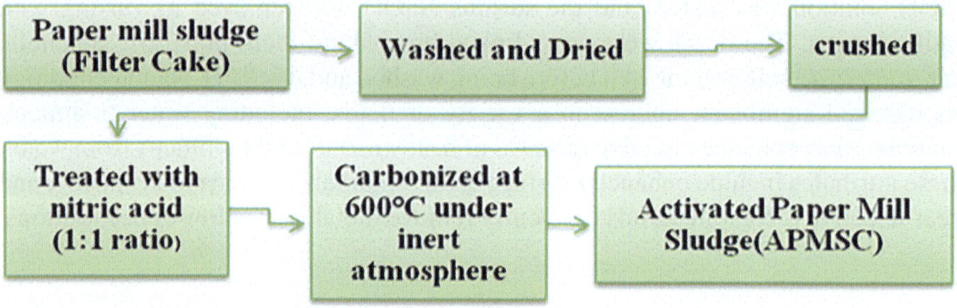

Fig. 2 Using flow chart to obtain APMSC

4.4 Sludge-Based Adsorbents (PMSC and APMSC)

Chemically treating industrial wastes including fly ash, waste iron, iron slags, and hydrous titanium oxide can increase the removal of metal from wastewater. Sludge-based activated carbon adsorbents, which are used in the treatment of wastewater, can be successfully made from industrial and domestic solid wastes. Paper mills generate a large amount of sludge throughout the manufacturing process, which causes major challenges with sludge disposal [10]. Paper mill sludge carbon and activated paper mill sludge carbon (PMSC and APMSC) are mesoporous sludge-based carbons that have been chemically activated using wet oxidizing agents like H_2SO_4, HNO_3, and H_2O_2. The potential for the spectrophotometric adsorption of heavy metal ions like Cr(VI) and Ni(II) has been compared with appear modified activated carbon.

Flowchart for getting PMSC and APMSC given in Figs. 1 and 2.

4.5 Water Treatment Sludge or Waterworks Sludge (WTRs)

Annual production of millions of tons of water treatment sludge (WTR's) makes efficient and responsible management difficult. The study of sorbent-based techniques is essential. The properties of heavy metals and metalloids sorption in SWTR's and

GWTR's are affected by time, pH, sorbent dose, *beginning* concentration, temperature, as well as the type and composition of the raw water. Recent studies emphasize WTRs as effective heavy metal sorbents, emphasizing their use in the removal of pollutants from aqueous solutions [11].

4.6 *Removing Heavy Metals from Industrial Effluents*

Heavy metal ion removal is crucial for maintaining a healthy ecosystem and human population. Chemical precipitation, flocculation, adsorption, ion exchange, and electrochemical plating are a few conventional techniques to eliminate contaminants from effluent. Using biosorbents, heavy metal removal from industrial effluents is summarized in Table 2. To treat electroplating wastewater, it is advised to use flocculation/precipitation after biosorption, since the suggested treatment has benefits over-existing procedures and uses a large amount less electricity. Chromium may be recovered from the sorbent with a high degree of purity and with less chemical additions [12]. Adsorption is the most effective and adaptable approach for removing heavy metals, even at very low concentrations. The expense of adsorbents, which is typically activated carbon, is regarded to be the biggest barrier to industrial application. The cost of using activated carbon to clean commercial effluent is prohibitive. To do this, research on heavy metal adsorption has moved its attention to plentiful natural materials (*biopolymers*) as well as wastes from commercial and agricultural activities. Biopolymers like lignocellulosic, amino cellulose, and sludge-based sorbents are economical, environmentally friendly alternatives for removing dangerous metals from water and wastewater [13].

5 Outlooks in Sorption Technology for Aqueous Toxic Metals

Household wastewater consists of non-manufacturing byproducts like sewage, bacteria, viruses, and sanitary waste. Untreated industrial wastewater is the primary source of water contamination, while industrial effluent contains toxins, endangering human health and the environment. The goal of this review is to draw attention to the significant and advantageous potential of inexpensive sorbents for wastewater treatment and to promote future investigation into their broad use.

5.1 Biosorption

Biosorption is a novel or inventive technique for removing heavy metals from wastewater. This process is thought to be an effective heavy metal detoxification technique even at low concentrations. The solid (*sorbent*) and liquid (*solvent*) stages of the biosorption adsorption process are both involved. To remove heavy metals, biological entities that are both viable and nonviable are desired. Biosorbents are an effective, affordable adsorbent that can be acquired as a waste material from a range of businesses. Functional groups like alcohol, aldehydes, ketones, carboxylic, ether, and phenolic enhance metal adsorption activity. Biomass adsorption process involves adsorption, ion exchange, complexation, surface precipitation, and contact time.

5.2 Factors Affecting Biosorption of Materials

- In batch mode sorption investigations, a variety of factors affect how well adsorbents remove heavy metals from wastewater. Starting temperature, concentration, adsorbent dosage, pH, contact time, and stirring rate are among the factors. Heavy metal removal increases with concentration, temperature, dose, contact time, stirring rate.
- The influence of pH was studied to determine the active species of the adsorbates and the degree of ionization during adsorption by adjusting the initial pH of the solution from 2 to 10 using the buffers HCl–KCl for the range 1–3, CH_3COOH-CH_3COONa for the range 4–6 and boric acid-NaOH for the range 8–10, maintaining a constant dosage of adsorbent and initial metal ion concentration.
- To evaluate the equilibrium time, the effect of contact time at varied initial metal ion concentrations was examined while keeping pH, temperature, adsorbent dosage, and agitation speed constant. Up until the adsorption equilibrium period, the metal ion concentration was observed at various time intervals.
- The optimal shaking speed and adsorption activation parameters were investigated using similar methods in relation to the impacts of agitation speed and temperature. Equilibrium isotherm investigations were also performed at constant temperature, agitation speed, and pH with varying initial metal ion concentrations.
- Spectrophotometric analysis was utilized at each time to determine the quantitative quantity of adsorption capacity (q_e) and percentage removal effectiveness (%) of the sorbent using Eqs. (1) and (2).

$$q_e = (C_o - C_e)V/m \tag{1}$$

$$\text{Percentage removal} = [(C_o - C)/C_o] \times 100 \tag{2}$$

C_o and C_e represent initial and equilibrium concentrations, 'V' represents solution volume, q_e represents adsorbed amount, m represents adsorbent weight, and 'C' represents remaining solution concentration after adsorption process.

5.3 *Isotherm Modeling in Biosorption of Chromium by SDCCB and ASDP*

Adsorption isotherms reflect the relationship between adsorbate concentration and adsorption degree on the adsorbent surface at constant temperature. As shown in Table 3, the equilibrium properties of chromium adsorption by SDCCB and ASDP were described using a variety of isotherm equations, including Langmuir, Freundlich, and Temkin (4). The Langmuir isotherm shows monolayer adsorption and surface uniformity. Additionally, separation factors (RL) are examined in this. For optimal adsorption, this value should be between 0 and 1. The Freundlich isotherm exhibits multilayer sorption and surface heterogeneity. The effects of adsorbent–adsorbate interaction are taken into consideration by the Temkin model. (i) Is presumable. Due to interactions between adsorbent and adsorbate, the heat of adsorption for each model in the layer would drop linearly with coverage. (ii) The distribution of the highest binding energy 'B' is uniform. The slope and intercept in Table 3 are used to create the linearized versions of isotherms and constants. Freundlich > Langmuir > Temkin is the order of the isotherm in both surfaces; hence, multilayer lateral adsorption is possible based on R^2 values. The adsorption process is therefore the same, but sawdust in composite beads has a higher maximum adsorption capacity (= 125 mg/g) than sawdust in powder [14].

Table 3 Linearized form of isotherms and the constants

Sl. No.	Isotherm model	Equation	Constants for SDCCB	Constants for ASDP
1	Langmuir isotherm	$C_e/q_e = 1/b \cdot \theta + C_e/\theta$ $K_L = (\theta \cdot b)$ $R_L = 1/1 + bc_0$	$\theta = 125$ mg/g $b = 0.04$ L mg^{-1} $K_L = 5$ L g^{-1} $R_L = 0.2$ $R^2 = 0.967$	$\theta = 45.5$ mg/g $b = 0.07$ L mg^{-1} $K_L = 5.9$ L g^{-1} $R_L = 0.13$ $R^2 = 0.979$
2	Freundlich isotherm	$\ln q_e = \ln K_F + 1/n \ln C_e$	$n = 1.8$ $K_F = 8.74$ L g^{-1} $R^2 = 0.974$	$n = 2.7$ $k_F = 8.63$ L g^{-1} $R^2 = 0.991$
3	Temkin isotherm	$q_t = B \ln K_T + B \ln C_e$	$B = 22.9$ $K_T = 0.58$ L g^{-1} $R^2 = 0.960$	$B = 7.05$ $K_T = 3.39$ L g^{-1} $R^2 = 0.967$

6 CNT Adsorbents

Due to its useful qualities and applications, it is frequently used in the heavy metal treatment procedure. Cylindrical graphite sheets that have been folded into a tube-like shape make up carbon nanotubes. SWCNTs are single-walled graphite sheets, while MWCNTs are made from multiple layers. The best-fit model for metal ion interaction with carbon nanotubes is shown in Table 4.

However, their utilization has been curtailed due to the high CNT concentration and dearth of functional groups. Increase CNT efficacy by modifying acid treatment, propagating functional groups, and saturating metals/metalloids (3). Freundlich isotherm has better-fitting model and higher correlation coefficient compared to Langmuir isotherm. As a result, this illustrates the usefulness of Cr(III) ion monolayer coverage on the adsorbent's surface. The significant Cr(III) ion adsorption to the surface of CNTs is further supported by the positive Freundlich isotherm correlation coefficient, which has led to the conclusion that CNTs have a significant potential for application as an adsorbent for the removal of Cr(III) ions in water treatment [15].

6.1 *Optimum Conditions for Biosorption onto WMR*

The optimal operating conditions identified by several research for the biosorption of heavy metals utilizing biosorbents [16] (Tables 5 and 6).

Table 4 CNTs' interaction mechanisms

Type of CNTs	Metals removed	pH	Adsorbent dosage	Interaction mechanism	Best-fitted model	R^2
Raw MWCNT's	Cr(VI)	3	75 mg	Electrostatic interaction	Langmuir	0.973
Raw CNT's	Cd(II)	7	50 mg	Electrostatic interaction	Langmuir	0.979
Alumina decorated	Cd(II)	7	50 mg	Electrostatic interaction, physical adsorption, surface	Langmuir	0.9972
MWCNT's				Precipitation, Van Der Waals interaction	Freundlich	0.9830
MnO_2-coated carbon nanotubes	Hg(II)	5–7	NA	Electrostatic interaction	Freundlich	0.997
COOH-MWCNTs	Hg(II)	4.3	NA	Electrostatic interaction/ complexation	Langmuir	0.990

Table 5 Biosorption parameters using watermelon rind

Adsorbate	~ pH	Sorbent dosage	Temp	Initial solute concentration (mg/L)	Max. sorption capacity (mg/g)	% Removal
Cu(II)	5	1 g/L	40 °C	10	31.25	88
Zn (II)	7	0.02 g	Room temp.	1	–	90.3
Ni(II)	5	2.5 g/L		100	18.4	–
Pb(II)	5	5 g/L	30 °C	300	116.2	–
Cd(II)	9.12	1–5 g/L	20 °C	500	40.16	
Cr(III)	3	1.5 g/L	30 °C	50	172.6	91
Fe(II)	7	0.04 g	Room temp.	2.2	4.98	98.3
Mn(II	6	0.05 g	Room temp.	0.7	1.37	98.9
Co(II)	6	2 g/L	Room temp.	50	23.3	57

Table 6 Demonstrates the way watermelon rind stacks up against other agricultural waste biosorbents in terms of maximal biosorption capabilities

Adsorbate	Maximum biosorption capacity (mg/g)	
	Watermelon rind	Other agricultural waste biosorbents
Cu(II)	39.2	17.42 (wheat shell)
Cu(II)	31.25	4.64 (barley straw
Zn(II)	22.55	80 (orange peel)
Ni(II)	38.98	39.75 (mango peel)
Pb(II)	98.06	52 (pomegranate peel)
Cd(II)	40.16	13.84 (rice straw)
Cr(III)	172.6	250 (mosambi peel)
Fe(II)	4.98	18.52 (pomegranate peel)
Mn(II)	1.37	28.6 (walnut shells)
Co(II)	23.3	6.10 (peanut shells)

6.2 *Kinetic Models of Cr(VI) and Ni(II) Using SDCCB and ASDP*

The assessment of the pseudo-first-order, pseudo-second-order, intra particle diffusion, and fundamental Elovich kinetic models was done to determine the rate of Cr(VI) adsorption by SDCCB and ASDP. Table 7 displays the linearized adsorption kinetics and associated constants. The best match for both surfaces is pseudo-second-order kinetics, as determined by the R^2 values. As a result of the validity of intra particle diffusion and the Elovich model of ASDP and SDCCB, adsorption happens via chemisorptions, which is the rate-limiting phase. The better-fitting Elovich kinetic data rather than the intra particle diffusion approach highlights the

Table 7 Summary of parameters derived for various kinetic models

S. No.	Kinetic model	Linear equation	Constants for SDCCB	Constants for ASDP
1	Pseudo-first-order	$\ln(q_e - q_t) = \ln q_e - K_{1_{ad}} t$	$K_{1ad} = 3.4 \times 10^{-2}$ min^{-1} $R^2 = 0.877$	$K_{1ad} = 1.9 \times 10^{-2}$ min^{-1} $R^2 = 0.955$
2	Pseudo-second-order	$\frac{t}{q_t} = \frac{1}{K_{2ad} \cdot q_e^2} + \frac{t}{q_e}$ $h = K_{2ad} \cdot q_e^2$	$q_e = 25$ mg/g $K_{2ad} = 4.3 \times 10^3$ g/mg min $h = 2.7$ mg/g min $R^2 = 0.994$	$q_e = 23.3$ mg/g $K_{2ad} = 8.41 \times 10^{-3}$ $h = 4.6$ mg/g min $R^2 = 0.999$
3	Simple Elovich	$q_t = \beta \ln t + \beta \ln(\alpha\beta)$	$\alpha = 1.2$ mg/g min $\beta = 3.62$ g/mg $R^2 = 0.984$	$\alpha = 1.09$ mg/g min $\beta = 1.85$ g/mg $R^2 = 0.978$
4	Intra particle diffusion	$q_t = K_{id} t^{1/2} + C$	$K_{id} = 0.867$ mg/g min$^{0.5}$ $R^2 = 0.958$	$K_{id} = 0.550$ mg/g min$^{0.5}$ $R^2 = 0.963$

importance of binding between chromium ions and ASDP and SDCCB activation sites [17].

6.3 Sorption Kinetic Analysis onto Sludge-Based Carbon

Table 8 displays *linear* plot parameters tested for adsorption rates of Ni(II) and Cr(VI) ions onto ASDP, PMSC, and APMSC using pseudo-first-order, second-order, simple Elovich, and intraparticle diffusion kinetics.

The two-phase adsorption of Ni(II) and Cr(VI) ions on inexpensive lignocellulosic adsorbents is revealed by the best-fitting pseudo-second-order model, with rapid adsorption occurring in the first stage and sluggish adsorption occurring subsequently. The q_t versus $t^{1/2}$ plot's early linear sections show a dual character that can be attributed to boundary layer diffusion effects, but the last linear component is a result of intraparticle diffusion effects. This is because the most practicable places on the ASDP and mesoporous surface of PMSC and APMSC were used first [18]. The intraparticle diffusion ($K_{id,2}$) played a significant influence but was not the rate-determining step in the multilinearity of the Webber and Morris model ($K_{id,1}$, $K_{id,2}$) because of the positive intercept values. Instead, the regulation of adsorption kinetics was aided by film diffusion ($K_{id,1}$). Adsorption behavior is influenced by ion exchange and H-bonding, with pseudo-second-order rate-limiting step suggesting chemisorption's. Simple Elovich and intraparticle diffusion models valid. Due to the huge volume of wastewater involved, batch adsorption studies are not appropriate for

Table 8 Various kinetic models are summarized in the following table

Kinetic model and Equation	Constant	Values for Cr(VI)			Values for Ni(II)		
		ASDP	PMSC	APMSC	ASDP	PMSC	APMSC
Pseudo-first-order $\ln(q_e - q_t) = \ln q_e - k_1 \cdot t$	K_1 (min^{-1})	1.96×10^{-2}	1.2×10^{-2}	3.3×10^{-2}	1.65×10^{-2}	1.3×10^{-2}	2.9×10^{-2}
	R^2	0.956	0.934	0.994	0.960	0.728	0.339
Pseudo-second-order $t/q_t = 1/K_2 \cdot q_e^2 + t/q_e$ $h = k_2 q_e^2$	K_2 (g/mg min)	8.21×10^{-3}	4.7×10^{-3}	4.8×10^{-3}	2.11×10^{-3}	3.9×10^{-3}	3.8×10^{-3}
	q_e (mg/g)	23.0	21.3	25.6	26.8	25	26.3
	h (mg/g min)	4.34	2.1	3.1	1.51	2.4	2.6
	R^2	0.998	0.996	0.998	0.998	0.997	0.997
Simple Elovich $q_t = \beta \ln t + \beta \ln \alpha\beta$	α (mg/g min)	6.14	249.5	1380	0.38	153.11	30.41
	β (g/mg)	1.85	1.75	1.84	4.36	2.12	2.63
	R^2	0.978	0.931	0.927	0.994	0.916	0.980
Intra-particle diffusion $q_t = K_{id} \cdot t^{0.5} + C$	$K_{id,1}$, $K_{id,2}$ (mg/g min)	1.02, 0.51	0.39, 0.61	0.53, 0.55	1.31, 0.49	0.40, 0.56	0.75, 0.17
	C_1, C_2	14.0, 16.4	14.7, 12.1	18.5, 17.1	10.0, 18.0	17.3, 16.3	16.3, 22.7
	R^2	0.985, 0.955	0.99, 0.99	0.94, 0.84	0.998, 0.829	0.97, 0.92	0.97, 0.77

industrial applications. When there is a lot of wastewaters to be treated, continuous treatment is faster. Because of its low running costs and the capacity of columns to adapt to diverse processes, the fixed bed adsorption process is more practical in real water treatment industries. For column studies to be successfully designed, the breakthrough curves' shape must be protected. Breakthrough curve timing and form are critical factors; effluent concentration at 0.1% influent concentration is breakthrough concentration. The term "*point of column exhaustion*" might refer to the point where the effluent concentration exceeds 99%. A packed bed column is therefore a useful tool for cyclic sorption. The concentration difference driving force for metal ion adsorption is maximized in this operating mode. Beginning at the inlet and finally bursting through the column, the saturated solid sorbent zone moves progressively across the column. The form and slope of the breakthrough curve are often governed by the equilibrium isotherm and mass transfer to and through the column's adsorbent, and the breakthrough record frequently produces a classic S-shaped breakthrough curve [18]. Table 8 gives the summary of parameters derived for various kinetic models.

7 Column Data Analysis

The performance of the column adsorption procedures is based on the breakthrough curve (7), which is derived by plotting C_t/C_o versus time (t), where C_t and C_o are the outlet and input metal ion concentrations in ppm, respectively. The fixed column loading behavior is expressed by breakthrough point (t_b), which is defined as the time required to approach a specific breakthrough concentration (1% of initial concentration, C_o), and exhaustion time (t_e), which is the time in minutes at which the effluent concentration reaches 99% of Co. The total column adsorption capacity in milligrams is calculated as follows:

q_{total} = area under breakthrough curve (A) × flow rate (Q) in mL/min.

$$q_{\text{total}} = \frac{QA}{1000} = \frac{Q}{1000} \int_{t=0}^{t=t_{\text{total}}} C_{\text{ad}} \mathrm{d}t \tag{3}$$

The Origin 8 program was used to determine the area (A) under the curve. Plotting the adsorbed concentration (C_{ad}, mg/L) versus time (t, min) allowed researchers to determine the area under the breakthrough curve. To determine the equilibrium adsorption capacity, use Eq. (4).

$$q_e = \frac{q_{\text{total}}}{m}, \tag{4}$$

where m represents the total dry weight of SDCCB in grams. Equation (5) is used to compute the quantity of metal ions that was injected into the column overall.

$$W_{\text{total}} = \frac{C_0 Q t_e}{1000}, \tag{5}$$

where W_{total} is the total amount of metal ion fed into the column, Eq. (6) is used to compute the total % removal of metal ions.

$$\text{Percentage removal (\%)} = \frac{q_{\text{total}}}{W_{\text{total}}} \times 100 \tag{6}$$

7.1 Column Sorption Modeling

The column performance was analyzed using three models (Thomas, Yoon-Nelson, and bed depth service time model). Various continuous adsorption theoretical models, including the Thomas model, the Yoon-Nelson model, and the bed depth service time model, were fitted with experimental data to investigate the solute interaction behavior in the system.

(i) **Thomas Model**: $\ln\left(\frac{C_o}{C_t}-1\right)=\frac{k_{Th}q_e m}{Q}-k_{Th}C_o(t)$.
(no external and internal diffusion limitations)
$\ln\{(C_o/C_t)-1\}$ versus time used to give q_e and k_{Th},
k_{Th}, Thomas rate constants (L/min mg),
q_e = equilibrium metal uptake (mg/g).

(ii) **Yoon-Nelson Model**: $\ln\left(\frac{C_t}{C_o-C_t}\right)=k_{YN}(t)-k_{YN}\tau$.
(rapid saturation behavior of column)
plot of $\ln[C_t/(C_o-C_t)]$ versus time (t) gives k_{YN} and τ from the slope and intercept,
k_{YN} = Uptake rate constant (1/min), τ = time required for 50% adsorbate breakthrough (min).

(iii) **Bed Depth Service Time Model**: $t_s=\frac{N_o}{C_o F}Z-\frac{1}{C_o K_a}\ln\left(\frac{C_o}{C_t}-1\right)$.
(Relation between Z and t_s) plot of t_s versus Z gives N_o and $K_a N_o$ is the adsorption capacity (ppm), K_a is the rate constant (L/mg min).

7.2 *Summary of Sorption Models*

For biosorption technologies to operate at their best, the sorption rate is essential. The rate is influenced by factors like residence duration, solution pH, temperature, and concentration. To explain the dynamic process of sorption systems, kinetic models including Thomas, Yoon-Nelson, Bed Depth Service Time, Clark, Adams–Bohart, and Wolborska have been researched. These models aid in comprehending the reaction pathways, associated processes, and dynamics of the biosorption process [19]. The *Redlich Peterson model, the BET model, and Lagergren's first-order equation are three prominent adsorption capacity models.* The interaction behavior between metal and biosorbents at equilibrium solute concentration and constant temperature is modeled using isotherms, such as those proposed by Langmuir and Freundlich. Understanding the biosorption mechanism and maximizing the usage of biosorbents are made possible by these models [20].

7.3 *Remediation of Heavy Metals on Microbes*

Depending on their chemical forms, total metal ion concentrations, and other factors including redox potential, the presence of heavy metals can either encourage or hinder microbes. Since more protons are available to saturate metal-binding sites at acidic pH levels, heavy metals frequently produce free ionic species [21]. The surface of the adsorbent becomes more positively charged at higher hydrogen ion concentrations, which decreases the adsorbent's attraction to metal cations and increases its toxicity [6]. Temperature is a key factor in the adsorption of heavy metals [22]. The rate of adsorbate diffusion via the external boundary layer increases as temperature rises. Heavy metals become more soluble at higher temperatures, increasing their bioavailability [23]. How the degradation process plays out depends on the substrate and several environmental parameters, which are compiled in Table 6.

7.4 Future Perspective and Challenges in Removal of Heavy Metals

The bio-adsorbents used in this study to remove heavy metals are affordable adsorbents that can take the place of sorbents that are manufactured commercially. According to several studies, the removal effectiveness of adsorbents for heavy metal removal from wastewater increases following modification. However, less research has been done in this area. As a result, our long-term goals are to improve the removal efficiency of bioabsorbents after modification (*with the least amount of acid, base, and heat*), regenerate sorbents, recover metal ions, and commercialize biosorbents [24]. The difficulty in removing heavy metals from wastewater is that considerable volumes of bio adsorbents and additional chemicals may be required to maintain a pH that allows for sorption. At pH 5–6, agricultural waste products have been demonstrated to be the most advantageous for the biosorption of most cationic metal species [25]. Temperature does not have a significant impact in the majority of biosorption processes involving agricultural waste materials [26]. Although a rise in temperature improves sorption capacity in most circumstances, a high temperature is avoided to avoid thermally induced structural damage. However, some undiscovered difficulties may occur because of the advancement in treatment technology. The great majority of the investigations covered in this article were conducted in laboratories. Research related to these future views provides the finest platform for commercial industrial applications, which must be analyzed to implement treatment processes on a pilot and industrial scale. To create large-scale treatment systems, this aspect must be further studied [27].

8 Conclusion

Most of the studies discussed in this article were carried out in lab settings. To execute treatment procedures on a pilot and industrial scale, research pertaining to these future views provides the right foundation for industrial applications at the commercial level. To create expansive therapeutic systems, this element needs to be further studied. Heavy metals clean-up can be accomplished using agricultural and industrial waste products, which are extremely effective, affordable, and renewable sources of biomass. Since it contains a lot of organic matter, heavy metals can be successfully removed from it. It is also widely known that the carboxyl group is the most important group for complexing with metallic ions and delaying dangerous ions through the sorption process. These biosorbents can also be modified for increased effectiveness, and several re-uses expand their industrial-scale viability. Ion exchange, complexation, physical and/or chemical adsorption, chelation, and inorganic microprecipitation are some of the mechanisms that can be involved in the biosorption process. Among these, the ionic exchange mechanism is thought to be the most important in the removal of metals by cellulosic biosorbents such as organic wastes—natural biopolymers, chemical composites, surface-modified biosorbents, and industrial solid wastes. As a result of their availability, chemical composition, and low cost in nature, various agricultural wastes and solid waste reuse are ideal precursors for the synthesis of activated carbon.

One of the numerous sectors where nanotechnology has gained significant attention is the treatment of water. Nanomaterials are more exciting possibilities for the elimination of metal ions and other contaminants from fluids due to their special features, which include enhanced surface area, extraordinary adsorption capacity, and superior selectivity. They are very effective and selective at removing metal ions, even at low concentrations. Adsorbent dosage, starting dye concentration, test solution pH, contact time, and break through curves were among the system variables whose impacts were investigated. Different isotherm models and kinetic behavior were fixed using experimental data to define the interaction of the solute with the adsorbent. Additionally, the sorption process's thermodynamic properties were calculated. One of the many sectors where nanotechnology has garnered a lot of interest is water treatment. Nanomaterials are more promising materials for removing heavy metal ions and other pollutants from water because of their special features, which include enhanced surface area, amazing adsorption capability, and superior selectivity. Acidic functional groups are introduced onto the surface of activated carbon after an acidic treatment. Because of their excellent selectivity, removal efficiency, enhanced stability, and low-cost energy needs, the separation of harmful metal ions using immobilized material is promise for improving environmental quality. Researchers frequently use spectroscopic methods to keep track of the presence of distinct functional groups and their association with metallic substances during the process of biosorption. Similar to this, it is believed that biosorption will be tremendously helpful in the recovery and purification of high-value items including proteins, steroids, medicines, and pharmaceuticals. These biomaterials have therefore proven to be capable of serving as metal biosorbents. A novel or creative method for eliminating heavy metals from wastewater is biosorption. This process is thought to be a powerful detoxifying approach for eliminating heavy metals, even at low concentrations. To remove heavy metals, biological entities that are both viable and nonviable are desired. In general, availability and cost effectiveness are the two most important factors to consider when choosing the most likely sorbent to treat heavy metals in wastewater. The treatment techniques discussed here can remove heavy metals from wastewater. The best method must be chosen, nevertheless, considering factors including wastewater characteristics, operational costs, and heavy metal content. Reusing organic waste that results from industrial or natural processes is another important step in today's efforts to prevent or minimize environmental degradation. Therefore, unconventional defined biosorbents would be very helpful in creating a technique that is appropriate for successfully removing harmful metals from water and polluted water.

References

1. B.E. Igiri, S.I.R. Okoduwa, G.O. Idoko, E.P. Akabuogu, A.O. Adeyi, I.K. Ejiogu, Toxicity and bioremediation of heavy metals contaminated ecosystem from tannery waste water. J. Toxicol. 1–16 (2018)
2. S. Babel, T.A. Kurniawan, Cr(VI) removal from synthetic wastewater using coconut shell charcoal and commercial activated carbon modified with oxidizing agents and/or chitosan. Chemosphere **54**(7), 951–967 (2004)
3. D. Park, Y.S. Yun, J.M. Park, The past, present, and future trends of biosorption. Biotechnol. Bioprocess Eng. **15**(1), 86–102 (2010)
4. Y. Bulut, Removal of heavy metals from aqueous solution by sawdust adsorption. J. Environ. Sci. **19**(2), 160–166 (2007)
5. C.Y. Yin, M.K. Aroua, W.M.A.W. Daud, Review of modifications of activated carbon for enhancing contaminant uptakes from aqueous solutions. Sep. Purif. Technol. **52**(3), 403–415 (2007)
6. N.M.I. Alhaji, K.T.M. Begum, Optimization and kinetic study for the removal of chromium (VI) ions by acid treated sawdust chitosan composite beads. Int. Res. J. Pure Appl. Chem. **5**(2), 160 (2015)
7. N. Xiong, P. Wan, G. Zhu, F. Xie, S. Xu, C. Zhu, A.S. Hursthouse, Sb (III) removal from aqueous solution by a novel nano-modified chitosan (NMCS). Sep. Purif. Technol. **236**, 116266 (2020)
8. K.M. Tajun Meera Begum, N.M.I. Alhaji, A. Ayeshamariam, V.S. Vidhya, M. Jayachandran, Removal of Ni (II) ions on to polymer loaded sawdust (PLSD)–batch adsorption studies. Fluid Mech. Open Acc. **3**(138), 2 (2016)
9. K.M.T.M. Begum, N.M.I. Alhaji, Spectrophotometric determination of the efficiency of ligno-cellulosic sludge-based adsorbents in the removal of Cr (VI) and Ni (II) ions from aqueous solution. Chem. Sci. **5**(3), 523–540 (2016)
10. M. Wołowiec, M. Komorowska-Kaufman, A. Pruss, G. Rzepa, T. Bajda, Removal of heavy metals and metalloids from water using drinking water treatment residuals as adsorbents: a review. Minerals **9**(8), 487 (2019)
11. V.K. Gupta, A. Nayak, S. Agarwal, Bioadsorbents for remediation of heavy metals: current status and their future prospects. Environ. Eng. Res. **20**(1), 1–18 (2015)
12. Renu, M. Agarwal, K. Singh, Methodologies for removal of heavy metal ions from wastewater: an overview. Interdiscipl. Environ. Rev. **18**(2), 124–142 (2017)
13. C. Liu, N. Fiol, J. Poch, I. Villaescusa, A new technology for the treatment of chromium electroplating wastewater based on biosorption. J. Water Process Eng. **11**, 143–151 (2016)
14. M. Agarwal, K. Singh, Heavy metal removal from wastewater using various adsorbents: a review. J. Water Reuse Desalin. **7**(4), 387–419 (2017)
15. M.A. Atieh, O.Y. Bakather, B.S. Tawabini, A.A. Bukhari, M. Khaled, M. Alharthi, M. Fettouhi, F.A. Abuilaiwi, Removal of chromium (III) from water by using modified and nonmodified carbon nanotubes. J. Nanomater. **2010**, 1–9 (2010)
16. C.F. Carolin, P.S. Kumar, A. Saravanan, G.J. Joshiba, M. Naushad, Efficient techniques for the removal of toxic heavy metals from aquatic environment: a review. J. Environ. Chem. Eng. **5**(3), 2782–2799 (2017)
17. K.M.T.M. Begum, N. Alhaji, Adsorption efficiency of sawdust in activated and surface-modified forms—a comparative study. Chem. Sci. Trans. **2**(4), 1364–1369 (2013)
18. C. Bhattacharjee, S. Dutta, V.K. Saxena, A review on biosorptive removal of dyes and heavy metals from wastewater using watermelon rind as biosorbent. Environ. Adv. **2**, 100007 (2020)

19. G.M. Gadd, Biosorption: critical review of scientific rationale, environmental importance and significance for pollution treatment. J. Chem. Technol. Biotechnol. **84**, 13–28 (2009)
20. H.A. Alalwan, M.A. Kadhom, A.H. Alminshid, Removal of heavy metals from wastewater using agricultural byproducts. J. Water Supply Res. Technol. AQUA **69**(2), 99–112 (2020)
21. Y. Na, J. Lee, S.H. Lee, P. Kumar, J.H. Kim, R. Patel, Removal of heavy metals by polysaccharide: a review. Polym.-Plast. Technol. Mater. **59**(16), 1770–1790 (2020)
22. G. Annadurai, L.Y. Ling, J.F. Lee, Adsorption of reactive dye from an aqueous solution by chitosan: isotherm, kinetic and thermodynamic analysis. J. Hazard. Mater. **152**(1), 337–346 (2008)
23. A. Hussain, S. Madan, R. Madan, Removal of heavy metals from wastewater by adsorption, *Heavy Metals—Their Environmental Impacts and Mitigation* (2021)
24. H.A. Hegazi, Removal of heavy metals from wastewater using agricultural and industrial wastes as adsorbents. HBRC J. **9**(3), 276–282 (2013)
25. B.A.M. Bandowe, M. Bigalke, L. Boamah, E. Nyarko, F.K. Saalia, W. Wilcke, Polycyclic aromatic compounds (PAHs and oxygenated PAHs) and trace metals in fish species from Ghana (West Africa): bioaccumulation and health risk assessment. Environ. Int. **65**, 135–146 (2014)
26. M. Ahemad, M. Kibret, Recent trends in microbial biosorption of heavy metals: a review. Biochem. Mol. Biol. **1**(1), 19–26 (2013)
27. T.M. Frederick, E.A. Taylor, J.L. Willis, M.S. Shultz, P.J. Woodruff, Chromate reduction is expedited by bacteria engineered to produce the compatible solute trehalose. Biotechnol. Lett. **35**, 1291–1296 (2013)

Goal 7: Affordable Clean Energy

Overview of the chapters:

Goal 7: Affordable and Clean Energy

(The goal should be to ensure that everyone has access to modern, affordable, sustainable, and clean energy)

Our expertise lies in the development and search for new sustainable nanomaterials, and we are very familiar with the current advanced energy technologies. Technology and human needs are met by the amount of effort required to create and maintain an object. Energy is essential to every living thing and every process in the universe. Among humans, animals, plants, and microorganisms on Earth, the sun provides the majority of the energy they need and can access. This energy arrives from either direct or indirect sources. In order to prevent biodiversity loss, clean and sustainable energy conversion systems must be developed in response to the exhaustion of fossil fuels. It is now essential to maintain a continuous power supply while utilizing this new energy source due to its limitations, such as its time and region restrictions. A significant increase in popularity, complexity, and power of portable electronics, particularly autonomous devices and medical devices, has occurred over the past ten years. It might be possible to make these futuristic devices run on lithium-ion microbatteries with low specific energy due to the low specific energy of thin-film, all-solid-state batteries. It is crucial that a device's weight-to-energy (w/e) and volume-to-energy (v/e) ratios are high in order to provide the needed energy. Despite the fact, organic materials are less costly than inorganic ones, and their chemical stability and low cost have led them to be used in modern technology more than inorganic materials. In addition, worldwide, the amount of electronic waste generated has risen alarmingly. It is estimated that global e-waste production will reach 74.7 million metric tons by 2030, according to Global E-Waste Monitor 2020. Several countries in Europe and the USA are leading contributors to e-waste, as well as primarily technology-consuming nations. E-waste can have grave environmental consequences if not properly managed and disposed of. Due to this, energy is a necessary ingredient for all universal processes, which is why it fuels human progress. Approximately 80% of Earth's energy comes from the sun. Fossil fuels are primarily responsible for the high energy consumption. Sustainable energy solutions are needed to keep pace with growing technology. Energy storage innovations are being driven by the urgency of pursuing renewable energy. Battery technology, capacitors, and more are covered in this part, as well as the advancements that will be made in the future to provide electrical energy storage.

Green Sustainable Nanocomposites for Energy Storage Devices

Ayesha Kausar and Ishaq Ahmad

Abstract Current advanced energy technologies are continuously developing and searching the novel sustainable nanomaterials. Green nanocomposites combine exclusive properties of sustainable polymers, nanoparticles, and nanomaterials. Green or environmental nanomaterials have advantages of inexpensiveness, light weightiness, environmentally friendliness, and competence for energy related technical applications. Ecological, biodegradable, and sustainable nanocomposites have been applied in number of energy devices including the fuel cells, batteries, light-emitting diodes, etc. Among all energy storage devices, supercapacitors have been acknowledged as the record competent charge and energy-storing systems. Therefore, the main focus of this chapter is to highlight the energy storage application of the green nanocomposites particularly focusing the supercapacitors. This episode initially covers the fundamentals of green sustainable nanocomposites. Subsequently, the energy storage connected solicitations of the eco–nanocomposites have been stated. Consequently, the developments of the multi-functional high-performance supercapacitors based on the ecological nanocomposites have been debated. Additionally, the challenges and forthcoming forecasts of the sustainable nanocomposites toward energy storage devices were deliberated.

1 Introduction

Green polymers have also been named as eco-polymers, environmental friendly polymers, ecological polymers and several related terms [1, 2]. These polymers mostly exist in nature or can be extracted using the renewable sources. The important properties of green polymers include biologically friendliness (*obviously*), sustainability,

A. Kausar (✉) · I. Ahmad
NPU-NCP Joint International Research Center on Advanced Nanomaterials and Defects Engineering, Northwestern Polytechnical University, Xi'an, China
e-mail: dr.ayeshakausar@yahoo.com

UNESCO-UNISA Africa Chair in Nanosciences/Nanotechnology, iThemba LABS, Somerset West, South Africa

K. Kasinathan et al. (eds.), *New Technologies for Energy Transition Based on Sustainable Development Goals*, https://doi.org/10.1007/978-981-97-2527-4_13

biodegradation, and compatibility with other green materials and systems [3]. Preliminary research has engrossed the adoption of natural green polymers [4]. Afterward, various green polymers have also been prepared synthetically. Moreover, the research progress has resulted in the development of the ecological friendly nanocomposites [5]. Variety of green polymers and nanofillers have been assumed to form the derived nanomaterials. The green nanocomposites have owned the properties of both the eco-polymers as well as the ecological nanoscale materials used. Accordingly, the green nanocomposites revealed several beneficial properties such as the structural stability, better mechanical properties, sustainability, and bio decomposability toward the high-performance applications [6]. In this regard, the certain non-green polymers or nanofillers have also been modified to attain the green nanocomposite structures [7]. If the nanocomposite is designed using both the eco-polymers and green nanofillers, then the resulting green nanocomposites have better sustainability and bio decomposability features [8]. Consequently, the obtained green nanocomposites have been applied in wide ranging high-efficiency technical applications ranging from the electronics or devices to the engineering structures as well as the biomedical and environmental applications [9, 10]. All such mythological fields have been turned toward the use of green materials due to the increasing demands of environmentally friendliness, sustainability as well as biodegradability features. Among synthetic green nanofillers, carbon nanoparticles, metal nanoparticles, metal oxides, nanoclays, and several other inorganic nanofillers. In the resulting green nanocomposites, the cooperative possessions of both the ecological polymers as well as the green nanofillers occasioned into numerous advantageous physical characteristics and high-performance applications, in addition to the eco-sociability and biodegradability characteristics [11]. Like non-green nanocomposites, essential features of the ecological nanocomposites have also been found to considerably enhance with the nanofiller addition. Furthermore, using the facile preparation routes, the interface formation and matrix–nanofiller interactions have been observed leading to several valuable properties of the nanomaterials. A very interesting use of the green nanocomposites have been observed for the energy storage devices and systems. The traditionally used non-green nanomaterials obviously have drawbacks of the non-biodegradability and environmental hazardousness [12]. Therefore, the research has enthused toward the use of sustainable biomaterials having enhanced electron conduction as well as the charge storage capabilities. The green nanocomposites offer fine replacements of the pollution producing energy materials. The capacitors ad related devices have been applied as the advanced energy storing maneuvers [13]. Especially, supercapacitors have been used as the effectual charge storage capacitor devices. In this regard, green nanomaterials have advantageous of the light weightiness, low cost, sturdiness, enhanced charge or energy storage, energy density, and specific capacitance features. In addition, the green nanomaterials own intrinsic biodegradable, sustainable, and recyclability properties [14]. The eco-nanocomposites have also been researched for improved engineering features for better use in the supercapacitors [15, 16]. However, the choice of appropriate designs and use in the high-performance advanced energy storage systems face several design to applicability challenges [17].

To the best of acquaintance, this chapter is innovative owing to the designed framework and the gathered literature regrading green nanomaterials for energy storage [18, 19]. Here, employment of the green polymeric matrices as well as the ecological nanofillers have been explored for the formation of the ecological nanocomposites. Otherwise, the final nanocomposite must have sustainability, regarding the nature of the individual components involved. The future of energy storage devices relies on using the sustainable nanocomposites. Hence, this chapter is comprehensive and purposes the indispensable methodical and marketable expansions of the green nanocomposites toward the energy sector.

2 Green Sustainable Nanocomposites

Environmentally friendly polymers are often also termed as eco-polymers or green polymers. The green polymers occur both naturally and can be synthesized using appropriate processes. Consequently, variety of natural or synthetic green polymers including the starch, cellulose, chitosan, nylons, poly(ethylene glycol), poly(ethylene oxide), polycarbonate, polyurethane, poly(lactic acid), poly(vinyl alcohol), and many more, have been adopted in materials science for technical purposes in Fig. 1 [20]. Green polymers are mostly biodegradable and ecological friendly in nature, therefore have much demand in the sustainability applications. Different techniques used for the formation of green polymers have been surveyed in literature [21]. The ecological polymers or green polymers have been technically applicable in the range of applications such as the membranes and coats [22], adhesives [23], pharmaceutical uses and drug delivery [24], and other biomedical applications [25].

For polymers or green polymers, nanofillers used can also be green or ecological friendly in nature. Green nanofillers also have biodegradability as a prominent and demanding feature. Figure 2 demonstrates some commonly applied green nanofillers or used after some modification to render them eco–nanomaterials. One important green nanofiller used is chitosan [26]. Although, chitosan has also been used as the matrix in green nanocomposites. Then, lignin has been functional as a green nanofiller, on the other hand, it can be applied as a green polymer matrix [27]. As nanofillers, the ecological nanomaterials have been added in the natural as well as synthetic polymers such as the nylons, cellulose, poly(lactic acid), polyester, polyethylene, polypropylene, polystyrene, etc. [28, 29]. An important nanofiller is the montmorillonite nanoclay, i.e., from wide the group of Phyllosilicate clay minerals [30]. As a green nanofiller, montmorillonite nanoclay has gained much attention [31, 32]. Montmorillonite nanoclay is a well biodegradable and environmental friendly nanofiller, so has been adopted for variety of the synthesized or green polymeric systems [33]. The carbon nanofillers, after some modifications, have also been applied in the green nanomaterial systems owing to remarkable properties. Among common carbon nanofillers used with the green nanomaterials include the graphene, graphene oxide, reduced graphene oxide, and other graphene derivatives, carbon nanotube, functional carbon nanotube, etc. [34, 35]. In addition to

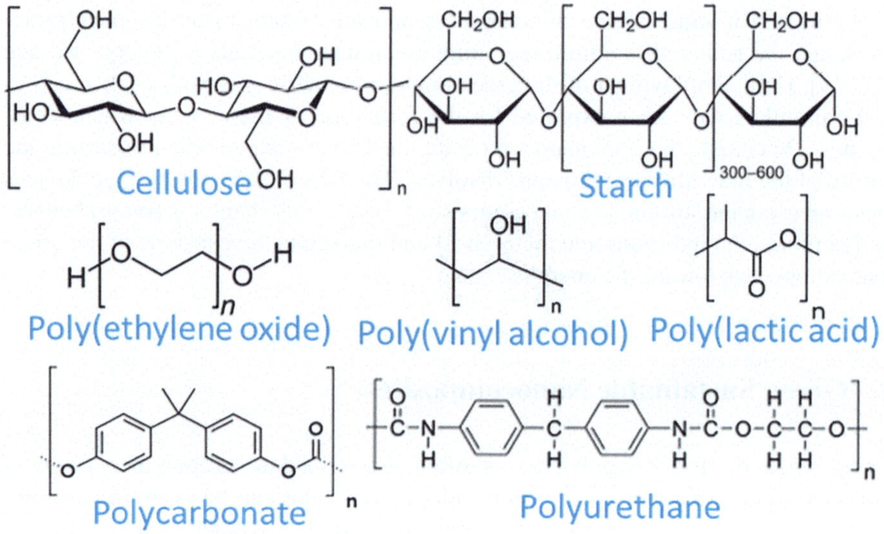

Fig. 1 Structures of some green natural or synthetic polymers

biodegradability properties, these green nanofillers have improved thermal constancy, mechanical stability, electron conduction, charge transportation, thermal conduction, antimicrobial nature, and several other crucial features of the green nanocomposites.

The structure chitosan is cationic (1–4)-2-amino-2-deoxy-β-D-glucan, i.e., can be easily partially acetylated [36]. A very common and handy way of attaining chitosan is the manufacturing from the marine chitin. Moreover, it has some very attractive benefits of non-toxicity, biocompatibility, biodegradability, etc. In the chitosan (as a matrix) derived nanocomposite, the nanoparticles such as nanocarbons, silver, gold, metal oxides, etc., have been effectively reinforced [37–39]. Carlo et al. [40] reported the chitosan and gold nanoparticles derived green nanocomposite. The colloidal chitosan matrix and gold nanofiller-based nanomaterial was obtained using a green technique. In this green method, $HAuCl_4$ salt precursor has been used to form the gold nanoparticles. Using an organic acid, it can reduce the Au^{III} to Au^{o} in the aqueous solution. In this attempt, the caffeic acid was used to reduce the gold precursor to colloidal gold nanoparticles, in the presence of chitosan solution as shown in Fig. 3. The reduction process was also found to alter the microstructure of the gold nanoparticles as well as the chitosan and gold-based nanomaterials. In the nanocomposite, the hydroxyl and amine functionalities of chitosan offered better interaction sited to the gold nanoparticles. Therefore, the biocompatibility and biodegradability properties of chitosan can be better combined with the features of the gold nanoparticles.

Starch is also an economical, processable and green matrix used for the ecological friendly nanomaterials [41]. On the other hand, starch has also been applied as a green nanofiller in various green or non-green matrices [42]. Cheviron et al. [43] designed the starch matrix and silver nanoparticles derived green nanocomposites. The ecological nanocomposites have been employed for the antibacterial applications, packages,

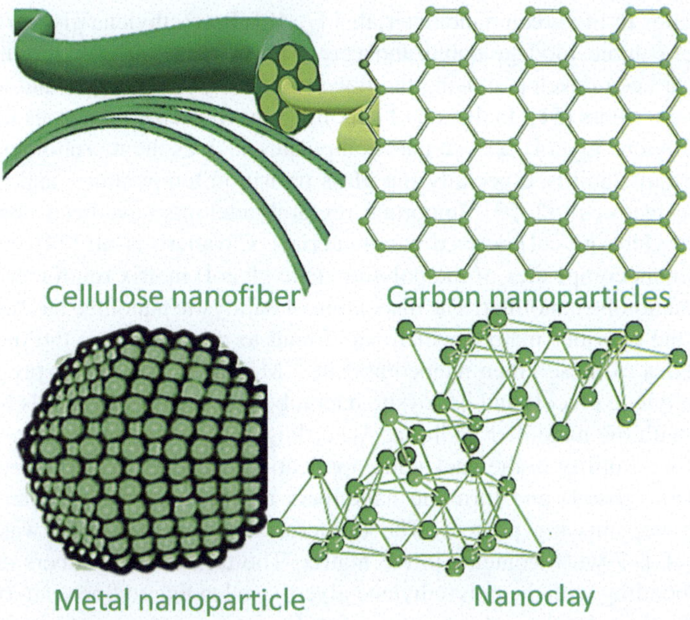

Fig. 2 An environmentally friendly nanofiller

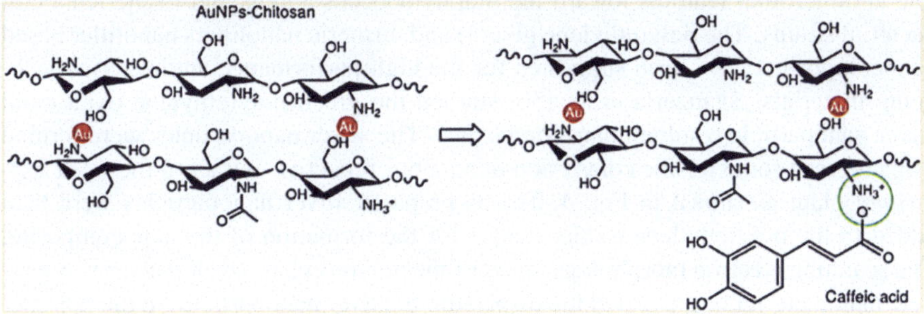

Fig. 3 A green method for the formation of chitosan and gold nanoparticles derived ecological nanocomposite [40] (Reproduced with permission from ACS)

and sensing purposes. Kaushik et al. [44] developed the nanofibrils of cellulose and applied to form the nanocomposites with the starch matrix. The starch with the cellulose nanofibrils nanofiller have been fabricated through the high shear mixing. The nanocomposites revealed fine mechanical properties as the tensile modulus varied in the range of 24–224 MPa. Similarly, among natural polymers, lignin has been significantly applied as a green polymer [45, 46]. Lignin has been easily obtained from natural sources. Like starch and cellulose, lignin can also be used as a matrix as well as a nanofiller in the green nanomaterials [47, 48]. Poly(ethylene glycol) is a non-toxic, water soluble, biodegradable, and a green polymer [49, 50]. Various nanocomposites have been developed using the poly(ethylene glycol) matrix and natural or synthetic nanofillers [51]. Inclusion of nanofillers have altered numerous features of the poly(ethylene glycol) derived nanocomposites such as the morphological properties, thermal stability especially the glass transition temperature, and important mechanical features [52, 53]. Important research attempts have been observed on the poly(ethylene glycol) derived nanomaterials. Cavallaro et al. [54] formed the ecological nanocomposites of the poly(ethylene glycol) matrix reinforced with the halloysite nanotube nanofiller. The dispersion of halloysite nanotube has been investigated in the polymer matrix, which was found to directly affect the mechanical characteristics of these green nanocomposites. Moreover, the barrier properties of the poly(ethylene glycol) and halloysite nanotube derived nanomaterials have been improved with the nanofiller addition. Accordingly, these ecological nanocomposites have found utility in the packaging application. Gopi et al. [55] fabricated the poly(ethylene glycol) and turmeric nanofibers resultant nanocomposites through a multiple step missing process. The turmeric nanofiber nanofiller was used in the range of 1–7 wt.% contents in the matrix. The turmeric nanofibers developed hydrogen bonding with the poly(ethylene glycol), so leading to fine scattering properties. Consequently, the poly(ethylene glycol) and turmeric nanofibers nanofiller based nanocomposites revealed high Young's modulus of 49.36 MPa, relative to neat poly(ethylene glycol) matrix (19.76 MPa). In addition, the nanocomposites showed fine antimicrobial features toward the Staphylococcus aureus and Escherichia coli bacterial strains. The poly(ethylene glycol) and turmeric nanofibers nanofiller based nanocomposites have been suggested for the high-performance antibacterial packaging materials. Sganzerla et al. [56] studied the green poly(ethylene oxide) and silver nanoparticle resultant nanocomposites. The silver nanoparticles were formed using a green route for the conversion of aqueous bioactive extract of the *Acca sellowiana* plant as shown in Fig. 4. The as prepared silver nanoparticles were then added to the poly(ethylene oxide) matrix for the formation of the nanocomposite. The scanning electron morphology study of the poly(ethylene oxide)/silver nanoparticle nanomaterial has revealed the dispersion of silver nanoparticles in the polymer matrix. The ecological friendly and biodegradable nanocomposites have revealed the antimicrobial and antioxidant characteristics due to the presence of silver nanoparticles. Moreover, addition of silver nanoparticles were found to enhance the elongation at break of the poly(ethylene oxide) leading to better film forming properties. Consequently, the applications of the poly(ethylene oxide) and silver nanoparticle subsequent nanocomposites have been observed for the food packaging application having the human health-friendly properties. Hong et al. [57] studied the molecular modeling of the silica nanoparticles grafted with the short chain poly(ethylene oxide) units. According to the molecular modeling studies, the silver nanoparticles were assumed as the solid spheres, while the polymer chains were reserved as the united-atom representation with the bending and torsional bonding interactions. The Green–Kubo relationships were applied to explore the viscosities of the polymer chain grafted silica nanoparticles and polymer chain non-grafted silica nanoparticles.

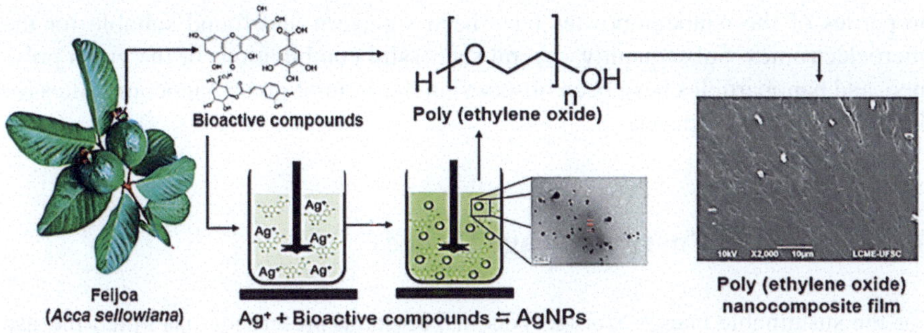

Fig. 4 Synthesis route and scanning electron microscopic imaged morphology of poly(ethylene oxide) and silver nanoparticles derived nanocomposites [56] (Reproduced with permission from Elsevier)

It has been observed that the rapid dynamics were observed for the polymer-grafted silver nanoparticles system, relative to the non-grafted.

Poly(lactic acid) has also been espoused as an important green and natural polymer [58, 59]. Like starch, poly(lactic acid) has biodegradability and softening properties. Krikorian et al. [60] designed the green poly(lactic acid) and green nanoclay-based ecological nanomaterials. In addition to biodegradability, crystalline and high storage modulus features were explored. Inclusion of nanoclay altered the X-ray diffraction patterns of the nanocomposites. Moreover, the addition of nanoclays to the poly(lactic acid) matrix improved the storage modulus in the range of 20–150 °C. It has observed that the storage modulus of the poly(lactic acid) matrix was enhanced up to 61% with the nanoclay additive of 15 wt.%. Wang et al. [61] formed the green nanocomposites based on poly(lactic acid) and nanocellulose crystal nanofiller. The addition of the nanocellulose crystals progress the crystallinity properties of the poly(lactic acid) matrix by 38% due to better polymer chain arrangements. Consequently, the 2 wt.% nanofiller addition improved the crystallinity up to 33%. The strength of the poly(lactic acid) was found to enhance up to 53.9 MPa with the nanofiller addition, relative to the neat polymer (41.9 MPa). Li et al. [62] developed the poly(lactic acid), poly(ethylene oxide), and cellulose nanocrystals resulting biodegradable green nanocomposites. The crystalline properties of the nanocomposites were enhanced with the inclusion of the nanoparticles due to better interface associations between the polymers and the nanofiller. The barrier properties of the nanomaterials were also studied and were found effective for important technical applications. Like other green and biodegradable polymers, poly(vinyl alcohol) is widely used matrix for the formation of nanocomposites [63]. In nanocomposite form, poly(vinyl alcohol) revealed several essential biomedical and ecological applications due to non-toxic nature such as anti-oxidants, antibacterial, and wound healers [64]. Morsi et al. [65] intended the poly(vinyl alcohol) and gold nanoparticle derived nanomaterials as green contenders. Zhao et al. [66] used the carboxymethyl chitosan nanofiller for the poly(vinyl alcohol) matrix. The nanocomposites revealed antimicrobial and non-toxicity features. Furthermore, the electron conduction and dielectric permittivity properties of the nanocomposites have been sightseen and found suitable for the microelectronics. Subsequently, several successful combinations of the green polymers and nanoparticles have been observed in the form of green nanocomposites for the technological purposes.

3 Sustainable Energy Storage Devices

For the sustainable energy storage systems, research preference has opted the use of the ecological materials, particularly the nanomaterials. History of using eco-materials in energy applications can be related to using these in the turbines and the wind turbine blades [67, 68]. The eco-materials have been designed having the low price, less heaviness, and superior strength properties. In this regard, the metallic materials in wind turbine blades have been replaced with the natural composites having eco-polymers or green materials [69]. Later, efforts have revealed the use of eco-materials in the optoelectronic field [70, 71]. Accordingly, the green polymers possess fine donor–acceptor behavior for light energy conversion applications [72]. The light-emitting diodes have also found the sue of the eco-polymers and related materials [73, 74]. Furthermore, the solar cell devices have found the use of green materials to replace the traditional non-green contents and enhance the light conversion efficiency [75]. Chen et al. [76] proposed the poly(dimethyl siloxane) and zinc derived green nanoparticles based nanomaterials for solar cells. The sustainable green nanocomposite depicted efficient luminescence spectrum [77, 78]. The eco–nanomaterial has been efficiently applied in the solar cell devices. Zhang et al. [79] formed the thermoelectric generators derived energy conversion devices using the green nanocomposites. In this concern, they developed the bio-based conjugated polymers and resulting nanocomposites. The ecological materials performed efficiently for the thermoelectric devices. Moreover, such green polymers and nanomaterials have been formed for the optoelectronics and solar cell devices [80, 81]. The green materials professionally industrialized the donor–acceptor energy conversion mechanism [82]. Incidentally, Zhuang et al. [83] also examined the ecological conducting polymers for the solar cells. The photophysical features of the materials were found suitable for the energy related devices and systems [84]. Hence, there has been increasing demand of the eco-materials in the energy devices counting the nanogenerators, supercapacitor, optoelectronics, solar cells, and several other related devices.

4 Nanocomposites in Supercapacitors

Continuous research efforts have focused the development of the energy storage devices aiming the electron, ion, and charge storage properties [85]. In this regard, numerous energy storage appliances have been explored with time including the capacitors (electrochemical or electrostatic), fuel cells, batteries, superconducting

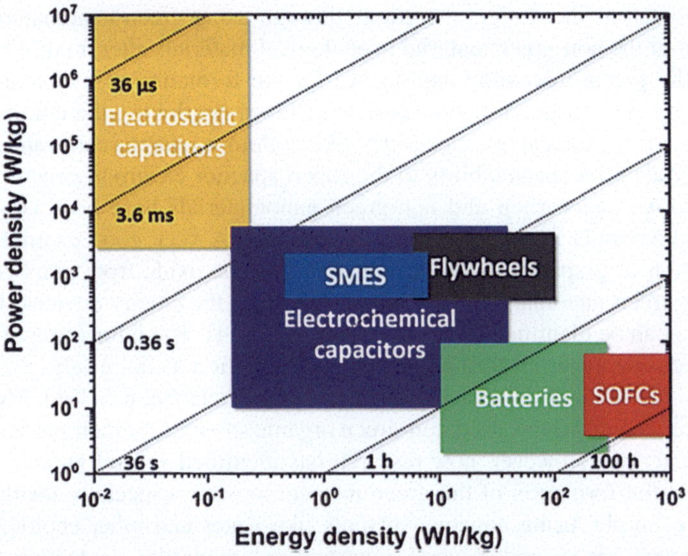

Fig. 5 Ragone plots based comparative behavior of the numerous energy storage devices studied so far, such as the capacitors, superconducting magnetic energy storage capacitors, flywheels, batteries, and solid oxide fuel cells [89] (Reproduced with permission from Elsevier)

magnetic energy storage devices, and so on [86, 87]. Figure 5 discloses the Ragone plots for the dissimilar energy storage devices [88]. The plots shoe the placements of the charge storage devices as per their performance. All these devices have been developed uninterruptedly over time, and according to the needs of the efficient charge storing properties. At this juncture, it has been considered essential to compare and analyze the best energy storing device according to the performance parameters. Apparently, capacitors have gained special stance in the Ragone plots of energy and power density. The performance was qualified to the better power density and charge/discharge properties. The capacitor devices have low price, low operating price and voltage, in addition to the high power density [90]. Consequently, the capacitor maneuvers have found essential solicitations including the power organizations, electronics, electronic vehicles, and green energy storage set ups.

5 Using Green Sustainable Nanocomposites in Supercapacitors

The supercapacitors have been found as the noteworthy performing charge or energy storing maneuvers. The energy storage devices can be converted into the green devices for the charge and energy storage depending upon the various factors. For the progress of green supercapacitor devices, some effective ways have been sorted out such as the design alterations through using green nanomaterials, then conversion of the non-green material to ecological materials after modifications and adopting the green-processing approaches for the formation of nanomaterials for green energy. Use of green nanomaterials involves employing the common green polymer matrices for energy electrodes like cellulose, starch, chitosan, etc. [91]. Such materials offer sustainability to the supercapacitor electrode structures. Similarly, operative nanocarbon and non-green nanomaterials have been converted to green nanomaterials for energy-storing electrodes. A very good example here is the formation of graphene oxide or reduced graphene oxide from graphene. More competent green nanomaterials can be obtained for the energy devices if suitable approaches can be identified for their end-uses [92–94]. The fabrication of supercapacitor electrode materials through the green fabrication tactics evades the environmental and health hazards related to organic synthesis techniques [95]. Most of the polymers and nanocarbons need non-green organic solvents for their processing into the nanomaterials. However, now research has identified several green routes and solvents for the formation of the green nanocomposites, especially for the energy uses. For example, using aqueous solvents like water and other ecological solutions have been discovered. Several important polymers like conducting polymers and epoxies can be easily processed in aqueous nonhazardous media. Nevertheless, replacing the whole supercapacitor electrode material with green components have been found challenging in literature due to limited research attempts in this sector. The entirely green nanocomposite electrode means having the green polymer, green nanofiller, green binder, and green substrate for the final design production. Mainly, the attainment of eco-friendliness along with very high charge or power density and capacitance values as well as robustness desire several intense focused future studies [96, 97].

The well-organized research efforts have been observed for the construction of the green materials for the energy storing devices. Graphene has been acknowledged as an extraordinary performing carbon nanomaterial [98]. One of the important applications of graphene has been observed for the formation of nanocomposites and polymeric nanocomposites [99, 100]. To the nanocomposites, including minor amounts of nanofillers may enhance the desired physical properties of the ensuing materials [101]. In energy-storage devices and systems, graphene and graphene-based nanocomposites have been effectively applied. One of the successful design combinations has been identified as the conducting polymer- and graphene-derived supercapacitor electrodes [102]. The conjugated polymer/graphene nanomaterials have inexpensiveness, structural stability, high surface area, capacitance, energy density, and charge/discharge performance [103–105]. Various polymers like polyaniline, polypyrrole, polythiophene, and polythiophene derivatives have been sued to form the supercapacitor electrodes with graphene nanofiller [106]. Here, eco-friendliness as well as sustainability of these materials have been the major concerns for the formation of energy storage components. Ezeigwe et al. [107] adopted green liquid phase exfoliation method for the formation of graphene and zinc oxide based nanocomposite electrodes for the supercapacitors. The nanomaterials have been fabricated using the green solvothermal technique with ethanol and water plus sonication steps.

The ecological graphene and zinc oxide derived nanocomposite electrodes possess specific capacitance of 236 Fg^{-1} at a scan rate of 10 mV s^{-1}. The efficient supercapacitance performance was attributed to the design of electrodes as well as the efficient green method adopted to form these materials. Nayak et al. [108] reported on a two electrode solid-state asymmetric supercapacitor cell based on the graphene supported tungsten oxide nanowires as the negative electrode. The performance of the two-electrode solid-state asymmetric supercapacitor cell was analyzed using the energy density, power density, and specific capacitance profiles, in addition to the real light-emitting diode-based application of these supercapacitors. Figure 6 displays the supple solid-state asymmetric energy storage device and the real asymmetric supercapacitor device for lightening the red colored light-emitting diodes. Moreover, Fig. 7 presents the Ragone plots for the trends of the energy density and power density profiles of the nanomaterials. In addition, the cyclic stability of specific capacitance for the given cycle numbers have been studied. After 4000 charge–discharge cycles, the energy density was maintained in the range of 6–25 $Whkg^{-1}$. Hence, the high specific capacitance, energy density, power density, and cyclic performance of the green derived nanomaterials have been found remarkable.

Arthisree et al. [109] formed the polyacrylonitrile, polyaniline, and graphene quantum dot derived efficient nanocomposite for supercapacitor electrodes. Figure 8 demonstrates the formation of an energy storage assembly obtained using an ecological route. The green nanocomposite electrode had nanofiller added in 1.5 wt.% contents. It was constructed by sandwiching the layers of nanomaterial, polymers, and Al foils. The device revealed output power of 1.4 V after 1 h of working. The supercapacitor depicted high specific capacitance in the range of > 100 to ~ 600 Fg^{-1}. It was suggested that the conducting nature of the conjugated polymers like polyaniline and polyacrylonitrile and their synergistic effects with the nanofiller were responsible for the enhanced supercapacitor performance [110]. Here, the nanocomposite electrodes can be further altered to incorporate the green materials in designing

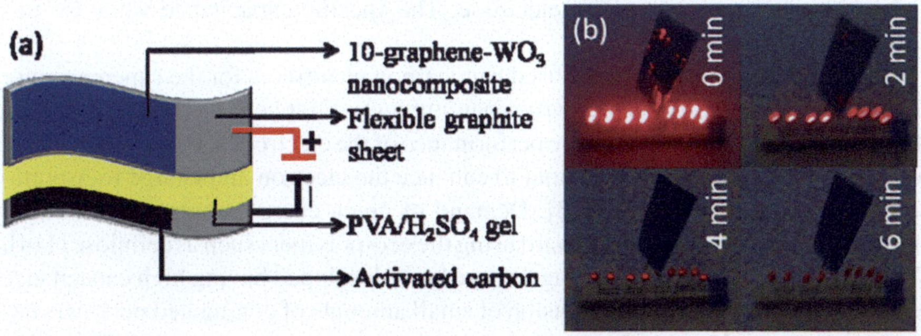

Fig. 6 **a** Representation of a flexible solid-state asymmetric supercapacitor device; and **b** diagram of a real ASC device to light red LED after charging [108]. *ASC* asymmetric supercapacitor, *LED* light-emitting diode (Reproduced with permission from ACS)

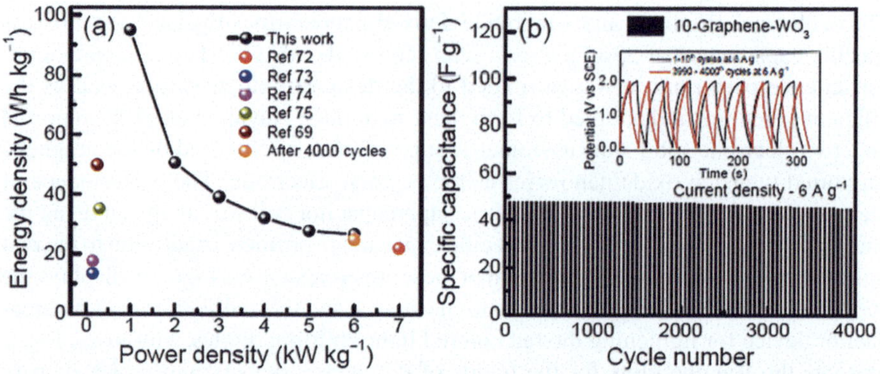

Fig. 7 **(a)** Ragone plots of energy density versus power density; and **(b)** cyclic stability pots of specific capacitance as a function of cycle numbers (current density 6.0 A g^{-1}) with a two-electrode solid-state ASC cell (graphene–WO_3 negative electrode and activated carbon positive electrode) [108]. *ASC* asymmetric supercapacitor (Reproduced with permission from ACS)

to attain sustainability of the charge storage devices [111]. Ciplak et al. [112] introduced the graphene oxide, reduced graphene oxide, and gold nanoparticle nano-reinforcements to the polyaniline matrix. An ecological approach was used to form the green supercapacitor electrodes. Here, graphene oxide was obtained using a facile ecofriendly route and then converted to reduce graphene oxide. On the other hand, the polyaniline and gold nanoparticle resultant nanocomposite were prepared through the in situ polymerization. In the nanocomposite structure, π–π stacking relations and electrostatic linking have been observed. Figure 9 expresses the relation profiles of the specific capacitance and the scan rates. Here, the performance of the neat polyaniline, polyaniline/graphene oxide-gold nanoparticles, and polyaniline/reduced graphene oxide-gold nanoparticles were compared. The specific capacitance of the polyaniline/reduced graphene oxide-gold nanoparticles was found around 64% higher than the neat polyaniline electrode. The specific capacitance value for this nanomaterial was 212.8 Fg^{-1}.

Henceforward, the commonly used nanocarbon nanofillers for the supercapacitor electrodes include the graphene, carbon nanotube, etc. Graphene has been found most efficient to enhance the capacitance performance of the electrodes. Furthermore using conjugated polymers has been found to enhance the electron and charge transportation through these materials [113]. Demand of green energy storage devices have moved the electrode designing toward using the eco-polymers such as cellulose [114]. The cellulose/graphene oxide aerogels have been developed having high capacitance performance [115–117]. The inclusion of small amounts of conjugated polymers has further increased the specific capacitance of these energy devices [118]. The green graphene oxide-based electrodes have also been modified using metal oxides such as the nickel oxide, magnesium oxide, nickel hydroxide, etc., for enhanced efficiency. In addition to conjugated polymers, the metal oxides and metal hydroxides have also been added to cellulose matrix for superior energy storing efficiency [119–121].

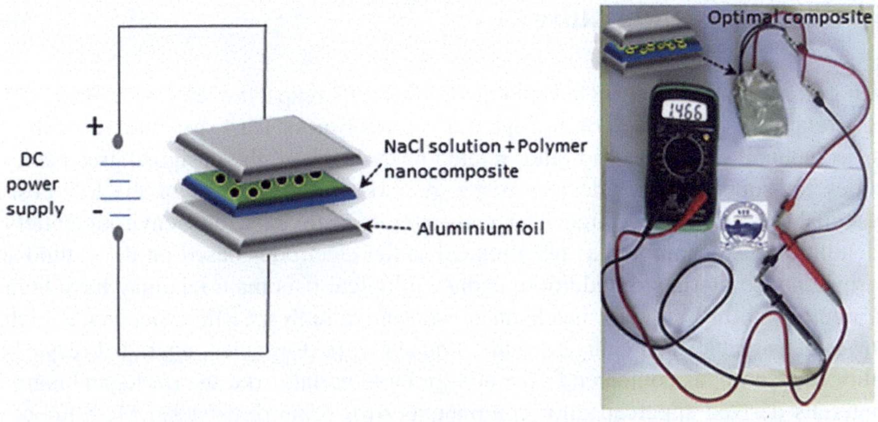

Fig. 8 Design of PAN/PANI@G-1.5 wt.% resulting supercapacitor along with the characteristic digital photograph of optimal nanocomposite for voltage generation [109]. *PAN/PANI@G* polyacrylonitrile/polyaniline/graphene quantum dot (Reproduced with permission from Elsevier)

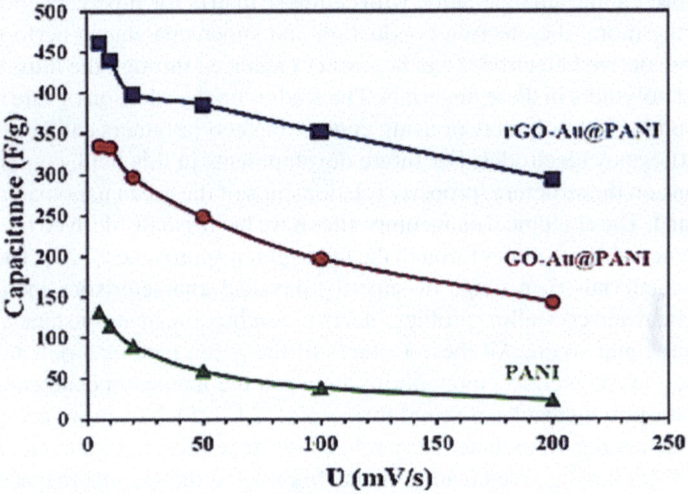

Fig. 9 The specific capacitance dependence on the scan rate (5–200 mVs^{-1}) for neat PANI, GO-Au@PANI, and rGO-Au@PANI [112]. *PANI* polyaniline, *GO-Au@PANI* graphene oxide-gold nanoparticle@polyaniline, *rGO-Au@PANI* reduced graphene oxide-gold nanoparticle@polyaniline (Reproduced with permission from Elsevier)

Into the bargain, numerous research efforts have been performed using the green biomaterials based energy storing electrodes depicting high surface area, capacitance, electrochemical, and charge density features for the supercapacitor application [122].

6 Challenges and Future

Using conventionally designed non-ecological supercapacitor electrode materials have related disadvantages of the high price, noxiousness, least structural reliability, environmental effluence, and other related hazards. Here, the designed green electrodes for supercapacitor devices reveal remarkable advantages of the low cost, sustainability, reliability, high heat constancy, non-toxicity, and environmentally friendliness. An example has been related to the electrodes based on the cellulose derived nanomaterials. In addition, using ecological nanomaterials may have high structural stability, superior mechanical robustness, high specific capacitance, high power density, elevated charge density, fine cyclic performance, upgraded recyclability, and so on, as compared to the non-green materials. Yet, the ecological nanomaterials derived supercapacitor components own some drawbacks. Most importantly, neat eco-polymers may have low charge storing properties. Therefore, these materials cannot be used without structural alterations or nanocomposite formation. Here, the combinations of the green materials and non-green nanomaterials have been suggested in literature. However, green carbon nanoparticles like graphene oxide can offer a fine amalgamation with cellulose matrix for supercapacitor components. Furthermore, the electron conduction and supercapacitance performance of the cellulose derived electrodes can be further enhanced through the introduction of conjugated polymers in these materials. The studies on the addition of safe non-toxic levels of conducting polymers or using conducting eco-polymers can be considered for high-efficiency electrodes. For future developments in this field, comprehensive explorations on the structure–property relationships of the green nanomaterials must be performed. The ecological nanocomposites have been usually derived from several priceless renewable resources through the facile green approaches [123]. The ensuing green nanomaterials own range of superior physical characteristics including the morphological and crystalline profiles, electron conduction, heat constancy, mechanical heftiness, and so on. All these features of the green nanomaterials rely on the fine interactions as well as compatibility between the nanocomposite components, especially the polymer and nano-reinforcements [124, 125]. Numerous eco-polymers employed in this regard include the starch, cellulose, chitosan, lignin, etc. and similarly nanofillers having green features led to high-performance supercapacitor materials [126]. Nevertheless, enhancing the conductivity and capacitance may reduce the structural stability of the nanomaterials. Consequently, design of the green nanocomposites must be focused for high-end devices and systems. Fewer studies have focused the mechanisms behind the charge storage and energy density capabilities of the supercapacitor electrodes and comprehensive research efforts have been desirable in this field. Accordingly, the matrix-nanofiller interface, hydrogen bonding associations, electrostatic interactions, covalent contacts, etc. must be considered along with the consistent nanoparticle dispersion in the green nanomaterials. Hence, the incipient research on the ecofriendly supercapacitors components has been desirable for the high-performance sustainable and green future of the energy storage devices

[127]. Besides the energy storage maneuvers, the foremost utility of the capacitors and supercapacitors devices can be observed in the electronics, microwaves, telecommunication, and related devices.

7 Summary

This chapter uniquely and comprehensively statuses the expansions in the arena of green nanocomposites and particulars of these materials focusing the energy storage solicitations. The green sustainable nanomaterials have been unceasingly functional for the supercapacitance applications. The investigations revealed the steadfastness and high charge storage recital of the ecological materials derived energy storage systems. The use of green nanomaterials along with the application of the green fabrication and utilization practices have been considerably found to improve the charge storage performance of the advanced supercapacitors. Both the ecological polymers as well as the ecological nanofillers have been found to grow the high-performance energy related devices and systems. Here, the choice of green polymers as well as green nano-reinforcements matters a lot for the better performance of the energy devices. Use of green processing methods have also played role for the enhancement of the charge storing performance of these materials and resulting high efficacy supercapacitors. Several essential characteristics of the supercapacitors have been enhanced using the green polymers, green nanofillers, as well as the green synthesis approaches including the crystalline nature, microstructure, matrix-nanoparticle associations, electron conduction, dielectric features, specific capacitance with better charge storing tendencies, charge density, power density, charge–discharge, cyclic performance, reliability, life cycle, and several other related features. The future research may extend the use of the discussed materials toward the green and sustainable photovoltaics, diodes, nanogenerators, microelectronics, and several other devices. Thus, this book section portrays the evolution in the stadia of ecological and sustainable energy storing materials. In this regard, various non-green polymers like thermosets, conjugated polymers, or thermosets and non–eco–nanofillers have been used after modifications to render them green features for the supercapacitor applications. Henceforth, all these approaches offer remarkable future opportunities toward the future green supercapacitor expertise. Besides, the continuous ongoing research efforts may resolve the underlying design, properties, and performance challenges in this field.

References

1. B.Y. Karlinskii, V.P. Ananikov, Recent advances in the development of green furan ring-containing polymeric materials based on renewable plant biomass. Chem. Soc. Rev. **52**(2), 836–862 (2023)
2. A. Kausar, I. Ahmad, Degradable green polymers, green nanopolymers and green nanocomposites derived from natural systems: statistics and headways. Nano-Horizons (2023). https://doi.org/10.25159/NanoHorizons.49f1ef7cdb71
3. Y. Zhong, P. Godwin, Y. Jin, H. Xiao, Biodegradable polymers and green-based antimicrobial packaging materials: a mini-review. Adv. Ind. Eng. Polym. Res. **3**(1), 27–35 (2020)
4. L. Fertier, H. Koleilat, M. Stemmelen, O. Giani, C. Joly-Duhamel, V. Lapinte, J.-J. Robin, The use of renewable feedstock in UV-curable materials—a new age for polymers and green chemistry. Prog. Polym. Sci. **38**(6), 932–996 (2013)
5. N.P.B. Tan, C.H. Lee, P. Li, Green synthesis of smart metal/polymer nanocomposite particles and their tuneable catalytic activities. Polymers **8**(4), 105 (2016)
6. A. Kausar, Progress in green nanocomposites for high-performance applications. Mater. Res. Innov. **25**(1), 53–65 (2021)
7. N. Sakthieswaran, M. Sophia, *Prosopis juliflora* fibre reinforced green building plaster materials—an eco-friendly weed control technique by effective utilization. Environ. Technol. Innov. **20**, 101158 (2020)
8. E. Bakhoum, G. Garas, M. Allam, Sustainability analysis of conventional and eco-friendly materials: a step towards green building. ARPN J. Eng. Appl. Sci. **10**(2), 788–796 (2015)
9. F. Chemat, N. Rombaut, A. Meullemiestre, M. Turk, S. Perino, A.-S. Fabiano-Tixier, M. Abert-Vian, Review of green food processing techniques. preservation, transformation, and extraction. Innov. Food Sci. Emerg. Technol. **41**, 357–377 (2017)
10. J. Liu, Y. Feng, Q. Zhu, J. Sarkis, Green supply chain management and the circular economy: Reviewing theory for advancement of both fields. Int. J. Phys. Distrib. Logis. Manag. (2018)
11. Y. Ahmadi, S. Ahmad, Surface-active antimicrobial and anticorrosive oleo-polyurethane/graphene oxide nanocomposite coatings: synergistic effects of in-situ polymerization and $\pi-\pi$ interaction. Prog. Org. Coat. **127**, 168–180 (2019)
12. S. Chen, A. Skordos, V.K. Thakur, Functional nanocomposites for energy storage: chemistry and new horizons. Mater. Today Chem. **17**, 100304 (2020)
13. R.M. Dell, D.A.J. Rand, Energy storage—a key technology for global energy sustainability. J. Power Sources **100**(1–2), 2–17 (2001)
14. T. Mohan, K. Kanny, *Green Nanofillers for Polymeric Materials: Green Nanomaterials* (Springer, Singapore, 2020), pp. 99–138
15. C. Yang, H. Wei, L. Guan, J. Guo, Y. Wang, X. Yan, X. Zhang, S. Wei, Z. Guo, Polymer nanocomposites for energy storage, energy saving, and anticorrosion. J. Mater. Chem. A **3**(29), 14929–14941 (2015)
16. S.S. Siwal, Q. Zhang, N. Devi, V.K. Thakur, Carbon-based polymer nanocomposite for high-performance energy storage applications. Polymers **12**(3), 505 (2020)
17. M. Mooney, A. Nyayachavadi, S. Rondeau-Gagné, Eco-friendly semiconducting polymers: from greener synthesis to greener processability. J. Mater. Chem. C **8**(42), 14645–14664 (2020)
18. X. Hu, L. Wei, R. Chen, Q. Wu, J. Li, Reviews and prospectives of Co_3O_4-based nanomaterials for supercapacitor application. Chem. Select **5**(17), 5268–5288 (2020)
19. H. Yang, A review of supercapacitor-based energy storage systems for microgrid applications, pp 1–5 (2018)
20. M.D. Tabone, J.J. Cregg, E.J. Beckman, A.E. Landis, Sustainability metrics: life cycle assessment and green design in polymers. Environ. Sci. Technol. **44**(21), 8264–8269 (2010)
21. A.R. Bagheri, M. Arabi, M. Ghaedi, A. Ostovan, X. Wang, J. Li, L. Chen, Dummy molecularly imprinted polymers based on a green synthesis strategy for magnetic solid-phase extraction of acrylamide in food samples. Talanta **195**, 390–400 (2019)

22. M. Marti, L. Molina, C. Aleman, E. Armelin, Novel epoxy coating based on DMSO as a green solvent, reducing drastically the volatile organic compound content and using conducting polymers as a nontoxic anticorrosive pigment. ACS Sustain. Chem. Eng. **1**(12), 1609–1618 (2013)
23. C. Yuan, M. Chen, J. Luo, X. Li, Q. Gao, J. Li, A novel water-based process produces eco-friendly bio-adhesive made from green cross-linked soybean soluble polysaccharide and soy protein. Carbohyd. Polym. **169**, 417–425 (2017)
24. H. Jahangirian, E.G. Lemraski, T.J. Webster, R. Rafiee-Moghaddam, Y. Abdollahi, A review of drug delivery systems based on nanotechnology and green chemistry: green nanomedicine. Int. J. Nanomed. **12**, 2957 (2017)
25. K. Kalantari, A.M. Afifi, H. Jahangirian, T.J. Webster, Biomedical applications of chitosan electrospun nanofibers as a green polymer—review. Carbohyd. Polym. **207**, 588–600 (2019)
26. B. Patanair, A. Saiter-Fourcin, S. Thomas, M.G. Thomas, P. Parathukkamparambil Pundarikashan, K. Gopalan Nair, V.K. Kumar, H.J. Maria, N. Delpouve, Promoting interfacial interactions with the addition of lignin in poly (lactic acid) hybrid nanocomposites. Polymers **13**(2), 272 (2021)
27. B. Fauzi, M.G.M. Nawawi, R. Fauzi, S.N.L. Mamauod, Physicochemical characteristics of sago starch-chitosan nanofillers film. BioResources **14**(4), 8324–8330 (2019)
28. A. Grossman, W. Vermerris, Lignin-based polymers and nanomaterials. Curr. Opin. Biotechnol. **56**, 112–120 (2019)
29. C.S. Lancefield, N.J. Westwood, The synthesis and analysis of advanced lignin model polymers. Green Chem. **17**(11), 4980–4990 (2015)
30. S. Jayrajsinh, G. Shankar, Y.K. Agrawal, L. Bakre, Montmorillonite nanoclay as a multifaceted drug-delivery carrier: a review. J. Drug Deliv. Sci. Technol. **39**, 200–209 (2017)
31. S.M.S. Hosseini, M. Mirzaei, Assessment of the colloidal montmorillonite dispersion as a low-cost and eco-friendly nanofluid for improving thermal performance of plate heat exchanger. SN Appl. Sci. **2**(10), 1–16 (2020)
32. H.R. Penchah, A. Ghaemi, H. Godarziani, Eco-friendly CO_2 adsorbent by impregnation of diethanolamine in nanoclay montmorillonite. Environ. Sci. Pollut. Res. **2021**, 1–17 (2021)
33. S. Mallakpour, M. Dinari, Synthesis and properties of biodegradable poly (vinyl alcohol)/organo-nanoclay bionanocomposites. J. Polym. Environ. **20**(3), 732–740 (2012)
34. Y. Kim, TEMPO-oxidized nanofibrillated cellulose film (NFC) incorporating graphene oxide (GO) nanofillers. Virginia Tech (2017)
35. Y.-Y. Wu, J. Zhang, C. Liu, Z. Zheng, P. Lambert, Effect of graphene oxide nanosheets on physical properties of ultra-high-performance concrete with high volume supplementary cementitious materials. Materials **13**(8), 1929 (2020)
36. H. Hamedi, S. Moradi, S.M. Hudson, A.E. Tonelli, M.W. King, Chitosan based bioadhesives for biomedical applications: a review. Carbohyd. Polym. **282**, 119100 (2022)
37. A. Konwar, N. Gogoi, G. Majumdar, D. Chowdhury, Green chitosan–carbon dots nanocomposite hydrogel film with superior properties. Carbohyd. Polym. **115**, 238–245 (2015)
38. M.B. Ahmad, M.Y. Tay, K. Shameli, M.Z. Hussein, J.J. Lim, Green synthesis and characterization of silver/chitosan/polyethylene glycol nanocomposites without any reducing agent. Int. J. Mol. Sci. **12**(8), 4872–4884 (2011)
39. F. Ansari, A. Sobhani, M. Salavati-Niasari, Green synthesis of magnetic chitosan nanocomposites by a new sol–gel auto-combustion method. J. Magn. Magn. Mater. **410**, 27–33 (2016)
40. G. Di Carlo, A. Curulli, R.G. Toro, C. Bianchini, T. De Caro, G. Padeletti, D. Zane, G.M. Ingo, Green synthesis of gold–chitosan nanocomposites for caffeic acid sensing. Langmuir **28**(12), 5471–5479 (2012)
41. N. Soykeabkaew, N. Laosat, A. Ngaokla, N. Yodsuwan, T. Tunkasiri, Reinforcing potential of micro-and nano-sized fibers in the starch-based biocomposites. Compos. Sci. Technol. **72**(7), 845–852 (2012)
42. D. Lu, C. Xiao, S. Xu, Starch-based completely biodegradable polymer materials. Exp. Polym. Lett. **3**(6), 366–375 (2009)

43. P. Cheviron, F. Gouanvé, E. Espuche, Green synthesis of colloid silver nanoparticles and resulting biodegradable starch/silver nanocomposites. Carbohyd. Polym. **108**, 291–298 (2014)
44. A. Kaushik, M. Singh, G. Verma, Green nanocomposites based on thermoplastic starch and steam exploded cellulose nanofibrils from wheat straw. Carbohyd. Polym. **82**(2), 337–345 (2010)
45. R. Vanholme, K. Morreel, C. Darrah, P. Oyarce, J.H. Grabber, J. Ralph, W. Boerjan, Metabolic engineering of novel lignin in biomass crops. New Phytol. **196**(4), 978–1000 (2012)
46. A.J. Ragauskas, G.T. Beckham, M.J. Biddy, R. Chandra, F. Chen, M.F. Davis, B.H. Davison, R.A. Dixon, P. Gilna, M. Keller, Lignin valorization: improving lignin processing in the biorefinery. Science **344**, 6185 (2014)
47. I. Norberg, Y. Nordström, R. Drougge, G. Gellerstedt, E. Sjöholm, A new method for stabilizing softwood kraft lignin fibers for carbon fiber production. J. Appl. Polym. Sci. **128**(6), 3824–3830 (2013)
48. A. Awal, M. Sain, Characterization of soda hardwood lignin and the formation of lignin fibers by melt spinning. J. Appl. Polym. Sci. **129**(5), 2765–2771 (2013)
49. N.A. Alcantar, E.S. Aydil, J.N. Israelachvili, Polyethylene glycol-coated biocompatible surfaces. J. Biomed. Mater. Res. **51**(3), 343–351 (2000)
50. S. Shu, X. Zhang, D. Teng, Z. Wang, C. Li, Polyelectrolyte nanoparticles based on water-soluble chitosan–poly (L-aspartic acid)-polyethylene glycol for controlled protein release. Carbohyd. Res. **344**(10), 1197–1204 (2009)
51. I.D. Gunbas, U. Aydemir Sezer, S. Gülce İz, I.S. Deliloğlu Gürhan, N. Hasirci, Semi-IPN chitosan/PEG microspheres and films for biomedical applications: characterization and sustained release optimization. Indus. Eng. Chem. Res. **51**(37), 11946–11954 (2012)
52. J.S. Jayan, B. Deeraj, A. Saritha, K. Joseph, Theoretical modelling of kinetics of glass transition temperature of PEG toughened epoxy. Plast. Rubber Compos. **49**(6), 237–244 (2020)
53. N. Ljungberg, B. Wesslen, Tributyl citrate oligomers as plasticizers for poly (lactic acid): thermo-mechanical film properties and aging. Polymer **44**(25), 7679–7688 (2003)
54. G. Cavallaro, G. Lazzara, S. Milioto, Sustainable nanocomposites based on halloysite nanotubes and pectin/polyethylene glycol blend. Polym. Degrad. Stab. **98**(12), 2529–2536 (2013)
55. S. Gopi, A. Amalraj, N. Kalarikkal, J. Zhang, S. Thomas, Q. Guo, Preparation and characterization of nanocomposite films based on gum arabic, maltodextrin and polyethylene glycol reinforced with turmeric nanofiber isolated from turmeric spent. Mater. Sci. Eng. C **97**, 723–729 (2019)
56. W.G. Sganzerla, M. Longo, J.L. de Oliveira, C.G. da Rosa, A.P. de Lima Veeck, R.S. de Aquino, A.V. Masiero, F.C. Bertoldi, P.L.M. Barreto, M.R. Nunes, Nanocomposite poly (ethylene oxide) films functionalized with silver nanoparticles synthesized with *Acca sellowiana* extracts. Colloids Surf. A Physicochem. Eng. Aspects 602, 125125 (2020)
57. B. Hong, A.Z. Panagiotopoulos, Molecular dynamics simulations of silica nanoparticles grafted with poly (ethylene oxide) oligomer chains. J. Phys. Chem. B **116**(8), 2385–2395 (2012)
58. H. Yamane, K. Sasai, Effect of the addition of poly (D-lactic acid) on the thermal property of poly (L-lactic acid). Polymer **44**(8), 2569–2575 (2003)
59. P. Simamora, W. Chern, Poly-L-lactic acid: an overview. J. Drugs Dermatol. JDD **5**(5), 436–440 (2006)
60. V. Krikorian, D.J. Pochan, Poly (L-lactic acid)/layered silicate nanocomposite: fabrication, characterization, and properties. Chem. Mater. **15**(22), 4317–4324 (2003)
61. K. Wang, J. Lu, R. Tusiime, Y. Yang, F. Fan, H. Zhang, B. Ma, Properties of poly (L-lactic acid) reinforced by l-lactic acid grafted nanocellulose crystal. Int. J. Biol. Macromol. **156**, 314–320 (2020)
62. L. Li, R.-Y. Bao, T. Gao, Z.-Y. Liu, B.-H. Xie, M.-B. Yang, W. Yang, Dopamine-induced functionalization of cellulose nanocrystals with polyethylene glycol towards poly (L-lactic acid) bionanocomposites for green packaging. Carbohyd. Polym. **203**, 275–284 (2019)

63. M.M. Ibrahim, A. Koschella, G. Kadry, T. Heinze, Evaluation of cellulose and carboxymethyl cellulose/poly (vinyl alcohol) membranes. Carbohyd. Polym. **95**(1), 414–420 (2013)
64. A. Joorabloo, M.T. Khorasani, H. Adeli, Z. Mansoori-Moghadam, A. Moghaddam, Fabrication of heparinized nano ZnO/poly (vinylalcohol)/carboxymethyl cellulose bionanocomposite hydrogels using artificial neural network for wound dressing application. J. Ind. Eng. Chem. **70**, 253–263 (2019)
65. M. Morsi, A. Oraby, A. Elshahawy, R. Abd El-Hady, Preparation, structural analysis, morphological investigation and electrical properties of gold nanoparticles filled polyvinyl alcohol/carboxymethyl cellulose blend. J. Market. Res. **8**(6), 5996–6010 (2019)
66. Y. Zhao, Y. Zhou, X. Wu, L. Wang, L. Xu, S. Wei, A facile method for electrospinning of Ag nanoparticles/poly (vinyl alcohol)/carboxymethyl-chitosan nanofibers. Appl. Surf. Sci. **258**(22), 8867–8873 (2012)
67. G.S. Mann, L.P. Singh, P. Kumar, S. Singh, Green composites: a review of processing technologies and recent applications. J. Thermoplast. Compos. Mater. **33**(8), 1145–1171 (2020)
68. K. Georgios, A. Silva, S. Furtado, Applications of green composite materials. Biodegrad. Green Compos. **16**, 312 (2016)
69. L. Mishnaevsky, K. Branner, H.N. Petersen, J. Beauson, M. McGugan, B.F. Sørensen, Materials for wind turbine blades: an overview. Materials **10**(11), 1285 (2017)
70. N.P. Sukumaran, S. Gopi, *Overview of Biopolymers: Resources, Demands, Sustainability, and Life Cycle Assessment Modeling and Simulation: Biopolymers and their Industrial Applications* (Elsevier, New York, 2021), pp. 1–19
71. Q.M. Jebur, A. Hashim, M.A. Habeeb, Structural, electrical and optical properties for (polyvinyl alcohol–polyethylene oxide–magnesium oxide) nanocomposites for optoelectronics applications. Trans. Electr. Electron. Mater. **20**(4), 334–343 (2019)
72. J. Guo, Y. Xu, S. Jin, L. Chen, T. Kaji, Y. Honsho, M.A. Addicoat, J. Kim, A. Saeki, H. Ihee, Conjugated organic framework with three-dimensionally ordered stable structure and delocalized π clouds. Nat. Commun. **4**(1), 1–8 (2013)
73. P. Perumal, P.C. Selvin, S. Selvasekarapandian, P. Sivaraj, K. Abhilash, V. Moniha, R.M. Devi, Plasticizer incorporated, novel eco-friendly bio-polymer based solid bio-membrane for electrochemical clean energy applications. Polym. Degrad. Stab. **159**, 43–53 (2019)
74. P.A. Haigh, F. Bausi, T. Kanesan, S.T. Le, S. Rajbhandari, Z. Ghassemlooy, I. Papakonstantinou, W. Popoola, A. Burton, H. Le Minh, A 20-Mb/s VLC link with a polymer LED and a multilayer perceptron equalizer. IEEE Photonics Technol. Lett. **26**(19), 1975–1978 (2014)
75. S.K. Ghosh, T.K. Sinha, B. Mahanty, S. Jana, D. Mandal, Porous polymer composite membrane based nanogenerator: a realization of self-powered wireless green energy source for smart electronics applications. J. Appl. Phys. **120**(17), 174501 (2016)
76. L. Chen, M.-C. Wong, G. Bai, W. Jie, J. Hao, White and green light emissions of flexible polymer composites under electric field and multiple strains. Nano Energy **14**, 372–381 (2015)
77. N.P. Holmes, M. Marks, J.M. Cave, K. Feron, M.G. Barr, A. Fahy, A. Sharma, X. Pan, D.A. Kilcoyne, X. Zhou, Engineering two-phase and three-phase microstructures from water-based dispersions of nanoparticles for eco-friendly polymer solar cell applications. Chem. Mater. **30**(18), 6521–6531 (2018)
78. S. Lee, D. Jeong, C. Kim, C. Lee, H. Kang, H.Y. Woo, B.J. Kim, Eco-friendly polymer solar cells: advances in green-solvent processing and material design. ACS Nano **14**(11), 14493–14527 (2020)
79. Q. Zhang, Y. Sun, W. Xu, D. Zhu, Organic thermoelectric materials: emerging green energy materials converting heat to electricity directly and efficiently. Adv. Mater. **26**(40), 6829–6851 (2014)
80. L. Cui, S. Yu, W. Gao, X. Zhang, S. Deng, C.-y Zhang, Tetraphenylenthene-based conjugated microporous polymer for aggregation-induced electrochemiluminescence. ACS Appl. Mater. Interfaces **12**(7), 7966–7973 (2020)
81. K. Yuan, P. Guo-Wang, T. Hu, L. Shi, R. Zeng, M. Forster, T. Pichler, Y. Chen, U. Scherf, Nanofibrous and graphene-templated conjugated microporous polymer materials for flexible chemosensors and supercapacitors. Chem. Mater. **27**(21), 7403–7411 (2015)

82. W. Feng, P. Long, Y. Feng, Y. Li, Two-dimensional fluorinated graphene: synthesis, structures, properties and applications. Adv. Sci. **3**(7), 1500413 (2016)
83. X. Hhuang, D. Gehrig, N. Forler, H. Liang, M. Wagner, M.R. Hansen, F. Laquai, F. Zhang, X. Feng, Conjugated microporous polymers with dimensionality-controlled heterostructures for green energy devices. Adv. Mater. **27**(25), 3789–3796 (2015)
84. W. Zhao, Y. Jiao, J. Li, L. Wu, A. Xie, W. Dong, One-pot synthesis of conjugated microporous polymers loaded with superfine nano-palladium and their micropore-confinement effect on heterogeneously catalytic reduction. J. Catal. **378**, 42–50 (2019)
85. X. Liu, K. Li, Energy storage devices in electrified railway systems: a review. Transp. Saf. Environ. **2**(3), 183–201 (2020)
86. A. Chatzivasileiadi, E. Ampatzi, I. Knight, Characteristics of electrical energy storage technologies and their applications in buildings. Renew. Sustain. Energy Rev. **25**, 814–830 (2013)
87. S. Koohi-Kamali, V. Tyagi, N. Rahim, N. Panwar, H. Mokhlis, Emergence of energy storage technologies as the solution for reliable operation of smart power systems: a review. Renew. Sustain. Energy Rev. **25**, 135–165 (2013)
88. T. Christen, M.W. Carlen, Theory of Ragone plots. J. Power Sources **91**(2), 210–216 (2000)
89. L. Yang, X. Kong, F. Li, H. Hao, Z. Cheng, H. Liu, J.-F. Li, S. Zhang, Perovskite lead-free dielectrics for energy storage applications. Prog. Mater Sci. **102**, 72–108 (2019)
90. K. Yao, S. Chen, M. Rahimabady, M.S. Mirshekarloo, S. Yu, F.E.H. Tay, T. Sritharan, L. Lu, Nonlinear dielectric thin films for high-power electric storage with energy density comparable with electrochemical supercapacitors. IEEE Trans. Ultrason. Ferroelectr. Freq. Control **58**(9), 1968–1974 (2011)
91. S. Liu, T. Yu, Y. Wu, W. Li, B. Li, Evolution of cellulose into flexible conductive green electronics: a smart strategy to fabricate sustainable electrodes for supercapacitors. RSC Adv. **4**(64), 34134–34143 (2014)
92. A.M. Alsaad, I.A. Aljarrah, A. Ahmad, Q.M. Al-Bataineh, A. Shariah, M.A. Al-Akhras, A.D. Telfah, The structural, optical, thermal, and electrical properties of synthesized PEO/GO thin films. Appl. Phys. A **128**(8), 1–11 (2022)
93. J. Singh, A. Dhaliwal, K. Sharma, R. Sehgal, V. Kumar, Conductive polymer-based composite photocatalysts for environment and energy applications, in *Conjugated Polymers for Next-Generation Applications*. (Elsevier, New York, 2022), pp. 505–538
94. Y.M. Albarqouni, S.P. Lee, G.A. Ali, A.S. Ethiraj, H. Algarni, K.F. Chong, Facile synthesis of reduced graphene oxide aerogel in soft drink as supercapacitor electrode. J. Nanostruct. Chem. **12**(3), 417–427 (2022)
95. K. Sun, J. Li, D. Wu, J. Jiang, Green synthesis of porous honeycomblike carbon materials for supercapacitor electrodes. Ind. Eng. Chem. Res. **59**(32), 14288–14295 (2020)
96. A. Borenstein, O. Hanna, R. Attias, S. Luski, T. Brousse, D. Aurbach, Carbon-based composite materials for supercapacitor electrodes: a review. J. Mater. Chem. A **5**(25), 12653–12672 (2017)
97. Z.S. Iro, C. Subramani, S. Dash, A brief review on electrode materials for supercapacitor. Int. J. Electrochem. Sci. **11**(12), 10628–10643 (2016)
98. A. Kausar, N-doped graphene and polymer sequent nanocomposite—nitty-gritties and scoping insights. Polym. Plastics Technol. Mater. **62**, 1–17 (2023)
99. A. Kausar, I. Ahmad, T. Zhao, O. Aldaghri, M. Eisa, Graphene in polymeric nanocomposite membranes—current state and progress. Processes **11**(3), 927 (2023)
100. A. Kausar, I. Ahmad, T. Zhao, O. Aldaghri, M. Eisa, Polymer/graphene nanocomposites via 3D and 4D printing—design and technical potential. Processes **11**(3), 868 (2023)
101. F. Xu, M. Gao, H. Wang, H. Liu, F. Yan, H. Zhao, Q. Yao, Polymer-based graphene composite molding: a review. RSC Adv. **13**(4), 2538–2551 (2023)
102. X. Zhu, S. Yu, K. Xu, Y. Zhang, L. Zhang, G. Lou, Y. Wu, E. Zhu, H. Chen, Z. Shen, Sustainable activated carbons from dead ginkgo leaves for supercapacitor electrode active materials. Chem. Eng. Sci. **181**, 36–45 (2018)

103. J. Zhang, X. Zhao, Conducting polymers directly coated on reduced graphene oxide sheets as high-performance supercapacitor electrodes. J. Phys. Chem. C **116**(9), 5420–5426 (2012)
104. S. Chandra, M.D. Patel, H. Lang, D. Bahadur, Dendrimer-functionalized magnetic nanoparticles: a new electrode material for electrochemical energy storage devices. J. Power Sources **280**, 217–226 (2015)
105. L.L. Zhang, X. Zhao, Carbon-based materials as supercapacitor electrodes. Chem. Soc. Rev. **38**(9), 2520–2531 (2009)
106. S.V. Selvaganesh, J. Mathiyarasu, K. Phani, V. Yegnaraman, Chemical synthesis of PEDOT-Au nanocomposite. Nanoscale Res. Lett. **2**(11), 546 (2007)
107. E.R. Ezeigwe, M.T. Tan, P.S. Khiew, C.W. Siong, One-step green synthesis of graphene/ZnO nanocomposites for electrochemical capacitors. Ceram. Int. **41**(1), 715–724 (2015)
108. A.K. Nayak, A.K. Das, D. Pradhan, High performance solid-state asymmetric supercapacitor using green synthesized graphene–WO_3 nanowires nanocomposite. ACS Sustain. Chem. Eng. **5**(11), 10128–10138 (2017)
109. D. Arthisree, W. Madhuri, Optically active polymer nanocomposite composed of polyaniline, polyacrylonitrile and green-synthesized graphene quantum dot for supercapacitor application. Int. J. Hydrogen Energy **45**(16), 9317–9327 (2020)
110. A. Sumboja, X. Wang, J. Yan, P.S. Lee, Nanoarchitectured current collector for high rate capability of polyaniline based supercapacitor electrode. Electrochim. Acta **65**, 190–195 (2012)
111. S. Chakraborty, N. Mary, Biocompatible supercapacitor electrodes using green synthesised ZnO/polymer nanocomposites for efficient energy storage applications. J. Energy Stor. **28**, 101275 (2020)
112. Z. Çıplak, A. Yıldız, N. Yıldız, Green preparation of ternary reduced graphene oxide-Au@ polyaniline nanocomposite for supercapacitor application. J. Energy Stor. **32**, 101846 (2020)
113. C. Zhong, Y. Deng, W. Hu, J. Qiao, L. Zhang, J. Zhang, A review of electrolyte materials and compositions for electrochemical supercapacitors. Chem. Soc. Rev. **44**(21), 7484–7539 (2015)
114. F. Ren, Z. Li, W.-Z. Tan, X.-H. Liu, Z.-F. Sun, P.-G. Ren, D.-X. Yan, Facile preparation of 3D regenerated cellulose/graphene oxide composite aerogel with high-efficiency adsorption towards methylene blue. J. Colloid Interface Sci. **532**, 58–67 (2018)
115. J. Tian, D. Peng, X. Wu, W. Li, H. Deng, S. Liu, Electrodeposition of Ag nanoparticles on conductive polyaniline/cellulose aerogels with increased synergistic effect for energy storage. Carbohyd. Polym. **156**, 19–25 (2017)
116. C. Wan, Y. Jiao, D. Liang, Y. Wu, J. Li, A geologic architecture system-inspired micro-/nano-heterostructure design for high-performance energy storage. Adv. Energy Mater. **8**(33), 1802388 (2018)
117. G. Zu, J. Shen, L. Zou, F. Wang, X. Wang, Y. Zhang, X. Yao, Nanocellulose-derived highly porous carbon aerogels for supercapacitors. Carbon **99**, 203–211 (2016)
118. W. Ouyang, J. Sun, J. Memon, C. Wang, J. Geng, Y. Huang, Scalable preparation of three-dimensional porous structures of reduced graphene oxide/cellulose composites and their application in supercapacitors. Carbon **62**, 501–509 (2013)
119. X. Yang, B. Fei, J. Ma, X. Liu, S. Yang, G. Tian, Z. Jiang, Porous nanoplatelets wrapped carbon aerogels by pyrolysis of regenerated bamboo cellulose aerogels as supercapacitor electrodes. Carbohyd. Polym. **180**, 385–392 (2018)
120. W. Tian, Q. Gao, L. Zhang, C. Yang, Z. Li, Y. Tan, W. Qian, H. Zhang, Renewable graphene-like nitrogen-doped carbon nanosheets as supercapacitor electrodes with integrated high energy–power properties. J. Mater. Chem. A **4**(22), 8690–8699 (2016)
121. L.F. Chen, Z.H. Huang, H.W. Liang, Q.F. Guan, S.H. Yu, Bacterial-cellulose-derived carbon nanofiber@ MnO_2 and nitrogen-doped carbon nanofiber electrode materials: an asymmetric supercapacitor with high energy and power density. Adv. Mater. **25**(34), 4746–4752 (2013)
122. K. Mensah-Darkwa, C. Zequine, P.K. Kahol, R.K. Gupta, 'Supercapacitor energy storage device using biowastes: a sustainable approach to green energy. Sustainability **11**(2), 414 (2019)

123. V. Zargar, M. Asghari, A. Dashti, A review on chitin and chitosan polymers: structure, chemistry, solubility, derivatives, and applications. ChemBioEng Rev. **2**, 204–226 (2015)
124. K. Shi, X. Yang, E.D. Cranston, I. Zhitomirsky, Efficient lightweight supercapacitor with compression stability. Adv. Func. Mater. **26**(35), 6437–6445 (2016)
125. K. Wang, L. Li, T. Zhang, Z. Liu, Nitrogen-doped graphene for supercapacitor with long-term electrochemical stability. Energy **70**, 612–617 (2014)
126. E.A. Olivetti, J.M. Cullen, Toward a sustainable materials system. Science **360**(6396), 1396–1398 (2018)
127. R. Sharma, S.M. Jafari, S. Sharma, Antimicrobial bio-nanocomposites and their potential applications in food packaging. Food Control **112**, 107086 (2020)

The Application of Planar Heterojunction Solar Cells Based on Au/BG/p–Si/Al in Thin Films in the Energy Sector

M. Benhaliliba, Y. S. Ocak, A. Ayeshamariam, and K. Kaviyarasu

Abstract Low-cost spin coating of *p*-type silicon substrates enables the fabrication of dye brilliant green (BG)-based devices. In order to achieve the front gold contact, BG thin film is deposited at high vacuum conditions. Various light intensities at room temperature are used to measure current and voltage in the dark. Electrical parameters of BG-based devices are extracted from experimental data, and they describe non-ideal diode behaviors. Using Cheung and Norde approximations and the thermionic emission theory, appropriate electrical parameters are calculated, including ideality factor (n), barrier height (Φ_b), series resistance (R_S), open-circuit voltage (V_{OC}), short-circuit current ($I/_{SC}$), and interface state density (D_{it}). Various characteristics of the Au/BG/p–Si/Al heterostructure are discussed at room and dark temperatures, including capacitance–voltage (C–V) and conductance–voltage (G–V). Further, the dark and illumination modes of operation of the diodes are influenced by resistance series and interface states.

M. Benhaliliba (✉)
Film Device Fabrication Characterization and Application (FDFCA), Research Group USTOMB, 31130 Oran, Algeria
e-mail: mbenhaliliba@gmail.com

USTOMB University, Mnaouer, BP1505, 31130 Oran, Algeria

Y. S. Ocak
Smart-Lab, 21280 Diyarbakir, Turkey

A. Ayeshamariam
Department of Physics, Khadir Mohideen College (Affiliated to Bharathidasan University, Thiruchirappalli), Adirampattinam, Tamilnadu 614701, India

K. Kaviyarasu
UNESCO-UNISA Africa Chair in Nanosciences/Nanotechnology Laboratories, College of Graduate Studies, University of South Africa (UNISA), Muckleneuk Ridge, P. O. Box 392, Pretoria, South Africa

K. Kasinathan et al. (eds.), *New Technologies for Energy Transition Based on Sustainable Development Goals*, https://doi.org/10.1007/978-981-97-2527-4_14

1 Introduction

Organic materials have been used as alternative candidates to inorganic ones because of the key properties, including low cost, high thermal and chemical stability, and ease of device fabrication in modern industry. Organic molecules can be used in many different ways. Numerous authors have studied the electrical and photo-electric properties of metal–semiconductor (MS) devices with thin organic interfacial layers, such as Schottky diodes, photovoltaic cells, and photodiodes [1]. The organic molecules have been previously used for the electrical device fabrication such as pentacene:poly[2-methoxy-5-(2-ethyl)hexoxy-phenylenevinylene]/p-Si, p-NiO/n-polyaniline/n-Si Schottky diode [2, 3]. In order to make organic thin films various deposition routes are already used like spray pyrolysis and spin coating [4–6]. Organic semiconductors offer several advantages like a lightweight for practical portability and flexibility, leading to less fragile than inorganic semiconductors, which must be deposited on flat and unbending substrates. Organic semiconductors are commonly easy and conservative to be created in the research facility. Although organic semiconductors based devices have certain restrictions, like the lifespan of organic display devices is shorter than that of traditional liquid crystal displays (LCDs), organic semiconductors meet several applications such as they are used OLEDs are organic light-emitting diodes developed in optoelectronics and devices competing with LCDs, the matrix of red, green, and blue pixels is easily manufactured either by a vacuum evaporation procedure or utilizing the inkjet printing technique. Organic dyes have been used as an active part of various electronic devices, including dye-sensitized solar cells (DSSCs), heterojunctions, and biosensors [7–9]. Brilliant green (BG) having the $C_{27}H_{33}N_2{\cdot}HO_4S$ chemical formula is characterized as one of the colors of triarylmethane, used to color silk and wool. It has been used to adsorption and photocatalytic behavior of various materials, in a light microscope, metal oxides, chalcogenides, and boron-doped diamonds are visible [10–13].

A p-type silicon substrate is used for the deposition of BG thin films. As BG/p–Si structure is vaporized at 10^{-6} Torr, the gold front contact is evaporating. The investigation of the electric characteristics of the organic diode-based brilliant green dye organic material is achieved. From the current–voltage characteristic under dark and room temperature environments, Cheung's and thermionic methods are programmed to extract electrical parameters, such as the ideality factor (n), and the resistance series (R_S). We measure capacitance–voltage (C–V) and conductance–voltage (G–V) at a number of different frequencies, making it possible to evaluate the carrier's concentration, diffusion potential, barrier height, and density interface. Optical properties of I–V in the presence of light are evidenced to determine the photovoltaic properties of the fabricated diodes. Up to our best knowledge, occasional papers take the electrical properties of such heterojunction-based BG dye molecules into account. No papers on brilliant green dye-based heterojunction diode and their electrical characterizations were reported in detail.

2 Experimental Technique

The cleaning is an essential step to obtain high-quality electronic devices. Thus, the Si wafers are degreased in a boiling trichloroethylene bath for 5 min. Removed grease and impurities from the surface of the substrate using ultrasonic vibration in acetone and methanol. A hydrofluoric acid bath is used to chemically etch silica wafers of their native oxide layers. The process is completed just before the deposition to form *H* passivated Si surface, which prevents oxide formation during the storage of the substrate. Experimental details are also gathered in Table 1. Before forming the device, thermally evaporating 200 nm Au into a vacuum at 10^{-6} Torr, on an unpolished side of a silicon wafer, using a mask size 1.5 mm, and annealing it at 570 °C for 3 min allows low-resistance ohmic contacts to be formed. The device based on the BG molecule is manufactured by the facile spin coating route. 24 mg of BG is diluted in 50 ml of methanol. BG thin films are deposited onto the p–Si substrates by a spin coater rotated at the speed of 2000 rpm for 5 min. The thickness of such thin film is of 673 nm, determined by using high-frequency capacitance–voltage measurements [14, 15]. Figure 1a shows the structure of Au/BG/p–Si/Al structures, showing the chemical structure of dye BG brilliant green. In Fig. 1b, we see a source meter unit Keithley Model 2400 which has been used in the fabrication and electrical characterization of diodes. Various intensities of dark and light are used for the current–voltage measurements at room temperature. The I–V characterization is a technique that makes it possible to learn about the electrical response of a given sample and indicates whether the electric current in the sensor exhibits an ohmic behavior or a rectifying behavior at the electrode contacts/semiconductor. It also allows calculating the ideality factor, the resistance of the material, and saturation current. Anode and cathode contacts are then placed on the diode after it is fixed in the test cell. An illumination device is used to test the solar cell, a solar cell test simulator (AM1.5) with an adjustable light output based on filters or silicon cells. At room temperature and in dark conditions, Agilent HP 4294A impedance analyzers are used to determine C–V characteristics.

Table 1 Experimental details

Materials	
Conditions of layer deposition	0.024 g BG powder
Precursor	
Mass	
Diluent	Methanol 50 ml
Deposition time	5 min
Distance	5 cm
Mask radius	1.5 mm
Substrate	Psi
Contact	Or (Au)

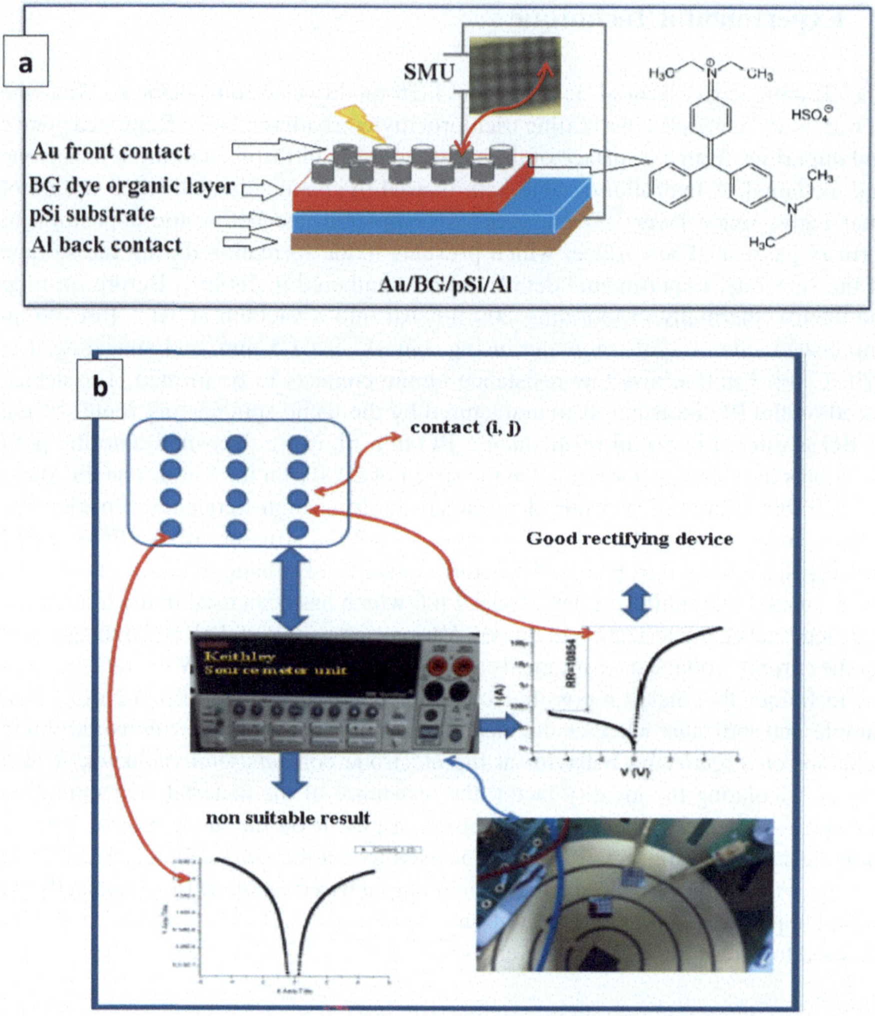

Fig. 1 Schematic diagram of the Au/organic BG/pSi/Al heterostructure diode and chemical structure of dye BG brilliant green is displayed at right (**a**), the test of contacts and Keithley set-up (**b**)

3 Results and Discussion

3.1 Characteristics of Current–Voltage in the Dark and Under Light Conditions

I–V (current–voltage) variations in dark rooms are depicted in Fig. 2 and light 60–120 mW/cm^2 conditions of Au/BG/p–Si/Al organic heterostructure. Using the experiment for Au/BG/p–Si/Al diodes, it is possible to visualize their semi-logarithmic reverse and forward biases. The rectifying ratio is found to be 78 in the dark and is affected by light. An organic heterostructure diode with Au/BG/p–Si/Al has three distinct characteristics in a dark environment and under a light environment, which are reverse, low voltage, and forward I–V, as shown in Fig. 2. The reverse bias region shows the effect of light on current while no substantial impact on current is visible in forward bias, especially in the low voltage ($V < 1$ V) region. It is noted that the photocurrent, as seen in Fig. 2, is strongly dependent on light intensity as [16];

$$I_{ph} = \beta P^{\gamma} \tag{1}$$

where γ is an exponent and β is a scaling constant. The value of γ, is calculated by the fitting of Eq. 1, is related to the density of the traps. By linear fitting Log I_{ph} versus Log P, the value of γ is found to be 1.003, ($0.55 + 1.003 \times P$) implies

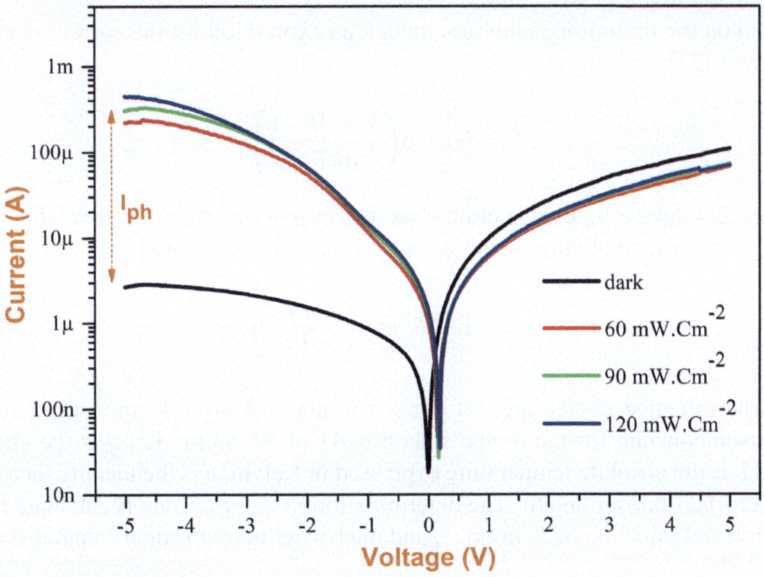

Fig. 2 Current–voltage semilog characteristics of Au/BG/pSi/Al organic heterostructure diode under dark–light intensity and room temperature environments

that the bimolecular recombination mechanism is dominant. Besides, the parameters are determined from current–voltage characteristics of Au/BG/pSi/Al at room temperature. Illuminated at 60, 90, and 120 mW/cm^2, the BG device exhibits the photovoltaic parameters such as I_{sc}, V_{oc}, I_m, V_m, P_{max}, FF, efficiency, I_{ph}, and photosensitivity as listed in Table 2. The geometrical configuration (*not shown here*) is a valuable tool for determining photovoltaic parameters of the as-fabricated heterojunction based on BG organic molecule. Figure 3 illustrates the photocurrent (I_{ph}) *vs* the light powerplots for BG/p–Si organic heterostructure, indicating an increase of I_{ph} with light power. The electrical parameters of the BG device liken, R_S and Φ_b are determined using several techniques at room temperature in darkness and light. The I–V characteristics of Au/BG/pSi/Al organic heterostructure with in 60–120 mW/cm^2 light power range are plotted in Fig. 4, describing a decay by linear segments. Photovoltaic parameters such as V_{oc} and I_{sc} in terms of light power of such device are plotted in double Y-axis as shown in Fig. 5. Similar studies have been performed by various research groups. For instance, Alsharari et al. [17] obtained Au/VONc/n–Si/Au planar heterojunction device deposition of nano-Vanadyl 2,3-naphthalocyanine (VONc) organic compound thin film on n-Si wafer. In addition to the photovoltaic parameters exhibited by the device, other measurements include open-circuit voltage (0.70 V), short circuit current density (11.30 mA/cm^2), fill factor (0.394), and power conversion efficiency (3.12%). In comparison to similar structures [18–21], this study obtained lower results. Field-dependent photogeneration and high series resistance of the organic layer are the main contributing factors to lower photovoltaic properties. Short-circuit current is reduced by the series resistance, which then reduces the fill factor and major power output [22].

Based on the thermionic emission model, an expression for the current would be as follows: [23];

$$I = I_0\left[\exp\left(\frac{V - \mathrm{IR_S}}{nkT}\right)\right] \tag{2}$$

In the BG device, I_o can be defined as the reverse saturation current when I and V are the experimental current and biasing voltage, respectively [23];

$$I_0 = AA^*T^2\exp\left(-\frac{q\Phi_b}{kT}\right), \tag{3}$$

where the effective device area of contact is noted A (0.071 cm^2), the effective Richardson constant for the p-type silicon is A^* of 32 A/cm^2 K^2, q is the electron charge, T is the absolute temperature expressed in Kelvin, n is the ideality factor, Φ_b is the zero-bias barrier height. The determination of the Φ_b value is calculated after extraction and injection of I_0 in Eq. 3, and the barrier height is then calculated using [24];

$$\Phi_b = \frac{kT}{q}\ln\left(\frac{AA^*T^2}{I_0}\right) \tag{4}$$

Table 2 Photovoltaic parameters of BG/Si organic heterojunction diode

Light (mW/cm^2)	I_{sc} (μA)	V_{oc} (mV)	I_m (nA)	V_m (mV)	P_{max} (nW)	FF	Efficiency	I_{ph} (μA)	Photo sensitivity
60	1.023	151.189	553.091	70.677	39.091	0.252744	0.0036	220.761	83.3788
90	1.280	171.271	595.148	90.425	53.816	0.24539	0.0033	304.177	114.506
120	1.451	191.062	703.805	90.404	63.628	0.229505	0.0030	447.074	167.83

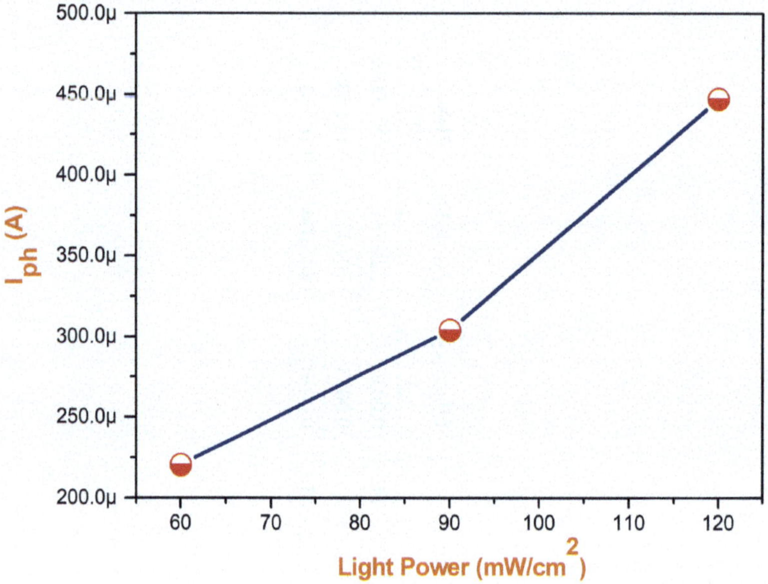

Fig. 3 Photocurrent profile against light power of Au/BG/pSi/Al organic heterostructure diode

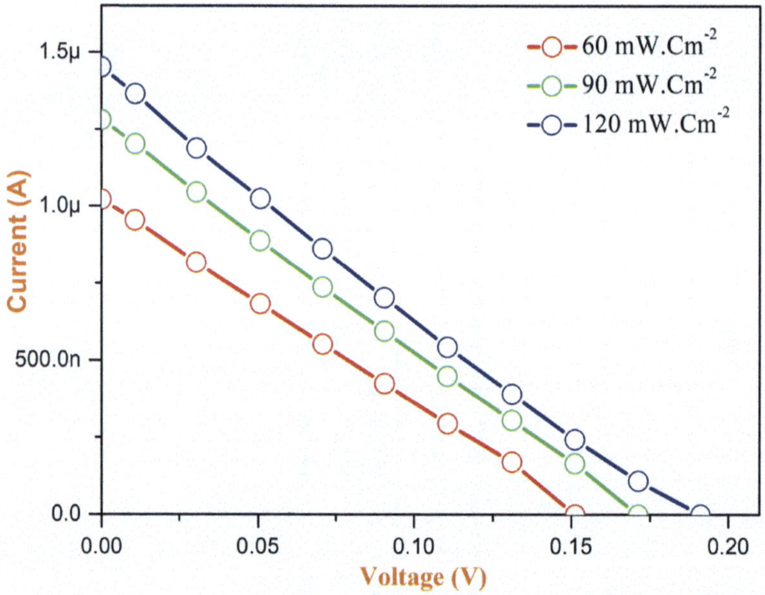

Fig. 4 I–V characteristics of Au/BG/pSi/Al-based heterostructures at $T = 300$ K in light exposure

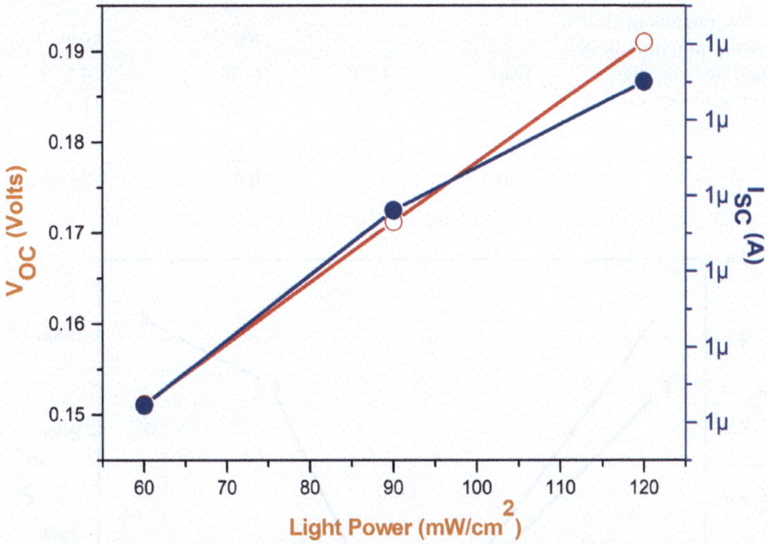

Fig. 5 V_{oc} (left) and I_{sc} (right) of Au/BG/pSi/Al-based heterostructures plots versus light intensity

As an alternative to ideal thermionic emission theory, experimental electrical characteristics have a higher *n* value versus thermionic emission theory's ideality factor of 1. Where *n* is expressed as follows [20];

$$n = \frac{e}{kT}\frac{\mathrm{d}V}{d(\ln I)} \tag{5}$$

Because of interface density and series resistance, experiences offer *n* values higher than unity, contrary to theoretical studies [19]. Parameters of BG/Si organic heterojunction diode determined by using the thermionic emission in dark and light conditions are listed in Table 3. Saturation currents and ideality factors relating to light power, in double *y*-axis, of BG-based device are sketched in Fig. 6. The non-homogeneous thickness of the organic film also contributes to such values of the ideality factor. The non-ideal behavior of organic molecule-based heterojunction like the Al/PEDOT:PSS/p-InP Schottky contact is reported before by Aydin [25]. It is noted that the rectification ratio (RR) is around 40 for BG-based heterojunction, and three regions within forward bias voltage are recorded. The first one, $0 < V < 0.1$ V, is related to the generation–recombination process describing by an exponential profile. For $V > 3kT/q$. An organic BG device's ideality factor and saturation current are determined by the semi-logarithmic curve. The third area highlights ($1 < V < 5$) the effect of series resistance to high bias voltages.

Series resistance is a significant factor that influences the electrical measurements of organic diodes. Cheung and Cheung defined equations to determine electrical parameters, including n, R_S, and Φ_b [9, 10], as given as follows [20];

Table 3 Parameters of BG/Si organic heterojunction diode determined by using the thermionic emission in dark and light conditions

	n	Φ_b (V)	I_s (nA)
Dark	4.29	0.68	144.72
60	5.75	0.69	110.257
90	5.83	0.69	109.351
120	6.33	0.69	128.009

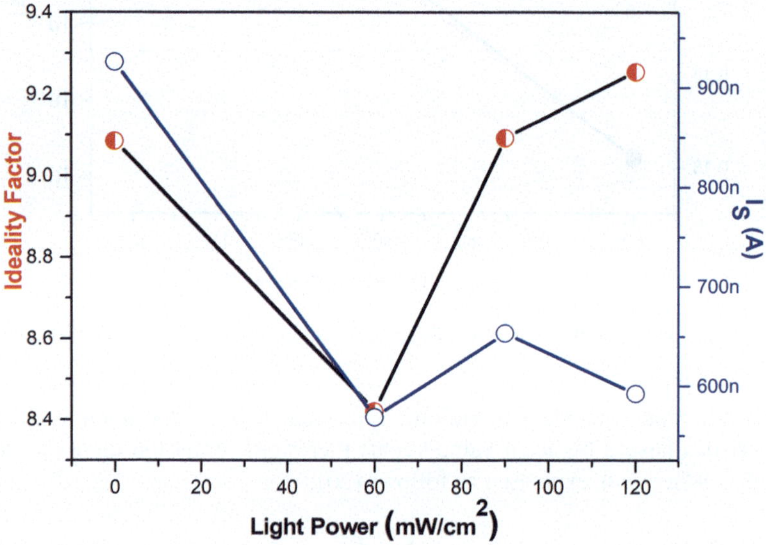

Fig. 6 The variation of ideality factor (left) and saturation current (right) of diode-based BG as a function of light power

$$\frac{dV}{d(\ln I)} = \mathrm{IR_S} + n\left(\frac{kT}{q}\right). \tag{6}$$

$$H(I) = V - n\left(\frac{kT}{q}\right)\ln\left(\frac{I}{AA^*T^2}\right) \tag{7}$$

And in another form, $H(I)$ is then

$$H(I) = \mathrm{IR_S} + n\Phi_b. \tag{8}$$

The Cheung equations are used to extract the ideality factor 'n' and the series resistance RS from the I–V data. The intercept and the slope are extracted using the linear part of I–V throughout the linear fitting of the dV/dLnI versus I (*not shown here*). By the mean of Eqs. 6–8, 'n' is deduced, and R_S and Φ_b values are easily determined. The obtained data using the Cheung method are gathered in Table 4. The found values from both equations of Cheung seem to be close to each other

Table 4 Parameters of BG/Si organic heterojunction diode determined using Cheung method in dark and light conditions

	n	R_{s1} by $dV/d\ln J$ (Ω)	R_{s2} by $H–I$ (Ω)	Φ_b (eV)	I_s (nA)	RR
Dark	9.08561	46,032.6	45,741.7	0.639	925.73	41.247
60	8.4217	81,786.5	81,515	0.652	568.799	0.313
90	9.09271	73,520.6	74,111.3	0.648	653.094	0.238
120	9.25526	70,215.1	69,005.7	0.651	592.579	0.157

and proved the suitability of this method. Another modified approach, defined as the Norde function, is used for the determination of R_S and Φ_b, as [26];

$$F(V) = \frac{V}{\delta} - \frac{kT}{q}\left(\frac{I(V)}{AA^*T^2}\right). \tag{9}$$

When the forward bias of the I–V characteristics is greater than 'n', the current and voltage can be determined. An organic device with such a barrier height has the following characteristics [27];

$$\Phi_b = F(V_0) + \frac{V_0}{\gamma} - \frac{kT}{q}. \tag{10}$$

As a result, the series resistance can be expressed as follows [28, 29];

$$R_s = \frac{kT(\gamma - n)}{qI}, \tag{11}$$

where $F(V_0)$ is the minimum value taken from the plot of $F(V)$ (*not shown here*), V_0 is the corresponding voltage, and I_{min} is the minimum value of current taken at $V = V_0$.

3.2 Electrical Transport and Conduction Mechanism in the Dark and Under Light Conditions

The I–V electrical measurements in dark and under light are used to identify the different conduction parameters and mechanisms. The permitted electronic bands are very narrow in organic materials, leading to low mobility of charge carriers (*low overlap between the molecular orbitals*). Thus, the load transport model can be described by a load jump model between localized states. Charge carriers cannot be treated as electrons or free holes but as small polaron-type quasiparticle, due to the localization of these carriers at molecular sites. Energy band calculations represent the excess electron's (or hole's) wave function by integrating the molecular orbitals of the excess electron, each orbital Φnis being calculated approximately by the linear combination atomic orbitals (LCAO) method from Huckel. The band method can be used when the intermolecular distances decrease. So, the application of this model to polymers is possible within a chain macromolecular. It is found that the ideality factor is much greater than 1. Therefore, the conduction mechanism is not thermionic emission. The plot of Log *I*–Log *V* allows the predominant conduction mechanism to be defined using the relation $I \alpha V^{\mathrm{m}}$ where conduction mechanism type is accurately described by the exponent '*m*'. The shape of the Log *I*–Log *V* curve is linear, the slope close to 2 and greater than 1, that the current obeys to ohmic or space charge limited current (SCLC) conduction mechanisms. Figure 7 shows the Log *I*–Log *V* plot of BG/p–Si organic heterostructure, indicating two regions in dark and illumination condition. The shape of the Log *I*–Log *V* curve is linear, the slope close to 2 and greater than 1, and then we conclude that the current follows an ohmic or SCLC regime. Under the light, the achievement of I–V characteristics by the measured of generated current by different diodes against voltage in dark and light is fulfilled. Such I–V curves lead us to evaluate the performances and electrical behavior of fabricated organic heterostructure diodes. By measuring the voltage versus current of the different diodes, the current versus voltage characteristics are determined. This characteristic allows us to assess the performance of our structures and their electrical behavior. The measurements taken in the dark do not show any short-circuit current or open-circuit voltage. The photoelectrical responses are observed after exposure of the diode to light. The current–voltage characteristics show that the reverse current increases with light intensity, especially for the Au/BG/p–Si/Al structure. In this model, a charge carrier surrounded by its electronic polarization is a polaron (*polaron transport is an intermediate state between molecular transport and transport in the band model*) [30].

In the case of a perfect insulator, that is to say, a dielectric without traps and mobility of charge carriers (μ) is independent of the electric field, the SCLC current density follows the Child law [31];

$$J_{\mathrm{SCLC}} = \frac{9\varepsilon_0\varepsilon_r\mu V^2}{8d^3}. \tag{12}$$

Where 'μ' is free charge carrier mobility, '*d*' is the thickness of the layer, and '*V*' is the potential that is applied across the sample. It corresponds to the maximum power that the diode can be delivered and is the product of the maximum voltage V_{max} and the maximum current I_{max}, where $0 < V_{\mathrm{max}} < V_{\mathrm{OC}}$ and $0 < I_{\mathrm{max}} < I_{\mathrm{SC}}$ [32];

$$FF = \frac{P_{\mathrm{max}}}{V_{\mathrm{oc}} x I_{\mathrm{SC}}} = \frac{I_{\mathrm{max}} V_{\mathrm{max}}}{V_{\mathrm{oc}} I_{\mathrm{SC}}} \tag{13}$$

$$\eta = \frac{I_{\mathrm{sc}} V_{\mathrm{OC}} FF}{P_{\mathrm{ins}}}. \tag{14}$$

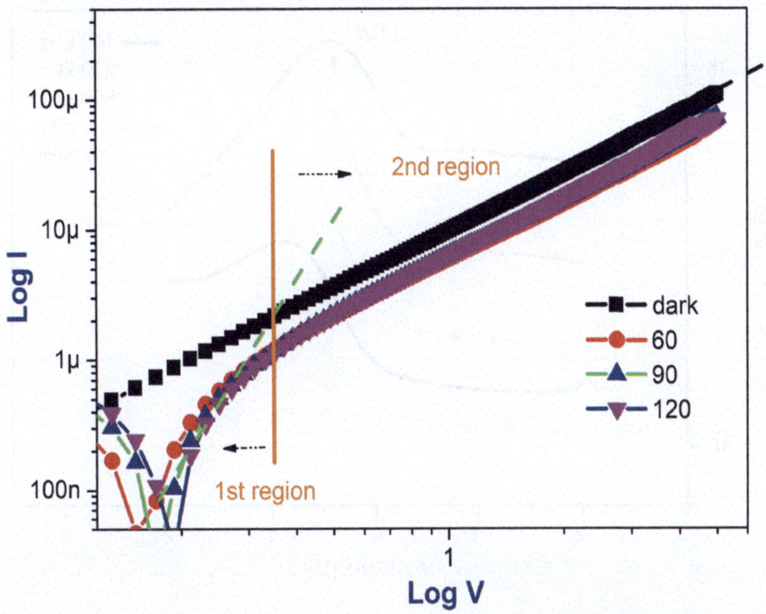

Fig. 7 Log *I*–Log *V* plot of BG/Si organic heterostructure **a** forward and **b** reverse bias voltage. Two linear segments, two regions, related to two different conduction mechanisms

The R_S–V profile describes a peak, and the peak position is removed by frequency towards higher voltage, and the peak level as frequency increases, it decreases, as displayed by arrows in Fig. 8. The highest point is recorded at 1.4 kΩ and declines gradually with increase in frequency to reach 55 Ω. The charges following the frequency of practical biasing voltage produce a shift in peak location. Anyway, AC signals with more prominent frequencies cannot be followed by interface states, so they don't consider a significant commitment to interface states. The R_S–V plots describe a peak at 1.493 kΩ, which is shifted towards higher voltage when frequency increases from 100 kHz to 2 MHz. It is noted that R_S peak intensity declines from 1.493 kΩ to 51.6 Ω as a result of frequency. It is consequently that R_S–V plots are frequency-dependent due to the occurrence of interface state. Besides, at the minor frequency, the interface state can follow the AC signal, which high resistance at 100 kHz is recorded, and an additional capacitance is stored. While at higher frequency 1–2 MHz, less value of $R_S < 100$ Ω due to the inability of the interface state to follow the Ac signal, no contribution to interface state can be attributed to the AC signal, similar behavior is reported [33].

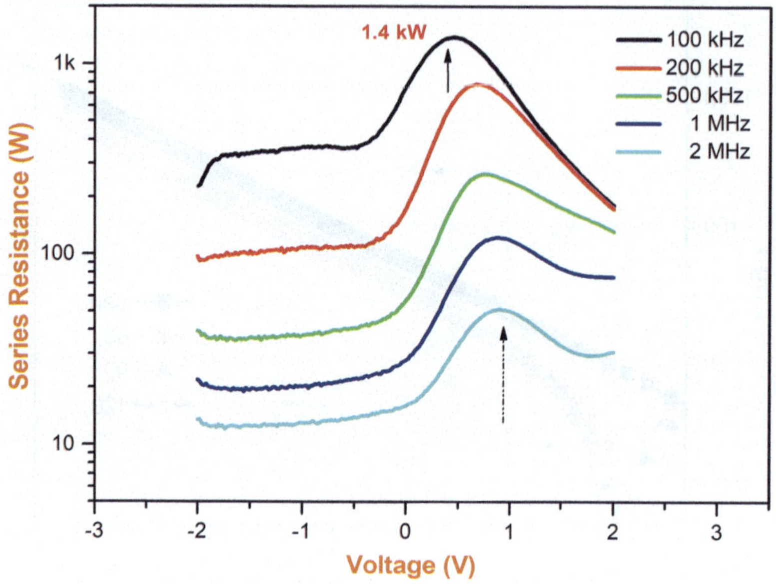

Fig. 8 Series resistance as a function bias voltage of Au/BG/pSi/Al organic heterostructure diode under dark and room temperature environments for various frequencies 100 k–2 MHz. The peak shift is indicated by arrows

3.3 Determination of the Capacitance–Voltage Characteristics

According to Fig. 9, the capacitance varies with frequency depending on the bias voltage, indicating *deep depletion (dd), depletion (d)* and *accumulation (A).* From this curve, it is noted that the capacitance varies only at low frequencies. The independence of capacitance from high-frequency voltage has been reported in the literature by several researchers. The phenomenon can be explained as a result of a large number of trapped charges that do not follow changes in voltage [32]. The C–V characteristics plotting show a lower value of capacitance > 200 pF which defines the deep depletion region within the reverse biasing voltage (− 2 V, 0 V). The capacitance increases rapidly within the forward voltage in the depletion region and attains a top around 1.1 nF, indicating an accumulation region of the organic device. The dielectric parameters such as N_a concentration, the diffusion potential V_d and the barrier height φ_B are deduced using the linear fit of $1/C^2$–V plot scheme for various experimental frequencies as illustrated in Fig. 10. Such extracted parameters from capacitance are determined from the $1/C^2$ as a function of bias voltage using the linear part approximation within the − 0.4 V, 0 V reverse voltage region, as seen in Fig. 10. In Fig. 11 describes the barrier height versus applied frequency of planar heterojunction based on BG molecule; further N_d and V_d parameters versus applied frequency, respectively, are portrayed in Fig. 12. The N_d, V_d, and Φ_b parameters at

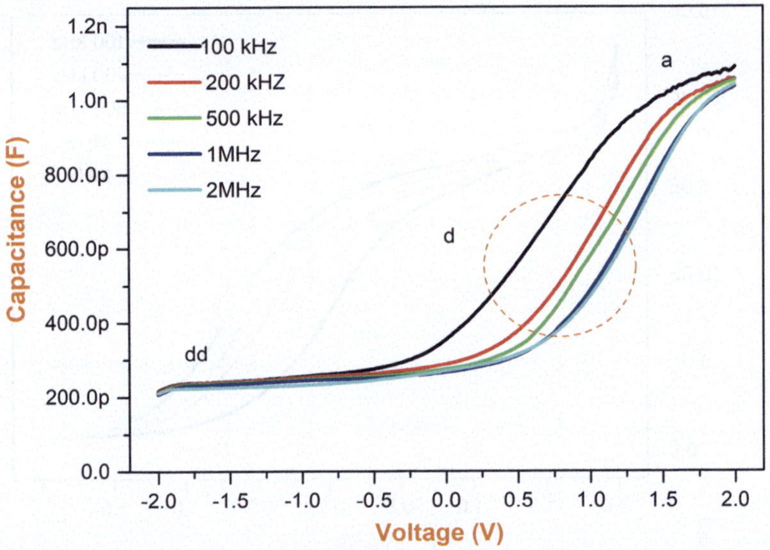

Fig. 9 Capacitance–voltage characteristics of Au/BG/pSi/Al organic heterostructure diode under dark and room temperature environments. Deep depletion (dd), depletion (d), and accumulation (a) regions are defined

various frequencies, 100–2000 kHz range, are extracted using the linear part of $1/C^2 - V$ plot within − 2 V, − 1 V range as indicated in Table 5. It is observed that N_a increases roughly with frequency, recording a high point of 2.29×10^{16} cm^{-3} at 2000 kHz. Both the diffusion potential and height barrier obey to the same increasing profile of N_d as seen in Table 5.

The G–V plots of Au/BG/pSi/Al organic heterostructure are drawn in Fig. 13, describing a sharp rise in 1–2 V range with an increase in frequency, and log G versus sqrt (V) is plotted in Fig. 14 according to the following equation;

$$J \alpha V \exp\left[\frac{q(-\phi_b + 2a\sqrt{V})}{kT}\right]. \tag{15}$$

In the case where 'a' $(q/4\pi\varepsilon_i d)^{1/2}$, voltage is written in terms of enough electric field as $V = E_i d$, and density of current J is expressed as $J = I/A = V/\text{RA} = V \times G/A$. Such plot exhibits roughly three linear regions according to $\log = -q\Phi_b/kT + q(2a/kT)(V)^{1/2}$. Such calculation gives the slope of $q2a/kT$ and intercept of $-q\Phi_b/kT$. For higher values of voltage (region C) of Fig. 14, intercept of − 8.84 and slope of 2.71 are recorded, then the parameter a is of 0.035. The Al/p-Si/P3HT/Ag organic MSP diode at 1 MHz device showed the same behavior as reported before by Meftah et al. but the higher value is recorded about 1500 nF but the BG device shows lower

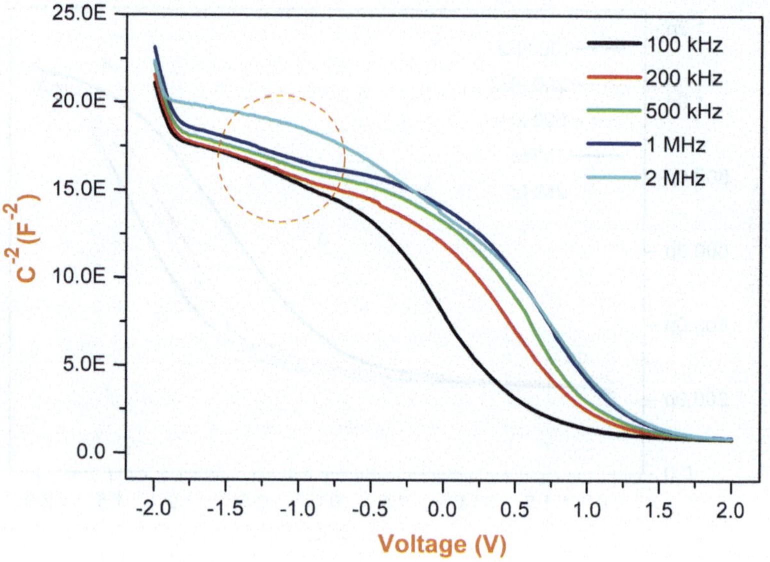

Fig. 10 $1/C^2 - V$ characteristics of Au/BG/pSi/Al organic heterostructure diode under dark and room temperature environments. Linear part is spotted by a dashed circle

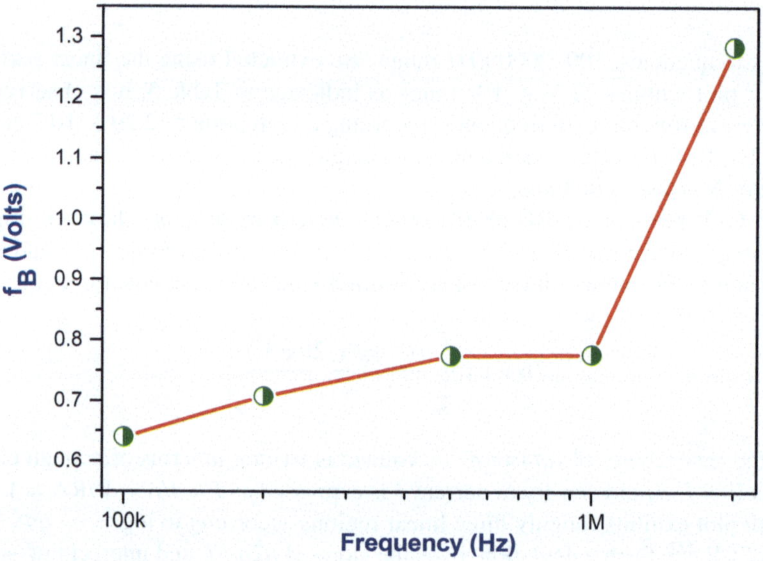

Fig. 11 Barrier height versus applied frequency of planar heterojunction-based BG molecule

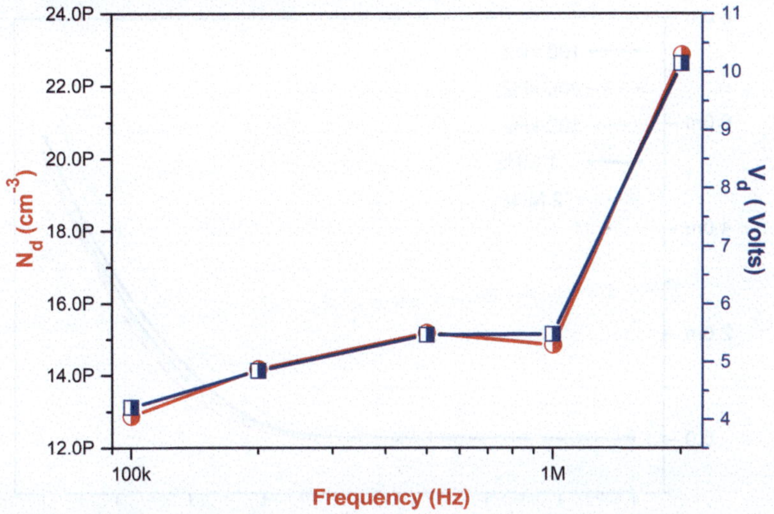

Fig. 12 The variation of N_d and V_d parameters of diode-based BG as a function of applied frequency

Table 5 Dielectric parameters of BG/Si organic heterojunction diode determined using C–V characteristics method at various experimental frequencies

f (kHz)	Na (10^{16} cm^{-3})	V_d (V)	Φ_b (eV)
100	1.28	4.21	0.64
200	1.42	4.84	0.71
500	1.52	5.46	0.77
1000	1.49	5.47	0.78
2000	2.29	10.14	1.28

capacitance around 1nF [20]. Lower values around 400 pF of C–V at 1 MHz are also reported by Yakuphanoglu et al. for the Au/methylene blue/n-Si diode [34]. The 1/C^2 versus voltage is given as follows [35];

$$\frac{1}{C^2} = \frac{2(V_D + V)}{q\varepsilon A^2 N}. \tag{16}$$

Taking the slope of the $1/CD^2$ curve as a basis, and extrapolating from its linear part, we can get the calculated potential VD by intersecting it with the voltage axis as $V_D = V_{bi} - kT/q$ [36]. Used constants are listed in Table 6. Nicollian and Goetzberger (NG) approximation methods give by the mean of adjusted capacitance and conductance the interface state density, and such NG functions express as [37, 38];

$$C_{adj} = C_m \frac{G_m^2 + (\omega C_m)^2}{a^2 + (\omega C_m)^2} \tag{17}$$

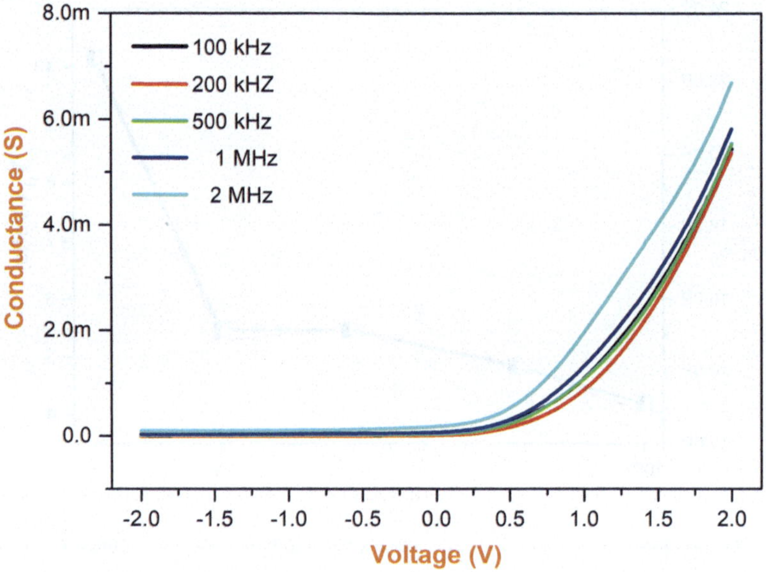

Fig. 13 Conductance–voltage characteristics of Au/BG/pSi/Al organic heterostructure diode under dark and room temperature environments

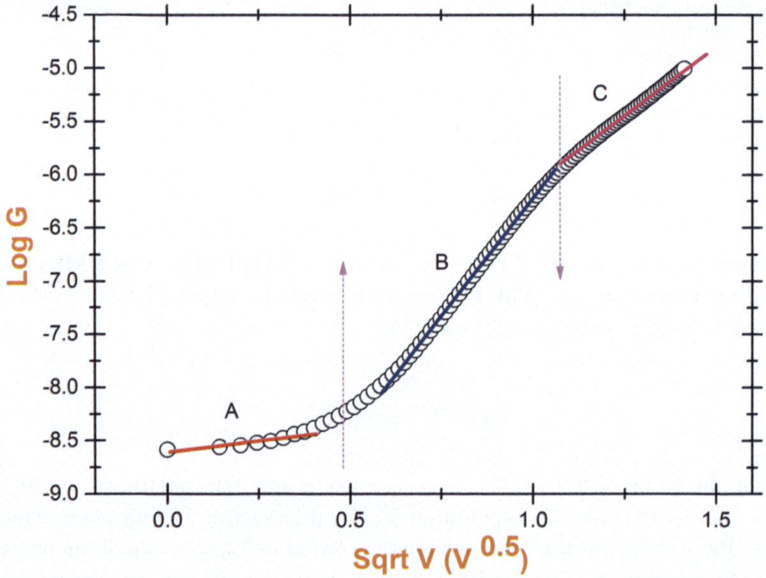

Fig. 14 Ln G–$V^{0.5}$ plot at 2 MHz of organic device-based BG

Table 6 Used constants

Used parameters	Values
Richardson constant p-Si A*	32 A K^{-2} cm^{-2}
Boltzmann constant (K)	1.38×10^{-23} J/K
Charge of electron (q)	1.6×10^{-19} C
Temperature (T)	300 K
Permittivity of vacuum (ϵ_0)	8.84×10^{-14} (F/cm)
Radius of contact	1.5 mm
Dielectric constant of silicon (ϵ_s)	11.8
Density of states in valence band (N_v)	1.82×10^{19} cm^{-3}

and,

$$G_{\text{adj}} = a\frac{G_{\text{m}}^2 + (\omega C_{\text{m}})^2}{a^2 + (\omega C_{\text{m}})^2}. \tag{18}$$

In this equation, C_{m} and G_{m} magnitudes refer to capacitance and conductance in strong accumulation regions, a is the corrected parameter, $\omega = 2\pi f$ is the f is the angular frequency, and R_s is the resistance in series [36];

$$a = G_{\text{m}} - \left[G_{\text{m}}^2 + (\omega C_{\text{m}}^2)\right]R_s. \tag{19}$$

Figure 15 shows that the adjusted conductance G_{adj} is plotted versus voltage as a function of voltage. A G_{adj} curve exhibits a narrow peak at the edge of forward bias voltage, which decreases with increasing frequency as shown in this figure. Furthermore, the peak is shifted towards low voltage as the frequency increases, as assigned by arrows in Fig. 15.

3.4 AC Impedance Spectroscopy and Interface State Density of Au/BG/pSi/Al

Through this research and as proved by results, the non-ideality of such heterostructure is due to occurrence of interface state density (D_{it}) written in terms of frequency ($f = \omega/2\pi$), C_{ox}, C_{m} and G_{m} as [37, 39];

$$D_{\text{it}} = \frac{2G_m/\omega}{qA[(G_m/\omega C_{\text{ox}})^2 + (1 - C_m/C_{\text{ox}})^2]} \tag{20}$$

A denotes the area of the diode, and C_{ox} represents the capacitance of the insulator layer. Control interface traps are affected by voltages induced changes in Fermi-level situations about the interface trap levels. It impacts MOS capacitance when this

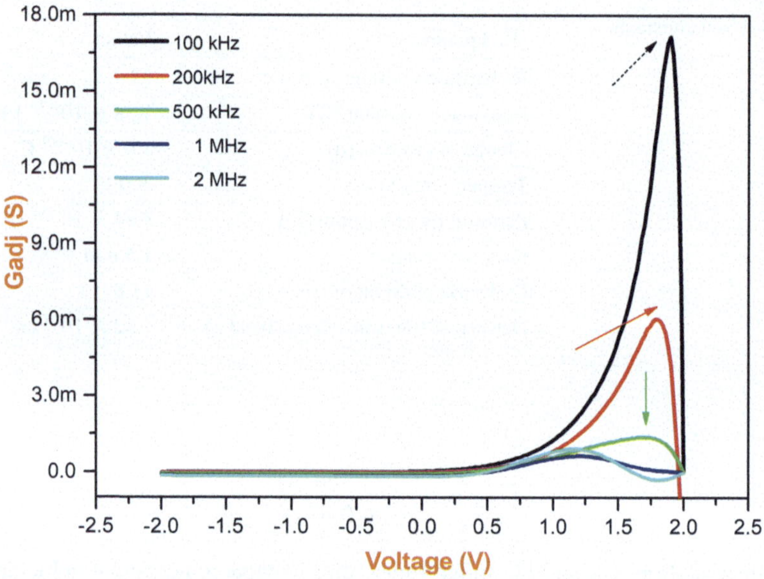

Fig. 15 G_{adj} versus bias voltage plot of (Au/BG/pSi/Al) at several experimental frequencies (100 k–2 MHz). The peaks are assigned by arrows

charge changes. In Fig. 16, the semiconductor capacitance is C_S, the interface trap capacitance is C_i, the oxide capacitance is C_{ox}, and the interfacial layer obstruction is R_i, and the equal circuit represents MOS capacitors. Equivalent circuits can be expressed as recurrence subordinate capacitances C_p, capacitances C_m, and conductances (G_m) with the equal part being converted into recurrence subordinate capacitances C_p and C_m. The study of the capacitance as a function of the frequency for our device allowed us to calculate the density of interface states using the equations developed above. The low-frequency regime helps to calculate this density using the plot of the modified conductance, named G_{adj}. It was shown in Fig. 17 that interface state density (D_{it}) of BG organic heterojunction is lowering with an increase of frequency from roughly 1×10^{13} 1/eV $\times$ cm^2) with frequency growth for measured 60, 90, and 120 mW/cm^2 light till a constant. The exploration of the different results of heterojunction-based BG leads us to explain the different calculated values. The Au/BG/p–Si/Al structure reveals an ideality factor n much greater than 1. The excessively high value of n can also result from the insulating layer inserted between the contact (Au) and the organic layer and strong series resistance (70 kΩ), a large interface concentration assumed. Several authors have linked the variation of the coefficient of ideality to the interface concentration. This effect is apparent in our diodes, the barrier height obtained by I–V matches that of C–V. The peaks present in the G–V characteristic show the existence of interface states, and their density decreases at high frequency because these states do not follow the high-frequency signal. The intensity of the light increases the short-circuit current and the voltage

V_{oc}. This finding proves that our structures operate in photodiode mode. As excitons are produced and dissociated at the barrier between BG and p–Si, a photocurrent is generated.

Low efficiency is observed for Au/BG/p–Si/Al due to a significant series resistance of 80 kΩ. It is noticed that the resistance of a diode affects its photovoltaic properties. The Au/BG/p–Si Al structure reveals an ideality factor '*n*' much greater than 1. The value too high of '*n*' may also result from the effect of the insulating layer inserted between the contact (Au) and the organic layer and strong series resistance (70 kΩ),

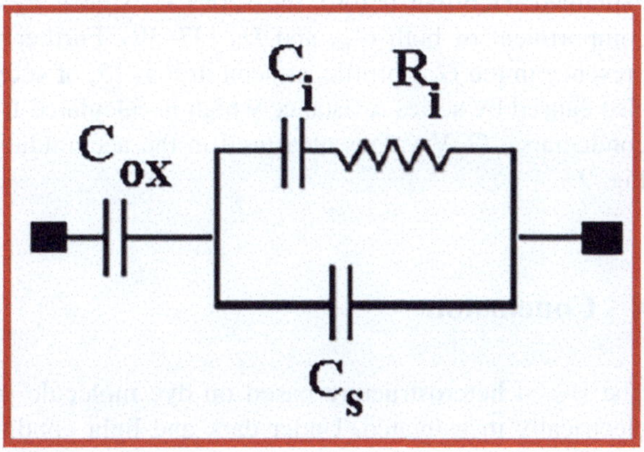

Fig. 16 Sketch of circuits for interface states: I—metal insulator semiconductor capacitor through interface trap (C_i and R_i)

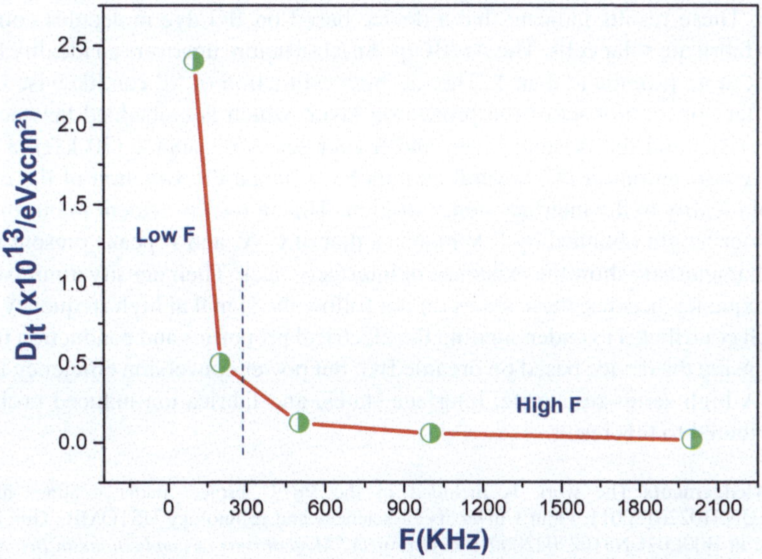

Fig. 17 Interface state density as a function of experimental frequency of BG based organic heterostructure

a large interface concentration says. Several authors have linked the variation in the coefficient of ideality to the interface concentration. This effect is apparent in our diodes, the barrier height obtained by I–V matches that of C–V for the diode. The peaks present in the G–V characteristic show the existence of interface states, and their density decreases at high frequency because these states do not follow the signal in high frequency. Whereas D_{it} as a function of the applied frequency of the Al/Ti–a:C/p–Si/Al photodiode is already studied and showed growth with frequency is reported [15]. There is a strong correlation between interface state densities and greater frequencies, resulting from the non-ideal performance of diodes (D_{it}). As explained Nicollian before, the series resistance R_S plays a role in affecting the comportment of both G_{adj} and D_{it} [37–39]. Further, it is discussed that the peak presence in the G_{adj} profile, as seen in Fig. 15, of such device based organic dye is also caused by series resistance which is calculated from the capacitance C–V and conductance G–V values measured in the accumulation region (a) as observed in Fig. 9.

4 Conclusion

The BG/Si heterostructure based on dye molecule is successfully fabricated and electrically investigated. Under dark and light conditions, BG/Si diodes are evaluated by their I–V characteristics, as well as their C–V characteristics at different frequencies. In some cases, the calculated parameters are in agreement with previous reports. These results indicate that a device based on BG dye molecules could be used to fabricate solar cells. The Au/BG/p–Si/Al structure uncovers an ideality factor 'n' a lot more prominent than 1. The too-high estimation of 'n' can likewise be the after effect of the impact of the protecting layer, which is embedded between the contact (Au) and the organic layer, and higher series resistance (70 kΩ), a large interface concentration D_{it}. Several authors have linked the variation of the coefficient of ideality to the interface concentration. This impact is evident in our diodes, the barrier height obtained by I–V matches that of C–V, and F peaks present in the G–V characteristic show the existence of interface states. Their density diminishes at high frequency because these states do not follow the signal at high frequency. This research contributes to understanding the electrical properties and conduction mechanism inside the device based on organic BG, but power conversion efficiency is still weak. A high series resistance, interface states, and fabrication-induced problems are attributed to this result.

Acknowledgements The work is included in the PRFU project under contract number B00L02UN310220180011, Oran University of sciences and technology USTOMB. This is also included in B00L02UN310220220001PRFU project "*Mechanisms of conductions in heterojunctions based on organic and inorganic materials for solar energy*" with USTOMB university and MESRS ministry.

Declaration of Competing Interest The authors declare that they have no known competing financial interests or personal relationships that could have appeared to influence the work reported in this paper.

References

1. H.J. Jang, J.Y. Lee, J. Ind. Eng. Chem. **93**, 388–393 (2021)
2. Z.A. Alahmed, Sh.A. Mansour, M.E. Aydın, F. Yakuphanoglu, Solid State Commun. **163**, 23–27 (2013)
3. S. Ameen, M.S. Akhtar, H.S. Shin, Electrochim. Acta **215**(10), 200–211 (2016)
4. Y. Yasuda, S. Kobayashi, T. Uchida, Y. Hosh, Thin Solid Films 698(31), 137868 (2020)
5. W. Cheng, H. Di, Z. Shi, D. Zhang, J. Alloys Compd. **831**(5), 154607 (2020)
6. V. Chaudhary, R.K. Pandey, R. Prakash, N. Kumar, A.K. Singh, Synth. Met. **258**, 116221 (2019)
7. P. Sanjaya, K. Deepab, J. Merline Shylab, J. Madhavanb, S. Senthil, Materials Today: Proceedings **8**, 130–135 (2019)
8. A. Ugur, A. GencerImer, Y.S. Ocak, Mater. Sci. Semicond. Process. **39**, 569–574 (2015)
9. A.M. Nawar, A.M. El-Mahalawy, J. Mater. Sci. Mater. Electron. **31**, 3256–3273 (2020)
10. F.L. Migliorini, J.R. Steter, R.S. Rocha, M.R.V. Lanza, M.R. Baldan, N.G. Ferreira, Diam. Relat. Mater. **65**, 5–12 (2016)
11. G.S. Sree, S.M. Botsa, B.M. Reddy, K.B. Ranjitha, Arab. J. Chem. **13**, 5137–5150 (2020)
12. P. Taneja, S. Sharma, A. Umar, S.K. Mehta, A.O. Ibhadon, S.K. Kansal, Mater. Chem. Phys. **211**, 335–342 (2018)
13. S. Kaur, S. Sharma, A. Umar, S. Singh, S.K. Mehta, S.K. Kansa, Superlattices Microstruct. **103**, 365–375 (2017)
14. T.T. Grove, M.F. Masters, R.E. Miers, Am. J. Phys. **73**, 52–56 (2005)
15. N. Aslan, M.M. Koc, A. Dere, B. Arif, M. Erkovan, A.G. Al-Sehemi, A.A. Al-Ghamd, F. Yakuphanoglu, J. Mol. Struct. **1155**, 813–818 (2018)
16. D.A. Aldemir, M. Benhaliliba, C.E. Benouis, Optik **222**, 16548 (2020)
17. A.M. Alsharari, S.I. Qashou, A.A.A. Darwish, M.M. El-Nahass, Thin Solid Films **704**, 137977 (2020)
18. M.M. Makhlouf, M.M. Shehata, Sol. Energy **214**, 326–336 (2021)
19. M.M. El-Nahass, W.M. Desoky, Optik **171**, 44–50 (2018)
20. S.E. Meftah, M. Benhaliliba, M. Kaleli, C.E. Benouis, C.A. Yavru, A.B. Bayram, Phys. B **593**, 412238 (2020)
21. E.M. El-Menyawy, H.H. Nawar, I.T. Zedan, Surf. Interf. **23**, 101040 (2021)
22. A.A. Darwish, E.A.A. El-Shazly, A.A. Attia, K.F. Abd El-Rahman, J. Mater. Sci. Mater. Electron. **27**(8), 8786–8792 (2016)
23. M.E. Aydin, F. Yakuphanoglu, Microelectron. Eng. **85**, 1836–1841 (2008)
24. M. Benhaliliba, Optik **217**, 164832 (2020)
25. M.E. Aydin, F. Yakuphanoglu, Microelectron. Reliab. **52**(7), 1350–1354 (2012)
26. H. Norde, J. Appl. Phys. **50**, 5052 (1979)
27. S.E. Meftah, M. Benhaliliba, M. Kaleli, C.E. Benouis, C.A. Yavru, A.B. Bayram, J. Electron. Mater. **593**, 412238 (2020)
28. J.W. Zhang, J.J. Qin, H.M. Yu, X.Q. Chen, X.Y. Hou, Org. Electron. **28**, 239–324 (2016)
29. F. Parlaktürk, A. Agasiev, A. Tataro, Gazi University J. Sci. **20**(4), 97–102 (2007)
30. M. Zoli, Phys. Rev. B **61**(21), 14523–14530 (2000)
31. C.D. Child, Phys. Rev. Ser. I **32**(5), 492 (1911)
32. K.K. Ng, S.M. Sze, *Physics of Semiconductor Devices*, 3rd edn. (Wiley, New Jersey, 2007)
33. S. Altındal, H. Uslu, J. Appl. Phys. **109**, 074503 (2011)
34. F. Yakuphanoglu, Y.S. Ocak, T. Kılıçoğlu, W.A. Farooq, Microelectron. Eng. **88**, 2951–2954 (2011)

35. F. Yakuphanoglu, M. Kandaz, B.F. Senkal, Thin Solid Films **516**, 8793–8796 (2008)
36. M. Benhaliliba, C.E. Benouis, D.A. Aldemir, Phys. B **572**, 175–183 (2019)
37. E.H. Nicollian, A. Goetzberger, Bell Labs Tech. J. **46**(6), 1055–1033 (1967)
38. S. Karatas, A. Turut, Vacuum **74**, 45 (2004)
39. M. Marín Genescà, J. García Amorós, R. Mujal Rosas, L.M. Vidal, X.C. Fajula, Polymers **12**, 2675 (2020)

Developing Energy Storage Applications for Next Generation

Hari Babu Pengonda, Naresh Kumar Rotte, Sampath Kumar Puttapati, and Subbareddy Yerramala

Abstract Energy fuels human progress, which is essential for all universal processes. The sun is Earth's primary energy source. High energy consumption, mainly from fossil fuels, has led to environmental issues. Growing technology demands sustainable energy solutions. The pursuit of renewable energy is urgent, driving innovations in energy storage. This chapter focuses on advancing electrical energy storage, including batteries, capacitors, and more, to meet future needs. Energy can be transformed, not stored indefinitely. Experts work on efficient energy storage for easy conversion to electricity. Storage involves internal, potential, or kinetic energy, managed through charging, storing, and discharging in energy storage systems. ESS types are: thermal, electrical, mechanical, and chemical. Advanced ESS offers quick dispatch and more extended backup. The comparison includes power, energy, discharge duration, output, cycle life, and efficiency. Effective ESS accelerates global energy transition, reducing fossil fuel reliance. Cutting-edge research is reshaping the landscape of rechargeable batteries. Focus areas encompass comprehending battery interfaces, self-healing battery technologies, monitoring battery health, refining manufacturing processes, and enhancing battery recyclability. These breakthroughs hold immense significance for applications ranging from electric vehicles and drones to biomedical devices. Thermal energy storage (TES) reduces reliance on conventional thermal energy through optimized storage. Supercapacitors offer high-power storage for electronics, while SMES offers lossless energy storage. Chemical energy storage uses bonds and electrolysis for sustainability. Fuel

H. B. Pengonda
Department of Chemical Engineering, Rajiv Gandhi University of Knowledge Technologies, Nirmal District, Basar, Telangana 504107, India

N. K. Rotte (✉)
M/S. Regal Enterprises, Eluru District, Kamavarapukota, Andhra Pradesh 534449, India
e-mail: nareshk.rotte@gmail.com

S. K. Puttapati
Department of Chemical Engineering, National Institute of Technology Warangal, Warangal, Telangana 506004, India

S. Yerramala
Department of Chemistry, Andhra Loyola College (Autonomous), Vijayawada, Andhra Pradesh 520008, India

K. Kasinathan et al. (eds.), *New Technologies for Energy Transition Based on Sustainable Development Goals*, https://doi.org/10.1007/978-981-97-2527-4_15

cells convert chemical energy to electricity using electrolytes, anodes, and cathodes. Stacking cells boost power, categorized by electrolyte and fuel types. This overview encapsulates these strides, underscoring their capacity to revolutionize industries and advance sustainable energy solutions.

1 Introduction

Energy measures how much effort must be made on an object to meet human needs and create technology. Every life and all processes that take place throughout the universe depend on energy [1]. The sun provides the most energy needed and available to Earth's humans, animals, plants, and microorganisms. This energy can be obtained either directly or indirectly from the sources. Energy sources could be classified as renewable and non-renewable [2, 3]. Human civilization has achieved very high development at the expense of high energy utilization. Electricity generation and consumption were once the global economies. Still, it is exhausting fossil fuels affecting global climate change and pollution significantly. With the rapid expansion of modern science and technology, electronic products have become a part of daily life, such as mobile phones, health trackers, microsensors, and many more, which comprise standard, intelligent, flexible, and miniaturized devices, and components to provide necessary power [4]. The search for green renewable energy has emerged and has received an unprecedented push as traditional energy sources are becoming limited. Now, new options are overturned as the world is running fast against time to create new energy-based sources to fulfill essential needs. As the industry adopts newer renewable energy, energy storage is promising. Figure 1 gave us insight into how the electrical power source progressed from the eighteenth century to the current decade by developing a series of rechargeable and non-rechargeable battery breakthroughs [2]. The figure underscores the paramount importance of lead-acid batteries on a global scale and highlights the substantial market share they command. It also points out that ongoing technological advancements have significantly enhanced the efficiency of these batteries over time. Furthermore, it draws attention to the current era, characterized by a burgeoning demand for batteries in consumer applications, which has spurred increased research efforts to provide viable solutions. Therefore, an attempt is made in this chapter to focus on the electrical energy storage devices not only on batteries, capacitors, and supercapacitors but beyond for upcoming generations.

2 Energy Storage Devices and Classifications

As per the fundamental principles of thermodynamics, it is known that energy cannot be stored indefinitely, but it can certainly be transformed or converted from one form to another. Experts in the field of science and technology are continuously

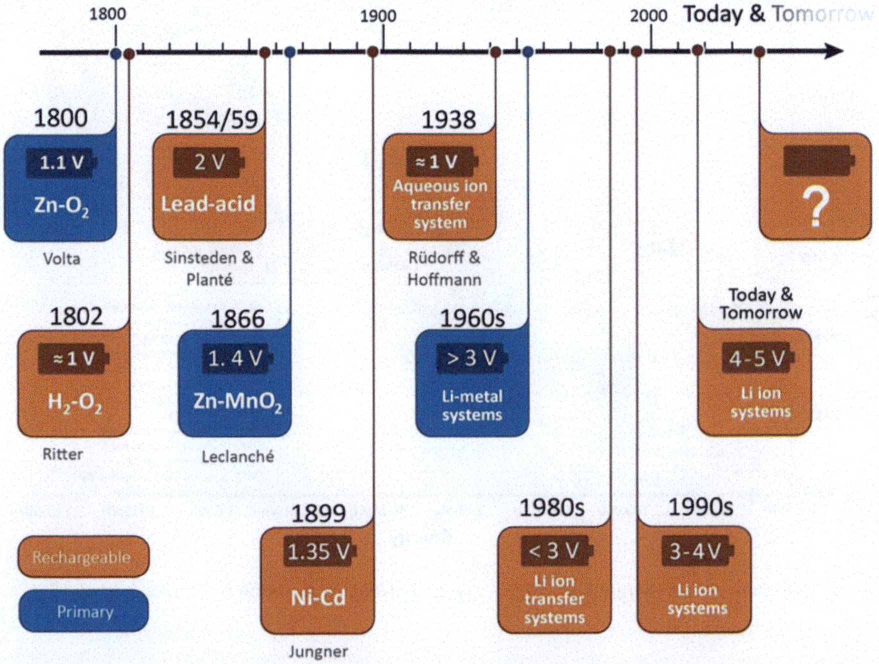

Fig. 1 Historical overview of the development of battery types by Placke et al. [2]

exploring new and innovative ways to store energy in a manner that minimizes loss, so that it can be effortlessly converted back into electricity as and when required. Energy storage can be accomplished through the storage of internal energy (U), potential energy (PE), or kinetic energy (KE). The storage process primarily involves three essential steps, namely charging, storing, and discharging. These processes are orchestrated by a series of energy converters, storage devices, and peripheral systems, which collectively comprise an energy storage system (ESS). ESS can be classified into thermal storage of energy (TSE), electrical storage of energy (EST), mechanical storage of energy (MSE), and chemical storage of energy (MCSE) [3]. Energy storage is applicable to many technologies such as pumped hydro storage (PHS), flywheels (FW), supercapacitors (SC), compressed air (CS), thermal energy storage (TES), and batteries. Advanced ESS can dispatch electricity at very high speeds and provide longer power backups. Energy storage systems can be compared and assessed over various parameters such as rated power, energy, discharge duration as shown in Fig. 2 [5], power output, cycle life, and efficiency of storage technologies. Energy storage technologies can be grouped into three categories. The first is for short discharge times (seconds to 1 min), including DLC, SMES, and FES, with fast activation and low energy-to-power ratios. The second category is for medium discharge times (minutes to hours) and includes FES, Li-ion, LA, and NaS batteries, which offer a balanced energy-to-power ratio. Finally, the third category is for long

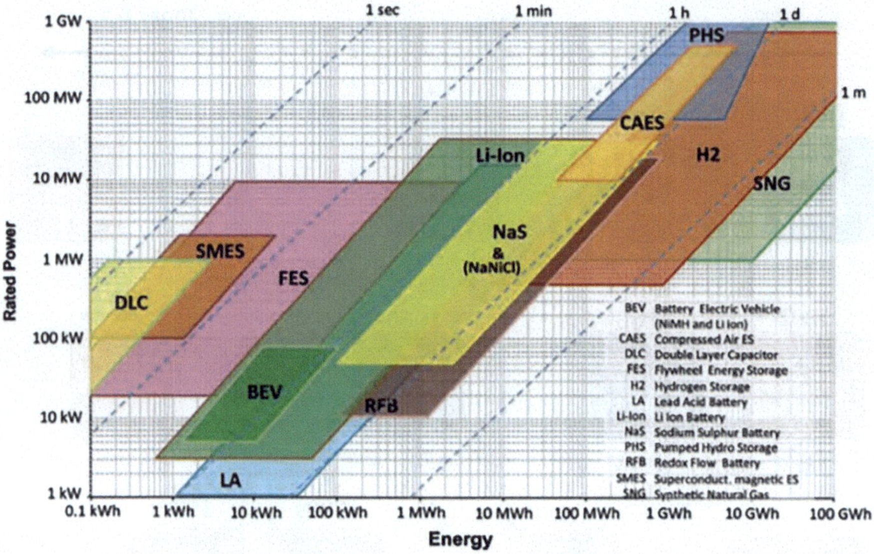

Fig. 2 ESS comparison by rated power, energy, and discharge duration by Carnegie et al. [5]

discharge times (days to months) and includes H_2 and SNG systems, which have a considerably high energy-to-power ratio, suitable for extended storage needs. This overview offers a comprehensive insight into various energy storage possibilities, with particular emphasis on hydrogen as a notable contender owing to its exceptional energy content. Nevertheless, it is crucial for current researchers to direct their efforts toward enhancing hydrogen's efficiency by a substantial margin, ideally by 30–40%. Effective ESS has shown a significant inclination to enhance and accelerate global energy transition thereby reducing reliance on fossil fuel-based generation.

2.1 *Mechanical Storage of Energy*

Pumped hydro storage (PHS), compressed air energy storage (CAES), and flywheels are major sectors of mechanical storage.

2.1.1 Pumped Hydro Storage (PHS)

PHS derives its electrical energy from water-sourced potential energy, usually from a reservoir upstream via a hydroelectric turbine that produces power. The above scheme supports energy storage for long durations whenever required. The installed capacity of electricity generated via PHS is approximately 165 Giga Watts (GW) worldwide [6]. Central Electricity Authority (CEA) reports a potential of more than 96.5 GW as of date. About 9 plants in India operate under PHS to yield about 4.8 GW. In addition to the above India plans to expand its plants at Tehri and Koyna, which are under progress, and Kundah, Malshej Ghat, Humbali, and Turg are in the planning stage. As of 2022, 28.75% accounts for hydroelectric schemes in operation, and 7.63% (110,835 MW) are under execution [7]. Loss of agricultural plains, forest lands, and forced relocation commonly leads to social and political issues that affect delays in PHS projects. Other reasons would be seasonal water flows, remote locations, and capital with engineering technology is undoubtedly challenged. Against the odds, pumped hydro storage is still considered the most common technology due to its high maturity, reliability, and influential contribution to integrating renewables into power systems by achieving optimal operation in combinations [8].

2.1.2 Compressed Air Energy Storage (CAES)

Air found in underground caverns, high-pressure containers, salt caverns, complex rock caverns, or saline aquifers, which are in a very highly compressive state is converted into electrical energy and is usually referred to as CAES [9]. In addition, offshore and underwater storage systems are emerging research areas [10]. By using the expansion turbine, the collected pressurized air temperature is elevated to the required to generate power when in demand. CAES systems can be categorized into diabatic, adiabatic, and isothermal CAES. Recent advancements are happening in integrated CAES systems to recover waste heat, reduce energy destruction, and have higher capacities with longer longevity [11]. Apart from natural gas, any fuel virtually can be used for combustion by using an air expander with an external combustion chamber and a bypass making it environmentally friendly is one of the latest innovative solutions [12].

2.1.3 Flywheel Energy Storage (FES)

Mechanical rotational energy of a heavy magnitude is sourced for FES [13]. FES systems usually comprise a sizeable rotational cylinder upheld by a stator. The squirrel cage inductor machine (SCIM), the double-fed inductor machine (DFIM), and the permanent magnet synchronous machine (PMSM) are added to increase efficiency and help in the implementation of its control process. FES requires a little work to keep in good condition and lower carbon footprint. Globally, 975 MW is the current grid-scale capacity. The use of composite materials, especially carbon fiber, has increased 5X in contrast with steel flywheels [14], and more electromagnetic and superconducting variants are in current ongoing research. In recent work, a full advance, based on synergetic theory control of PMSM-FES unified with a hybrid PV-wind system is established to ensure more stable electrical power production [15].

2.2 *Electrochemical Storage Energy (ECSE)*

ECSE incorporates various battery technologies by using the reactivity nature of chemical compounds, which could release or absorb energy, which can be turned into electric power. Electrochemical energy is typically based on chemical reactions which largely consists of the electrode and electrolyte materials and the various research organization is performing basic-level studies to optimize the electrochemistry, and derive composites and assemblies [16]. It is estimated that a total of 1TWh globally installed storage capacity is required to run the application in both stationary and mobile sectors, adding to further challenges [17]. Researchers have rapidly developed many ECSEs expedients like lithium-metal, lithium-ion batteries, supercapacitors and zinc-ion batteries, multivalent-ion batteries, metal-sulfur/air batteries, etc., [18], and many more combinations to improve energy and power density, longevity, and fast charging capabilities and to overcome security flaws [19], reducing the thermal runaway [20] to alleviate the huge ingesting of fossil fuels and mounting claim for energy globally [21]. New approaches are needed to overcome future requirements with novel solutions that can rush research to build more efficient materials. In today's technologically driven world, batteries play a crucial role in powering our devices and enabling our everyday activities. From smartphones to electric vehicles, the ability to store and discharge energy efficiently is essential. Two prominent types of battery technologies, solid rechargeable batteries and flow batteries have emerged as key contenders in the quest for efficient energy storage. Apart from the above two, a second energy carrier is also in research focusing on storage and energy carriers. These systems can support grid power, transportation, and for extensive needs [22]. We have gone from the era of lead-sulfuric acid batteries to the latest innovation in fuel cells stacked like batteries. In recent years, continuous effort has been focused on rational materials and electrode designs to improve the performance and conversion of devices.

2.2.1 Rechargeable Batteries

Current state-of-the-art research is happening around rechargeable batteries, to study battery interface genome, self-healing batteries, improving battery health, optimizing and manufacturability of batteries, and recyclability of the batteries where they find numerous applications coming onto the marketplace from EVs, flying drones, biomedical grafts, etc., [23]. An attempt is made to elaborate on the latest happenings from the history of rechargeable batteries to the modern era.

2.3 Lead-Acid Batteries (LAB)

The LAB has made significant contributions to the development of batteries and is widely recognized as a top electrochemical energy storage system. Extensively seen its applications in starters, electric drives, and stationary batteries for the quickest emergency need of electricity. High-performance variations of LABs can be studied under advanced lead acid which has longer longevity with better performance.

2.4 Advanced Lead-Acid Batteries (ALAB)

There are typically two classifications for ALABs, namely lead carbon type which uses carbon additives, and the bipolar lead acid type to increase the efficiency of LABs. Nowadays, carbon finds use in LAB as an additive to the negative active mass to improve its electrochemical properties, use of carbon in the current collectors is increasing their low specific energy and finds vast applications in solar farms [24].

2.5 Bipolar Lead-Acid Battery (BLAB)

The chemistry involved with a BLAB is like the conventional lead-acid battery. Both these types, involve the participation of lead compounds in reversible electrochemical reactions on the positive and negative plates of the cell. These electrochemical reactions involve charge exchange reactions, specifically reduction–oxidation reactions. In a bipolar design, the battery utilizes bipolar substrates. These substrates are electronically conducting and serve to join the active materials of opposite polarities within the battery. The bipolar substrates help ensure efficient electrical conductivity and facilitate the movement of charge between the two opposite polar plates. By utilizing bipolar substrates, the bipolar lead-acid battery design enhances the overall performance and efficiency of the battery. This design allows for a more compact and lightweight battery, making it suitable for various applications where space and weight are critical factors. The use of bipolar substrates in the bipolar design improves the battery's efficiency and overall performance [25]. In addition to having higher specific energy and energy density, this type also has a greater footprint reduction and recyclable materials. Recently, graphite/polytetrafluoroethylene emulsion has been used as a composite electrode for bipolar plates to improve electrolyte diffusion channels [26].

2.6 Lithium-Ion Batteries (LiB)

Unlike traditional batteries, LiB's are advantageous as they have the unique characteristic of being light-weighted with higher energy density potential. Their portability makes them ideal for portable applications such as EVs and E-devices. LiB stands for lithium-ion battery, and it falls into two categories: systems with metallic lithium and systems without. In systems containing metallic lithium, the Li ions can solidly embed in electrodes. There are several variations of this type of LiB, depending on the internal chemistry [27]. Some examples include:

- Lithium Cobalt Oxide (LCO)
- Lithium Titanate Oxide (LTO)
- Lithium Manganese Oxide (LMO)
- Lithium Iron Phosphate (LFP)
- Lithium Nickel Manganese Cobalt Oxide (NMC)
- Lithium Nickel Cobalt Aluminum Oxide (NCA).

LiBs have witnessed rapid progress in research and development in recent years, which has led to a vertical fall in prices due to production scale-up. One key advantage of LiB's over lead-acid batteries is the increasing energy densities, which means they can store more energy in a smaller and lighter package. Refer to Fig. 3 [28] for a Ragone plot which compares dissimilar rechargeable batteries like lithium-based (e.g., lithium-ion) and nickel-based (e.g., nickel-metal hydride) batteries based on their specific energy and specific power. Lithium-based batteries excel in specific energy, ideal for energy-dense applications like smartphones and electric vehicles. In contrast, nickel-based batteries offer higher specific power, suited for quick bursts of energy, as in power tools and hybrid cars. The plot helps decision-makers choose the right battery for their application, considering trade-offs between energy and power.

Rechargeable LiBs store energy by allowing Li ions to be inserted and removed from functioning materials through an electrochemical redox reaction known as intercalation. In recent years, there have been developments in using teardrop-shaped tin phosphide particles to increase energy densities in LiBs. These particles are amalgamated from the reaction of tin acetate in a mixed solution of trioctyl phosphine and trioctylphosphine oxide. By incorporating these particles into the LiB, it is possible to improve the energy storage capacity and overall performance of the battery [29]. Graphite-encapsulated Fe_3O_4 composites in lithium dual-ion batteries have helped enhance electrochemical performance [30]. Attempts are made to under the latest automotive Li-ion chemistries such as $LiMn_2O_4$ (LMO), $LiFePO_4$ (LFP), $LiNi_{0.8}Co_{0.15}Al_{0.05}O_2$ (NCA), and layered $LiNi_{1-x-y}Mn_y Co_zO_2$ (NMC), Ni-rich NMC cathode materials, silicon-based anode materials and innovative battery chemistries including solid-state batteries are next-generation upcoming EV applicable systems [31]. The electric vehicle industry has garnered a lot of attention, resulting in a surge in demand for lithium batteries. It is projected that the worldwide capacity for lithium batteries will quadruple by the end of the decade.

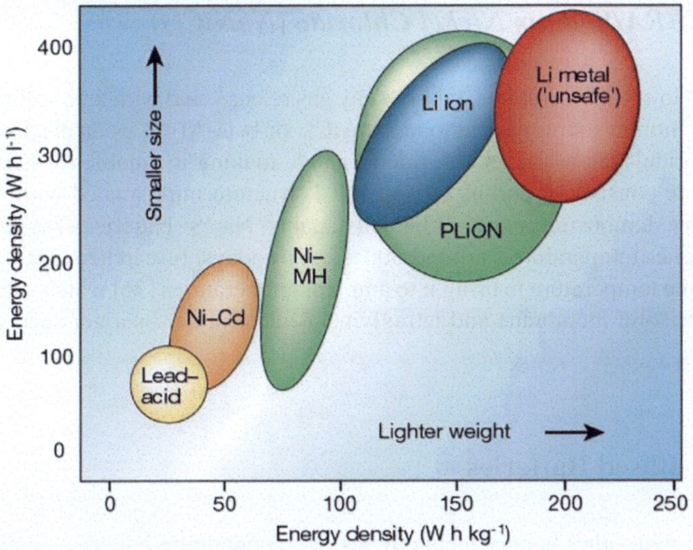

Fig. 3 Ragone plot for dissimilar rechargeable batteries

3 High-Temperature Sodium (HTS)

HTS batteries propose all-solid-state batteries consisting of electrolytes of inorganic or polymer matter above 300 °C showing longer life cycles, higher thermal stability, and reliability [32]. Ceramic or polymer electrolytes are possible, solid electrolytes because of their properties of high elastic moduli of ceramics and low for polymers [33]. Beta-alumina to late NASICON-type glass and glass ceramics have been studying electrolytes over the last decades under the HTS systems. Broadly HTS be classified into sodium sulfur (NaS) and sodium nickel chloride ($NaNiCl_2$), elaborated below.

3.1 Sodium Sulfur (NaS)

It is assembled with molten mass an anode and liquid 'S' as a cathode bounded inside a cylindrical cell and enclosed by a steel casing made up of sodium beta-alumina tube as a solid electrolyte. On discharge, sodium-polysulfides are formed, and reactions get switched back during charge [34]. There are currently 200 installations of NaS batteries available for grid applications globally, which account for a total of 315 Mega Watt. These batteries are mainly used for stationary storage purposes [35]. NTPC has tested the feasibility of NaS batteries in Indian grid solar conditions as per the reports of the Intelligent Grid India Forum (ISGF).

3.2 *ZEBRA/Sodium Nickel Chloride ($NaNiCl_2$)*

It works at lower temperatures, without any gas release, and with zero self-discharge [26], with molten sodium as cathode, $NaAlCl_4$ or beta-Al_2O_3 as an electrolyte, and metal chloride such as nickel chloride as anode making it suitable for hybrid EVs. The cathode consists of porous Ni and NaCl structure impregnated with $NaAlCl_4$. This battery demonstrates several benefits against Na-'S' batteries. These batteries require higher temperatures, around 300 °C. Nowadays, researchers are attempting to reduce the temperature to bring it to ambient temperatures [36] which uses a dense beta Al_2O_3 solid membrane and tetraglyme as electrolyte separators and catholyte solvents.

4 Zinc-Based Batteries

Compared with other battery technologies, zinc-containing batteries combine zinc with a variety of chemicals to produce a battery that is younger in development. The types of zinc batteries differ from one another according to the composition of their cathodes and their electrolytic charge carriers. In the realm of zinc batteries, there are various classifications that are determined by the materials utilized for the cathode and the electrolytes that function as charge carriers. This distinction is crucial as it impacts the overall performance and efficiency of the battery. Understanding the different types of zinc batteries can aid in selecting the appropriate battery for specific applications. The zinc-based chemistry previously used in zinc batteries could not be recharged, but manufacturers are currently working on the development of fully rechargeable zinc-based chemistry (RZBs). Attempts were made to understand systems which include mild electrolyte RZBs and alkaline RZBs, Zn-ion, and Zn-air capacitors/batteries [37]. This advanced technological advancement boasts a low carbon footprint making it an apt battery.

4.1 *Zinc Air Battery*

The zinc air fuel cells, which are also called zinc oxide fuel cells (ZAFs) are designed to oxidize zinc with oxygen through the radical oxidation of polyacrylamide (PAM) hydrogel, and the rate at which the reaction takes place is controlled by adjusting airflow. Many high-performance non-precious metals-based oxygen reduction reaction (ORR) electrocatalysts have been studied and developed recently, and much research is done on CO_2 management [38].

4.2 Zinc Manganese Battery

Rechargeable $ZnMnO_2$ batteries have recently been developed with a 2–8 h discharge duration. Attempts were made to create a highly reversible and stable 2-electron transfer solid–liquid reaction based on MnO_2 and soluble Mn $(CH_3COO)_2$ $(Mn(AcO_2))$ under a neutral medium showing CE of 99% and EE of 78% at 40 $mAcm^{-2}$ [39]. As a safer, nontoxic replacement to lithium batteries, these batteries contain no lead, heavy metals, or flammable electrolytes, making them a more cost-effective alternative to lithium batteries.

5 Flow Batteries

To attain carbon-zero goals, renewable energy utilization has become necessary. This requires an efficient storage medium for a stable energy supply of the sporadic nature of renewable energy sources [40]. Flow batteries are one such proficient energy storage device. A timeline of the development of flow batteries is provided in Fig. 4 [41]. It depicts the flow battery technology, which began in the late 1960s, has evolved significantly. It encompasses aqueous and non-aqueous systems and hybrid configurations. Aqueous flow batteries, like the VRB, excel in grid-scale energy storage due to reliability. Non-aqueous options, such as organic flow batteries, offer higher energy density and cost-effectiveness. Hybrid flow battery systems, combining aqueous and non-aqueous chemistries, are gaining traction. They promise optimized performance for large-scale energy storage, especially in renewable integration. Key moments include Tesla's Powerwall in 2015 and Cambridge's quinone-based flow battery in 2017. Ongoing research drives advancements in these diverse flow battery technologies. Flow batteries are distinct from traditional batteries as they store energy within the electrolyte. Canisters contain electrolyte solutions that flow within a shared compartment. This compartment is separated by a membrane that enables the movement of electrons, which allows for electricity to pass between the electrolytes. There are several types of flow batteries available in the market today, such as vanadium redox flow batteries, zinc-iron flow batteries, and zinc-bromine batteries. Ongoing research and development have resulted in the creation of new types of flow batteries, such as zinc-iron and hydrogen-bromine batteries. Zinc batteries have variations based on their cathode materials and electrolytic charge carriers. It is worth noting that the technology of grid-connected flow batteries has been successfully commercialized worldwide, with a total of 326 MW being installed across 108 projects as of 2018. However, in India, this technology is still in its early stages of adoption and is limited to test trials. In 2015, a microgrid powered by a 30 kW vanadium redox battery (VRB) was installed, representing a significant milestone in the country's exploration of this technology. Furthermore, researchers at the Indian Institute of Science in Bangalore are currently developing a new type of flow battery known as the soluble lead-acid flow battery, which shows promising potential.

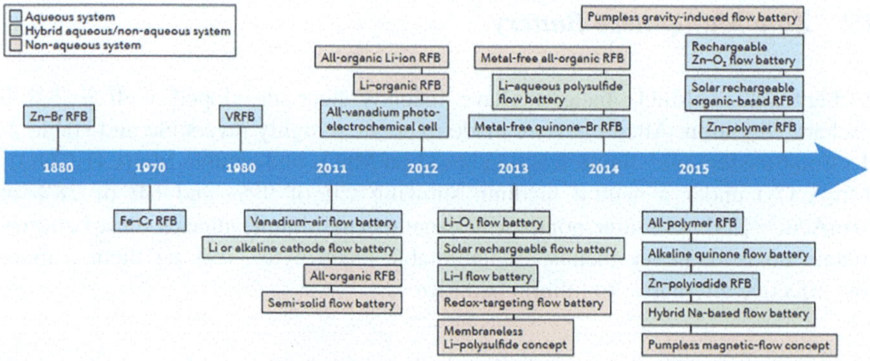

Fig. 4 Timeline of key developments around flow-battery systems

Vanadium redox flow batteries (VRFBs) are one of the types of flow batteries that can be utilized for grid storage applications. Though VRFBs have advantages, it also struggles to meet the desired efficiency [40]. Studies show that improving both the physical and electrochemical characteristics of deep eutectic solvent electrolytes, along with improving rate of reaction, can be applicable in non-aqueous FBs that can have an optimistic effect on oxidation and reduction peaks of deep eutectic solvent electrolyte [42]. Redox flow batteries have also found their application in both electrolysis synthesis and secondary batteries by encompassing a three-layer system with bi-continuous microemulsions as the middle phase [43]. A lot of progress is under research to increase the efficiency of carbon electrodes in vanadium redox flow batteries through thermal activation [44]. Similarly, the stable discharge capacity of aqueous redox flow batteries (ARFBs) is ensured by using a 30 μm thick polybenzimidazole (PBI) membrane that shows excellent performance [45].

5.1 *Aqueous Batteries*

Aqueous batteries (Abs) based on water have minimal application at large scale but have a lot of scope with the advent of advanced materials and emerging electrochemistry [46], leading to the successful fabrication of Ni–Fe, Ni–Cd, and Pd-acid ABs. $LiMn_2O_4/VO_2$ Li AB has shown ~ 55 Wh kg^{-1} energy density built on the notion of aqueous rocking-chair battery chemistries which competed with the lead-acid batteries [47]. The energy densities have increased over the past by the induction of low-cost salts which include $NaClO_4$, $ZnCL_2$, lithium and potassium acetate's which are the forerunners in research exploration under ABs [48].

5.2 *Current Pathways of Research in Rechargeable Batteries*

The batteries research communal is concentrating on many trails of research to substitute critical elements of the batteries with more sustainable solutions with improvements in the chattels of the energy devices. Below are some of the mentioned strategies played around the globe.

1. Looking forward for naturally abundant elements, which lead to the manufacturing of low Co-containing NMC materials, many other relevant systems have been developed based on Co-free cathode having higher capacities which includes Li–Ni–Mn oxide-systems or Li-iron phosphate (LFP), some of them are in commercial stages [47].
2. On the anode front, post-Li-systems are being developed as charge carriers. Recently, Li is getting plated on the current collector itself which in turn is saving lithium metal as a resource [49].
3. Li-chalcogenide combination which includes Li-sulfur [27] or Li-oxygen [49] is being studied under conversion-based energy systems.
4. Liquid electrolyte is replaced by a solid-ion conductor leading to the development of solid-state batteries [50].

The development of solid rechargeable batteries and flow batteries has opened new possibilities for efficient energy storage. As we strive toward a more sustainable future, these battery technologies hold immense promise in various industries, including transportation, renewable energy, and grid-scale energy management. By understanding the composition, working principles, advantages, and applications of these battery technologies, we can better appreciate their potential to revolutionize our daily lives and contribute to a cleaner and more energy-efficient world.

6 Thermal Energy Storage (TES)

Any improvement in areas of thermal energy storage would cut down the huge dependency on thermal energy consumption. Effective thermal energy management involves a range of functions, and one of the most crucial ones is thermal energy storage (TES). TES encompasses a variety of technologies such as ice-based, hot and chilled water, molten salt, and rock storage technologies. These systems operate by storing thermal energy during off-peak hours and then releasing it during peak hours to meet the energy demands of the building or facility. By utilizing TES, energy managers can optimize energy usage, reduce energy costs, and increase the overall efficiency of the building's energy systems in all the above schemes, excess thermal energy is poised together for future demand. TES has advantages over electrochemical storage systems where there is a requirement for low-grade electricity, CSP plants or fissionable reactors would be more advantageous in such situations, whereas TES systems may not be suitable for storing electricity in the form of grid due to lesser round-trip efficiency of 50% [51]. The TES has an improved life cycle and stays longer than the typical cycling-induced degradation across ECSS. Higher latent heat and specific heat with better conductivity are essential thermophysical properties. Other desired properties would include low cost, thermal stability, nontoxic, compatible melting and low flammability, etc., required by TES materials [52]. TES systems based on the types of TES materials can be categories being selected for heat or cold storage, which are briefly described below.

6.1 Sensible Heat Storage (SHT)

The available energy is stored as a temperature within a material without any change in the phase, which can be used whenever required to raise or reduce the temperature of a particular application. The heat stored would be proportional to many factors including density, V, temperature, and material storage properties. Some of these include molten salt storage (integrated with CSP), hot water storage, and chilled water storage for domestic purposes utility. Though the SHT systems have lower specific heat, they still find many applications because of their larger workable temperature range and higher energy densities.

6.2 Hot Water Storage and Chilled Storage/Thermal Oils

Water acts both as heat transfer carrier materials and efficient thermal energy as a source. As the nature of the water can transform into any form like ice-liquid-stream, it can be easily used for home warming, preservation of food and beverages, and hot water usage. Hot stream and thermal oil can also be accumulators in steel tanks for energy storage which can withstand higher pressures [53]. Thermal oils such as non-edible vegetable oils remain liquid even at higher temperatures with lower vapor pressure having a heat transfer coefficient in the range of 1000–3500 $Wm^{-2}\ K^{-1}$ [54]. A lot of work is in progress with adding nanoadditives [55]. Thermal oils have degrading properties at higher temperatures and lead to the corrosion of containers and pipes, which must be addressed.

6.3 Molten Salt Storage

Thermal oils can be replaced with molten salts as these resources can endure higher heat, reaching up to 565 °C [56] and are primarily used in concentrating solar plants and nuclear power plants as a thermal energy storage medium. Pure salts or salt eutectics which include ternary salt mixture and binary salt mixture solar salt are commonly deployed as molten salt storage medium [57]. Apart from the above two storage mediums, attempts are being made to study liquid metals such as Na, lead–bismuth eutectic to be used as coolant medium in nuclear reactors [54, 58]. And Earth metals, such as rocks, sand, and gravel are used as filler materials in thermocline single storage tanks. Quartzite among the rock types and stones, pebbles, and gravels are applied as packing materials in packed beds with high content of SiO_2. Ceramics, dry cement powder, and coal are used in suspended nature to recover the heat emitted from CSP [59].

6.4 Latent Heat Storage (LHS)

Whenever a phase change occurs between solid and liquid materials, this change has the potential to hold the energy and use it when it's in demand. In both the conditions, like from solids to liquid or liquids to solids, and from liquid to gases or gases to liquids, liquid–gas phase change emerges out as the highest latent heat carrier [60]. LHS are 50–100 X times larger than SHS due to the higher energy density. Metals and alloys, salts and their hydrates, organic and inorganic, paraffin and non-paraffin such as glycols, alcohols, esters, and fatty acids are some of the well-known latent heat storage mediums.

6.5 Thermochemical Storage (TCS)

The chemical reaction can be in forward or backward direction which depends upon the given standard characters such as pressure and temperature, these reactions have the potential to store thermal energy. TCS is categorized under two groups; (i) thermochemical sorption storage system applied for low-grade heat applications but can be overcome by using chemical heat pumps and (ii) chemical reaction storage system employed for higher temperature applications. Hydration and dehydration of metal oxide are the most favorable reactions among TCS systems, depending on the reaction temperatures [61]. Nickel oxide, cobalt oxide (200 °C), magnesium oxide (300 °C), strontium oxide and barium oxide (700 °C) [62] are the potential forerunners candidates under TCS systems, which vary depending upon the temperatures applied. TCS systems have an added advantage over SHS and LHS, which can reflect higher potential to store heat without any thermal losses. Silica gel/water, magnesium sulfate/water, lithium bromide/water, lithium chloride/water, and NaOH/water are some of the prominent working pairs [63].

7 Electrical Storage Supercapacitors (SCs) and Superconducting Magnetic Energy Systems (SMES)

SMES has the potential to store electricity in the form of electric and electromagnetic form with minimalist energy losses. The characteristic of energy and power of SCs shows advance performance when compared with conventional capacitors or batteries. SCs find their applications in memory backups, portable tools, distribution and storage, decoupling of power needs, and regeneration devices [64]. Intensive development of new technologies and new materials, such as the hybrid electrode, which combines double-layer capacitance and pseudo-capacitance, is used to increase the number of charges and decrease the discharge cycle [26]. With the advent of nanotech, materials such as graphene, CNT, and Fullerene show promising results in carbon-based electrode supercapacitors [65]. Attempts are underway on perovskite oxides based on lanthanum, strontium, and cerium being researched [66] apart from natural copper minerals, primarily chalcosine [67] and coveline [68]. Conductive polymers containing double bonds are also of great interest, under the pseudocapacitive reactions by increasing their conductivity and porosity [69]. SMES store direct current transmitting magnetic field traveling via a superconducting coil with zero electrical resistance. The current continues to loop continuously until it is needed and discharged. Metal alloys, mercury, vanadium, and niobium-titanium are cooled to close to zero to achieve the superconducting stage. SMES systems are now commercially available and can find applications in microchip fabrication facilities.

8 Chemical Storage (CES)

CES systems store energy in their chemical bonds between atoms and molecules of the materials and release energy whenever these bonds are broken and new one forms when a chemical reaction occurs. According to CES systems, various materials including crude petroleum gas, coal, and gasoline are classified as derivatives. These fuels undergo a process of transformation, where they are converted into mechanical storage energy before being converted into power. Among the advanced CES systems, there is a process called electrolysis of water, which produces H_2 [70], a storage material that can be used for various energy sources and can be used as a transportation fuel. This process of electrolysis is highly efficient and sustainable, making it a promising solution for the future of energy production.

9 Electrolyzes/Power to Gas (PtG)

Power to gas is a prominent opportunity which shows the transformation of an electrical source of energy from the release of chemical conversion of combustible gas. An electrolyzer splits water using electric power into two components O_2 and H_2. Methanation process utilizes hydrogen, which largely depends on the various embedding energy systems. Alkaline electrolysis (AE), proton exchange membrane electrolysis (PEMC), and solid oxide electrolysis (SOE) cells are well-known electrolyte systems. AE mostly often used aqueous alkaline electrolytes to produce hydrogen with a fair conversion of 65–70% with the lowest investment cost but a larger plant footprint. PEM increases the efficiency to 85% which uses a combination of polymeric electrolytic solution, but it is very expensive as it is usually rare metals such as platinum and iridium. SOECs, efforts are made to commercialize these concepts, using ceramics as electrolytes that can handle high-temperature streams as a heat source.

10 Fuel Cells (FC)

Fuel cells fall under the electrochemical segment in which, chemical reaction energy is transformed into electrical energy [71]. This device comprises of electrolytes with the support of anode and cathode electrodes, these devices are classified [72] based on the type of electrolytes and fuels used, Fig. 5 illustrates the design of a stacked fuel cell (FC) configuration, where multiple individual fuel cells are connected in series and separated by bipolar plates to achieve a higher overall power output [71]. These fuel cells can be categorized based on the type of electrolyte they utilize. Stacking fuel cells in series is a common approach to increase the voltage and power output of a fuel cell system, making it suitable for various applications, such as stationary power generation and transportation. The choice of electrolyte type can vary depending on the specific fuel cell technology being used, with different types offering various advantages and suitability for different operating conditions and applications. Proton exchange membrane or polymer electrolyte membrane (PEMFC), high temperature PEMFC, direct methanol fuel cell (DMFC), molten carbonate fuel cell (MCFC), phosphoric acid fuel cell (PAFC), solid oxide fuel cell (SOFC) and alkaline fuel cell (AFC) are some of the well-known fuel cells where extensive research is undergoing. The power of battery technology lies not only in the devices we use daily but also in the potential to shape a brighter and more sustainable future for generations to come.

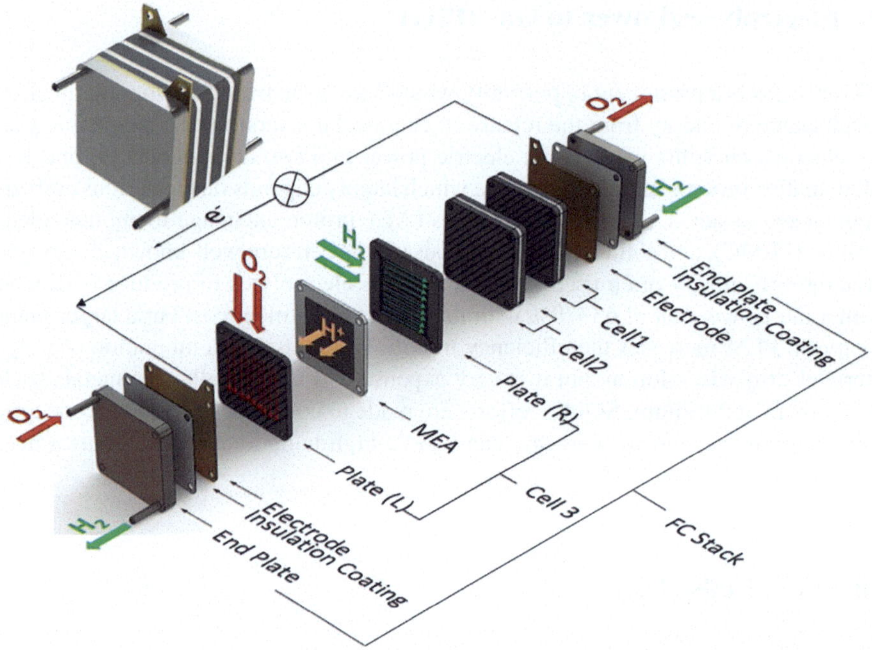

Fig. 5 Fuel cell stack operations and components

References

1. A. Ometov et al., A survey on wearable technology: history, state-of-the-art and current challenges. Comput. Netw. **193**, 108074 (2021)
2. T. Placke, R. Kloepsch, S. Dühnen, M. Winter, Lithium ion, lithium metal, and alternative rechargeable battery technologies: the odyssey for high energy density. J. Solid State Electrochem. **21**, 1939–1964 (2017)
3. G.L. Kyriakopoulos, G. Arabatzis, Electrical energy storage systems in electricity generation: energy policies, innovative technologies, and regulatory regimes. Renew. Sustain. Energy Rev. **56**, 1044–1067 (2016)
4. X. Lu et al., Polyethylene glycol/carbon black shape-stable phase change composites for peak load regulating of electric power system and corresponding thermal energy storage. Eng. Sci. **9**(4), 25–34 (2020)
5. R. Carnegie, D. Gotham, D. Nderitu, P.V. Preckel, Utility-scale energy storage systems. State Util. Forecast. Group. Purdue Univ. **1** (2013)
6. M.S. Guney, Y. Tepe, Classification and assessment of energy storage systems. Renew. Sustain. Energy Rev. **75**, 1187–1197 (2017). https://doi.org/10.1016/j.rser.2016.11.102
7. Central Electricity Authority Ministry of Power Government of India, CEA Annual Report 2021–22 (2022)
8. R.J. Mahfoud, N.F. Alkayem, Y. Zhang, Y. Zheng, Y. Sun, H.H. Alhelou, Optimal operation of pumped hydro storage-based energy systems: a compendium of current challenges and future perspectives. Renew. Sustain. Energy Rev. **178**, 113267 (2023). https://doi.org/10.1016/J.RSER.2023.113267
9. X. Luo, J. Wang, M. Dooner, J. Clarke, C. Krupke, Overview of current development in compressed air energy storage technology. Energy Procedia **62**, 603–611 (2014)

10. J.D. Hunt et al., Isothermal deep ocean compressed air energy storage: an affordable solution for seasonal energy storage. Energies **16**(7), 3118 (2023)
11. A.M. Rabi, J. Radulovic, J.M. Buick, Comprehensive review of compressed air energy storage (CAES) technologies. Thermo **3**(1), 104–126 (2023)
12. K. Kosowski, M. Piwowarski, W. Włodarski, P. Ziemiański, G. Pawlak, Technical and economic analysis of energy storage in the compressed air technology with low capacity for the production plant. Energy Convers. Manag. **282**, 116872 (2023)
13. S. Nemsi, S.A. Belfedhal, L. Barazane, Role of flywheel energy storage system in microgrid. J. Eng. Res. Technol. **3**(3) (2016)
14. R. Sebastián, R.P. Alzola, Flywheel energy storage systems: Review and simulation for an isolated wind power system. Renew. Sustain. Energy Rev. **16**(9), 6803–6813 (2012)
15. M. Beghdadi, K. Kouzi, A. Ameur, Robust fully synergetic control of PMSM-flywheel energy storage system FESS integrated in standalone hybrid PV-wind system. Rom. J. Inf. Technol. Autom. Control **33**(1), 67–80 (2023)
16. S.P.S. Badwal, S.S. Giddey, C. Munnings, A.I. Bhatt, A.F. Hollenkamp, Emerging electrochemical energy conversion and storage technologies. Front. Chem. **2**, 79 (2014)
17. Z. Zhongming, L. Linong, Y. Xiaona, Z. Wangqiang, L. Wei, Tight battery market is next test for EVs after chip crisis (2021)
18. J.B. Goodenough, Energy storage materials: a perspective. Energy Storage Mater. **1**, 158–161 (2015)
19. L. Kong, Y. Li, W. Feng, Strategies to solve lithium battery thermal runaway: from mechanism to modification. Electrochem. Energy Rev. **4**(4), 633–679 (2021)
20. X. Feng, M. Ouyang, X. Liu, L. Lu, Y. Xia, X. He, Thermal runaway mechanism of lithium ion battery for electric vehicles: a review. Energy Storage Mater. **10**, 246–267 (2018)
21. M. Zhong, M. Zhang, X. Li, Carbon nanomaterials and their composites for supercapacitors. Carbon Energy **4**(5), 950–985 (2022)
22. S.T. Revankar, Chemical energy storage, in *Storage and Hybridization of Nuclear Energy* (Elsevier, 2019), pp. 177–227
23. M. Fichtner et al., Rechargeable batteries of the future—the state of the art from a BATTERY 2030+ perspective. Adv. Energy Mater. **12**(17), 2102904 (2022)
24. J. Lach, K. Wróbel, J. Wróbel, P. Podsadni, A. Czerwiński, Applications of carbon in lead-acid batteries: a review. J. Solid State Electrochem. **23**, 693–705 (2019)
25. S.K. Pradhan, B. Chakraborty, Substrate materials and novel designs for bipolar lead-acid batteries: a review. J. Energy Storage **32**, 101764 (2020)
26. R. Lakra, R. Kumar, P.K. Sahoo, D. Thatoi, A. Soam, A mini-review: Graphene based composites for supercapacitor application. Inorg. Chem. Commun. **133**, 108929 (2021)
27. N. Nitta, F. Wu, J.T. Lee, G. Yushin, Li-ion battery materials: present and future. Mater. today **18**(5), 252–264 (2015)
28. J.-M. Tarascon, M. Armand, Issues and challenges facing rechargeable lithium batteries. Nature **414**(6861), 359–367 (2001)
29. Y. Kim, H. Hwang, C.S. Yoon, M.G. Kim, J. Cho, Reversible lithium intercalation in teardrop-shaped ultrafine SnP0. 94 particles: an anode material for lithium-ion batteries. Adv. Mater. **19**(1), 92–96 (2007)
30. T.T. Salunkhe, A.N. Kadam, J. Hur, I.T. Kim, Green and sustainably designed intercalation-type anodes for emerging lithium dual-ion batteries with high energy density. J. Energy Chem. **80**, 466–478 (2023)
31. Y. Ding, Z.P. Cano, A. Yu, J. Lu, Z. Chen, Automotive Li-ion batteries: current status and future perspectives. Electrochem. Energy Rev. **2**, 1–28 (2019)
32. K.B. Hueso, M. Armand, T. Rojo, High temperature sodium batteries: status, challenges and future trends. Energy Environ. Sci. **6**(3), 734–749 (2013)
33. V. Palomares, P. Serras, I. Villaluenga, K.B. Hueso, J. Carretero-González, T. Rojo, Na-ion batteries, recent advances and present challenges to become low cost energy storage systems. Energy Environ. Sci. **5**(3), 5884–5901 (2012)

34. B. Dunn, H. Kamath, J.-M. Tarascon, Electrical energy storage for the grid: a battery of choices. Science (80-.). **334**(6058), 928–935 (2011)
35. M.M. Thackeray, C. Wolverton, E.D. Isaacs, Electrical energy storage for transportation—approaching the limits of, and going beyond, lithium-ion batteries. Energy Environ. Sci. **5**(7), 7854–7863 (2012)
36. X. Lu et al., Advanced intermediate-temperature Na–S battery. Energy Environ. Sci. **6**(1), 299–306 (2013)
37. C. Xie, H. Zhang, W. Xu, W. Wang, X. Li, A long cycle life, self-healing zinc–iodine flow battery with high power density. Angew. Chemie **130**(35), 11341–11346 (2018)
38. Y. Li, H. Dai, Recent advances in zinc–air batteries. Chem. Soc. Rev. **43**(15), 5257–5275 (2014)
39. C. Xie, T. Li, C. Deng, Y. Song, H. Zhang, X. Li, A highly reversible neutral zinc/manganese battery for stationary energy storage. Energy Environ. Sci. **13**(1), 135–143 (2020)
40. R.K. Sankaralingam, S. Seshadri, J. Sunarso, Parametric optimisation using data model to improve the energy efficiency of vanadium redox flow batteries. J. Energy Storage **64** (2023). https://doi.org/10.1016/j.est.2023.107163
41. M. Park, J. Ryu, W. Wang, J. Cho, Material design and engineering of next-generation flow-battery technologies. Nat. Rev. Mater. **2**(1), 1–18 (2016)
42. L. Lu et al., Synergistic effect of ultrasonic and magnetic fields on the performance of non-aqueous iron-vanadium redox flow battery with deep eutectic solvent electrolyte. J. Energy Storage **63** (2023). https://doi.org/10.1016/j.est.2023.107036
43. K. Nakao et al., Electrochemistry in bicontinuous microemulsions derived from two immiscible electrolyte solutions for a membrane-free redox flow battery. J. Colloid Interface Sci. **641**, 348–358 (2023). https://doi.org/10.1016/j.jcis.2023.03.060
44. K. Köble et al., Multimodal characterization of carbon electrodes' thermal activation for vanadium redox flow batteries. J. Power Sources **569** (2023). https://doi.org/10.1016/j.jpowsour.2023.233010
45. G. Park, S. Eun, W. Lee, D. Henkensmeier, Y. Kwon, Polybenzimidazole membrane based aqueous redox flow batteries using anthraquinone-2,7-disulfonic acid and vanadium as redox couple. J. Power Sources **569** (2023). https://doi.org/10.1016/j.jpowsour.2023.233015
46. D. Chao et al., Roadmap for advanced aqueous batteries: from design of materials to applications. Sci. Adv. **6**(21), eaba4098 (2020)
47. W. Li, J. R. Dahn, D.S. Wainwright, Rechargeable lithium batteries with aqueous electrolytes. Science (80-.). **264**(5162), 1115–1118 (1994)
48. D. Kundu, B.D. Adams, V. Duffort, S.H. Vajargah, L.F. Nazar, A high-capacity and long-life aqueous rechargeable zinc battery using a metal oxide intercalation cathode. Nat. Energy **1**(10), 1–8 (2016)
49. T. Liu, J.P. Vivek, E.W. Zhao, J. Lei, N. Garcia-Araez, C.P. Grey, Current challenges and routes forward for nonaqueous lithium–air batteries. Chem. Rev. **120**(14), 6558–6625 (2020)
50. A.J. Louli, M. Coon, M. Genovese, J. deGooyer, A. Eldesoky, J.R. Dahn, Optimizing cycling conditions for anode-free lithium metal cells. J. Electrochem. Soc. **168**(2), 20515 (2021)
51. H. Singh, R.P. Saini, J.S. Saini, A review on packed bed solar energy storage systems. Renew. Sustain. Energy Rev. **14**(3), 1059–1069 (2010). https://doi.org/10.1016/j.rser.2009.10.022
52. G. Alva, L. Liu, X. Huang, G. Fang, Thermal energy storage materials and systems for solar energy applications. Renew. Sustain. Energy Rev. **68**, 693–706 (2017)
53. E. González-Roubaud, D. Pérez-Osorio, C. Prieto, Review of commercial thermal energy storage in concentrated solar power plants: steam vs. molten salts. Renew. Sustain. Energy Rev. **80**, 133–148 (2017)
54. H. Benoit, L. Spreafico, D. Gauthier, G. Flamant, Review of heat transfer fluids in tube-receivers used in concentrating solar thermal systems: Properties and heat transfer coefficients. Renew. Sustain. Energy Rev. **55**, 298–315 (2016)
55. T.T. Beyene et al., Concentrated dual-salt electrolyte to stabilize Li metal and increase cycle life of anode free Li-metal batteries. J. Electrochem. Soc. **166**(8), A1501–A1509 (2019)
56. R. Jacob, M. Belusko, A.I. Fernández, L.F. Cabeza, W. Saman, F. Bruno, Embodied energy and cost of high temperature thermal energy storage systems for use with concentrated solar power plants. Appl. Energy **180**, 586–597 (2016)

57. A. Gil et al., State of the art on high temperature thermal energy storage for power generation. Part 1—concepts, materials and modellization. Renew. Sustain. Energy Rev. **14**(1), 31–55 (2010)
58. J. Pacio, A. Fritsch, C. Singer, R. Uhlig, Liquid metals as efficient coolants for high-intensity point-focus receivers: implications to the design and performance of next-generation CSP systems. Energy Procedia **49**, 647–655 (2014)
59. H. Zhang et al., Particle circulation loops in solar energy capture and Storage: gas–solid flow and heat transfer considerations. Appl. Energy **161**, 206–224 (2016)
60. B. Cárdenas, N. León, High temperature latent heat thermal energy storage: phase change materials, design considerations and performance enhancement techniques. Renew. Sustain. Energy Rev. **27**, 724–737 (2013)
61. S.T. Kim, H. Takasu, Y. Kato, Thermal energy storage in the transport sector, in *Advances in Thermal Energy Storage Systems* (Elsevier, 2021), pp. 717–723
62. C. Milone, Y. Kato, E. Mastronardo, Thermal energy storage with chemical reactions. Recent Adv. Mater. Syst. Therm. Energy Storage An Introd. Exp. Charact. Methods 15–32 (2019)
63. S. Kalaiselvam, R. Parameshwaran, *Thermal Energy Storage Technologies for Sustainability: Systems Design, Assessment and Applications* (Elsevier, 2014)
64. J. Castro-Gutiérrez, A. Celzard, V. Fierro, Energy storage in supercapacitors: focus on tannin-derived carbon electrodes. Front. Mater. **7**, 217 (2020)
65. A.G. Olabi, M.A. Abdelkareem, T. Wilberforce, E.T. Sayed, Application of graphene in energy storage device–a review. Renew. Sustain. Energy Rev. **135**, 110026 (2021)
66. Y. Cao et al., Recent advances in perovskite oxides as electrode materials for supercapacitors. Chem. Commun. **57**(19), 2343–2355 (2021)
67. Z. Stević, M. Rajčić-Vujasinović, Chalcocite as a potential material for supercapacitors. J. Power. Sources **160**(2), 1511–1517 (2006)
68. M. Rajčić-Vujasinović, Z. Stević, S. Bugarinović, Electrochemical characteristics of natural mineral covellite. Open J. Met. **2**(3), 60–67 (2012)
69. I. Shown, A. Ganguly, L. Chen, K. Chen, Conducting polymer-based flexible supercapacitor. Energy Sci. Eng. **3**(1), 2–26 (2015)
70. A.K. Worku et al., Energy storage technologies; recent advances, challenges, and prospectives, in *Planning of Hybrid Renewable Energy Systems, Electric Vehicles and Microgrid: Modeling, Control and Optimization* (Springer, 2022), pp. 125–150
71. J. Larminie, A. Dicks, M.S. McDonald, *Fuel Cell Systems Explained*, vol. 2 (J. Wiley Chichester, UK, 2003)
72. U. Lucia, Overview on fuel cells. Renew. Sustain. Energy Rev. **30**, 164–169 (2014)

Circular Economy and the Recycling of E-Waste

Mohammad Mahdi Sarkhoshkalat, Ali Afkham, Majid Bonyadi Manesh, and Maryam Sarkhosh

Abstract E-waste, also referred to as electronic waste, is a term used to describe discarded or obsolete electronic devices and equipment. It includes a wide range of electronic items such as computers, laptops, smartphones, televisions, printers, cameras, and other consumer electronics. E-waste also encompasses electronic components, circuit boards, cables, and accessories. The proliferation of technology and rapid advancement in electronic devices has led to a significant surge in global e-waste production. The constant evolution of electronic products, driven by technological progress, results in their frequent replacement, raising valid environmental concerns. The management and disposal of e-waste pose substantial environmental and health hazards. Hazardous elements like lead, mercury, cadmium, brominated flame retardants, and several others are present in various electronic devices. These hazardous components can leach into the soil, water, and air when e-waste is discarded in landfills or incinerated, causing ecological harm and posing risks to public health. In response to these challenges, numerous countries have enacted regulations and initiatives to foster responsible e-waste management. The core pillars of effective e-waste management involve recycling and proper disposal. Recycling processes for electronic devices encompass the recovery of valuable metals (such as gold, silver, and copper), polymers, and glass. These actions are pivotal in ensuring efficient e-waste management, mitigating environmental impact, and promoting sustainable practices. The extraction and processing of raw materials can be minimized by using these resources in the production of new electrical devices.

Additionally, some organizations and manufacturers have established e-waste collection programs and take-back initiatives to ensure the safe disposal of electronic devices. These initiatives strive to avert the deposition of e-waste in landfills and the

M. M. Sarkhoshkalat
Department of Mechanical Engineering, Islamic Azad University, Mashhad, Iran

A. Afkham · M. Sarkhosh (✉)
Department of Environmental Health Engineering, School of Health, Mashhad University of Medical Sciences, Mashhad, Iran
e-mail: marya.sarkhosh@yahoo.com

M. Bonyadi Manesh
Department of Oral and Maxillofacial Prosthodontics, School of Dentistry, Mashhad University of Medical Sciences, Mashhad, Iran

K. Kasinathan et al. (eds.), *New Technologies for Energy Transition Based on Sustainable Development Goals*, https://doi.org/10.1007/978-981-97-2527-4_16

illicit exportation of such waste to developing nations with lenient environmental standards. Additionally, endeavors are underway to enhance public consciousness regarding the significance of e-waste recycling, alongside advocating for the creation of electronic products characterized by prolonged usability and simplified recyclability. This includes encouraging manufacturers to adopt environmentally friendly practices, such as using fewer hazardous materials, designing products for easy disassembly and recycling, and providing repair options to extend the lifespan of electronic devices. In conclusion, e-waste refers to discarded electronic devices and equipment. Proper management, recycling, and responsible disposal of e-waste are crucial to minimize environmental damage and health risks associated with hazardous materials present in electronic devices. The notion of a circular economy presents a hopeful blueprint for tackling these predicaments and advocating for sustainable methodologies. Within this context, this chapter delves into the interconnection between e-waste and the circular economy, accentuating the conceivable advantages and pivotal factors entailed in executing a circular strategy for the management of electronic waste.

1 Introduction

1.1 Generation and Consequences of E-Waste

Electronic waste, often referred to as "e-waste," encompasses discarded electronic devices such as computers, smartphones, televisions, and appliances. The proliferation of technology and escalating consumer demand for electronic products has propelled e-waste into a substantial global predicament. The generation of electronic waste has surged at an alarming rate on a worldwide scale. In 2019, the Global E-Waste Monitor 2020 [1] reported a global e-waste production of 53.6 million metric tonnes, and this figure is projected to escalate further to 74.7 million metric tonnes by 2030. Predominantly technology-consuming nations, including the USA, China, and various European countries, serve as the primary generators of e-waste. The improper management and disposal of e-waste have dire environmental implications. Hazardous substances like lead, mercury, cadmium, and brominated flame retardants, coupled with various toxic chemicals, are present within e-waste. Incorrect handling of these materials can lead to contamination of soil, water bodies, and the atmosphere, posing substantial threats to ecosystems and human well-being. Moreover, e-waste contributes to the depletion of natural resources. This is due to the fact that electronic devices harbor valuable metals and rare materials, which are often left unrecovered.

1.2 Trends and Statistics in E-Waste Generation

Electrical and electronic equipment (EEE) has developed into essential components of our daily lives, enabling us to enjoy a higher standard of living. However, the current methods of manufacturing, using, and disposing of e-waste are unsustainable and unacceptable. According to the United Nations University's Electronic Waste Management Outlook 2021 report [1], global e-waste production is projected to have surged to 53.6 million metric tonnes in 2019, with estimations foreseeing a further increase to 74.7 million metric tonnes by 2030. As per the report, the Asia–Pacific region led e-waste generation in 2019 with 24.9 million metric tonnes, trailed by the Americas with 13.1 million metric tonnes and Europe with 12 million metric tonnes. The study underscores the immediate necessity for effective e-waste management strategies, as the growth in e-waste volume is surpassing both global population expansion and GDP growth. Additional insights into the origins and handling of global e-waste are offered in the Global E-Waste Monitor 2020, a collaborative publication by the International Telecommunication Union, International Solid Waste Association, and United Nations University [1]. According to the research, primary contributors to e-waste include household appliances, information, and communication technology (ICT) equipment, and medical devices. The study also highlights that a mere 17.4% of e-waste generated in 2019 underwent proper collection and recycling, leaving the majority to be either landfilled, incinerated, or informally recycled, thereby generating significant environmental and health repercussions. In essence, the compiled data and statistics concerning e-waste production accentuate the pressing requirement for holistic and effective e-waste management policies and practices. These measures are crucial for tackling the escalating quantities of discarded electronic devices and appliances and the associated environmental and health ramifications. The COVID-19 pandemic has significantly impacted our reliance on electronics and digital solutions, as people have increasingly used these devices to stay connected with colleagues, healthcare professionals, family, and social circles. The use of smartphones for online education among children has both positive and negative effects. While they gain access to progressive learning, furthermore, they find themselves susceptible to the adverse influences emanating from social media and online gaming. Notably, there was a 4.6% year-on-year escalation in the demand for personal computers and tablets within the European Union, underscoring the confluence of the pandemic's impact and the burgeoning utilization of artificial intelligence (AI). This trend has prompted concerns relating to infringed privacy, diminished physical activity, health-related issues, and decreased social interactions.

As per the latest global e-waste statistics [1], an astounding 53.6 million tons of e-waste were generated globally in 2019, reflecting a substantial 21% upsurge over the preceding five years. The anticipated trajectory is equally alarming, with the worldwide volume of e-waste predicted to nearly double within a mere 16-year span, from 2014 to 2030, reaching a staggering 74 million tons. Consequently, e-waste represents one of the fastest-growing categories of residential waste on a global scale. The hastening pace of product obsolescence, swift technological advancements, shifts in consumer behaviors, and limited options for repairing electrical and electronic equipment (EEE) stand out as key drivers of this trend. In terms of regional contributions, Asia led the pack in e-waste generation for 2019, accounting for 24.9 million tons, trailed by the Americas (13.1 million tons), Europe (12 million tons), Africa (2.9 million tons), and Oceania (0.7 million tons). A mere 17.4% (equivalent to 9.3 million tons) of e-waste underwent proper collection and effective management, highlighting a stark contrast with the remaining 82.6% (approximately 44.3 million tons) that encountered improper disposal methods or exhibited geographically disparate impacts on the environment and human well-being. E-waste, a latent resource, conceals an array of unexplored raw and refined commodities, including precious metals like platinum, base metals, plastics, and other non-metallic components. These valuable constituents hold an estimated value of around USD 65 billion, akin to the gross domestic product (GDP) of numerous nations. Regrettably, rather than being systematically processed within formal facilities to recover these resources as secondary raw materials and protract their life cycle, a substantial portion of e-waste finds itself subjected to open dumping or incineration. The COVID-19 pandemic has underscored our increased reliance on electronics and digital solutions for connectivity and functionality (Fig. 1).

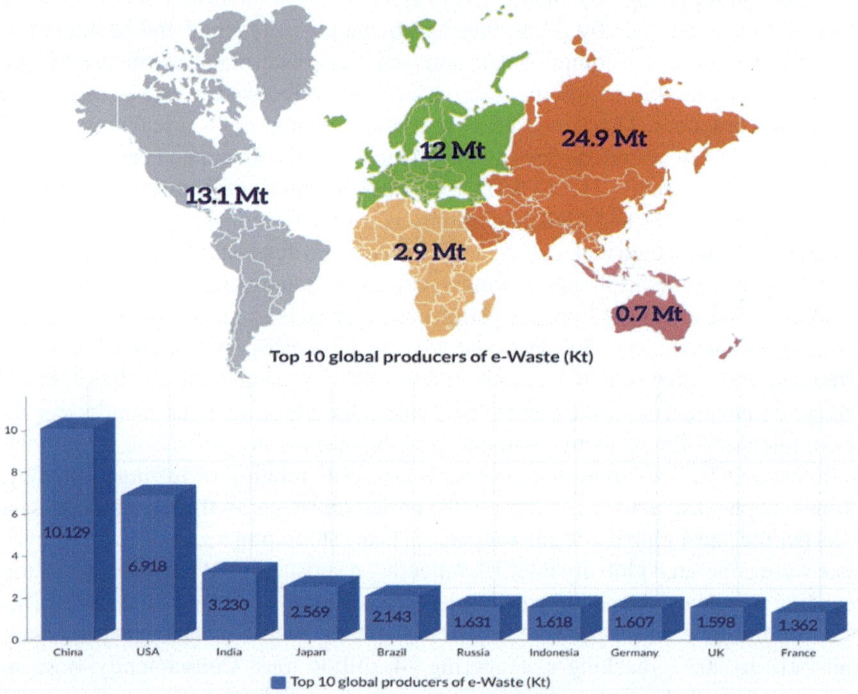

Fig. 1 Information about the largest global manufacturers of electronic waste [2]

Approximately 12.9 million women engaged in the informal sector are at risk of exposure to e-waste due to its hazardous nature, as highlighted in the inaugural report by the World Health Organization (WHO) on this matter [3]. This peril extends to their unborn children, compounding health hazards for both generations. Adding to this concern, the realm of informal waste processing employs an estimated 18 million minors and adolescents, some as young as five years old. Adults and adolescents, particularly susceptible to the dangers of e-waste exposure, face potential risks from toxic gases, particulate matter inhalation, corrosive substances, chemicals contacting their skin, and the consumption of contaminated food and beverages. Such exposures can lead to a variety of health repercussions, potentially compromising their overall metabolic well-being. In order to protect the millions of lives that are at risk, especially those of children, teenagers, and expectant mothers who are most vulnerable as a result of the informal management of e-waste, immediate and effective obligatory interventions are needed. Natural resource depletion is happening at a frightening rate. The development of secondary production techniques through urban mining is necessary due to the dearth of metals extracted from the Earth's core. The finished products can be reintroduced into the supply chain by putting in place a methodical process and strict regulations, which will result in a circular economy that is sustainable. Utilizing recycled materials and reducing dependency on new resources are the goals of this strategy (Table 1).

To conclude, it is of utmost importance to prioritize the monitoring of e-waste quantity and movement, as this plays a crucial role in safeguarding our natural resources and nurturing a sustainable society. Ongoing monitoring not only enables informed decision-making but also pinpoints avenues for enhancement, guaranteeing the enforcement of suitable strategies for environmentally responsible e-waste management.

Table 1 Natural resource depletion rate as a result of new mining [2]

Valuable metal	Amount accessible in subterranean deposits (tons)	Yearly utilization rate (tons)	Exhaustion interval (years)	Principal geographic origins
Gold	50,000	4400	11	South Africa, USA, Indonesia
Silver	503,000	33,000	15	Peru, Australia, Poland
Platinum	69,000	252	19	South Africa, Russia, Zimbabwe
Copper	870,000,000	28,000,000	31	Chile, Mexico, USA, Indonesia

1.3 Ecological and Health Consequences of E-Waste

Electrical and electronic equipment (EEE) encompasses diverse components, a significant portion of which contain hazardous substances. If not handled appropriately, these substances can pose threats to both human health and the environment [4]. The improper treatment of e-waste yields consequences that reverberate beyond immediate health impacts [5]. These repercussions encompass secondary health effects resulting from the consumption of crops cultivated in soil contaminated by e-waste [6, 7]. Illicit dumping engenders groundwater pollution that infiltrates the food chain, thus contributing to the propagation of various diseases. The disposal of metals into rivers can contaminate water sources, undermining their suitability for consumption [8]. Uncontrolled incineration and acid leaching of e-waste discharge toxic emissions into the atmosphere, inflicting respiratory harm and potential fatalities upon workers involved in these processes and local residents alike [9–11]. The atmospheric pollution emanating from e-waste also contributes to far-reaching concerns such as the depletion of the warming global and ozone layer [12, 13]. Moreover, apart from impairing soil fertility and disrupting the food chain, heavy metal poisoning can induce genotoxicity, damaging genetic material and elevating the risk of diseases like cancer. Among those most adversely affected by these illicit practices are children, expectant mothers, and employees working at e-waste processing facilities. The persistence of such unlawful activities in both developing and developed countries can be attributed to an array of factors, including but not limited to cheap labor availability, high illiteracy rates, lax policy enforcement, inadequate data on e-waste generation, deficient formal infrastructure, limited awareness among stakeholders, insufficient individual and societal responsibility, and the prevalence of informal sectors (Table 2).

1.4 Legal and Regulatory Frameworks for E-Waste Management

International policies and legislation play a crucial role in addressing the challenges posed by e-waste through the formulation of guidelines that guide the conduct of stakeholders across public and private sectors [15]. This section will delve into the array of laws implemented to mitigate the repercussions stemming from improper e-waste management. The establishment of the Basel Convention in 1992 marked a significant stride in overseeing the cross-border movement of hazardous waste and ensuring its proper treatment and environmentally responsible disposal [16]. With 175 member states, this convention is designed to shield public health and the environment from the detrimental consequences of hazardous waste generation, illicit transit, and mismanagement [17]. Despite the European Union's pioneering adoption of these provisions within its member nations, the USA has yet to ratify the

Table 2 Environmental effects and potential health risks [14]

Electronic waste element	Approach adopted by informal industries	Possible health risk	Potential ecological consequence
Cathode ray tubes (CRT)	Fracturing of the copper yoke, extracting and subsequently disposing of it in landfills	Silicosis, also referred to as miner's phthisis Cuts from CRT glass Exposure to phosphor containing cadmium or other metals through inhalation or skin contact	The infiltration of heavy metals like lead and barium, among others, into groundwater through leaching, coupled with the discharge of toxic phosphorus
Printed circuit boards (PCBs)	Removing solder and segregating computer chips	Breathing in lead and tin particles Possible inhalation of brominated dioxins, cadmium, mercury, and beryllium	Likely release of the same substances into the atmosphere
Processing of disassembled PCBs	Uncontrolled burning of PC boards	Hazard posed by the inhalation of toxic substances such as tin, lead, brominated dioxins, beryllium, cadmium, and mercury to both employees and residents in the vicinity	Inhalation of brominated dioxins, beryllium, cadmium, and mercury, alongside contamination of the surrounding environment, including surface and groundwater, by tin and lead
Gold-coated materials and additional microchips	Application of nitric and hydrochloric acid for chemical removal of contaminants along riverbanks	Direct contact of acid with the eyes or skin can result in lasting harm Inhaling acid mists, fumes, chlorine, and sulfur dioxide gases can provoke respiratory irritation, leading to severe outcomes like circulatory failure, pulmonary edema, and even fatality	Direct discharge of hydrocarbons, heavy metals, brominated compounds, and additional pollutants into the river and its adjacent areas Inducing river acidification, resulting in the demise of aquatic life and vegetation
Extraction of plastics from computers and peripheral devices	Melting at reduced temperatures and shredding	Individuals residing in the proximity of the incineration operations may have been subjected to hydrocarbons, brominated dioxins, and PAH exposure	Emissions of brominated dioxins, heavy metals, and hydrocarbons into the atmosphere

(continued)

Table 2 (continued)

Electronic waste element	Approach adopted by informal industries	Possible health risk	Potential ecological consequence
Processing of secondary steel, copper, and valuable metals through smelting	Steel or copper is reclaimed from the waste within the furnace	Workers might face exposure to dioxins and heavy metals	Dioxins and heavy metals are discharged into the environment, leading to air contamination
Wires	Utilization of open incineration to recover copper	Individuals residing near the incineration operations may have faced potential exposure to brominated and chlorinated dioxins as well as PAHs	Release of hydrocarbons and ash, which may contain PAHs, into the air, water, and soil

agreement. Regrettably, illicit transboundary transport of e-waste persists, underscoring challenges in curtailing hazardous waste movement even in the face of the Basel Ban's initiatives to curtail such activities [9–18]. The United Nations, along with its member states, embraced the 2030 Agenda for Sustainable Development in 2015. This comprehensive framework comprises 17 Sustainable Development Goals (SDGs) and 169 targets [19]. Its objectives are to end poverty, safeguard the environment, and advance prosperity for all during a 15-year period. [19]. A unique SDG indicator, 12.5.1, has been identified for tracking the national recycling rate and tons of material recycled in order to monitor the increase of e-waste and manage its potential risks and residual value [19]. SDG indicator 12.4.2, which focuses on the treatment, creation, and management of hazardous waste, emphasizes the significance of tackling e-waste [19]. In order to achieve these objectives within the 15-year time frame, precise rules established for monitoring are used to define and measure these targets and indicators.

Policymakers and stakeholders possess the potential to collaborate in fostering ethical conduct, mitigating the detrimental impact of e-waste on human health and the environment, and contributing to the attainment of sustainable development objectives by aligning e-waste management with the UN Sustainable Development Goals. At present, a noteworthy 71% of the global population has embraced policies pertaining to e-waste management, albeit with variances discernible across distinct regions. However, a remaining 29% still lack a legislative framework to regulate and champion sustainable approaches in e-waste management (Fig. 2) [15]. It's imperative to acknowledge that even within nations boasting e-waste legislation, illicit practices, and unauthorized imports continue to persist.

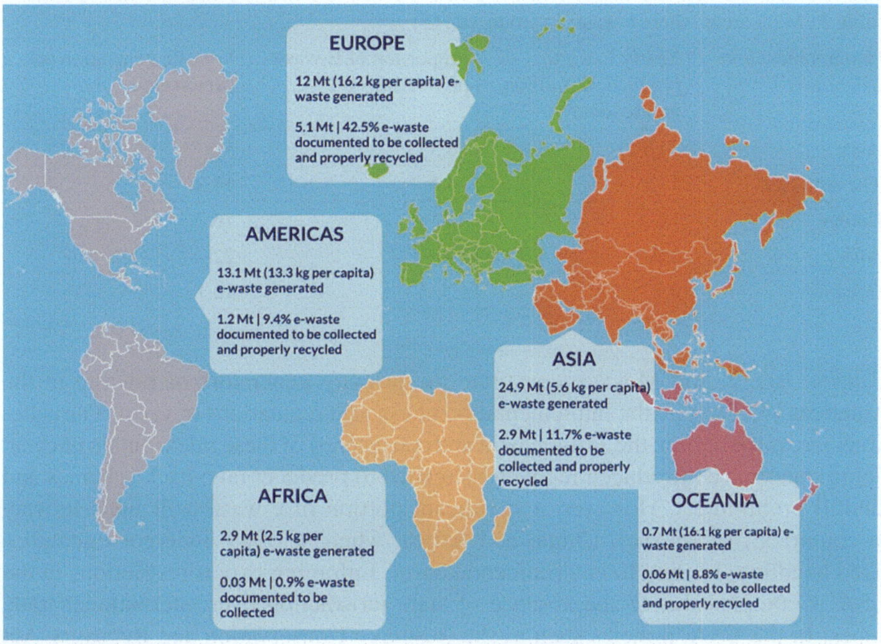

Fig. 2 Currently available statistics with formal recycling and E-waste documentation [15]

1.4.1 Asia

The domestic production of electronic trash and its management are frequently impacted by national economic situations. China, recognized as both the largest e-waste producer in Asia and the world, emerges as a primary recipient of imported e-waste (Table 3). Over the years, various Asian nations have grappled with the challenge of illicit e-waste importation and processing [20]. Nations in South Asia, including China, Japan, India, Korea, and Singapore, have taken proactive measures by enacting legislation to oversee e-waste management, underlining the imperative of effective e-waste treatment. Drawing insights from the European Union's (EU) directives, these countries have forged their own legal frameworks for e-waste management. Their approaches are geared toward advancing circular economy principles, bolstering extended producer responsibility (EPR), establishing product take-back systems, adhering to Restriction of Hazardous Substances (RoHS) regulations, and instituting approvals and licenses for ethical e-waste management.

1.4.2 The Americas

North and South America together with the Americas have been producing a lot of e-waste (Table 3). Higher-income nations, particularly those such as the USA,

Table 3 Main areas where e-waste is produced [15]

Geographical area	Yearly E-waste production (million metric tons)	Proportion of E-waste production (%)	E-waste generation per person (kg)
Asia	24.9	46.5	5.6
The Americas	13.1	24.4	13.3
Europe	12	22.4	16.2
Africa	2.9	5.4	2.5
Oceania	0.7	1.3	16.1

Canada, Brazil, and Mexico, constitute the primary generators of e-waste in the Americas [9]. Despite the implementation of e-waste regulations in various jurisdictions across the region, the enforcement and consistency of these rules remain uneven. Some states have introduced regulations related to producer take-back schemes and landfill prohibitions. However, a significant portion of e-waste still finds its way to countries like Mexico, China, and Africa, where it often undergoes unauthorized handling, leading to environmental harm. Enforcing export restrictions in the USA is complex due to the absence of state jurisdiction over international trade zones. In the USA, policies such as the Resource Conservation and Recovery Act (RCRA), regulations concerning cathode ray tube (CRT) recycling, and the National Strategy for Electronics Stewardship framework are employed to manage domestic e-waste. Furthermore, the US Environmental Protection Agency (EPA) launched the Sustainable Materials Management (SMM) Electronics Challenge in collaboration with retailers and original equipment manufacturers (OEMs). This initiative aims to gather used electronics and channel them to licensed recyclers for proper recycling [20, 21]. Only a few nations in Latin America have implemented special legislation, and the region's progress in regulating e-waste has been very modest. Along with Mexico, other nations like Costa Rica, Colombia, and Peru are working to enhance their current e-waste treatment infrastructures. In 2020, Brazil and Chile have set the foundation for the establishment of a structured regulatory framework for e-waste [9, 15].

1.4.3 Europe

The EU Commission put into effect the Waste Shipment Regulation (WSR), in the wake of the Basel Treaty's ratification in 1993, a prohibition on shipping hazardous e-waste to non-Organization for Economic Cooperation and Development (OECD) nations was instituted [22]. Subsequently, the Restriction of Hazardous Substances Directive (2012/95/EC) was introduced a decade later [23]. This directive's core aim is to elevate the recycling rate of waste electrical and electronic equipment (WEEE) originating from households. It emphasizes the incorporation of environmentally sustainable materials in product design and packaging, thereby curbing the use of hazardous substances. The primary legislative framework governing e-waste management in Europe is the WEEE Directive (2012/19/EU) [24], designed to ensure consistency among its member states. This comprehensive guideline governs the collection, recycling, and resource recovery processes for e-waste. It underscores the significance of methodical and separate collection of used electrical and electronic equipment (EEE) to enhance recycling efficiency and the retrieval of reusable e-waste components. As stipulated by the directive, processed e-waste must be documented and reported to the National Enforcement Authority. Member states are also encouraged to foster the production of EEEs that are easy to disassemble and recyclable. The directive sets treatment standards for diverse e-waste materials and components and outlines guidelines for storage facilities to ensure environmentally responsible processing. Additionally, it incorporates the concept of extended producer responsibility (EPR), obligating manufacturers to recycle their end-of-life products [9, 25].

1.4.4 Africa

While Africa accounts for approximately 5.4% of the global e-waste volume, numerous African nations experience cross-border flows of waste electrical and electronic equipment (WEEE). Similar to other continents, Africa lacks comprehensive government control and enforcement mechanisms, resulting in a prevalence of unofficial collectors and recyclers responsible for e-waste disposal. Only a handful of African countries, such as Egypt, Madagascar, Ghana, Rwanda, Nigeria, Cameroon, South Africa, and Côte d'Ivoire, have introduced e-waste regulations. However, the implementation of these regulations often poses challenges. For instance, Uganda established an Electronic Waste Management Policy in 2012, while Rwanda recently enacted legislation to govern e-waste management. In pursuit of establishing sustainable e-waste management systems, countries like Rwanda, Tanzania, Burundi, Uganda, South Sudan, and Kenya have collaborated to formulate a regional e-waste plan. Despite these initiatives, many African nations continue to lack comprehensive e-waste legislation essential for effective management [1].

1.4.5 Oceania

Oceania encompasses Australia, New Zealand, and the Pacific Island region. Among these nations, Australia stands as the sole country to have enacted dedicated legislation addressing the management of e-waste [27]. Australia's Product Stewardship Act (2011) is particularly notable, with a focus on recycling televisions and computers. This act serves as an exemplar for other countries in the region, showcasing producer responsibilities in e-waste management. Over 291 kt of TV and computer e-waste have been collected and recycled since its enactment. New Zealand and several Pacific Island nations are taking cues from Australia's lead, embarking on the implementation of product stewardship policies. Many of the smaller island governments in the vicinity have enacted environmental protection laws that encompass protocols for e-waste handling. In certain cases, Pacific countries are collaborating with the European Union to regulate hazardous waste [1]. Despite the presence of multiple laws and regulations, the informal sector continues to dominate e-waste management in major regions, perpetuating illegal practices. Inadequate enforcement of stringent laws, limited understanding of formalizing the informal sector, inadequate infrastructure and expertise, and a lack of awareness about global best practices contribute to the persistence of these illicit practices.

2 A Framework for the Circular Economy

The concept of the circular economy is rapidly gaining prominence as a sustainable departure from the conventional linear economic approach. Within a linear economy, resources are extracted, transformed into goods, utilized, and subsequently discarded as waste. In contrast, the circular economy strives to establish a closed-loop system that sustains the utilization of resources for as long as possible, while minimizing waste and maximizing value extraction. Embracing the circular economy involves cultivating an economic model that prioritizes the ongoing utilization of resources and materials, facilitated by practices such as recycling, reuse, repair, and remanufacturing. This approach seeks to disentangle economic growth from resource consumption and environmental deterioration. Fundamentally, the circular economy's principles encompass eradicating waste and pollution, prolonging the use of products and materials, and rejuvenating natural systems. By decreasing resource depletion, mitigating environmental effects, and fostering economic development, the circular economy provides a method to achieve sustainability goals. It makes it possible to use resources more effectively, lowers the amount of waste produced, and lessens the need for virgin materials. By lowering greenhouse gas emissions linked to resource extraction and manufacturing, the circular economy also helps to combat climate change. The circular economy offers methods for maximizing the value of electronic devices throughout their life cycles in the context of managing e-waste. These tactics include making things durable and repairable, putting take-back and recycling procedures in place, encouraging remanufacturing and refurbishing, and making it possible to recover important materials. The circular economy requires a shift in business models from the traditional linear model to a more circular approach. Businesses can adopt strategies such as product-as-a-service, sharing platforms, and leasing arrangements, which promote product longevity, sharing, and reuse. Furthermore, adopting circular supply chains and fostering collaboration with stakeholders throughout the value chain are essential components for effectively incorporating circular practices. Facilitating the shift toward a circular economy necessitates the implementation of enabling policies and regulatory frameworks. Governments hold a pivotal position in promoting circular practices by offering incentives such as extended producer responsibility, tax benefits, and eco-design regulations. Collaboration between governments, businesses, academia, and civil society is vital to create an enabling environment for circular economy implementation. Measuring and assessing the circularity of products, processes, and systems is essential to monitor progress and inform decision-making. Metrics such as material circularity, product lifespan, and resource efficiency can be used to evaluate the circularity of a product or system. Tools and frameworks like life cycle assessments and circularity indicators assist in quantifying the environmental and economic benefits of circular approaches.

To promote sustainable consumption and production habits, reduce waste. In order to safeguard resources for future generations, it is imperative to grasp the concept of the circular economy. Societies can advance toward a more sustainable and rejuvenating economic framework by embracing the principles and strategies inherent in the circular economy. The circular economy plays a crucial role in achieving sustainability goals by addressing the interconnected challenges of resource depletion, environmental degradation, and economic inefficiencies. The circular economy aims to minimize resource consumption by promoting the efficient use and reuse of materials. Through the adoption of approaches like recycling, remanufacturing, and repair, the circular economy lessens the reliance on extracting new resources. This preservation of resources plays a vital role in mitigating environmental repercussions, including habitat degradation, water and air contamination, and deforestation. By minimizing the generation of waste and maximizing the utilization of resources, the circular economy actively contributes to safeguarding natural resources for the well-being of future generations. Traditionally, economic growth has been closely tied to resource consumption, resulting in unsustainable practices. The circular economy seeks to decouple economic growth from resource use by focusing on the value derived from resources rather than their sheer quantity. Through strategies like product durability, modular design, and sharing platforms, the circular economy promotes the longevity of products and their components. This shift toward a more service-oriented and circular business model allows for economic growth while reducing environmental impacts. By lowering greenhouse gas (GHG) emissions throughout the product lifespan, the circular economy can make a substantial difference in the effort to combat climate change. The circular economy lowers the carbon footprint associated with resource extraction, manufacturing, and waste management by giving priority to renewable energy sources and energy-efficient activities. Moreover, strategies like recycling and remanufacturing require less energy compared to producing goods from virgin materials. By minimizing energy consumption and GHG emissions, the circular economy helps combat climate change and achieve sustainability goals. Transitioning to a circular economy creates new economic opportunities and jobs. The shift toward recycling, remanufacturing, and repair activities generates employment in sectors like waste management, resource recovery, and product refurbishment. The circular economy also fosters innovation and entrepreneurship by encouraging the development of sustainable business models and eco-friendly technologies. By creating a circular economy ecosystem, countries can enhance their economic resilience, promote social inclusion, and foster sustainable development.

The circular economy enhances resilience by reducing dependence on scarce and non-renewable resources. By closing resource loops and promoting circular supply chains, countries can minimize their exposure to resource price volatility and geopolitical risks associated with resource extraction. Diversifying resource inputs and adopting circular practices strengthen resource security, contributing to long-term economic stability. The circular economy necessitates active collaboration and participation from a diverse array of stakeholders, encompassing governments, businesses, consumers, and civil society organizations. By fostering partnerships and multi-stakeholder engagement, the circular economy can drive systemic change and overcome barriers to implementation. Collaborative efforts can facilitate knowledge sharing, innovation, and the development of supportive policies and regulatory frameworks. In conclusion, the circular economy plays a vital role in achieving sustainability goals by promoting resource conservation, reducing waste, decoupling economic growth from resource consumption, mitigating climate change, creating economic opportunities, enhancing resilience, and fostering stakeholder collaboration. By embracing circular practices, societies can facilitate the shift toward a more sustainable and regenerative economic system that balances environmental stewardship, social well-being, and economic prosperity. Circular economy strategies for e-waste management encompass a range of approaches aimed at maximizing the value of electronic products throughout their life cycle and minimizing the environmental and social impacts associated with e-waste. One key strategy is designing electronic products with durability and repairability in mind. This involves using high-quality materials and components that can withstand extended use and frequent repairs. Designing products for disassembly and easy access to components facilitates repair and enables the replacement of faulty parts, thereby extending the product's lifespan. Modular design involves creating electronic devices with interchangeable components, allowing for easy upgrades, repairs, and replacements. Component standardization ensures that components can be used across different product models, reducing the need for specialized parts and simplifying repair and recycling processes. Implementing responsible materials selection involves using recyclable and environmentally friendly materials in the production of electronic devices. This makes it easier to recover valuable resources from e-waste during recycling processes. Additionally, designing products with minimal hazardous substances ensures safer handling and reduces environmental contamination during disposal or recycling. Establishing product take-back programs enables consumers to return their end-of-life electronic devices to manufacturers or designated collection points. Effective reverse logistics systems ensure the proper handling, transport, and processing of returned products. This enables the recovery of valuable components and materials from e-waste, reducing the need for virgin resources and minimizing waste generation. Refurbishment involves repairing and restoring used electronic devices to a functional state, enabling their reuse. Remanufacturing takes refurbishment a step further by disassembling products, replacing worn-out components, and restoring them to "like-new" condition. Refurbished and remanufactured electronics can be sold as affordable alternatives, extending their lifespan and reducing the demand for new products. Effective e-waste recycling involves separating and processing electronic devices to recover valuable materials, such as precious metals (e.g., gold, silver) and rare earth elements. Advanced recycling technologies, including mechanical and hydrometallurgical processes, enable the extraction and purification of these resources for reuse in manufacturing new products. Responsible recycling also ensures the proper treatment and disposal of hazardous substances. Educating consumers about the importance of responsible e-waste management and encouraging their active participation is crucial. Providing information on proper disposal methods, repair options, and the benefits of recycling can help raise awareness and drive behavior change. Consumer awareness campaigns can also highlight the economic and environmental benefits of choosing repair, refurbishment, and recycling over disposal. Implementing circular economy strategies for e-waste management requires collaboration among stakeholders. Governments, manufacturers, recyclers, retailers, and consumers need to work together to develop and implement effective policies, regulations, and infrastructure. Public–private partnerships can help establish collection networks, develop recycling facilities, and foster innovation in e-waste management practices.

By implementing these circular economy strategies, e-waste management can be transformed from a linear, resource-intensive process to a more sustainable and resource-efficient system. This approach aids in the transition toward a more circular and sustainable economic model, simultaneously curtailing the environmental impact of e-waste, creating economic prospects, and safeguarding resources. The management of Waste Electrical and Electronic Equipment (WEEE) can present intricate challenges due to the presence of highly toxic substances. However, through collaborative endeavors and the establishment of universally benchmarked technology and infrastructure, this task becomes achievable. The progression toward a sustainable and circular economy hinges on the practice of urban mining, an emerging trend in e-waste processing, which must be executed within a structured framework [27]. Over the past decade, the annual demand for gold has been nearly met through a combination of new mining and recycling. Despite recent demand reductions, primary mining still persists. It's imperative to assess whether the price paid to satiate this demand is truly justifiable. Conversely, adhering to industry-recognized optimal recovery techniques yields invaluable, premium materials while simultaneously contributing to ecosystem preservation and the conservation of natural resources [28, 29]. Much like gold, silver demand has been satisfied to some extent, yet mining activities endure, leading to ecological harm. Conversely, the global supply has yet to meet the demand for platinum and palladium. A systematic process can be employed to recover a range of endangered rare earth metals and strategic elements utilized in defense equipment, including guiding, targeting, electronic warfare, and communication systems [30, 31] (Fig. 3). In conclusion, more than 15% of the demand for precious metals is being satisfied despite just 17.4% of the world's e-waste being handled in formal industries. By imagining a time when all electronic trash is formally treated, we can meet 85–95% or more of the demand. Virgin mining shouldn't be considered a good way to accomplish sustainable development [32].

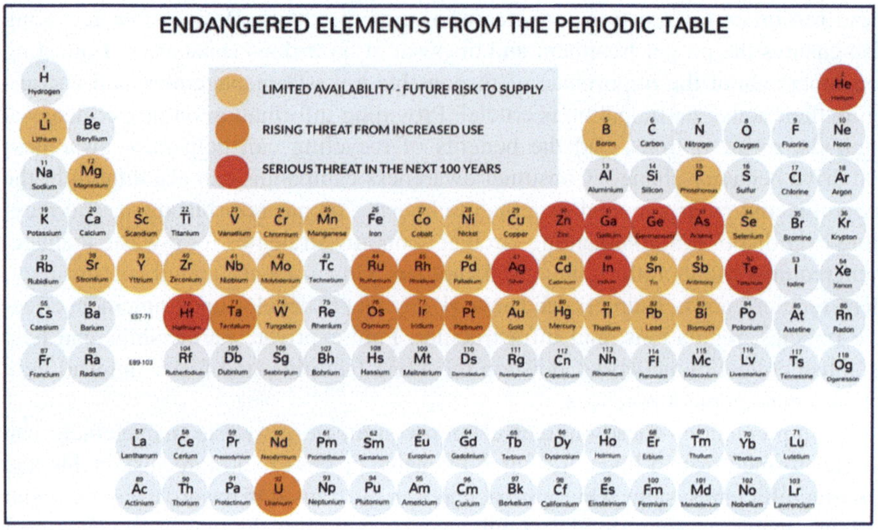

Fig. 3 Fundamental essence highlighted in the periodic table (ACS Green Chemistry Institute)

3 Extending Product Life Cycles

Extending product life cycles is an essential concept in addressing the issue of e-waste and promoting sustainable consumption. It involves maximizing the use and value of electronic products by prolonging their lifespan through various strategies and actions.

- Design for Durability: Manufacturers can design electronic products with durability in mind, using high-quality materials and components that are less prone to wear and tear. This approach ensures that devices can withstand extended use without malfunctioning or becoming obsolete quickly.
- Repair and Maintenance: Encouraging and enabling repairability is crucial for extending product life cycles. Manufacturers can design devices that are easily repairable, with accessible parts and modular designs. Additionally, supporting repair services and providing access to repair manuals and spare parts can empower consumers to fix their devices instead of discarding them.
- Software Updates: Regular software updates can improve device performance, security, and compatibility with new technologies. By providing software updates for older models, manufacturers can extend the usable life of electronic devices and enhance user experience without the need for immediate replacement.
- Upgrades and Modular Design: Modular design allows users to upgrade specific components of electronic devices, such as RAM, storage, or processors, instead of replacing the entire device. This approach reduces electronic waste and enables consumers to adapt their devices to changing needs without fully discarding functional parts.

- Consumer Education: Educating consumers about the benefits of extending product life cycles and responsible electronic consumption is essential. Providing information on repair options, maintenance tips, and the environmental impact of e-waste can empower individuals to make informed choices and actively contribute to sustainable practices.
- Second-Hand Markets: Facilitating the trade of used electronic devices through second-hand markets or certified refurbishment programs promotes the reuse of functional products. By extending the life of electronics through resale, fewer devices end up in landfills, reducing e-waste generation.
- Recycling and Responsible Disposal: Despite efforts to extend product life cycles, there will come a time when devices reach the end of their usable life. Implementing effective recycling programs and ensuring responsible disposal of e-waste is crucial to minimize environmental harm. Recycling facilities can recover valuable materials from discarded devices, reducing the need for virgin resource extraction.

Through the application of these tactics and the advocacy of circular economy principles, which emphasize the prolonged utilization of resources, the lifespan of electronic products can be substantially elongated. This approach not only mitigates the environmental consequences of e-waste, but also preserves resources, curtails energy usage, and fosters sustainable patterns of consumption.

3.1 Repair and Refurbishment Programs

Repair and refurbishment initiatives hold a significant position in the Life Cycle Analysis (LCA) of e-waste supply. LCA represents a comprehensive methodology that evaluates the environmental implications of a product or system across its entire life cycle, encompassing raw material extraction to eventual disposal. In the context of e-waste, life cycle assessment serves to appraise the ecological effects of diverse waste management choices and informs the adoption of more sustainable approaches. Repair and refurbishment programs are particularly relevant in the "end-of-life" phase of electronic products. Instead of immediately discarding devices, these programs aim to extend their lifespan by repairing and upgrading them for continued use. Here's how repair and refurbishment programs contribute to the LCA of e-waste:

- Environmental Benefits: Repairing and refurbishing electronic devices reduce the need for new product manufacturing. Prolonging the lifespan of existing devices circumvents the environmental ramifications tied to raw material extraction, energy use, and manufacturing procedures. LCA aids in quantifying and contrasting the environmental advantages of repair and refurbishment in comparison to alternative waste management strategies, such as recycling or disposal.
- Resource Conservation: Repairing and refurbishing electronic devices conserve valuable resources, including rare metals and minerals present in these products.

LCA assesses the reduction in resource depletion achieved through repair and refurbishment programs, providing insights into the overall resource efficiency of the e-waste management strategy.

- Energy Savings: Repairing and refurbishing devices generally require less energy compared to manufacturing new products. LCA evaluates the energy savings associated with repair and refurbishment activities, including energy requirements for component replacement, testing, and reconditioning. This information helps understand the energy efficiency gains achieved through these programs.
- Emissions Reduction: By avoiding the production of new electronic equipment, repair and refurbishment programs help reduce emissions of greenhouse gases and other air pollutants associated with manufacturing processes. LCA analyzes avoided emissions from repairs and renovations and enabled a comprehensive assessment of the environmental benefits in terms of air quality and climate protection.
- Waste Minimization: Repair and refurbishment programs help minimize e-waste generation by keeping devices functional and in use for a longer time. LCA assesses the reduction in waste generation achieved through these programs, offering insights into waste management strategies and the potential for a more circular economy.

By incorporating repair and refurbishment programs into the LCA of e-waste management, decision-makers can make informed choices that prioritize sustainable practices and optimize environmental outcomes. These programs not only extend the life of electronic devices but also contribute to resource conservation, energy efficiency, emissions reduction, and waste minimization, aligning with the principles of sustainable consumption and production.

3.2 *Consumer Education and Awareness for Responsible Product Use*

Enhancing consumer education and awareness holds a pivotal role in advocating responsible product usage and maximizing the effectiveness of life cycle assessment (LCA) within e-waste management. LCA evaluates the environmental impacts of products throughout their life cycle, and consumer behavior significantly influences these impacts. By educating and raising awareness among consumers, we can encourage responsible practices that minimize e-waste generation and enhance sustainability. Here's how consumer education and awareness contribute to LCA in e-waste management:

- Extended Product Lifespan: Educating consumers about the importance of prolonging the lifespan of electronic devices can significantly reduce e-waste. By providing information on proper maintenance, handling, and care, consumers can ensure that their devices function optimally for a longer period. This reduces the need for frequent replacements and subsequent waste generation.

- Responsible Disposal: Proper disposal of electronic devices is crucial to prevent environmental harm. Educating consumers about the potential risks associated with improper disposal, such as toxic substance leakage and groundwater contamination, encourages responsible disposal practices. This includes promoting the use of authorized recycling facilities or collection programs to ensure e-waste is managed appropriately.
- Sustainable Consumption: Consumer education can raise awareness about sustainable consumption practices, such as considering the environmental impact of products before purchase. By providing information about eco-labels, energy efficiency ratings, and product certifications, consumers can make informed choices that prioritize products with lower environmental footprints. This encourages manufacturers to produce more sustainable and durable devices.
- Recycling and Take-Back Programs: Educating consumers about the availability and benefits of recycling and take-back programs helps divert e-waste from landfills. Consumers can be informed about convenient collection points, drop-off locations, or manufacturer-sponsored programs that facilitate proper recycling of electronic devices. This ensures valuable resources are recovered and reused, reducing the environmental impact of new product manufacturing.
- Repair and Upcycling: Consumer education can highlight the advantages of repair and upcycling as alternatives to disposal. By providing information on repair services, spare part availability, and upcycling ideas, consumers are empowered to extend the life of their equipment through creative repurposing and repair. This reduces e-waste generation and promotes a more circular economy.
- Eco-conscious Behavior: Consumer education can encourage eco-conscious behaviors, such as reducing energy consumption, minimizing standby power, and using energy-saving features of electronic devices. These activities improve a product's environmental performance over the course of its life cycle and increase energy efficiency and carbon emissions reduction.
- Through the cultivation of consumer education and awareness, stakeholders have the ability to empower individuals with knowledge to make informed choices and embrace conscientious practices throughout the electronic product life cycle. This collective effort subsequently enhances the sustainability of e-waste management, harmonizing with LCA principles and promoting the adoption of a circular economy paradigm.

4 Recycling and Resource Recovery

4.1 E-Waste Recycling Technologies and Processes

The utilization of e-waste recycling technologies and processes is integral to the effective handling and treatment of electronic waste. These technologies aim to recover valuable resources, minimize environmental impacts, and ensure the safe handling of hazardous materials present in e-waste. Here's an overview of some commonly used e-waste recycling technologies and processes:

- Sorting and Segregation: The first step in e-waste recycling is the sorting and segregation of different types of electronic devices. This entails segregating items like computers, laptops, mobile phones, televisions, and other electronic devices according to their material composition and recycling necessities.
- Dismantling: Dismantling involves the careful disassembly of electronic devices into their individual components and parts. Skilled technicians use specialized tools to remove batteries, circuit boards, wires, connectors, screens, and other components. This process aims to extract valuable materials for recycling, such as precious metals, copper, aluminum, and plastics.
- Shredding and Crushing: After dismantling, the remaining components are often shredded or crushed into smaller pieces. This process facilitates further separation and liberation of materials. Shredding may involve the use of large shredders or crushers that reduce the size of e-waste to more manageable fragments.
- Magnetic Separation: The process of magnetic separation is utilized to segregate ferrous metals from shredded e-waste. Powerful magnets attract and remove ferrous materials, such as iron and steel, leaving behind non-magnetic materials for further processing.
- Eddy Current Separation: Eddy current separation is a technique used to separate non-ferrous metals, including aluminum, copper, and brass, from the shredded e-waste. An eddy current separator creates a rotating magnetic field that induces electric currents in conductive materials, causing them to be repelled and separated from the waste stream.
- Air Classification: Air classification is employed to separate lighter materials, such as plastics and foam, from heavier materials in e-waste. This process utilizes controlled air streams and mechanical separation techniques to separate materials based on their density and size.
- Chemical Leaching: Chemical leaching is harnessed to extract valuable metals from components of e-waste. Different chemical solutions are applied to dissolve and isolate metals, including gold, silver, palladium, and copper, from circuit boards and other electronic elements. These metals can then be further refined for reuse.

- Precious Metal Recovery: Precious metal recovery processes, such as smelting or hydrometallurgical methods, are employed to extract and refine valuable metals from e-waste. These metals, including gold, silver, and platinum, are recovered in a purified form and can be reintroduced into the manufacturing industry.
- Plastics and Non-metallic Recycling: Plastics and non-metallic materials recovered from e-waste can undergo processes such as shredding, washing, and granulating to convert them into reusable plastic pellets or other forms suitable for manufacturing new products.
- Disposal of Hazardous Materials: Proper disposal of hazardous materials or substances, such as batteries or components containing mercury, that cannot be recycled or recovered are disposed of safely using appropriate techniques to prevent environmental contamination.

It's important to note that e-waste recycling technologies and processes may vary depending on the type and complexity of the electronic devices being recycled, as well as the available infrastructure and regulations in different regions. Sustainable e-waste recycling practices aim to maximize resource recovery, minimize waste generation, and prioritize environmental and human health and safety considerations throughout the entire process.

4.2 Challenges and Opportunities in the Context of E-Waste Recycling

Challenges and opportunities in e-waste recycling are interconnected and arise from various factors, including technological, environmental, economic, and regulatory aspects. Understanding these challenges and seizing the opportunities is crucial for the development of effective e-waste recycling systems. Electronic devices consist of a wide range of materials and components, including hazardous substances, which make their recycling technically challenging. Constantly evolving technologies and product designs pose difficulties in keeping up with the changing composition and recycling requirements of e-waste.

- *Environmental Concerns*: E-waste contains hazardous substances like lead, mercury, cadmium, and brominated flame retardants, which pose significant environmental risks and potential health hazards if not effectively managed. E-waste's improper disposal or ineffective recycling results in the loss of important resources and makes it more necessary to mine new materials, which degrades the environment.
- *Economic Challenges*: Developing efficient and cost-effective recycling technologies for different types of e-waste is a challenge, as it requires significant investment in research and infrastructure. E-waste recycling can be economically challenging due to the fluctuating prices of recovered materials, high operational costs, and the need for extensive processing and sorting.

- *Regulatory and Policy Challenges*: Many regions lack comprehensive legislation and regulations specifically addressing e-waste recycling, resulting in inconsistent practices and ineffective management. Weak enforcement of existing regulations and limited awareness of e-waste recycling obligations among stakeholders hinder the proper implementation of recycling practices. E-waste encompasses valuable resources such as precious metals, rare earth elements, and plastics that have the potential to be reclaimed and reused. This practice not only lessens dependence on virgin materials but also advances the principles of a circular economy. E-waste recycling presents opportunities for job creation and economic growth, especially in developing countries, where the informal sector plays a significant role. Formalizing and improving recycling practices can lead to the development of a skilled workforce and the growth of sustainable businesses. Technological advancements offer opportunities for the development of innovative recycling processes and equipment, improving efficiency, resource recovery rates, and reducing environmental impacts. Implementing EPR policies and regulations that enforcing accountability on manufacturers for the complete life cycle of their products can incentivize the adoption of sustainable e-waste management practices. It encourages product design for recyclability and promotes investment in collection and recycling infrastructure. Increasing awareness among consumers about the importance of responsible e-waste management and the benefits of recycling can drive behavioral change and promote participation in recycling programs. Addressing the challenges and leveraging the opportunities in e-waste recycling requires collaboration among governments, manufacturers, recyclers, consumers, and other stakeholders. It involves investment in research and development, infrastructure development, policy improvements, and the promotion of sustainable practices throughout the e-waste management chain.

4.3 Extracting Valuable Resources from E-Waste: Metals, Plastics, and Rare Materials

Resource recovery from e-waste refers to the process of extracting and recovering valuable materials from electronic waste for reuse or recycling. E-waste contains various materials that can be recovered, including metals, plastics, and rare materials. Here's an explanation of the resource recovery process for these materials:

- ***Metals***: E-waste recycling is a significant source of valuable metals such as gold, silver, copper, palladium, and platinum. The recovery of metals from e-waste involves several steps: (a) *Collection and Sorting*: E-waste is collected and sorted to separate different types of electronic devices and components. (b) *Dismantling*: The devices are then dismantled to access the internal components and separate them based on material composition. (c) *Shredding and Size Reduction*: The components are shredded and mechanically processed to reduce their size and facilitate further separation. (d) *Physical Separation*: Methods such as gravity separation, magnetic separation, and eddy current separation are employed to isolate metals from other substances by capitalizing on their distinct physical characteristics. (e) *Chemical Processing*: Chemical processes, such as leaching or smelting, may be employed to extract metals from complex components or circuits. (f) *Refining*: The extracted metals undergo refining processes to obtain high-purity metals suitable for reuse in various industries.

- ***Plastics***: E-waste recycling also focuses on recovering plastics, which can be reused or processed into new plastic products. The plastic recovery process typically involves (a) *Sorting*: E-waste is sorted to separate plastic components from other materials. (b) *Shredding and Size Reduction*: The plastic components are shredded and reduced in size to facilitate further processing. (c) *Separation*: Mechanical processes like sieving, flotation, or density separation are used to separate different types of plastics based on their properties. (d) *Washing and Cleaning*: The separated plastics undergo washing and cleaning processes to remove contaminants, such as labels, adhesives, and dirt. (e) *Palletization*: The cleaned plastics are melted and converted into small pellets or granules, which can serve as raw materials for producing novel plastic goods.
- ***Rare Materials***: E-waste recycling also provides a chance to reclaim rare and precious materials, including rare earth elements (REEs) and strategic elements. The recovery of these materials involves (a) *Component Identification*: E-waste components containing rare materials are identified and separated during the dismantling and sorting stages. (b) *Extraction Techniques*: Various techniques, such as hydrometallurgical processes or solvent extraction, may be used to extract and separate rare materials from other components. (c) *Purification*: The extracted materials undergo purification processes to remove impurities and obtain high-purity rare materials suitable for reuse or resale. (d) *Reuse or Resale*: Recovered rare materials can be reused in new electronic devices, sold to manufacturers for specific applications, or stockpiled for future use.

Resource recovery from e-waste not only conserves valuable resources but also reduces the need for primary extraction of raw materials, which has environmental and social impacts. It promotes the circular economy by reintroducing recovered materials into the production cycle, reducing waste generation, and minimizing the reliance on virgin resources.

5 Promoting Reuse and Secondary Markets

Encouraging the reuse and development of secondary markets for electronic waste is a crucial element of sustainable e-waste management. It involves extending the lifespan of electronic devices by facilitating their reuse and creating viable markets for second-hand electronics. Here's an explanation of the importance and benefits of promoting reuse and secondary markets for e-waste:

- *Environmental Advantages*: Reuse substantially diminishes the environmental repercussions linked to the manufacturing and disposal of electronic devices. By extending the lifespan of electronics, it helps conserve natural resources, reduce energy consumption, and minimize electronic waste generation. Reusing devices also avoids the release of hazardous substances that can occur during the recycling or disposal process.
- *Conservation of Resources*: Promoting reuse helps preserve valuable resources embedded in electronic devices, such as metals, plastics, and rare materials. By keeping devices in circulation for longer, there is less demand for extracting and processing virgin resources, leading to reduced environmental degradation and energy consumption associated with resource extraction.
- *Economic Opportunities*: Reuse and secondary markets for e-waste create economic opportunities at various levels. It allows individuals and businesses to save money by purchasing used electronics at lower prices compared to new devices. Additionally, it supports local repair businesses and refurbishers, creating jobs and contributing to local economies. The secondary market for used electronics also enables affordable access to technology, particularly in developing regions or for marginalized communities.
- *Access to Technology*: Promoting reuse ensures that functional electronic devices reach individuals and communities that may not have access to new or expensive technology. This can bridge the digital divide and enhance digital inclusion by providing affordable devices for educational purposes, communication, and improving livelihoods.
- *Reduction of E-Waste*: Reuse diverts electronic devices from the waste stream, reducing the volume of e-waste that requires recycling or proper disposal. By keeping devices in use for longer, it decreases the need for frequent upgrades and reduces the accumulation of obsolete electronics. This, in turn, eases the burden on recycling infrastructure and minimizes the associated environmental and health risks.
- *Circular Economy Approach*: Encouraging reuse aligns with the principles of the circular economy, wherein products and materials are maintained in circulation for extended periods, minimizing waste and resource consumption. It fosters a more sustainable and resource-efficient approach to electronic consumption by shifting from a linear "take-make-dispose" model to one that emphasizes the reuse, repair, and repurposing of electronic devices.

To promote reuse and secondary markets for e-waste, various strategies can be implemented, including establishing dedicated e-waste collection centers, supporting refurbishment and repair businesses, implementing buy-back or trade-in programs, raising consumer awareness about the benefits of buying used electronics, and fostering collaborations between manufacturers, retailers, and the second-hand electronics market. These initiatives contribute to a more sustainable and responsible handling of e-waste, simultaneously maximizing the value and longevity of electronic devices.

5.1 *Reselling and Refurbishing Electronic Devices*

Reselling and refurbishing electronic devices play crucial roles in the realm of sustainable e-waste management. They involve giving a second life to used electronic devices by preparing them for reuse and making them available for sale in the secondary market. Reselling refers to the act of selling used electronic devices to new users or customers. It involves various steps, such as collecting used devices, assessing their condition, and preparing them for sale. Reselling can be done through online platforms, dedicated second-hand electronics stores, or other resale channels. Refurbishing involves the repair, restoration, and upgrading of used electronic devices to improve their functionality and extend their lifespan. Refurbishment may include repairing hardware components, replacing faulty parts, upgrading software, cleaning, and cosmetic enhancements. The goal is to bring the device to a condition where it performs reliably and meets the quality standards expected by the end-users. Benefits of Reselling and Refurbishing Electronic Devices:

- *Environmental Benefits*: Reselling and refurbishing electronic devices contribute to reducing electronic waste and its associated environmental impact. By giving used devices a second life, it avoids their premature disposal, reducing the need for raw material extraction, manufacturing, and energy-intensive production processes associated with new devices. This helps conserve resources and reduce greenhouse gas emissions and pollution.
- *Resource Conservation*: Reselling and refurbishing extend the useful life of electronic devices, conserving the valuable resources embedded in them. By maximizing the utilization of existing devices, it reduces the demand for new raw materials, including metals, plastics, and rare materials. This supports a more sustainable approach to resource management and reduces the environmental footprint of electronic consumption.
- *Cost Savings*: Reselling and purchasing refurbished devices offer cost savings compared to buying new electronics. Consumers can often find used devices at lower prices, making technology more affordable and accessible. Refurbished devices, which have undergone quality checks and repairs, provide a more budget-friendly alternative to new devices without compromising functionality.
- *Digital Inclusion*: Reselling and refurbishing contribute to bridging the digital divide by making technology more accessible to individuals and communities with limited financial resources. Affordable used devices allow people to participate in the digital world, access educational resources, job opportunities, and online services, thereby promoting digital inclusion and reducing the digital divide.
- *Job Creation and Economic Opportunities*: Reselling and refurbishing operations create employment opportunities within the repair and refurbishment industry. Small businesses or entrepreneurs involved in device repair, resale, and refurbishment contribute to local economies by generating jobs and supporting local communities.
- *E-Waste Minimization*: Through reselling and refurbishing processes, the quantity of e-waste destined for landfills or recycling centers is diminished. This not only

eases the burden on e-waste management systems but also mitigates the ecological and health hazards linked to inadequate disposal or recycling methods.

To promote reselling and refurbishing, it is important to establish partnerships between electronics manufacturers, retailers, refurbishes, and the second-hand market. Creating awareness among consumers about the benefits of purchasing used electronics, providing warranties or guarantees for refurbished devices, and implementing reliable quality control processes during refurbishment are also key factors in fostering trust and acceptance of refurbished products. The practices of reselling and refurbishing electronic devices play a pivotal role in fostering a more sustainable and circular approach to electronic consumption. By extending the operational lifespan of devices, these processes contribute to the conservation of valuable resources and the reduction of electronic waste. This not only benefits the environment but also promotes responsible and conscious consumption patterns.

5.2 Social and Economic Benefits of Secondary Markets

Secondary markets, in the context of e-waste, refer to the resale and reuse of electronic devices that have been previously owned or used. These markets provide social and economic benefits to various stakeholders. Secondary markets contribute to environmental sustainability by extending the lifespan of electronic devices. The reselling and refurbishing of electronic devices lead to a significant reduction in the demand for new device production. This, in turn, mitigates the necessity for raw material extraction, minimizes energy consumption, and curtails greenhouse gas emissions associated with the manufacturing process. The promotion of reuse through secondary markets contributes to the preservation of valuable natural resources and plays a pivotal role in diminishing the overall environmental footprint linked to the production and disposal of electronic devices. Secondary markets offer cost savings to consumers. Refurbished or used devices are typically sold at lower prices compared to new ones. This makes electronic devices more affordable, allowing individuals with limited budgets to access technology and bridge the digital divide. Lower prices also enable consumers to upgrade their devices more frequently, keeping up with technological advancements without bearing the full cost of brand-new devices. Secondary markets provide greater access to technology, particularly in underserved communities or developing countries where access to new devices may be limited due to financial constraints. By offering affordable options, secondary markets ensure that a wider population can benefit from the advantages of electronic devices, including access to information, communication, education, and employment opportunities. Secondary markets contribute to job creation and economic growth. The refurbishment, repair, and resale of electronic devices require skilled technicians, refurbishers, and retail personnel. This generates employment opportunities and supports the growth of a local circular economy. Additionally, secondary markets can stimulate economic activity by creating a value chain that includes device collection, testing, repair, refurbishment, distribution, and sales. Secondary markets play a crucial role in promoting social and digital inclusion. Access to affordable technology allows marginalized communities, low-income individuals, and disadvantaged groups to participate more fully in society, education, and the digital economy. It helps bridge the digital divide, empowering individuals with tools for communication, information access, online learning, and economic opportunities. By facilitating the reuse and resale of electronic devices, secondary markets contribute to the reduction of e-waste. Instead of devices being discarded or improperly disposed of, they find new owners and purposes. This reduces the volume of e-waste generated and alleviates the burden on recycling and waste management systems. Secondary markets provide opportunities for innovation and knowledge transfer. Refurbishers and repair technicians acquire a wealth of expertise in diagnosing and rectifying a diverse range of electronic devices. This accumulation of knowledge often results in the evolution of novel repair techniques and innovative solutions. Knowledge sharing within the secondary market ecosystem helps improve the overall quality and reliability of refurbished devices. In summary, secondary markets offer social and economic benefits by promoting environmental sustainability, cost savings, access to technology, job creation, economic growth, social inclusion, reduction of e-waste, and opportunities for innovation and knowledge transfer. These markets contribute to a more equitable and sustainable society by extending the lifespan and value of electronic devices.

6 Collaboration and Stakeholder Engagement

Collaboration and stakeholder engagement are vital aspects of addressing the challenges posed by e-waste effectively. They involve bringing together various parties, including government entities, industry players, recycling organizations, NGOs, consumers, and communities, to work collectively toward sustainable e-waste management. Collaboration allows for the sharing of knowledge, expertise, and best practices among stakeholders. Different organizations and individuals bring unique perspectives, experiences, and insights to the table. Through collaboration, stakeholders have the opportunity to share insights, exchange information regarding successful e-waste management strategies, and foster ongoing enhancements in their operational approaches. E-waste management involves multiple stages, including collection, transportation, recycling, and disposal. Collaboration enables stakeholders from different stages of the e-waste management life cycle to work together. By considering the entire life cycle and engaging stakeholders at each stage, it becomes possible to develop comprehensive and integrated solutions that address the challenges holistically. Effective e-waste policies and regulations necessitate collaborative efforts among stakeholders to ensure their development and successful implementation. By engaging government agencies, industry representatives, and NGOs, policies can be designed to ensure compliance, promote responsible practices, and establish standards for e-waste management. Stakeholder engagement also allows for feedback and input, ensuring that policies are practical, feasible, and reflective of the needs of various stakeholders. The establishment of e-waste collection centers, recycling facilities, and proper disposal infrastructure greatly relies on collaboration among various stakeholders. Government bodies, recycling organizations, and industry players can work together to invest in and develop the necessary infrastructure to support the collection, processing, and recycling of e-waste. Stakeholder engagement helps identify gaps in infrastructure and find solutions to address them effectively. Collaboration enables capacity building initiatives to enhance the knowledge and skills of stakeholders involved in e-waste management. Organizing training programs, workshops, and educational campaigns can effectively educate stakeholders about the environmental and health consequences of e-waste, proper handling methods, recycling techniques, and the significance of responsible consumption. This capacity building approach enhances stakeholders' practices and plays a pivotal role in fostering a sustainable e-waste management ecosystem. Stakeholder engagement plays a crucial role in raising awareness among consumers, businesses, and communities about the importance of responsible e-waste management. Through collaborative efforts, stakeholders can develop educational campaigns, public awareness programs, and initiatives that promote responsible consumer behavior, such as device repair, refurbishment, and proper disposal. Engaging consumers and communities fosters a sense of responsibility and encourages them to participate actively in e-waste management efforts. Collaboration encourages research and innovation in e-waste management technologies and processes. By bringing together academia, industry experts, and research institutions, stakeholders can explore and develop innovative solutions for e-waste recycling, material recovery, and sustainable product design. Collaborative research efforts can lead to the development of new technologies, methodologies, and business models that improve the efficiency and effectiveness of e-waste management. Through collaboration and engagement with stakeholders, a culture of continuous improvement in e-waste management practices is nurtured. This collective effort drives innovation, knowledge sharing, and the implementation of more effective strategies to address the challenges posed by electronic waste. Regular dialogues, feedback mechanisms, and evaluation processes allow stakeholders to assess the effectiveness of existing approaches, identify areas for improvement, and implement necessary changes. By working together, stakeholders can adapt to emerging challenges, address gaps in the system, and strive for continuous progress in e-waste management.

In summary, collaboration and stakeholder engagement play a pivotal role in achieving sustainable and effective e-waste management. By harnessing the collective knowledge, resources, and expertise of diverse stakeholders, it becomes possible to develop comprehensive strategies and continuously enhance practices for better e-waste management.

7 Economic Opportunities in the Circular Economy

The circular economy offers significant economic opportunities in the context of e-waste management. Transitioning to a circular economy model for e-waste management can generate new job opportunities across the value chain. The recycling and refurbishment sectors, in particular, require skilled workers for tasks such as disassembly, repair, component extraction, and material processing. The growth of these sectors can lead to job creation, contributing to local economies and providing employment opportunities. E-waste encompasses valuable resources, including precious metals, rare earth elements, and high-quality plastics. Adopting circular economy principles enables the extraction and recovery of these valuable materials through recycling and refurbishment processes. By recovering and reusing these resources, businesses can reduce their reliance on virgin raw materials, resulting in cost savings and improved resource efficiency. The concept of the circular economy promotes the creation of secondary markets for reconditioned electronic devices and recycled materials. Refurbished electronics can be sold at lower prices, making them more affordable for consumers and expanding the market for these products. Similarly, recycled materials can be supplied to manufacturers, reducing the demand for virgin materials and creating a market for recycled content. Implementing strategies such as repair, refurbishment, and remanufacturing extends the lifespan of electronic devices. This approach allows businesses to generate additional revenue streams by offering repair services, selling refurbished products, or leasing devices through product-as-a-service models. Prolonging the use of products reduces the need for frequent replacements, benefiting consumers and creating economic value. The circular economy drives innovation in e-waste management technologies and processes. The circular economy approach stimulates the advancement of more efficient and ecologically sustainable recycling techniques, cutting-edge sorting technologies, and innovative methods for extracting valuable materials. Allocating resources to research and development in this domain can yield technological breakthroughs, opening avenues for companies to provide pioneering solutions and attain a competitive advantage. Embracing the circular economy demonstrates a commitment to sustainability and responsible business practices. By implementing effective e-waste management strategies, companies can enhance their brand reputation, attract environmentally conscious consumers, and build trust in their products and services. Positive brand perception can translate into increased customer loyalty, market differentiation, and long-term profitability. E-waste regulations are becoming more stringent worldwide. Embracing the circular economy helps businesses comply with these regulations and mitigate the associated risks. By adopting responsible e-waste management practices, companies can avoid fines, legal liabilities, and reputational damage that may arise from improper handling or illegal disposal of e-waste. The circular economy fosters the development of innovative business models, such as leasing, sharing, or subscription-based services. These models provide opportunities to reach new customer segments, increase market penetration, and explore alternative revenue streams. Additionally, adopting circular economy principles can help businesses access markets with sustainability-focused procurement policies or certification requirements.

In conclusion, the circular economy approach to e-waste management brings forth numerous economic opportunities. It enables job creation, cost savings through resource recovery, market development for refurbished products and recycled materials, extended product life cycles, innovation, brand reputation enhancement, regulatory compliance, and access to new markets and business models. By embracing these opportunities, businesses can contribute to a sustainable and prosperous economy while addressing the challenges of e-waste effectively.

7.1 Job Creation and Economic Growth in the E-Waste Sector

The e-waste sector presents significant opportunities for job creation and economic growth. E-waste recycling and refurbishment require a skilled workforce to carry out activities such as dismantling, sorting, component extraction, material processing, repair, and refurbishment. As the volume of e-waste continues to increase, there is a growing demand for workers in these areas. The establishment of recycling facilities, refurbishment centers, and repair shops creates employment opportunities for both skilled and unskilled workers. In many regions, e-waste management is predominantly carried out by the informal sector, consisting of informal collectors and recyclers. Formalizing this sector by providing training, infrastructure, and regulatory support can lead to the creation of formal jobs and improve working conditions. Furthermore, it guarantees the appropriate management of e-waste, fosters the retrieval of resources, and diminishes the ecological and health hazards linked to unregulated methods. E-waste encompasses valuable materials such as precious metals like gold, silver, and copper, alongside rare earth elements and premium-grade plastics. Recovering and extracting these resources through recycling processes creates economic value. The establishment of recycling facilities and advanced technologies for resource recovery opens up business opportunities and generates revenue streams. The refurbishment and resale of electronic devices offer economic opportunities. Refurbished devices, when properly tested and repaired, can be sold at lower prices, making them affordable to a broader consumer base. This stimulates the secondary market for electronics and generates revenue for refurbishment businesses. The e-waste sector drives innovation in recycling technologies, material recovery processes, and sustainable business models. Allocating resources to research and development endeavors paves the way for the emergence of novel technologies, methodologies, and equipment aimed at enhancing the efficiency and efficacy of e-waste recycling. This dynamic environment not only fuels innovation but also garners investments, ultimately fostering substantial economic growth within the sector. E-waste management facilities and recycling plants contribute to local economic development. These facilities require infrastructure, machinery, and equipment, which are sourced locally, leading to economic benefits for suppliers and service providers. Additionally, the revenue generated from e-waste management activities circulates within the local economy, supporting other businesses and creating multiplier effects. The e-waste sector perfectly aligns with the objectives of the green economy and sustainable development. It champions the cause of conserving resources, curbing reliance on pristine materials, and mitigating the ecological degradation stemming from improper disposal practices. The burgeoning e-waste industry serves as a crucial contributor to the overarching sustainability blueprint, eliciting investments from conscientious businesses and forward-thinking investors. Extending the operational life of electronic devices via repair, refurbishment, and remanufacturing not only curbs the need for frequent replacements but also translates into cost savings for consumers. This, in turn, fosters a heightened demand for repair services and generates fertile ground for enterprises operating within the circular economy paradigm. The extension of product life cycles concurrently lessens the environmental and financial load associated with the production of new devices.

In conclusion, the e-waste sector offers substantial economic opportunities, including job creation, value recovery from e-waste, development of secondary markets, innovation, local economic development, and contributions to the green economy. Embracing sustainable practices, formalizing the sector, and investing in recycling infrastructure and technologies can further unlock the economic potential of e-waste management while addressing environmental and social challenges.

7.2 *Innovation and Entrepreneurship in Sustainable Electronics*

Innovation and entrepreneurship are pivotal driving forces in advancing sustainable electronics and effective e-waste management. Innovation in sustainable electronics begins with product design. Entrepreneurs and innovators are developing eco-friendly and energy-efficient electronic devices that are designed with materials that are easier to recycle or repair. These endeavors emphasize environmental sustainability across the entire life cycle of products, from raw material acquisition to end-of-life disposal. Entrepreneurs are actively exploring novel business models that align with circular economy principles within the electronics industry. One notable approach is the concept of "product-as-a-service," where consumers lease or rent devices instead of purchasing them outright. This encourages manufacturers to design products for longevity, ease of repair, and potential upgrades, fostering a closed-loop system that curtails waste generation and optimizes resource utilization. Continuous innovation is evident in the field of recycling technologies. Entrepreneurs are pioneering advanced methods like hydrometallurgical processes, bioleaching, and cutting-edge sorting technologies to extract valuable materials from electronic waste more efficiently. These breakthroughs make way for the retrieval of precious metals, rare earth elements, and other valuable resources, presenting economic opportunities while mitigating environmental impacts. Emerging enterprises and innovative minds are actively crafting imaginative strategies for the effective management of e-waste. They are in the process of setting up dedicated e-waste collection hubs, cutting-edge recycling complexes, and intricate networks for reversing the supply chain. Through the adept utilization of cutting-edge technologies such as the Internet of Things (IoT), sophisticated data analytics, and secure blockchain systems, these initiatives are seamlessly orchestrating the entire cycle of e-waste management.

Through the integration of IoT connectivity, devices are meticulously tracked and monitored, ensuring their efficient collection and seamless transportation. The power of data analytics offers insights into consumption patterns and disposal behaviors, which in turn fuels informed decision-making. Simultaneously, the application of blockchain technology guarantees transparency and security throughout the supply chain, enhancing traceability and accountability. By amalgamating these advanced tools, startups and visionaries are not merely simplifying the intricate process of e-waste collection, tracking, and processing; they are also cementing adherence to proper disposal practices. This holistic approach propels effective resource reclamation, mitigating the environmental burden triggered by electronic waste and fostering a future defined by sustainability and responsibility. The entrepreneurial landscape also embraces the repair and refurbishment of electronic devices. This movement not only reduces electronic waste but also extends the lifespan of gadgets, reducing the need for frequent replacements and conserving resources. These innovative initiatives collectively contribute to a more sustainable electronics industry and effective e-waste management.

They offer convenient and affordable repair options, extending the lifespan of products and reducing electronic waste. These businesses play a vital role in educating consumers about the benefits of repair and providing access to skilled technicians and spare parts. Innovators are developing platforms and tools to increase consumer awareness and education about sustainable electronics. They create online platforms, mobile apps, and informational resources to help consumers make informed decisions about product purchases, repair options, and responsible disposal of electronic devices. These initiatives empower consumers to choose sustainable electronics and contribute to reducing e-waste. Entrepreneurs in the sustainable electronics space often collaborate with various stakeholders, including manufacturers, governments, NGOs, and academic institutions. Such collaborations foster knowledge sharing, access to resources, and collective efforts to address sustainability challenges. Partnerships also enable entrepreneurs to scale their innovations and expand their impact. The growth of sustainable electronics entrepreneurship is supported by funding and investment from impact investors, venture capitalists, and government initiatives. Funding opportunities specifically targeting sustainable technology startups and e-waste management ventures help drive innovation, scale businesses, and create a sustainable electronics ecosystem.

In conclusion, innovation and entrepreneurship are essential drivers in the transition to sustainable electronics and e-waste management. Through sustainable product design, circular economy business models, advanced recycling technologies, and collaborative efforts, entrepreneurs and innovators are creating economic opportunities while addressing environmental challenges. Their efforts contribute to a more sustainable and responsible electronics industry.

7.3 Economic Feasibility and Financial Mechanisms for Circular Practices

Economic feasibility and financial mechanisms play a crucial role in promoting circular practices in the e-waste sector. Conducting a cost–benefit analysis is essential to determine the economic feasibility of circular practices. This analysis compares the costs associated with implementing circular practices, such as recycling and resource recovery, against the potential benefits, including cost savings from reduced raw material extraction and waste disposal, revenue generation from recycled materials, and environmental benefits. It aids in the assessment of the economic feasibility of circular strategies and provides valuable insights for decision-making. An impactful financial mechanism known as Extended Producer Responsibility (EPR) plays a pivotal role in holding producers accountable for the entire life cycle of their products. This approach mandates manufacturers to not only fund but also execute systems for the collection, recycling, and proper disposal of their electronic goods at their end of life. By redistributing the financial responsibility from governmental bodies and taxpayers to producers, EPR acts as a driving force compelling manufacturers to infuse circular design principles into their products and invest in robust recycling infrastructure. This mechanism effectively ensures that the expenses linked to e-waste management are inherently embedded within the product's life cycle. Product stewardship programs, on the other hand, encompass agreements—both voluntary and mandatory—forged between manufacturers, retailers, and other stakeholders. These agreements entail taking holistic responsibility for the entire life cycle of electronic products. Such programs might encompass a range of strategies, including the provision of financial incentives or fees to foster the collection, refurbishment, and recycling of electronic waste. This collaborative framework extends the responsibility beyond the point of purchase, channeling a collective commitment to the sustainable handling of electronic products throughout their life cycle.

By establishing cooperative frameworks and sharing the costs and benefits among stakeholders, product stewardship programs create a sustainable financial mechanism for circular practices. Market-based instruments, such as environmental taxes, levies, or tradable permits, can be used to internalize the environmental costs associated with e-waste generation and promote circular practices. For example, levies imposed on the sale of new electronic products can be used to fund e-waste collection and recycling programs. Tradable permits can create a market for recycled materials, incentivizing their recovery and reuse. These instruments provide economic incentives for circular practices and help finance the necessary infrastructure and initiatives. E-waste contains valuable resources, including precious metals, rare earth elements, and other reusable materials. The adoption of effective recycling technologies and streamlined processes enables the reclamation of these precious resources, which can subsequently be sold or reintegrated into the production of novel electronic goods. Revenue generated from the sale of recovered materials can offset the costs of e-waste management and create economic opportunities within the circular economy. Circular practices in the e-waste sector create new business opportunities and economic growth. Entrepreneurs and innovators can develop innovative technologies and processes for e-waste recycling, refurbishment, and remanufacturing. These innovations can lead to the creation of new companies, job opportunities, and revenue streams. Additionally, the development of secondary markets for refurbished electronic devices contributes to economic activity and consumer choice. Access to financing and investment is crucial for scaling up circular practices in the e-waste sector. Governments, financial institutions, and impact investors can provide funding, grants, and loans to support the establishment of e-waste recycling facilities, research and development of innovative technologies, and capacity building initiatives. Investment in circular practices not only generates financial returns but also contributes to sustainable development and environmental protection. Effective circular practices thrive on the collaborative efforts of various stakeholders, encompassing governments, enterprises, financial institutions, and civil society organizations, which is essential for their triumph. Public–private partnerships can leverage resources, expertise, and networks to drive the economic feasibility of circular initiatives. Collaboration also helps share risks and costs associated with e-waste management, enhances knowledge sharing, and fosters innovation in financial mechanisms for circular practices.

In conclusion, economic feasibility and financial mechanisms are essential for the successful implementation of circular practices in the e-waste sector. By conducting cost–benefit analyses, implementing extended producer responsibility and product stewardship programs, using market-based instruments, focusing on value recovery from e-waste, fostering innovation and entrepreneurship, and facilitating access to financing and investment, circular practices can be financially viable, promote sustainable economic growth, and contribute to a more resource-efficient and ecologically sustainable e-waste management system.

References

1. V. Forti, C.P. Baldé, R. Kuehr, G. Bel, *The Global E-waste Monitor 2020: Quantities, flows and the circular economy potential* (United Nations University (UNU)/United Nations Institute for Training and Research (UNITAR)—Co-hosted SCYCLE Programme, International Telecommunication Union (ITU) & International Solid Waste Association (ISWA), Bonn/Geneva/Rotterdam, 2020)
2. V. Murthy, S. Ramakrishna, A review on global E-waste management: urban mining towards a sustainable future and circular economy. Sustainability **14** (2022)
3. *Children and Digital Dumpsites: E-Waste Exposure and Child Health* (World Health Organization, Geneva, 2021). Licence: CC BY-NC-SA 3.0 IGO

4. A. Arain, R.L. Neitzel, A review of biomarkers used for assessing human exposure to metals from E-waste. Int. J. Environ. Res. Publ. Health. **16**, 15–16 (2019)
5. P. Roychowdhury, J.M. Alghazo, B. Debnath, S. Chatterjee, O.K.M. Ouda, Security threat analysis and prevention techniques in electronic waste, in *Waste Management and Resource Efficiency* (Springer: Singapore, 2018)
6. M.S. Sankhla, M. Nandan, S. Mohril, G.P. Singh, B. Chaturvedi, R. Kumar, M. Kumari, Effect of electronic waste on environmental & human health—a review. IOSR J. Environ. Sci. Toxicol. Food Technol. **10**, 98–104 (2016)
7. K. Tiller, Heavy metals in soils and their environmental significance, in *Advances in Soil Science*, 13 (Springer, Berlin/Heidelberg, Germany, 1989), 113–142
8. V.N. Pinto, E-waste hazard: the impending challenge. Indian J. Occup. Environ. Med. **12**, 65–70 (2008)
9. R.A. Patil, S. Ramakrishna, A comprehensive analysis of e-waste legislation worldwide. Environ. Sci. Pollut. Res.Pollut. Res. **27**, 14412–14431 (2020)
10. B. Ramesh Babu, A.K. Parande, C. Ahmed Basha, Electrical and electronic waste: a global environmental problem. Waste Manag. Res. **25**, 307–318 (2007)
11. P. Kiddee, R. Naidu, M.H. Wong, L. Hearn, J.F. Müller, Field investigation of the quality of fresh and aged leachates from selected landfills receiving E-waste in an arid climate. Waste Manag. **34**, 2292–2304 (2014)
12. C. Gangwar, R. Choudhari, A. Chauhan, A. Kumar, A. Singh, A. Tripathi, Assessment of air pollution caused by illegal E-waste burning to evaluate the human health risk. Environ. Int. **125**, 191–199 (2019)
13. R. Rautela, S. Arya, S. Vishwakarma, J. Lee, K.-H. Kim, S. Kumar, E-waste management and its effects on the environment and human health. Sci. Total. Environ. **733**, 145623 (2021)
14. F.-C. Mihai, M.-G. Gnoni, C. Meidiana, C. Ezeah, V. Elia, Chapter 1—waste electrical and electronic equipment (WEEE): flows, quantities, and management—a global scenario, in *Electronic Waste Management and Treatment Technology*, eds. by M.N.V. Prasad, M. Vithanage (Butterworth-Heinemann, Kidlington, UK, 2019), 1–34
15. C.P. Baldé, S. van den Brink, V. Forti, A. van der Schalk, F.T. Hopstaken, D.W.E.E.E. Flows, *What Happened Between 2010 and 2018* (United Nations University (UNU)/United Nations Institute for Training and Research (UNITAR), Tokyo, Japan, 2020), p.2020
16. D.P. Hackett, Assessment of the Basel convention on the control of transboundary movements of hazardous wastes and their disposal. Am. Univ. J. Int. Law Pol. **5**, 31 (1989)
17. United Nations Environment Programme (UNEP), Report on the Basel Convention on the Control of Transboundary Movements of Hazardous Wastes. http://www.basel.int/TheConvention/Overview/tabid/1271/Default.aspx
18. E-Waste Country Assessments. www.sprep.org/attachments/Publications/WMPC/Summary_Rep_PacWaste_TAP_E-waste_Country_Assess.pdf
19. UN Sustainable Development Agenda 2021: Sustainable Development Goals (SDGs). https://www.in.undp.org/content/india/en/home/sustainable-development-goals.html
20. J. Alghazo, O. Ouda, A. Elhassan, E-waste environmental and information security threat: GCC countries vulnerabilities. Euro-Mediterr. J. Environ. Integr. **3**, 5–9 (2018)
21. C.P. Baldé, V. Forti, V. Gray, R. Kuehr, P. Stegmann, *The Global E-Waste Monitor 2017* (United Nations University (UNU), International Telecommunication Union (ITU), and International Solid Waste Association (ISWA), 2017). https://collections.unu.edu/eserv/UNU:6341/Global-E-waste_Monitor_2017__electronic_single_pages_.pdf
22. J. Jupille, The European Union and international outcomes. Int. Organ. **53**, 409–425 (1999)
23. European Commission (2012/95/EU), *Restriction of Hazardous Substances in Electrical and Electronic Equipment (RoHS)*. https://ec.europa.eu/environment/topics/waste-and-recycling/rohs-directive_en
24. European Community Directive 2012/19/EU, *Waste Electrical and Electronic Equipment Directive (WEEE Directive)*. https://eur-lex.europa.eu/legal-content/EN/TXT/PDF/?uri=CELEX:32012L0019&from=EN

25. F. Cucchiella, I. D'Adamo, S.L. Koh, P. Rosa, A profitability assessment of European recycling processes treating printed circuit boards from waste electrical and electronic equipments. Renew. Sustain. Energy Rev. **64**, 749–760 (2016)
26. R.A. Patil, S. Ramakrishna, *A Comprehensive Analysis of E-Waste Legislation Worldwide* (Springer, Berlin, Germany, 2020), 5–10
27. S. Ramanayaka, S. Keerthanan, M. Vithanage, Urban mining of E-waste: treasure hunting for precious nanometals, in *Handbook of Electronic Waste Management*, eds. by M.N.V. Prasad, M. Vithanage, A. Borthakur (Butterworth-Heinemann, Kidlington, UK, 2019), 19–54
28. R. Ahirwar, A.K. Tripathi, E-waste management: a review of recycling process, environmental and occupational health hazards and potential solutions. Environ. Nanotechnol. Monit. Manag. **15**, 100409 (2021)
29. Step: Solving the E-Waste Problem. https://www.step-initiative.org/organisation-rev.html
30. A. Chatterjee, J. Abraham, Efficient management of E-wastes. Int. J. Environ. Sci. Technol. **14**, 211–222 (2016)
31. D. Pant, D. Joshi, M.K. Upreti, R.K. Kotnala, Chemical and biological extraction of metals present in E waste: a hybrid technology. Waste Manag. **32**, 979–990 (2012)
32. K.A. Shagun, A. Arora, Proposed solution of E-waste management. Int. J. Future Comput. Commun. **2**, 490–493 (2013)

Lithium-Doped Cobalt Oxide ($LiCoO_2$) Thin Film Can Be Applied as a Storage Medium for Energy

A. Sudha, N. M. I. Alhaji, A. Ayeshamariam, M. Ragamathunnisa, M. Ismail Fathima, and M. Sivabharathy

Abstract The $LiCoO_2$ powder was synthesized using a simple co-precipitation process in order to create thin films. After annealing the ITO substrate to 200 °C, this powder was deposited on an indium tin oxide (ITO) substrate using the electron beam evaporation technique (E-beam). The structure and morphology of the $LiCoO_2$ thin film formed on an ITO substrate were examined using structural, scanning electron microscopy with EDAX analysis, and transmission electron microscopy (TEM). The performance of the $LiCoO_2$ thin film was calculated and plotted extensively for cyclic voltammetry (CV) studies. Therefore, the prepared $LiCoO_2$ thin film was found to be suitable for cyclic voltammetry studies and their findings are reported in detail here.

A. Sudha · N. M. I. Alhaji
Department of Chemistry, Khadir Mohideen College (Affiliated to Bharathidasan University, Tiruchirappalli), Adirampattinam, Tamil Nadu, India

A. Sudha
Department of Chemistry, Kunthavai Nachiyar Government Arts College for Women (Affiliated to Bharathidasan University, Tiruchirappalli), Thanjavur, Tamil Nadu, India

A. Ayeshamariam (✉)
Department of Physics, Khadir Mohideen College (Affiliated to Bharathidasan University, Tiruchirappalli), Adirampattinam, Tamil Nadu 614701, India
e-mail: ayeshamariamkmc@gmail.com

M. Ragamathunnisa
PG & Research Department of Physics, Periyar EVR College (Autonomous), Tirchy, Tamil Nadu 620023, India

M. Ismail Fathima · M. Sivabharathy
Department of Physics, Sethu Institute of Technology, Pulloor, Kariyapatti, Tamil Nadu 626115, India

K. Kasinathan et al. (eds.), *New Technologies for Energy Transition Based on Sustainable Development Goals*, https://doi.org/10.1007/978-981-97-2527-4_17

1 Introduction

Over the past ten years, portable electronics have substantially increased in popularity, complexity, and power, raising new issues, notably in the fields of autonomous and medical devices. The low specific energy of thin film, all-solid-state lithium-ion micro batteries may enable these futuristic devices to run. For a device to provide the required energy, both the weight-to-energy (*w*/*e*) ratio and the volume-to-energy (*v*/*e*) ratio must be high. These needs can be met by lithium-ion batteries. In smaller dimensions, packaging considerations become more important as lithium-ion batteries are miniaturized. The integrated batteries concept was built on this foundation [1]. In order to reduce the weight of portable gadgets, there is a growing demand for smaller, thinner batteries. Thin-film rechargeable batteries have gotten a lot of attention as a way to minimize battery size. In lithium secondary batteries, $LiCoO_2$ has been widely used as a cathode material since it has a high specific capacity, high operating voltage, and a prolonged cycle life. Lithium ions are located on the octahedral 3*b* interstices, cobalt ions are found on the 3*a* sites, and oxygen anions are found in cubic close packing in crystalline $LiCoO_2$. This multilayer arrangement provides a Li-ion diffusion route that is accessible in two dimensions. The fundamental reason for this is that battery materials and technology have advanced significantly. In comparison with other commercial rechargeable batteries, lithium-doped batteries are more advanced. The electrical and solar cell industries partnered effectively to create a wide range of Li-ion devices with voltages triple times greater than Ni, Cd, Co, and iron batteries. In a battery module with a set of output voltages, the specific cell voltage could reduce the number of cells required, as well as the need for associated electronics, resulting in enhanced battery module dependability due to fewer parts [2]. Due to its affordability, availability, and environmental friendliness, doped lithium oxide has garnered a lot of interest. Cobalt oxide thin films doped with lithium oxide thin films showed high electrical conductivity and intriguing electrochemical performance. However, its capacity for energy storage encourages the creation of high energy density batteries, Li-ion devices, for example, and the use of novel electrochemical performance as a cathode in battery production. The thermal effect in single-cell Li-ion batteries triggers chain reactions, which can lead to disastrous consequences in energy storage devices or electric cars. As a result, it's critical to comprehend the components that could contribute to a reduction in single-cell thermal issues [3]. There have been several cathode materials suggested and implemented [2]. The layered structure of such cathode materials is usual. John B. Goodenough described and explored $LiCoO_2$ as among the standard cation exchange constituent grouping for laser-induced breakdown spectroscopy (LIBs) for the first time in 1980 [4]. Transition metals frequently experience a process of oxidation to a somewhat high oxidation state when Li^+ ions are removed from such cathode materials.

Following lithium recovery from $LiCoO_2$, there is often a progressive change from the hexagonal to the monoclinic phase as well as Co disintegration in the highly oxidized state. It is equivalent to a 50% reduction of Li from $LiCoO_2$, resulting in significant performance degradation. $LiCoO_2$ surface modifications, such as a protective coating, have been required to improve the material's availability at larger threshold potentials (4.4–4.5 eV). Because such techniques can successfully restrict changes in the lattice constant and the cycling phase shift from hexagonal to monoclinic. Uncertainty still surrounds the mechanism of improved $LiCoO_2$ capacity [5]. The Dahn group attributed the limited power consistency of $LiCoO_2$ at the greater discrimination potential (4.5 eV) to side reactions creating an increase in impedance at the $LiCoO_2$ surface. A cathode with a larger threshold potential still has a thermal stability issue, providing a safety risk, even though a higher threshold potential can lead to a larger reversal performance. In addition, Co in delighted $Li_{1x}CoO_2$ has a high oxidizing strength and can be conveniently induced by organic electrolytes to produce additional side reactions. As a result, $LiCoO_2$ is mostly used in low-power devices such as consumer electronics [6]. To build thin films of $LiCoO_2$, several methods can be used, including radio frequency sputtering, pulsed laser deposition, spray pyrolysis, chemical vapor deposition, and cobalt metal reactions [7]. An ITO-coated substrate is used for the deposition of thin-film $LiCoO_2$ cathodes. Even though thin films of $LiCoO_2$ have been examined for their electrochemical properties and fabrication parameters [8], the results show the $LiCoO_2$ thin films behavior because of the high nature of the active material for battery application. The details of the cyclic voltammetry technique, as well as the experimental findings of these investigations suggest the $LiCoO_2$ films' suitability for device applications [9]. We present a thin film of $LiCoO_2$ prepared by a simple co-precipitation process as a cathode material for lithium-ion batteries in this investigation. An array of electrochemical experiments is employed in order to investigate the structural and electrochemical properties of mesoporous $LiCoO_2$ films, including high angle XRD (XRD), scanner electron microscopy (SEM), electronic dispersive X-ray spectroscopy (EDAX), and transmission electron microscopy (TEM).

2 Experimental

Lithium oxide and cobalt oxide are the beginning materials (of purity 99.99%). The parent material lithium chloride is a 90% (2.689 g) pure powder of cobalt chloride and a 10% (0.749 g), which is synthesized using a simple co-precipitation process. The produced powder was analyzed using XRD to establish the presence of lithium and cobalt oxide in a required proportion. To counteract the nonuniform deposition of thin films in the electron beam evaporation process, a rotating sample holder is commonly utilized. In this experiment, the rate of evaporation was kept within a 0.05 nm/s range. With a pressure of 500 kg/cm^2, we use the hydraulic method to obtain pellets of these powders. ITO substrate subjected to ionic bombardment for 5 min of deposition, calibrated by chromel alumel thermocouple inserted inside bell jar of electron beam evaporation process is shown in Fig. 1. Electron beam evaporation was used to make lithium cobalt oxide thin films, with an indium tin oxide conducting plate as the substrate. The evaporation rate was changed by changing the electron beam current in the range of 40–50 mA while keeping the voltage at 5 kV using the reactive electron beam evaporation technique at a chamber pressure of 0.5×10^{-5} mbar with oxygen. When the electron beam evaporation process for $LiCoO_2$ film on ITO substrate is done, the degree of crystallization with respect to annealing temperature promotes the crystallization of thin film. For these deposited thin films, the cyclic voltagrams show highly defined redox peaks. This thin film is electrochemically active, with a discharge capability of about 48 μAh/(cm^2 μm) [9].

Fig. 1 Flowchart of experimental section of powder and thin film on ITO plate

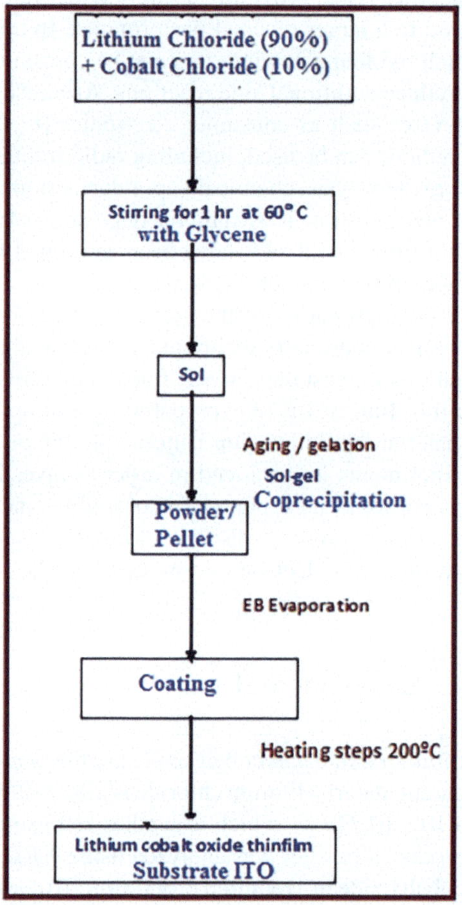

3 Results and Discussion

3.1 XRD Analysis

X-ray diffraction was performed along CuK_α emission in a 2θ range of 10°–80° at 15 mA and 40 kV using a diffractometer. VASP software was used to examine the XRD patterns. The crystalline structure was confirmed by the 2D refinement, which revealed free patterns. Using the collected data and Scherrer's formula, using the prepared sample, the crystallite size was calculated. Formula Debye–Scherrer, $D = 0.94\lambda/(\beta\cos\theta)$; where, λ—wavelength of CuK_α radiation = 1.5406 Å, β—FWHM in radian, and θ—Bragg's angle. The XRD study of $LiCoO_2$ indicated that it had an average crystal size of 37.29 nm, with prominent films of (101), (201), and (211) that were well-matched with JCPDS file number #78-1519 is shown in Fig. 2a. The lattice constant $a = 6.692$ Å and $c = 4.710$ Å with tetragonal structure were closely aligned with our reported JCPDS file. The textures, increasing in intensity (101), indicate nanoindentation of our prepared thin film [10], according to the two-dimensional (2D) refined analysis shown in Fig. 2b–c.

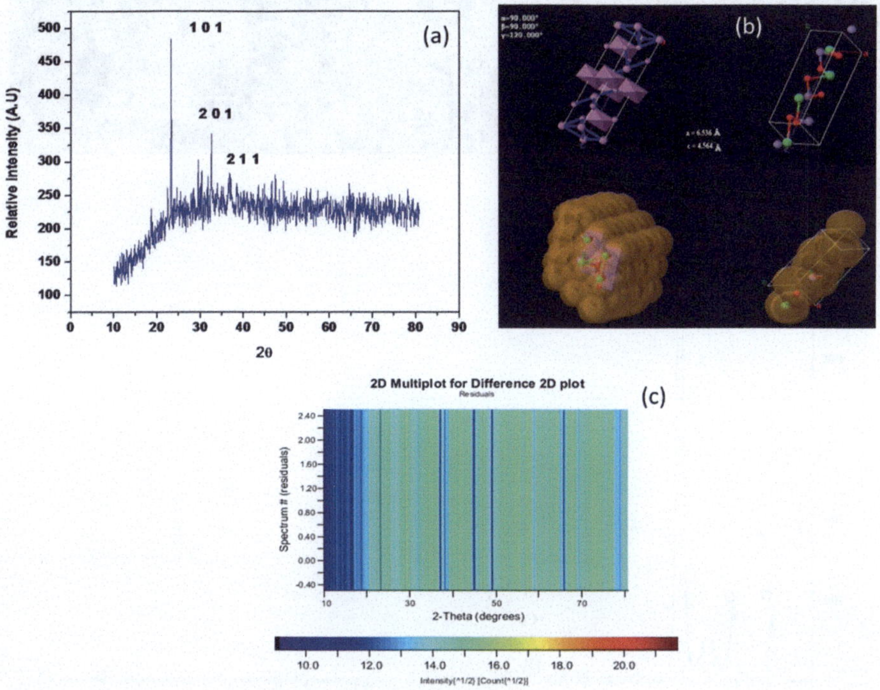

Fig. 2 **a** XRD of lithium cobalt oxide thin films (JCPDS File No.-78-1519); **b–c** Unit cell analysis and 2D view of lithium cobalt oxide thin films

3.2 SEM with EDAX Analysis

The SEM micrograph for thin film is presented in Fig. 3(a-b). Due to the thin film's lack of agglomeration, it was established that it maintained the same crystal structure with uniform particle sizes ranging from 2 to 2.5 nm and an average particle size of about 1.75 nm. This image shows a SEM micrograph with a homogeneous granular fresh surface. It revealed that the sample is thick, with no visible grain development or porosity. As a result, our material has a density of 2.79 g/cm^3, which corresponds to a high relative density of theoretical value (2.749 g/cm^3) [11]. Elemental analysis of the composition of thin film as shown in Fig. 3(c) respectively.

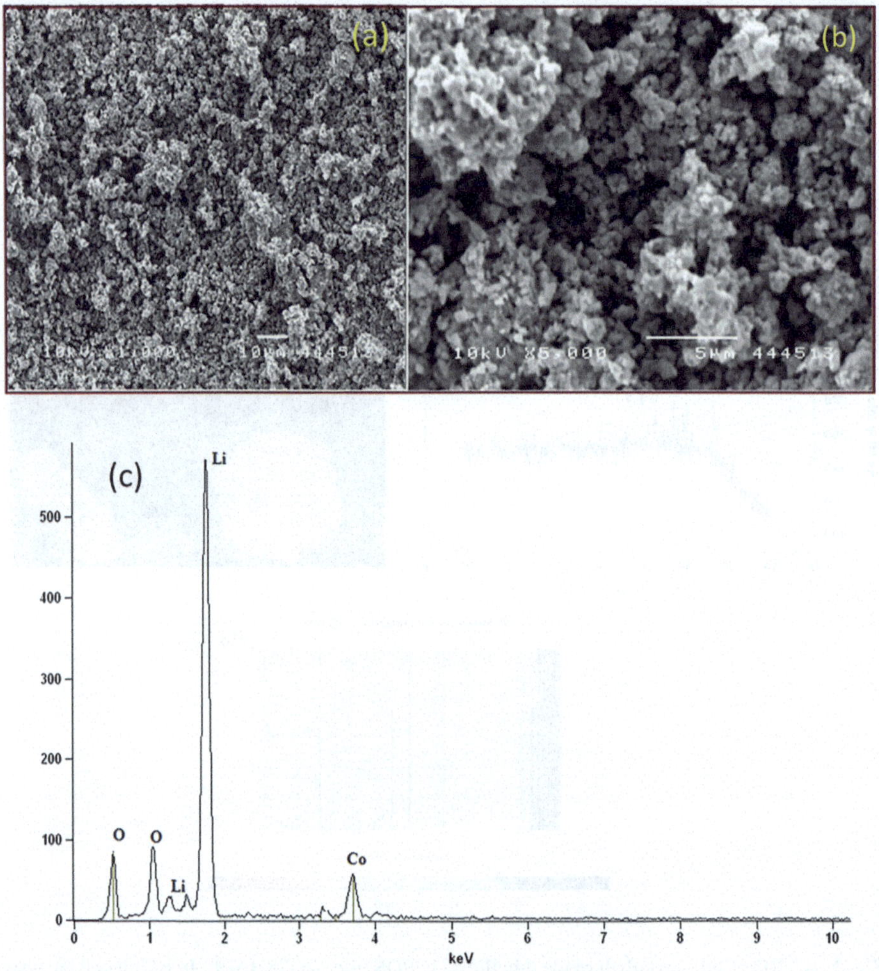

Fig. 3 a–c SEM and EDAX analysis of lithium cobalt oxide thin films

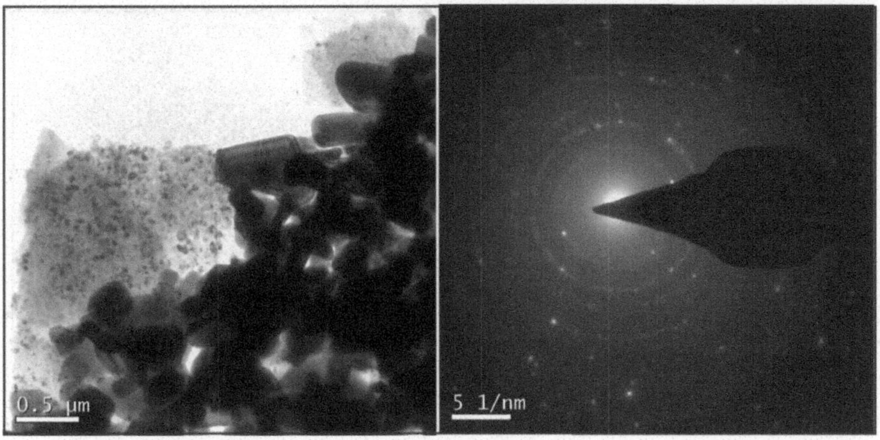

Fig. 4 **a–b** TEM analysis of lithium cobalt oxide thin films

3.3 TEM Analysis

A thin layer of film TEM images was implanted in thin film and based on the pattern shown in the TEM images and its diffraction planes are reported here as in Fig. 4a–b, For the plane of (101) planes, crystalline patterns with a spacing of 2.3791 nm were computed. The development of materials with changeable energy gaps is required for increased applications in electrochromic liquid crystal displays and optoelectronic devices. As a result, the absorption and transmission properties of the material were thoroughly investigated to compute bandgap fluctuations and related optical features. The bandgap of severely disordered evaporated $LiCoO_2$ films ranged between 3.2 and 3.4 eV reported earlier [12] is well-matched with SEM analysis. According to photo electrochemical tests, monoclinic, triclinic, and orthorhombic crystalline anodic films with varying degrees of hydration had a bandgap of 2.55–2.79 eV [9].

3.4 AFM Measurements

Figure 5a–c shows a representative AFM image of a 2 mN indent. Because of a raised piece at the lower right side of the indentation, our thin film has a minor crack. This dense crystalline $LiCoO_2$ exhibits nearly deeper indenter pile-up around two-dimensional (2D) and three-dimensional (3D) images of observed thin films with sizes ranging from 1 to 2 cm, this is consistent with the results of the SEM micrograph.

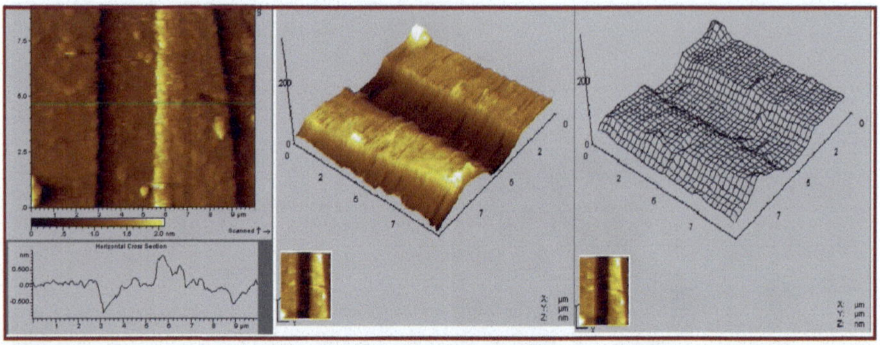

Fig. 5 **a–c** AFM analysis of lithium cobalt oxide thin films

4 Cyclic Voltammetry (CV) Measurements

For CV investigations on electrochromic thin films, a typical three-electrode setup was used. At this time, electrochemical cell approach is more or less standardized. Before the actual construction and performance evaluation of the cell, extensive research into the materials preparation, evaluation, determination of the electrochromic properties, and research on the electrode's and electrolytes' long-term stability was done. The right selection of constituent pieces is required for the electrochemical cell to perform properly [13]. CV is a straightforward but effective method for determining the electrochemical characteristics of a thin film, substance, biological material, or electrode surface. Since the pattern of the CV wave is affected by all aspects of the electrochemical cell, it is a strong approach. A schematic illustration of a cyclic voltammetry setup can be found in Fig. 6. Working electrode, reference electrode, and counter electrode are the three electrodes that make up CV configuration. The electrode power between the working and counter electrodes is changed in a suitable manner around a stable reference electrode system in electrochemical impedance. The current flow between the working and counter electrodes is monitored, same as in polarography [14]. After being linearly modulated from an initial to a final potential value, the voltage is swept back to the original potential at the same sweep rate. Figure 7 depicted the CV curves of a thin film that had been created. At a scan rate of 0.05 mV S^{-1}, our thin film's cyclic voltammetry curves within 0.1 and 3 V. A cathodic peak at 0.67 V may be attributed to embedded values on the electrode surface, while an anodic peak at 1.65 V may arise from Li^{2+} oxidation [11, 15, 16]. A polarization of the electrode materials may have led to the cathodic peak shifting to a bigger positive potential in later cycles. The CV curves virtually match after the first cycle, showing that the $LiCoO_2$ thin-film electrode holds its capacity well [17].

An example of a cyclic voltammogram (CV) in 0.1 M KOH electrolyte is as shown in Fig. 8. The CV spectra of typical CVs were recorded at a scan rate of 50 mV/s over multiple cycles (100–500 cycles). Voltage voltammograms change to a positive peak as the number of cycles increases, which causes a decrease in the current density

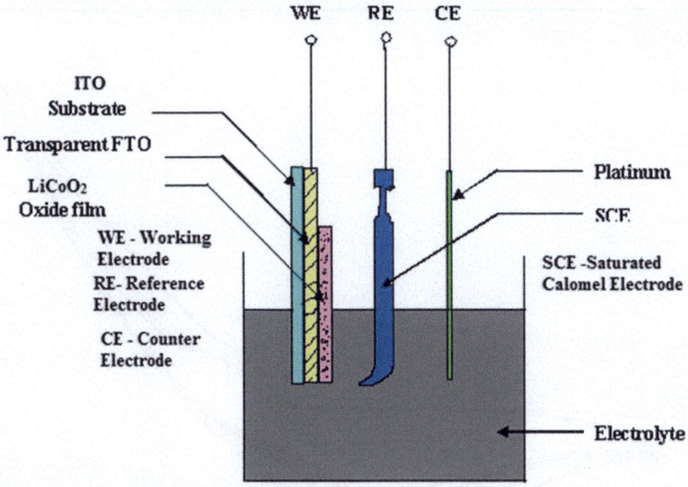

Fig. 6 Schematic representation of how a CV is set up

Fig. 7 Analysis of lithium cobalt oxide thin films is presented in the CV curve

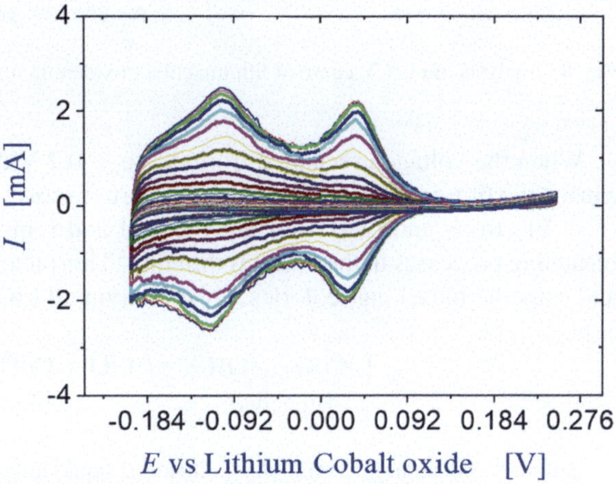

and an increase in potential value [18]. The reason for this is that $LiCoO_2$ thin films on ITO substrates degrade as permeable electrodes with increasing porosity [19]. A deep brown color is obtained in the film as a result of the oxidative reaction between Li(II) and Co(III), which occurs during the anodic scan, which scans from − 0.2 to 0.7 V. Anodic peaks are observed between 370 and 395 mV (E_a). As will be discussed below [20], oxygen evolution is the mechanism that causes the sudden spike in current that was noticed at the end of the anodic sweep.

$$2OH^- \rightarrow H_2O + {}^1\!/_2 O_2 \uparrow + 2e^- \quad (1)$$

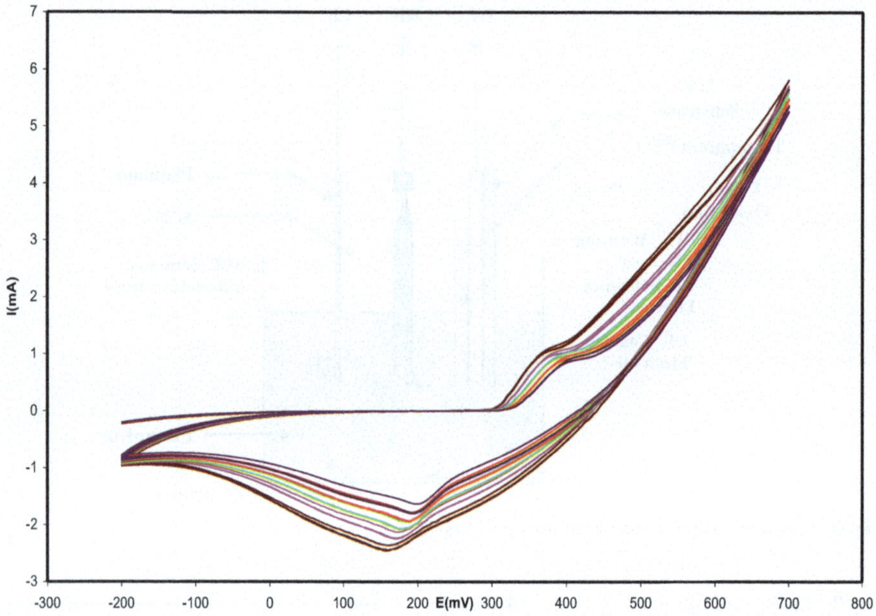

Fig. 8 Analysis of the CV curve of lithium cobalt oxide thin films for different cycling conditions

When the voltage is scanned from 0.7 to − 0.2 V during a cathodic scan, which causes the film to bleach, cathodic peaks are discovered between 200 and 155 mV (E_c). Electrons and OH^- ions are inserted and removed during the coloring and bleaching processes in the $LiCoO_2$ thin film. This pattern of anodic coloring (brown) and cathodic bleaching (colorless) is present in all $LiCoO_2$ thin films [21].

$$\underset{\text{Bleached}}{LiCoO_2 + OH^-} \leftrightarrow \underset{\text{Colored}}{LiO + CoOH + e^-} \tag{2}$$

Between 200 and 155 mV (E_c), cathodic peaks are observed during the cathodic scan, which scans the voltages from 0.7 to − 0.2 V, which causes bleaching of the film. During the coloring and bleaching processes of $LiCoO_2$ thin films, electrons and OH^- ions are inserted and removed. There is no difference in the coloration of $LiCoO_2$ thin films along the anodic and cathodic lines [21, 22]. Lithium cobalt oxide films do not require hydrogen to have electrochromic properties, therefore there are no fundamental requirements. Colored states occur on lithium and cobalt atoms when their valence bands are partially filled, or when electron vacancies are present. Bleached metal has a completely filled valence band. $LiCoO_2$ thin films with high oxygen content have been shown to exhibit p-type conductivity as a result of coloration induced by anodic scan [22, 23]. In $LiCoO_2$ thin films, the deep brown coloration can be attributed to oxidation peak detected before the current rise owing to oxygen evolution. Anodic electrochromism has been observed in thin films made of lithium cobalt oxide [24, 25], but the bleaching process appears to be linked to a cathodic reduction peak.

5 Conclusion

In addition to being energy-efficient and long-lasting, lithium-based batteries possess high energy density and long shelf lives. In order to grow lithium cobalt oxide thin films by co-precipitation, it is necessary to use a simple method for obtaining a reaction mixture. The XRD analysis of $LiCoO_2$ confirmed that it has average crystal size of 37.29 nm having predominant films of (101), (201), and (211) and crystallite size was found to decrease with doping concentration. The micro size uniform granular surface with a small crystalline size and larger surface area shows high reversible specific capacity and greater capacity persistence, the electrochemical studies explored the prepared thin film was electrochemically active and its discharge capacity is nearly 48 μAh/(cm^2 μm), respectively. The findings indicate that $LiCoO_2$ thin films could have a great prospect in lithium-ion batteries.

References

1. Y. Nishi, Lithium-ion secondary batteries; past 10 years and the future. J. Power Sources **100**(1–2), 101–106 (2001)
2. H. Li et al., Research on advanced materials for Li-ion batteries. Adv. Mater. **21**(45), 4593–4607 (2009)
3. C. Liu, Z.G. Neale, G. Cao, Understanding electrochemical potentials of cathode materials in rechargeable batteries. Mater. Today **19**(2), 109–123 (2016)
4. J.B. Goodenough, K.S. Park, The Li-ion rechargeable battery: a perspective. J. Am. Chem. Soc. **135**(4), 1167–1176 (2013)
5. Z. Chen, J.R. Dahn, Methods to obtain excellent capacity retention in $LiCoO_2$ cycled to 4.5 V. Electrochim. Acta **49**(7), 1079–1090 (2004)
6. T. Liu, X. Li, S. Sun, X. Sun, F. Cao, T. Ohsaka, J. Wu, Analysis of the relationship between vertical imparity distribution of conductive additive and electrochemical behaviors in lithium-ion batteries. Electrochim. Acta **269**, 422–428 (2018)
7. W. Mei, H. Chen, J. Sun, Q. Wang, Numerical study on tab dimension optimization of lithium-ion battery from the thermal safety perspective. Appl. Therm. Eng. **142**, 148–165 (2018)
8. D.W. Murphy, J.N. Carides, F.J. Di Salvo, C. Cros, J.V. Waszczak, Cathodes for nonaqueous lithium batteries based on VS2. Mater. Res. Bull. **12**(8), 825–830 (1977)
9. Y. Wang, Z.W. Fu, X.L. Yue, Q.Z. Qin, Electrochemical reactivity mechanism of Ni_3 N with lithium. J. Electrochem. Soc. **151**(4), E162 (2004)
10. A. Zhao, J. Liu, X. Ai, H. Yang, Y. Cao, Highly selective and pollution-free electrochemical extraction of lithium by a polyaniline/$LixMn_2O_4$ cell. ChemSusChem **12**(7), 1361–1367 (2019)
11. H.L. Wang, A. Gopalan, T.C. Wen, A novel lithium single ion based polyurethane electrolyte for light-emitting electrochemical cell. Mater. Chem. Phys. **82**(3), 793–800 (2003)
12. Y. Takahashi, T. Yamashita, D. Takamatsu, A. Kumatani, T. Fukuma, Nanoscale kinetic imaging of lithium-ion secondary battery materials using scanning electrochemical cell microscopy. Chem. Commun. **56**(65), 9324–9327 (2020)

13. A. Patil, V. Patil, D.W. Shin, J.W. Choi, D.S. Paik, S.J. Yoon, Issue and challenges facing rechargeable thin film lithium batteries. Mater. Res. Bull. **43**(8–9), 1913–1942 (2008)
14. H. Kitao, T. Fujihara, K. Takeda, N. Nakanishi, T. Nohma, High-temperature storage performance of Li-ion batteries using a mixture of Li-Mn spinel and Li-Ni-Co-Mn oxide as a positive electrode material. Electrochem. Solid State Lett. **8**(2), A87 (2004)
15. F. Kaneko, S. Wada, M. Nakayama, M. Wakihara, J. Koki, S. Kuroki, Capacity fading mechanism in all solid-state lithium polymer secondary batteries using PEG-borate/aluminate ester as plasticizer for polymer electrolytes. Adv. Func. Mater. **19**(6), 918–925 (2009)
16. H.C. Shin, J.A. Corno, J.L. Gole, M. Liu, Porous silicon negative electrodes for rechargeable lithium batteries. J. Power Sour. **139**(1–2), 314–320 (2005)
17. M.Z. Xue, Z.W. Fu, Electrochemical reaction of lithium with nanostructured thin film of antimony trioxide. Electrochem. Commun. **8**(8), 1250–1256 (2006)
18. K.J. Chee, V. Kumar, C.V. Nguyen, J. Wang, P.S. Lee, Polymer light-emitting electrochemical cell blends based on selection of lithium salts, LiX [X = trifluoromethane sulfonate, hexafluorophosphate, and bis(trifluoromethylsulfonyl) imide] with low turn-on voltage. J. Phys. Chem. C **120**(21), 11324–11330 (2016)
19. W. Choi, H.C. Shin, J.M. Kim, J.Y. Choi, W.S. Yoon, Modeling and applications of electrochemical impedance spectroscopy (EIS) for lithium-ion batteries. J. Electrochem. Sci. Technol. **11**(1), 1–13 (2020)
20. R. Senthil, G. Vijayaragavan, A. Ayeshamariam, K. Kaviyarasu, Nonlinear optical properties of single crystal of L-OOMHCL incorporation with glycine oxalic acid (GOA) with high chemical stability for optoelectronic applications. Surf. Interfaces **18**, 100417 (2020)
21. G.J. Wang, Q.T. Qu, B. Wang, Y. Shi, S. Tian, Y.P. Wu, R. Holze, Electrochemical behavior of $LiCoO_2$ in a saturated aqueous Li_2SO_4 solution. Electrochim. Acta **54**(4), 1199–1203 (2009)
22. K. Hoang, M. Oh, Y. Choi, Electronic structure and properties of lithium-rich complex oxides. ACS Appl. Electron. Mater. **1**, 75−81 (2019)
23. S. Waki, K. Dokko, T. Itoh, M. Nishizawa, T. Abe, I. Uchida, High-speed voltammetry of Mn-doped $LiCoO_2$ using a microelectrode technique. J. Solid State Electrochem. **4**(4), 205–209 (2000)
24. Y.H. Rho, K. Kanamura, Li+-ion diffusion in $LiCoO_2$ thin film prepared by the poly (vinylpyrrolidone) sol-gel method. J. Electrochem. Soc. **151**(9), A1406 (2004)
25. E.S. Lin, F.S. Chao, C.J. Liang, C.J. Chang, A. Fang, C.K. Lin, Y.C. Chang, C.Y. Tsay, J.J. Wu, C.Y. Chen, Hydrothermal synthesis of Co_3O_4/$ZnCo_2O_4$ core-shell nanostructures for high-performance supercapacitors. J. Electrochem. Soc. **168**(12), 123502 (2021)

In the Construction of Energy Storage Supercapacitors, Nanostructures Play a Key Role

S. John Sundaram, J. Bosco Franklin, S. Sachin, and A. Dhayal Raj

Abstract Efforts must be made to develop cleaner and more sustainable energy conversion systems in response to the exhaustion of fossil fuels and the resulting ecological problems. Supercapacitors and batteries have become essential for maintaining a continuous and balanced power supply in the face of limitations of these new energy sources, such as their time and location limitations. A novel electrode material is being explored by researchers to overcome these challenges. In an advanced technique, conductive carbon matrix is incorporated with electrochemically active oxide metals to form an electrode hybrid. Li-ion batteries (LIB) and supercapacitors have already demonstrated significant electrical performance improvements using metal oxide/graphene composites despite being relatively new research fields. Aside from amplified capacity/capacitance, enhanced rate capabilities, and improved cycling stability, these improvements also make it possible for the device to deliver higher energy and power densities. Energy storage technologies have been revolutionized because of graphene's remarkable properties, and further research is being undertaken in this area because of its potential.

1 Introduction

The exhaustion of fossil fuels and the resulting ecological issues must spur the innovation of clean and sustainable energy conversion systems. To address the limitations of these new energy sources, such as their time and region restrictions, energy storage systems like supercapacitors and batteries have become essential for maintaining a continuous and balanced power supply [1]. Despite having lower energy density compared with batteries, supercapacitors have garnered significant attention over the years because their remarkable power density and fast charging capacities [2]. Various areas applications were found, including powering hybrid electric vehicles, in transferrable electronics acting as power holdup, load-leveling

S. John Sundaram (✉) · J. Bosco Franklin · S. Sachin · A. Dhayal Raj
Department of Physics, Sacred Heart College (Autonomous), Tirupattur, Tamil Nadu 635601, India
e-mail: johnsundaram@shctpt.edu

K. Kasinathan et al. (eds.), *New Technologies for Energy Transition Based on Sustainable Development Goals*, https://doi.org/10.1007/978-981-97-2527-4_18

facilitating, and supporting other continuous energy needs. With the ever-increasing energy demands, researchers are actively searching for latest materials and refining developments to create extremely advanced devices that store energy. The scientific counsel is currently intensive on enhancing energy and power density of supercapacitors by developing materials for novel electrode, innovative electrolytes, and device designs that are improved [3]. Regarding their "*mechanism of energy storage*" and the properties of the conductor material, supercapacitors may be categorized into three main varieties. The first type is electrical double-layer capacitors (EDLCs), at boundary between the surface of the electrode and the electrolyte, the electrostatic collects charge. Secondly pseudocapacitors, where quick and reversible redox process involved (Faradaic processes) on the electrode surface [4]. Lastly, there are hybrid capacitors that combine both the EDLC and pseudocapacitor mechanisms as a one device.

Supercapacitors, specifically EDLCs, store electrical energy without involving Faradic charge transfer through ion adsorption, resulting in remarkable features such as extreme power density, rapid charge–discharge progressions, and admirable cycling stabilities. Though, the energy densities of conventional electrode materials remain limited as a result of their surface area and constrained pore size distribution [5]. To overcome this limitation, pseudocapacitors have been established to enhance energy density. In recent years, significant advancements in nanotechnology and nanoscience have provided a driving power for the creation of novel and superior supercapacitor electrode structures. The reduced size of nanomaterials plays an important role in significantly enlightening the utilization of active materials, and the transport path ions and electrons are shortened [6]. Despite being in the form of nanomaterials, single-phased nanomaterials still exhibit performance limitations due to their inherent feebler material properties, including cycling stability, conductivity, and mechanical stability. As researchers continue their efforts, the aim is to address these challenges and further optimize the performance of nanomaterial-based supercapacitors [7]. Higher energy density offered by metal oxides compared with conventional carbon materials exhibit good cycling stability comparing with polymer materials. However, they have low conductivity, which is a significant drawback. High cycle life and high-power density are demonstrated by carbon materials, but their broader applications are limited by their poor specific capacitance [8]. To overcome these challenges, researchers are exploring novel electrode materials. An advanced approach involves hybridizing electrode through incorporating metal oxides that are electrochemically active to a conductive carbon matrix. This entails loading and/or integrating metal oxide nanoparticles into nanostructured materials, like carbon nanotubes (CNT), mesoporous carbons, carbon aerogels, graphene, and others. The uses of active materials are enhanced, by creating metal oxide/carbon composites, while simultaneously improving the electrical conductivity and mechanical strength of the materials. This advanced approach holds potential in addressing the limitations of both carbon materials and metal oxides, leading to improved supercapacitor performance and broader practical applications.

Based on calculations, electrode material costs approximately 90% of a supercapacitor's cost. For instance, cost of ruthenium oxide poses a practical obstacle despite its exceptional properties. To address this issue, several studies have aimed on combining ruthenium oxide together with more economical metal oxides, aiming to reduce the reliance on precious metals [5]. In parallel, researchers have dedicated significant efforts to exploring alternative materials that are both cost-effective and environmentally friendly while exhibiting electrochemical behavior comparable with that of ruthenium oxide. Some of these other materials are like MnO_2, Fe_2O_3, Fe_3O_4, TiO_2, SnO_2, NiO, and V_2O_5. These ongoing endeavors aim to develop supercapacitors with improved cost-efficiency and sustainability, without compromising on performance. Power density has been actively focused on exploring electrode materials capable of enhancing specific capacitance and achieving extreme power density in supercapacitors simultaneously. In modern days, graphene has emerged as a wonderful material for supercapacitor electrode, owing to its remarkable characteristics. In a honeycomb like crystal lattice graphene, an ordered arrangement of carbon atoms bonded in the sp^2 configuration, forming a sheet that is just one atom thick, and which was thinnest material known till date [9]. Graphene exhibits distinct electronic behavior because of the Dirac equation defining charge carriers within it, stemming from its specific electronic properties. Graphene also showcases extraordinary electronic attributes, including an astonishingly fast electron mobility of approximately 15,000 cm^2/V s and a super high mobility of temperature-independent charge carriers at 200,000 cm^2/V s. Additionally, it boasts a productive Fermi velocity of 106 m/s in room temperature, making a noteworthy comparison with the speed of light. Graphene exhibits outstanding optical, mechanical, thermal, and electrochemical properties further than its electric properties, which surpass various carbon allotropes like fullerene, graphite, diamond, and carbon nanotubes (CNTs) [10]. Single-layer and graphene boasts outstanding mechanical strength with stiffness 130 GPa, 1.0 TPa Young's modulus, offering around 97.7% optical transmittance and an impressive thermal conductivity around 5000 W m^{-1} K^{-1}. Its extreme electrical conductivity coupled with exceptional flexibility further enhances its appeal. As a result of these exceptional properties, graphene is predicted to overtake metals, CNTs, graphite, and semiconductors in various applications, whether utilized as a separate material or as a component within hybrid. The potential of graphene in advancing supercapacitor technology has spurred widespread interest and ongoing research [11].

In energy storage applications graphene offers several unique advantages over carbon nanotubes (CNT's) and graphite. One of its prominent strengths is its significantly higher theoretic specific surface area, measuring 2600 m^2/g, exceeding the performance of (CNT's) (1301 m^2/g) and graphite (10–20 m^2/g). The surface area enlarged and provided a plethora of electrochemical reaction active sites, making graphene an excellent material for energy storage [12]. Furthermore, the flexibility of graphene sets it apart from brittle graphite, making it well-suited for constructing flexible energy storage devices. Graphene's flexibility offers practical advantages in certain applications. Additionally, graphene eliminates the issue of residual metallic impurities that can be challenging to remove, a problem often encountered with CNTs, and this obstacle hinders various practical applications, including storing energy. Graphene offers another crucial advantage due to the significant profusion of oxygen-incorporated functional groups, which are composed of various elements, present on the edges and surface of reduced graphene oxide (rGO) and graphene oxide (GO). These functional groups exert a potent influence on the morphology, dimensions, and dispersion of metal oxide particles on graphene. Graphene serves as an exceptional scaffold for the consistent initiation, expansion, or arrangement of metal oxides, featuring established characteristics of form, size, and crystal structure. It proficiently mitigates alterations in volume and the tendency of metal oxides to aggregate during the entirety of the charge–discharge procedure. Despite being a relatively recent area of research, metal oxide/graphene composites have already shown significant improvements for Li-ion batteries (LIBs) and supercapacitors as electrodes in electrochemical performance. These improvements include amplified capacity/capacitance, enhanced rate capability, enhanced cycling stability, and higher energy and power densities. The remarkable properties of graphene have demonstrated its potential in revolutionizing energy storage technologies and are driving further investigations in this field. Reduced graphene oxide (rGO) and chemically modified graphene (CMG) are particularly intriguing due to their accessible and cost-effective chemical method, resulting in a material with conductivity comparable with pristine graphene. Consequently, rGO has gained extensive usage as a highly conductive material for electrode in various applications, including supercapacitors. Understanding the mechanisms of reducing graphene oxide (GO) becomes crucial in the improvement of efficient materials for supercapacitor electrode. GO can be transformed into graphene by employing diverse reduction methods such as chemical, thermal, electrochemical, photothermal, laser sonochemical, photocatalytic, and microwave approaches [13]. These reduction techniques play a pivotal role in converting GO into rGO, leading to improved electrical conductivity, and making it a valuable component for constructing high-performance electrode for supercapacitor. The wide range of methods used to reduce rGO provides the opportunity to modify its properties to match specific requirements for storing energy. This capability contributes to driving significant strides forward and fostering innovative developments within the field of supercapacitor research.

Standard wet-chemistry approaches, such as employing chemical sol–gel method and hydrothermal process, were utilized in the crafting of a multitude of composite materials that integrate metal oxides and graphene. Commencing with a dispersed solution containing suspended graphene, these composites are created, wherein graphene acts as a precursor within an interconnected framework supporting discrete metal nanoparticles. In the initial stages, both graphene oxide (GO) and reduced graphene oxide (rGO) are scattered in aqueous or organic solvents, facilitated by electrostatic stabilization and chemical functionalization, as evidenced by Wu et al., in their 2012 study [14]. Notably, graphene assumes a crucial and indispensable role in effectively suppressing the undesirable agglomeration tendencies exhibited by metal oxide nanoparticles within the composite structure. The focal point of this chapter revolves around emphasizing the paramount significance attributed to nanostructured materials, which play a pivotal role in substantially enhancing the overall performance and capabilities of supercapacitors, a theme extensively discussed and explored in the context of Ref. [15].

2 Why Nanostructured Nanomaterials Are Used in Supercapacitors?

The researchers around the world have done enormous amount of work to obtain the suitable material for renewable energy resources, because of the growing energy demand, global warming, increasing oil and coal prices and its deficit in suppliers. The exploration of novel nanocrystalline materials and innovative technological solutions is poised to make a substantial impact on the progress of energy systems [16]. In nanocrystals, a significant proportion of atoms, often one in ten or more, reside on their surfaces. Consequently, nano-sized metals exhibit a considerable reduction in melting points, primarily attributed to the substantial influence of the total free energy. The phenomenon of size quantization in semiconductor nanoparticles, also known as nanocrystals, nanoclusters, and quantum dots, has sparked a thriving field of research in material science. Nanostructure exploration has led to the development of smaller, lighter, more affordable, and quicker devices with enhanced functionality, all while conserving raw materials and energy, thanks to the advancements in nanotechnology. Operating on scales as minuscule as a one in million compared with meter, nanotechnology represents the nearly invisible science to construct as shown in Fig. 1. Nanocrystal's optical characteristics are dictated by their size and surface chemistry, setting them apart significantly from those bulk solids [17]. Nanomaterials exhibit an array of highly exotic properties, contributing to their ever-increasing demand in various industries.

2.1 Key Properties of Nanomaterial

- Size-dependent properties

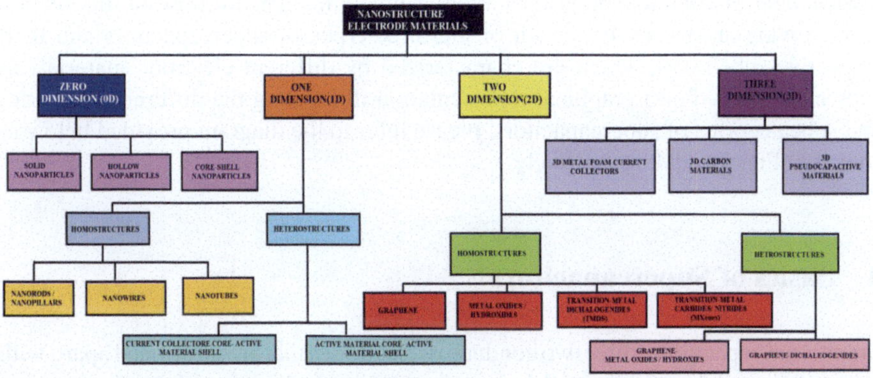

Fig. 1 Wide variety of highly exotic properties are demonstrated by the schematic mechanism of nanomaterials, contributing to their ever-increasing demand

- Extraordinary surface—volume ratio and surface energy
- Extraordinary mechanical strength and micro hardness
- High specific heat
- Huge thermal expansion
- High catalytic action
- Improved self-diffusion
- Extraordinary magnetic susceptibility
- Extraordinary sintering rate
- Other unique assets of nanomaterials.

As a result of incorporating nanomaterials, a high surface-volume ratio occurs, increasing the surface area and, therefore, the capacitance.

3 Taxonomy of Supercapacitors

Depending on recent research and development, supercapacitors are categorized by three main categories.

- Electrochemical double-layer capacitors
- Pseudocapacitors, and
- Hybrid capacitors.

Within each category, there exists a unique charge storage mechanism, which can be categorized as Faradaic, non-Faradaic, or a hybrid combination of both. The Faradaic reaction entails charge transfer occurring within the electrode and electrolyte, frequently encompassing oxidation and reduction processes. Conversely, non-Faradaic mechanisms operate independently of chemical processes, contributing to the multifaceted landscape of charge storage mechanisms across various classes; instead, charges are spread on surfaces through physical processes without the involvement of chemical bond formation or breaking. The purpose of this section is to provide an overview of each of the three types of supercapacitors and their respective subclasses, which are characterized by different electrode materials as shown in Fig. 2. For a graphical representation illustrating the different categories and subcategories of supercapacitors, please refer to the diagram provided below as shown in Fig. 3 and 4, respectively.

4 Basics of Supercapacitors

Traditional capacitors utilize two conductive electrodes that are positioned apart, with an insulating dielectric material in between. Upon applying a voltage, contrasting charges amass on the surface of each electrode. A standard supercapacitor, however,

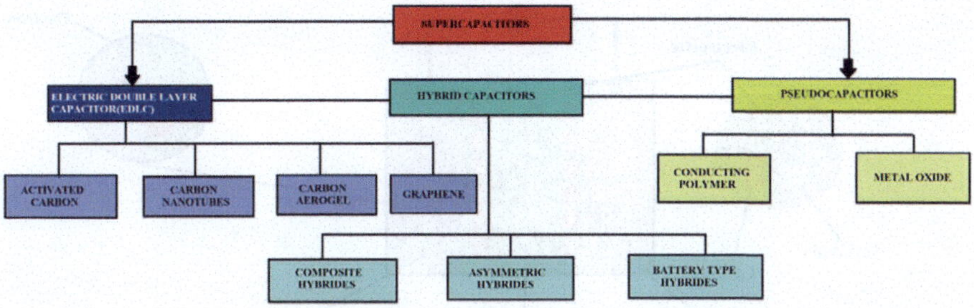

Fig. 2 Supercapacitor can be classified into a variety of types based on the materials used for its electrodes

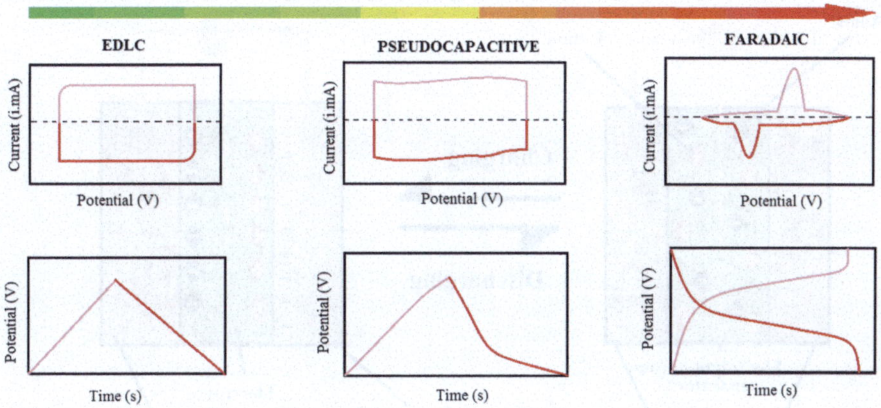

Fig. 3 Pseudocapacitor physiology distinguishes between Faradaic and non-Faradaic charge storage mechanisms, as well as a hybrid combination

comprises three crucial elements: electrodes, electrolyte, and separator. Collaborating seamlessly, these components efficiently store and release electrical energy. Figure 4 depicts a schematic representation of the basic pseudocapacitor mechanism. The effectiveness of supercapacitors is significantly influenced by the intrinsic physical characteristics of both the electrode and the electrolyte materials. The electrode's role is particularly crucial, actively storing and releasing charges, which directly shapes the supercapacitor's energy and power densities. Scientists often analyze the electrochemical performance of a supercapacitor using cyclic voltammetry and galvanostatic charge–discharge measurements [18–21]. These techniques offer valuable insights into the capabilities and operational characteristics of the supercapacitor [22, 23].

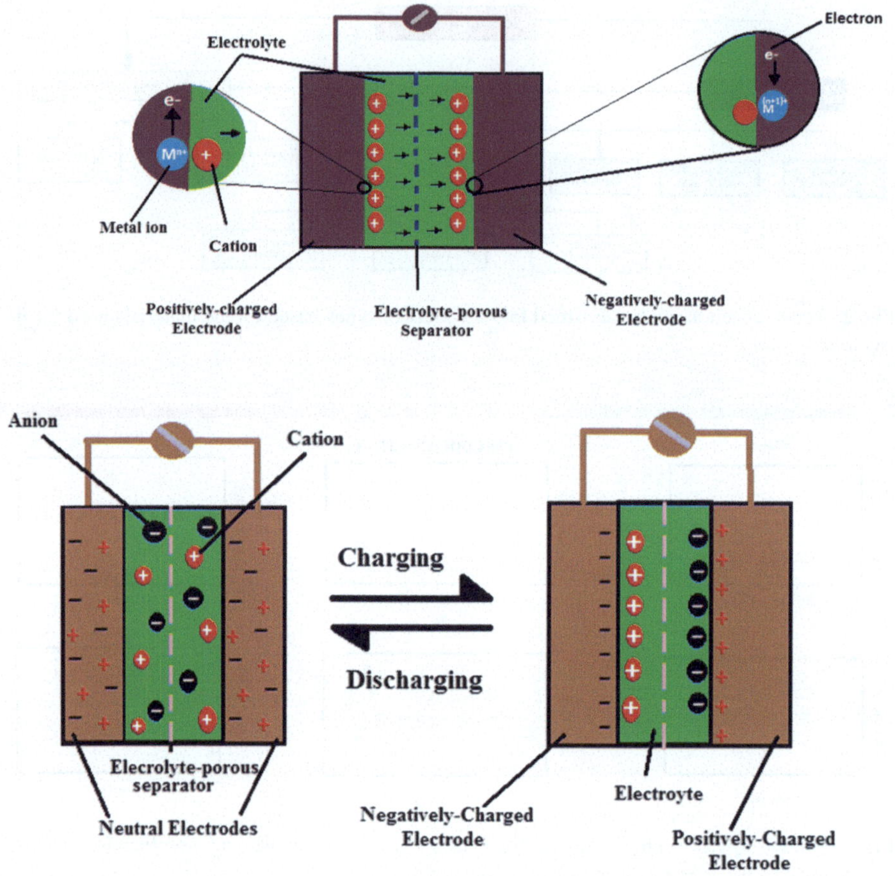

Fig. 4 Schematic illustration visually represents the mechanism of a pseudocapacitor

5 Charge Storage Mechanism in Supercapacitors

Sanket et al. [24] have discussed about the way supercapacitors store charges. The group proposed that analyzing the behavior of the supercapacitor could be done by investigating its energy density, and power densities involve studying the definitions of these parameters. Energy density (E) and power density (P) can be characterized by the following definitions:

$$E = \frac{V^2 * C}{2} \quad \text{and} \quad P = \frac{V^2}{4R}, \tag{1}$$

where

C—capacitance (Farad),

V—operating potential (volt),

R—equivalent series resistance (ohm).

Enhancing both the device's capacitance and its operational potential range is essential to achieve higher energy density in a supercapacitor. Simultaneously, expanding the potential range and reducing series resistance within the system contribute to improving power density. The capacitor's capacitance is intricately regulated by the selection of electrode material and the precise charge storage mechanism employed. In supercapacitor devices, the utilization of two primary charge storage mechanisms is evident: electric double-layer capacitance (EDLC) and pesudocapacitance. Additionally, a hybrid approach frequently combines EDLC-type electrodes with battery-type electrodes, by incorporating precise material loading, the goal is to attain elevated energy and power densities within the ultimate cell. Cyclic voltammetry (CV) and galvanostatic charge–discharge (CD) tests are frequently utilized to assess the capacitance of electrode materials. The specific capacitance (C_{sp}) can be calculated from the cyclic voltammetry data using the following expression.

$$C_{sp} = \frac{I\Delta t}{m\Delta V}, \tag{2}$$

where

I—discharge current (amp),

Δt—discharge time (sec),

m—the mass of active material contained within the electrode (grams),

ΔV—voltage change during discharge (volt).

Greater duration of discharge time signifies enhanced capacitance of the material. Cyclic voltammetry (CV) and galvanostatic charge–discharge (CD) curves showcase discernible and individualistic features when observed in the context of electric double-layer capacitors (EDLCs), pseudocapacitors, and hybrid capacitors. A thorough analysis of the charge storage mechanism is presented for each type of capacitor electrode material, with special attention given to nanostructured materials [25–29].

6 Role of 0D, 1D, 2D and 3D Materials in Supercapacitors

Division	Subdivision	Material/precursor	Electrochemical behaviors	References
1D	1D hemostructures	$NiCo_2S_4$ nanotube and sheets	11.2 F cm^{-2} (3227.94 F/g) at CD 2 mA cm^{-2} (2 A g^{-1})	[29]
		One-dimensional Ag_x/ TiO_2 nanostructure	Low power density	[30]

(continued)

(continued)

Division	Subdivision	Material/precursor	Electrochemical behaviors	References
		Ag/c-CN@CF//AC@CF supercapacitor	A remarkable 86.1% of the initial capacitance is retained even after undergoing 2500 cycles, high power, and high energy density	[31]
		Wood sawdust Activated carbon with KCl and KOH	The specific capacitance is recorded as 285 F g^{-1} at a current density of 1 A g^{-1}, while the energy density attains 5.60 Wh kg^{-1}. Remarkably, a power density of 9210 W kg^{-1} is achieved. Furthermore, an impressive capacitance retention of 99.8% is maintained even after undergoing an extensive 10,000 cycles	[32]
		Electrode $MoS_2/Cu(OH)_2$/CFP	At a current density of 1 mA cm^{-2}, a substantial capacitance of 1120 mF cm^{-2} is attained. 98.3% capacitance retention after 10,000 cycles	[33]
1D	1d heterostructures	CoNi-ZIF/*N*-CNT-2 fabrication of heterostructure 1D/2D nanostructures	Extreme specific capacitance up to 1110 F g^{-1} at 1 A g^{-1}	[34]
		$MnCo_2O_4$@NiCo-LDH core-sheath	An impressive specific capacitance of 4550.0 F/g is achieved at a current density of 1 Ag^{-1}	[35]

(continued)

(continued)

Division	Subdivision	Material/precursor	Electrochemical behaviors	References
		Co_3O_4@$NiMoO_4$/$CoMoO_4$ heterostructures	An elevated energy density of 53.9 Wh/kg is attained, coupled with a notable retention rate of 84.5% even after undergoing 1000 cycles	[36]
2D	2D hemostructures	2D/2D CoAl LDH/g-C_3N_4	High capacitance	[37]
		Amorphous cobalt doped nickel phosphate	An exceptional specific capacity of 1214 F g^{-1} is achieved at a current density of 1 A g^{-1}	[38]
		NiCo/g-C_3N_4 heterostructures	The composite demonstrates impressive values, reaching as high as 1100 and 46 mg/g at 30 °C	[39]
		CoAl LDH/g-C_3N_4	Wide linearity range	[37]
		2D ordered mesoporous carbon/MXene heterostructures	High capacitance around 240 F/g at CD 0.2 A/g	[40]
		2D/2D MXene/BCN heterostructure	A significant specific capacitance of 1170 F/g (equivalent to 1876 C g^{-1}) is achieved at a current density of 2 A g^{-1}	[41]
		Two-dimensional siloxene—graphene heterostructure	Energy density (55.79 Wh/kg)	[42]

(continued)

(continued)

Division	Subdivision	Material/precursor	Electrochemical behaviors	References
		$Cu(OH)_2$ nanosheet	An impressive retention rate of 86.2% is maintained over 6000 cycles, with a specific capacitance of 190.53 F g^{-1} at a current density of 1 A g^{-1}. Furthermore, the energy density reaches approximately 60.16 Wh/kg, accompanied by a power density of 749.98 W/kg	[43]
3D	3D hemostructures	PANI@CNF/CN	The system exhibits a commendable energy density of 65.3 Wh kg^{-1} at a power density of 800 W kg^{-1}, showcasing its robust performance. Furthermore, an impressive retention rate of 98% is sustained even after undergoing 5000 cycles at 10 A g^{-1}	[44]
		Two-dimensional (2D) graphene sheets in conjunction with three-dimensional (3D) carbon nanospheres	An elevated specific capacitance of 235.3 F g^{-1} is achieved at a current density of 0.2 A g^{-1}. The system also demonstrates remarkable cyclic stability, maintaining a capacitance retention of 91% at 5 A g^{-1} even after 5000 cycles	[45]
		Cobalt-doped nickel phosphate 3D flower-like microstructure	An impressive specific capacity of 1214 F g^{-1} is achieved at a current density of 1 A g^{-1}	[38]

(continued)

(continued)

Division	Subdivision	Material/precursor	Electrochemical behaviors	References
		Hollow tube-sheets with a $NiCo_2S_4$ core–shell nanoarray structure	An outstanding capacitance of 11.3 F cm^{-2} (equivalent to 3220.94 F/g) is achieved	[29]
3D	3D hemostructures	CuCe-MOF@PANI-1	At a current density of 1 A/g, the system exhibits a high capacitance of 724.4 F g^{-1} and 713.2 F g^{-1}, highlighting its exceptional performance	[46]
		$Ti_3C_2T_X$@NiO	Operating at a current density of 1 A g^{-1}, the specific capacitance achieves a notable level of up to 966 F g^{-1}	[47]
		Three-dimensional (3D) nanosheets composed of FeNi-layered double hydroxides (LDH) and $Ti3C_2T_x$-MXene	Energy densities of 94.1 Wh kg^{-1} and a power density of 7431.8 W kg^{-1} are realized, showcasing exceptional system capabilities	[48]

7 The Role of Conducting Polymers in Supercapacitors

The primary objective of this endeavor is to harness the potential of conducting polymers to construct and develop supercapacitor electrodes and devices, thereby advancing the realm of energy storage and capacitive technology. This effort aims to establish a connection between current carbon-based supercapacitors and batteries, effectively bridging the gap and yielding units that exhibit a moderate level of specific energy. In their research endeavors, scientists have examined prominent conducting polymer materials, including polypyrrole, polyaniline, polythiophene, and various polythiophene derivatives. Furthermore, these researchers have probed the synergistic amalgamation of these substances with carbon nanotubes and inorganic battery constituents [49]. Another variety of supercapacitor is known as a pseudocapacitor. It attains its capacitance by storing charge within the bulk of a redox material through a redox process. Exhibiting rapid redox reactions akin to capacitance, this phenomenon is aptly labeled as "*pseudocapacitance*" [50–52].

7.1 Polyaniline

Polyaniline (PANI) is prominently recognized as one of the most auspicious conducting polymers, extensively used as an economical positive electrode material in energy storage (ES) applications. This can be attributed to its versatile oxidation state and exceptional doping-de-doping characteristics, resulting in a remarkably high specific pesudocapacitance [53–55]. PANI can be synthesized through both chemical and electrochemical oxidation techniques, utilizing diverse substrates such as nickel, carbon, stainless steel, flexible substrates, or in combination with various materials for ES applications. Multiple literature studies on supercapacitors based on polyaniline (PANI) have showcased a diverse range of specific capacitance values that are observed, spanning from 30 to 3000 F/g. This significant variability can be ascribed to several aspects, including the structural morphology, dopant concentration, ionic diffusion length, and polymerization process, of the electroactive material. To develop an ideal PANI-based pseudocapacitor, two essential requirements are nanoscale structures and with high surface area, this achievement is possible by introducing various dopants under improved conditions [56]. Polyaniline exhibits a notable breadth of capacity, showcasing a wide spectrum that spans from a lower limit of 44 mAh/g to a higher limit of 270 mAh/g [57]. The difference in capacity is attributed to several influences, such as polymer morphology, the chosen synthetic route, type and quantity of adhesives and supplements, as well as the width of the electrode. Among conducting polymers, polyaniline demonstrates the highest variable specific capacitance, with electro-deposited polyaniline achieving even higher specific capacitance compared with chemically synthesized forms [49]. As a positive electrode, researchers have employed polyaniline, while carbon serves as the negative electrode in the developed device [58]. This device is categorized as a hybrid electrochemical capacitor and based on Park et al. [58]. Drawing upon the foundation of polythiophene derivatives, this approach yields comparable or greater standards than those of recently reported Type III supercapacitors.

7.2 Polypyrrole

Polypyrrole (PPy) is deemed exceptionally promising for Faradaic pseudocapacitor applications due to its impressive characteristics. These include, rapid charge–discharge mechanism, excellent heat stability, high conductivity cost-effectiveness, and substantial energy storage capacity [55]. PPy is commonly infused with both single and multiple-charge anions [59], further enhancing its electrochemical properties. In comparison with many conducting polymers, Polypyrrole grants a higher degree of adaptability in electrochemical treatment [60], making it a subject of significant interest for supercapacitor and battery electrode development [61, 62]. On the downside, polypyrrole lacks the ability to be *n*-doped like its thiophene derivatives counterparts, limiting its usage primarily to cathode materials [49]. However, polypyrrole compensates for this drawback with its higher density, leading to an increased capacitance within each unit volume, (400–500 F/cm^3) [63]. However, it should be noted that a decline in capacitance ensues when there is an occurrence of thicker and dense polypyrrole development on the current collector. This diminution can be attributed to the constriction of dopant accessibility to the internal sites of the polymer pillar, thereby hindering effective charge storage. In recent years, considerable research in the field of conducting polymers has been focused on polypyrrole for developing pseudocapacitor electrodes using both chemical and electrochemical methods. Chemical oxidation-polymerization allows the synthesis of various PPy nanostructures with the presence of excess surfactants. However, creating electrodes for energy storage (ES) implementing applications through this methodology proves to be intricate because of the elevated internal resistance induced by the existence of binders [64]. In comparison with alternative nanostructures, PPy nanosheets showcase a notably superior specific capacitance. To assess the conductance steadiness of PPy-coated paper under various curves, the current at a potential of 0.5 V was monitored. Notably, the conductance of the PPy-coated paper exhibited remarkable stability, remaining nearly unchanged even after enduring 100 instances of bending. This technique holds significant potential as a cost-effective method for the mass production of conductive paper-based electrodes, which find utility in energy storage applications and flexible electronics. Consequently, it has sparked intense interest in similar fields. Although this material shows promise, it has not yet reached high-performance levels. Consequently, substantial progress is necessary to achieve a significantly high specific capacitance suitable for real-world applications [55].

7.3 *Polythiophene*

Polythiophene (PTh) and its analogues showcase dual attributes, encompassing both p-type and *n*-type characteristics as conducting polymers (CPs). In their normal state, polythiophene displays weak conductivity. Conversely, exceptional stability is demonstrated by p-doped polymers even when exposed to air and humid conditions. While the specific capacitance of PTh-dependent electrodes tends to be comparatively lower than that of PAn or PPy, a significant edge emerges from their capacity to effectively operate within a relatively broader potential range, approximately 1.2 V [65], when *n*-doped, polythiophene requires much lower potentials [66, 67]. However, in this state, it becomes less stable to water and oxygen, and its conductivity decreases distinguished with the *p*-doped state [68].

8 The Role of Metal Oxides in Supercapacitors

Supercapacitors, alternatively labeled as ultracapacitors or electrochemical capacitors. They have attracted considerable attention within the realm of energy-related studies and endurance because of their exceptional electrochemical performance. They exhibit excellent specific capacitance, fast charge and discharge rates, more specific power, and an extended durability exceeding 10^5 times [69]. These features set them apart from other energy storage devices, making them highly desirable for various applications. A conventional full-cell supercapacitor configuration comprises two electrodes, a separator, and an electrolyte. The construction of a complete full-cell supercapacitor can take on two configurations: symmetric or asymmetric. In both cases, the configuration involves the placement of two distinct electrodes, which are effectively isolated from each other by a separator material. The separator plays a crucial role in facilitating ion transportation with its excellent ion permeability properties [70]. In the quest for advancing high-performance energy storage systems, the focus on augmenting progressive capabilities has led to significant interest in transition metal oxides (TMOs) as potential pseudocapacitor electrodes. This is attributed to their ability to deliver markedly superior specific energy output when compared with materials used in electric double-layer capacitors (EDLCs). Transition metal oxides (TMO's), including manganese oxide (MnO_2), ruthenium oxide (RuO_2), nickel oxide (NiO), vanadium oxide (V_2O_5), and cobalt oxide (Co_3O_4) are frequently adopted in supercapacitors due to their Faradaic charge-exchanging process, enabling them to achieve a substantial theoretical specific capacitance [71–76].

8.1 Ruthenium Oxide

Because of its impressive theoretical specific capacitance, ruthenium oxide (RuO_2) is indeed recognized as a vital electrode material for oxide-dependent supercapacitors. This attribute contributes to the attainment of advanced capacitive properties, further enhancing its significance (ranging from 750 to 2000 F/g) [77]. Electrical conductivity is boasted excellently with extraordinary rate capability. Though the synthesis process of RuO_2 is costly, it holds great potential as a promising supercapacitor material, capable of significantly enhancing specific energy distinguished with EDLC and other conducting polymer-based electrodes [78].

8.2 Manganese Oxide

MnO_2 has emerged as a brilliant Faradic material, offering a viable alternative to TMOs like RuO_2 because of their excellent environmental suitability [79, 80]. It possesses an extraordinary notional specific capacity (1370 F/g) [81–83], a prolonged working potential range, and low cost [80, 84]. Though, MnO_2 faces significant restrictions for commercial applications, mainly related to poor electrical (10^{-5} to 10^{-6} S cm^{-1}) and ionic conductivities (10^{-13} S/cm) [85]. As a result of this limitation, the achievable capacitance and rate capability are impeded. To overcome these challenges, a promising strategy involves the thoughtful design of balanced MnO_2 nanoarchitectures. By leveraging the pseudo capacitance of MnO_2 and the ample available surface area of conductive fiber resources, this strategy aims to fully achieve the potential of MnO_2 for enhanced performance in supercapacitor applications.

8.3 Nickel Oxide

Nickel oxide (NiO) emerges as a highly promising electrode candidate for supercapacitors in comparison with other transition metal oxides (TMOs) because of its excellent conceptual charge storage capacity (2584 F/g) [80, 85], ample surface space [80], less toxic nature, and cost-effective preparation [80]. Various NiO nanomorphologies, such as nanosheets, nanoribbons, nanowires, nanoflakes, and nanoparticles can be easily synthesized using different methods like chemical immersion, hydrothermal, solvothermal, microwave hydrothermal, and electrodeposition techniques. In recent times, researchers have been focusing on creating sustainable fibers that incorporate an enhanced NiO structure, leading to significant benefits in terms of fast electron transfer [86]. This approach aims to augment the fibers' specific surface area (SSA) while utilizing the upgraded NiO architecture to further enhance its performance as a supercapacitor material. The proposed method involves decorating and utilizing a hydrothermal technique, 3D NiO nanowalls are synthesized, with subsequent refinement through an annealing process at 600 °C. The amalgamation of these factors synergistically contributes to the seamless growth process of carbon nanofibers (CNFs) onto metal wires (MWs). This is effectively realized through the utilization of a chemical vapor deposition process (CVD), which ensures the controlled and uniform deposition of CNFs onto the MWs. This innovative approach not only enhances the structural integrity of the resulting composite but also enables the precise engineering of CNF-MW assemblies for diverse applications. By enhancing the base surface with pre-deposited three-dimensioned NiO nanowalls, the resulting CNFs/MW (3D) NiO structure evidencing a remarkable stride forward, it displays a notable augmentation in terms of areal capacitance, boasting an impressive value of 12.5 mF/cm^{-2}. Furthermore, this material demonstrates a high degree of resilience in cyclability, with an outstanding 91% retention rate achieved over an extensive span of 3000 cycles. This exceptional performance underscores its potential for prolonged and reliable utilization in energy storage applications, owing to the substantial capacitance contributions from the CNFs. To further validate the high performance (3D) NiO demonstrates distinct characteristics as a supercapacitor, showcasing remarkable energy storage capabilities. To leverage its potential, a novel strategy employs a sequential arrangement of two such devices. This innovative setup is then utilized to regulate the operation of a light-emitting diode (LED), exemplifying the CNFs/MW three dimension (3D) NiO supercapacitor's adaptability and versatility in various practical scenarios [80].

8.4 Cobalt Oxide

This highly promising energy storage solution holds immense promise across a broad range of applications thanks to the multitude of advantages that cobalt oxide-based supercapacitors offer. With their remarkable energy density and swift charge/discharge capabilities, they can effectively accumulate and release substantial amounts of energy in rapid succession [87]. The extended cycle life of these supercapacitors ensures their reliability and durability, enabling them to withstand numerous charge and discharge cycles with minimal degradation. Additionally, their adaptability to varying temperatures makes them well-suited for deployment in diverse environmental conditions. Furthermore, they boast superior safety compared with lithium-ion batteries, owing to their non-flammable electrolytes. Moreover, their eco-friendly nature, devoid of toxic heavy metals, makes them a more environmentally conscious choice. With flexibility in size and design, they can be tailored to fit specific energy storage needs. Their exceptional power density makes them an excellent fit for applications requiring quick energy delivery, such as regenerative braking in electric vehicles [86]. These supercapacitors also exhibit low self-discharge rates, ensuring efficient energy retention over extended periods. Furthermore, their compatibility with other energy storage technologies allows for hybrid solutions that optimize overall efficiency and performance. However, in choosing an appropriate energy storage solution, careful consideration of factors such as cost, performance, and sustainability is crucial, as cobalt extraction may raise environmental and ethical concerns [88–90]. To address these challenges, researchers are actively working on advancing supercapacitor technology and exploring alternative materials.

References

1. S. Yang, Y. Lin, X. Song, P. Zhang, L. Gao, Covalently coupled ultrafine H-TiO_2 nanocrystals/nitrogen-doped graphene hybrid materials for high-performance supercapacitor. ACS Appl. Mater Interfaces **7**(32), 17884–17892 (2015)
2. C. Kim, S. Kim, J. Lee, J. Kim, J. Yoon, Capacitive and oxidant generating properties of black-colored TiO_2 nanotube array fabricated by electrochemical self-doping. ACS Appl. Mater. Interfaces **7**(14), 7486–7491 (2015)
3. D. Wang, Q. Wang, T. Wang, Controlled synthesis of mesoporous hematite nanostructures and their application as electrochemical capacitor electrodes. Nanotechnology **22**, 135604 (2011)
4. G.M. Zhou, D.W. Wang, F. Li, L.L. Zhang, N. Li, Z.S. Wu, L. Wen, G.Q. Lu, H.M. Cheng, Graphene-wrapped Fe_3O_4 anode material with improved reversible capacity and cyclic stability for lithium ion batteries. Chem. Mater. **22**, 5306–5313 (2010)
5. G. Wang, L. Zhang, J. Zhang, A review of electrode materials for electrochemical supercapacitors. Chem. Soc. Rev. **41**, 797–828 (2012)

6. H. Zhou, Y. Zhang, Electrochemically self-doped TiO_2 nanotube arrays for supercapacitors. J. Phys. Chem. C **118**, 5626–5636 (2014)
7. J.S. Chen, Z. Wang, X.C. Dong, P. Chen, X. W. Lou Graphene-wrapped TiO_2 hollow structures with enhanced lithium storage capabilities. Nanoscale **3**, 2158–2161 (2011)
8. K. Lee, S. Deng, H.M. Fan, S. Mhaisalkar, H.R. Tan, E.S. Tok, K.P. Loh, W.S. Chin, C.H. Sow, α-Fe_2O_3 nanotubes-reduced graphene oxide composites as synergistic electrochemical capacitor materials. Nanoscale **4**, 2958 (2012)
9. Q. Ke, C. Tang, Y. Liu, H. Liu, J. Wang, Intercalating graphene with clusters of Fe_3O_4 nanocrystals for electrochemical supercapacitors. Mater. Res. Express **1**, 025015 (2014)
10. Q. Wang, L. Jiao, H. Du, Y. Wang, H. Yuan, Fe_3O_4 nanoparticles grown on graphene as advanced electrode materials for supercapacitors. J. Power. Sources **245**, 101–106 (2014)
11. S. Yang, X. Song, P. Zhang, J. Sun, L. Gao, Self-assembled α-Fe_2O_3 mesocrystals/graphene nanohybrid for enhanced electrochemical capacitors. Small **10**(11), 2270–2279 (2014)
12. M. Pumera, Electrochemistry of graphene: new horizons for sensing and energy storage. Chem. Rec. **9**, 211–223 (2009)
13. C. Hu, T. Lu, F. Chen, R. Zhang, A brief review of graphene–metal oxide composites synthesis and applications in photocatalysis. J. Chin. Adv. Mater. Soc. **1**(1), 21–39 (2013)
14. Z.S. Wu, G. Zhou, L.-C. Yin, W. Ren, F. Li, H.-M. Cheng, Graphene/metal oxide composite electrode materials for energy storage. Nano Energy **1**, 107–131 (2012)
15. Z. Zhang, F. Xiao, Y. Guo, S. Wang, Y. Liu, One-pot self-assembled three-dimensional TiO_2-graphene hydrogel with improved adsorption capacities and photocatalytic and electrochemical activities. ACS Appl. Mater. Interfaces **5**(6), 2227–2233 (2013)
16. P. Balaya, Size effects and nanostructured materials for energy applications. Energy Environ. Sci. **1**(6), 645–654 (2008)
17. S. Mohapatra, A. Acharya, G.S. Roy, The role of nanomaterial for the design of supercapacitor. Lat. Am. J. Phys. Educ. **6**(3), 380–384 (2012)
18. M. Winter, R.J. Brodd, What are batteries, fuel cells, and supercapacitors? Chem. Rev. **104**(10), 4245–4270 (2004)
19. M.D. Stoller, R.S. Ruoff, Best practice methods for determining an electrode material's performance for ultracapacitors. Energy Environ. Sci. **3**, 1294–1301 (2010)
20. P. Simon, Y. Gogotsi, Materials for electrochemical capacitors. Nat. Mater. **7**, 845–854 (2008). https://doi.org/10.1038/nmat2297
21. L. Zhang, X.S. Zhao, Carbon-based materials as supercapacitor electrodes. Chem. Soc. Rev. **38**, 2520–2531 (2009)
22. E.E. Miller, Y. Hua, F. Handan Tezel, Materials for energy storage: review of electrode materials and methods of increasing capacitance for supercapacitors. J. Energy Storage **20**, 30–40 (2018)
23. F. Shi, L. Li, X.-l. Wang, C.-d. Gua, J.-p. Tu, Metal oxide/hydroxide-based materials for supercapacitors. RSC Adv. **4**, 41910–41921 (2014)
24. S. Bhoyate, P.K. Kahol, R.K. Gupta, Nanostructured materials for supercapacitor applications. Nanoscience **5**, 1–29 (2018)
25. R. Burt, G. Birkett, X.S. Zhao, A review of molecular modelling of electric double layer capacitors. Phys. Chem. Chem. Phys. **16**(14), 6519–6538 (2014)
26. W. Liu, J. Li, K. Feng, A. Sy, Y. Liu, L. Lim, G. Lui, R. Tjandra, L. Rasenthiram, G. Chiu, A. Yu, Advanced Li-ion hybrid supercapacitors based on 3D graphene–foam composites. ACS Appl. Mater. Interfaces **8**(39), 25941–25953 (2016)
27. X. Zhao, B.M. Sánchez, P.J. Dobson, P.S. Grant, The role of nanomaterials in redox-based supercapacitors for next generation energy storage devices. Nanoscale **3**(3), 839–855 (2011)
28. P. Yang, W. Mai, Flexible solid-state electrochemical supercapacitors. Nano Energy **8**, 274–290 (2014); E. Mitchell, J. Candler, F. De Souza, R.K. Gupta, B.K. Gupta, L.F. Dong, High performance supercapacitor based on multilayer of polyaniline and graphene oxide. Synth. Met. **199**, 214–218 (2015)
29. J. Zhou, Y. Wang, J. Zhou, K. Chen, L. Han, Well-defined hollow tube@ sheets $NiCo_2S_4$ core–shell nanoarrays for ultrahigh capacitance supercapacitor. Dalton Trans. **50**(42), 15129–15139 (2021)

30. N. Al Suliman, C. Awada, A. Alshoaibi, N.M. Shaalan, Simple preparation of ceramic-like materials based on 1d-agx (X = 0, 5, 10, 20, 40 mm)/TiO_2 nanostructures and their photocatalysis performance. Crystals **10**(11), 1024 (2020)
31. W. Hao, F. Teng, Molten-state cubic cage-templated uniform C_3N_4 hollow nanocubes and improved electrochemical performance of asymmetric micro-supercapacitor. J. Alloy. Compd. **909**, 164790 (2022)
32. L. Yang, J. Qiu, Y. Wang, S. Guo, Y. Feng, D. Dong, J. Yao, Molten salt synthesis of hierarchical porous carbon from wood sawdust for supercapacitors. J. Electroanal. Chem. **856**, 113673 (2020)
33. G. Zhang, R. Zhu, R. Zhang, D. Zhang, C. Sun, Z. Long, Y. Li, 3D hetero-nanostructured electrode constructed on carbon fiber paper with 2D 1T-MoS2/1D $Cu(OH)_2$ for flexible asymmetric solid-state supercapacitors. J. Power. Sources **523**, 231031 (2022)
34. A. Zhang, H. Zhang, B. Hu, M. Wang, S. Zhang, Q. Jia, L. He, Z. Zhang, The intergrated nanostructure of bimetallic CoNi-based zeolitic imidazolate framework and carbon nanotubes as high-performance electrochemical supercapacitors. J. Colloid Interface Sci. **608**, 1257–1267 (2022)
35. Y. Wang, Z. Wang, X. Zheng, X. Teng, L. Xu, Y. Yuan, X. Liu, A. Fu, Y. Li, H. Li, Core-sheath heterostructure of $MnCo_2O_4$ nanowires wrapped by NiCo-layered double hydroxide as cathode material for high-performance quasi-solid-state asymmetric supercapacitors. J. Alloy. Compd. **904**, 164047 (2022)
36. Hierarchically Hybrid Porous Co3O4@NiMoO4/CoMoO4 Heterostructures for High-Performance Electrochemical Energy Storage
37. Y. Chen, L. Xu, M. Yang, Y. Jia, Y. Yan, J. Qian, F. Chen, H. Li, Design of 2D/2D CoAl LDH/g-C_3N_4 heterojunction-driven signal amplification: fabrication and assay for photoelectrochemical aptasensor of ofloxacin. Sens. Actuators B Chem. **353**, 131187 (2022)
38. R. Ma, Z. Xiao, Y. Li, X. Peng, X. Chen, H. Chai, Microwave assisted synthesis of amorphous cobalt doped nickel phosphate as electrode materials for supercapacitor. Available at SSRN 3967327
39. X. Zhang, P. Yang, NiCo-layered double hydroxide/g-C3N4 heterostructures with enhanced adsorption capacity and photoreduction of Cr(VI). Appl. Surf. Sci. **556**, 149772 (2021)
40. Z. Liu, H. Xiong, Y. Luo, L. Zhang, K. Hu, L. Zhang, Y. Gao, Z.A. Qiao, Interface-induced self-assembly strategy toward 2D ordered mesoporous carbon/MXene heterostructures for high-performance supercapacitors. Chemsuschem **14**(20), 4422–4430 (2021)
41. K. Nasrin, V. Sudharshan, K. Subramani, M. Karnan, M. Sathish, In-Situ Synergistic 2D/2D MXene/BCN heterostructure for superlative energy density supercapacitor with super-long life. Small **18**(4), 2106051 (2022)
42. K. Krishnamoorthy, P. Pazhamalai, V.K. Mariappan, S. Manoharan, D. Kesavan, S.J. Kim, Two-dimensional siloxene–graphene heterostructure-based high-performance supercapacitor for capturing regenerative braking energy in electric vehicles. Adv. Funct. Mater. **31**(10), 2008422 (2021)
43. H. Liu, S. An, X. Han, X. Sun, J. Cui, Y. Zhang, W. He, 3D heterostructure of 2D Y-doped $Cu(OH)_2$ nanosheet supported by nickel foam as advanced electrodes for high performance supercapacitor. Vacuum **187**, 110106 (2021)
44. S. Liu, K. Wan, C. Zhang, T. Liu, Polyaniline-decorated 3D carbon porous network with excellent electrolyte wettability and high energy density for supercapacitors. Compos. Commun. **24**, 100610 (2021)
45. Z. Huang, H. Guo, C. Zhang, Assembly of 2D graphene sheets and 3D carbon nanospheres into flexible composite electrodes for high-performance supercapacitors. Compos. Commun. **12**, 117–122 (2019)
46. P.P. Sun, Y.H. Zhang, H. Shi, F.N. Shi, Controllable one step electrochemical synthesis of PANI encapsulating 3d–4f bimetal MOFs heterostructures as electrode materials for high-performance supercapacitors. Chem. Eng. J. **427**, 130836 (2022)
47. W. Chen, Y. Peng, Z. Qiu, X. Zhang, H. Xu, 3D hierarchical $Ti_3C_2T_X$@ NiO-reduced graphene oxide heterostructure hydrogel as free-standing electrodes for high performance supercapacitor. J. Alloy. Compd. **901**, 163614 (2022)

48. R. Zhang, J. Dong, W. Zhang, L. Ma, Z. Jiang, J. Wang, Y. Huang, Synergistically coupling of 3D FeNi-LDH arrays with $Ti_3C_2T_x$-MXene nanosheets toward superior symmetric supercapacitor. Nano Energy **91**, 106633 (2022)
49. G.A. Snook, P. Kao, A.S. Best, Conducting-polymer-based supercapacitor devices and electrodes. J. Power. Sources **196**(1), 1–12 (2011)
50. M. Wu, G.A. Snook, V. Gupta, M. Shaffer, D.J. Fray, G.Z. Chen, Electrochemical fabrication and capacitance of composite films of carbon nanotubes and polyaniline. J. Mater. Chem. **15**(23), 2297–2303 (2005)
51. M. Mastragostino, F. Soavi, C. Arbizzani, W. van Schalkwijk, B. Scrosati, *Advances in Lithium-ion Batteries* (Springer, 2002)
52. K. Lota, V. Khomenko, E. Frackowiak, Capacitance properties of poly (3,4-ethylenedioxythiophene)/carbon nanotubes composites. J. Phys. Chem. Solids **65**(2–3), 295–301 (2004)
53. K.S. Ryu, K.M. Kim, N.-G. Park, Y.J. Park, S.H. Chang, Symmetric redox supercapacitor with conducting polyaniline electrodes. J. Power. Sources **103**, 305–309 (2002)
54. V. Khomenko, E. Frackowiak, F. Beguin, Determination of the specific capacitance of conducting polymer/nanotubes composite electrodes using different cell configurations. Electrochim. Acta **50**, 2499–2506 (2005)
55. I. Shown, A. Ganguly, L.C. Chen, K.H. Chen, Conducting polymer-based flexible supercapacitor. Energy Sci. Eng. **3**(1), 2–26 (2015)
56. Y. Li, X. Zhao, Q. Xu, Q. Zhang, D. Chen, Facile preparation and enhanced capacitance of the polyaniline/sodium alginate nanofiber network for supercapacitors. Langmuir **27**, 6458–6463 (2011)
57. S.R. Sivakkumar, R. Saraswathi, Performance evaluation of poly (N-methylaniline) and polyisothianaphthene in charge-storage devices. J. Power. Sources **137**(2), 322–328 (2004)
58. J.H. Park, O.O. Park, Hybrid electrochemical capacitors based on polyaniline and activated carbon electrodes. J. Power. Sources **111**(1), 185–190 (2002)
59. S. Suematsu, Y. Oura, H. Tsujimoto, H. Kanno, K. Naoi, Conducting polymer films of cross-linked structure and their QCM analysis. Electrochim. Acta **45**, 3813–3821 (2000)
60. M. Hughes, G.Z. Chen, M.S. Shaffer, D.J. Fray, A.H. Windle, Controlling the nanostructure of electrochemically grown nanoporous composites of carbon nanotubes and conducting polymers. Compos. Sci. Technol. **64**(15), 2325–2331 (2004)
61. G.A. Snook, G.Z. Chen, D.J. Fray, M. Hughes, M. Shaffer, Studies of deposition of and charge storage in polypyrrole–chloride and polypyrrole–carbon nanotube composites with an electrochemical quartz crystal microbalance. J. Electroanal. Chem. **568**, 135–142 (2004)
62. G.A. Snook, G.Z. Chen, The measurement of specific capacitances of conducting polymers using the quartz crystal microbalance. J. Electroanal. Chem. **612**(1), 140–146 (2008)
63. K. Naoi, Y. Oura, H. Tsujimoto, in *Proceedings of the Symposium on Electrochemical Capacitors Ii*, ed. by F.M. Delnick, D. Ingersoll, X. Andrieu, K. Naoi (Electrochemical Society Inc, Pennington, 1997), pp. 120–126
64. H. An, Y. Wang, X. Wang, L. Zheng, X. Wang, L. Yi et al., Polypyrrole/carbon aerogel composite materials for supercapacitor. J. Power. Sources **195**, 6964–6969 (2010)
65. G.A. Snook, G.Z. Chen, The measurement of specific capacitances of conducting polymers using the quartz crystal microbalance. J. Electroanal. Chem. **612**, 140–146 (2008)
66. M. Skompska, J. Mieczkowski, R. Holze, E. Heinze, J. Electroanal. Chem. **577**, 9 (2005)
67. M.D. Levi, Y. Gofer, D. Aurbach, M. Lapkowski, E. Vieil, J. Serose, J. Electrochem. Soc. **147**, 1096 (2000)
68. A. Laforgue, P. Simon, J.F. Fauvarque, M. Mastragostino, F. Soavi, J.F. Sarrau, P. Lailler, M. Conte, E. Rossi, S. Saguatti, J. Electrochem. Soc. **150**, A645 (2003)
69. X. Shi, S. Zheng, Z.-S. Wu, X. Bao, Recent advances of graphene-based materials for high-performance and new-concept supercapacitors. J. Energy Chem. **27**(1), 25–42 (2018)
70. W.H. Low, P.S. Khiew, S.S. Lim, C.W. Siong, E.R. Ezeigwe, Recent development of mixed transition metal oxide and graphene/mixed transition metal oxide based hybrid nanostructures for advanced supercapacitors. J. Alloy. Compd. **775**, 1324–1356 (2019)

71. J.H. Zheng, R.M. Zhang, P.F. Yu, X.G. Wang, Binary transition metal oxides (BTMO) (Co-Zn, Co-Cu) synthesis and high supercapacitor performance. J. Alloy. Compd. **772**, 359–365 (2019)
72. Z. Zhang, Z. Xu, Z. Yao, Y. Meng, Q. Xia, D. Li, Z. Jiang, Ultrahigh capacitance of TiO_2 nanotube arrays/C/MnO_2 electrode for supercapacitor. J. Alloy. Compd. **805**, 396–403 (2019)
73. S. Mothkuri, S. Chakrabarti, H. Gupta, B. Padya, T.N. Rao, P.K. Jain, Synthesis of MnO_2 nano-flakes for high performance supercapacitor application. Mater. Today: Proc. (2019)
74. T.-F. Yi, J. Mei, B. Guan, P. Cui, S. Luo, Y. Xie, Y. Liu, Construction of spherical NiO@MnO_2 with core-shell structure obtained by depositing MnO_2 nanoparticles on NiO nanosheets for high-performance supercapacitor. Ceram. Int. (2019)
75. J. Lin, Y. Yan, H. Wang, X. Zheng, Z. Jiang, Y. Wang, J. Qi, J. Cao, W. Fei, J. Feng, Hierarchical Fe_2O_3 and NiO nanotube arrays as advanced anode and cathode electrodes for high-performance asymmetric supercapacitors. J. Alloy. Compd. **794**, 255–260 (2019)
76. P. Wang, H. Zhou, C. Meng, Z. Wang, K. Akhtar, A. Yuan, Cyanometallic framework-derived hierarchical Co_3O_4-NiO/graphene foam as high-performance binder-free electrodes for supercapacitors. Chem. Eng. J. **369**, 57–63 (2019)
77. C.-C. Hu, K.-H. Chang, M.-C. Lin, Y.-T. Wu, Design and tailoring of the nanotubular arrayed architecture of hydrous RuO_2 for next generation supercapacitors. Nano Lett. **6**(12), 2690–2695 (2006)
78. L.Y. Chen, Y. Hou, J.L. Kang, A. Hirata, T. Fujita, M.W. Chen, Toward the theoretical capacitance of RuO_2 reinforced by highly conductive nanoporous gold. Adv. Energy Mater. **3**(7), 851–856 (2013)
79. J. Qu, L. Shi, C. He, F. Gao, B. Li, Q. Zhou, H. Hu, G. Shao, X. Wang, J. Qiu, Highly efficient synthesis of graphene/MnO_2 hybrids and their application for ultrafast oxidative decomposition of methylene blue. Carbon **66**, 485–492 (2014)
80. Y. Yang, F. Yang, H. Hu, S. Lee, Y. Wang, H. Zhao, D. Zeng, B. Zhou, S. Hao, Dilute NiO/carbon nanofibre composites derived from metal organic framework fibres as electrode materials for supercapacitors. Chem. Eng. J. **307**, 583–592 (2017)
81. M. Toupin, T. Brousse, D. Bélanger, Charge storage mechanism of MnO 2 electrode used in aqueous electrochemical capacitor. Chem. Mater. **16**(16), 3184–3190 (2004)
82. Y. Liu, M. Liu, P. Zheng, D. Ge, L. Yang, Controllable hydrogel-thermal synthesis of Mn_2O_3/CNT aerogels: shape evolution, growth mechanism and electrochemical properties. Mater. Des. **182**, 108022 (2019)
83. A.V. Radhamani, M. Krishna Surendra, M.S. Ramachandra Rao, Tailoring the supercapacitance of Mn_2O_3 nanofibres by nanocompositing with spinel-$ZnMn_2O_4$. Mater. Des. **139**, 162–171 (2018)
84. W. Wei, X. Cui, W. Chen, D.G. Ivey, Manganese oxide-based materials as electrochemical supercapacitor electrodes. Chem. Soc. Rev. **40**(3), 1697–1721 (2011)
85. N. Yu, H. Yin, W. Zhang, Y. Liu, Z. Tang, M.-Q. Zhu, High-performance fibre-shaped all-solid-state asymmetric supercapacitors based on ultrathin MnO_2 nanosheet/carbon fibre cathodes for wearable electronics. Adv. Energy Mater. **6**(2), 1501458 (2016)
86. S.G. Kandalkar, D.S. Dhawale, Chang-Koo Kim, C.D. Lokhande, Chemical synthesis of cobalt oxide thin film electrode for supercapacitor application (2010)
87. G. Zhu, J. Chen, Z. Zhang, Q. Kang, X. Feng, Y. Li, Z. Huang, L. Wang, Y. Ma, NiO nanowall-assisted growth of thick carbon nanofibre layers on metal wires for fibre supercapacitors. Chem. Commun. **52**(13), 2721–2724 (2016)
88. H. Wang, Z. Xu, H. Yi, H. Wei, Z. Guo, X. Wang, One-step preparation of single-crystalline Fe_2O_3 particles/graphene composite hydrogels as high performance anode materials for supercapacitors. Nano Energy **7**, 86–96 (2014)

89. S.H. Yang, X. Song, P. Zhang, J. Sun, L. Gao, Self-assembled α-Fe_2O_3 mesocrystals/graphene nano hybrid for enhanced electrochemical capacitors. Small **10**(11), 2270–2279 (2014)
90. X. Yu, G. Wang, T. Zhai, M. Yu, J. Gan, Y. Tong, Y. Li, Hydrogenated TiO_2 nanotube arrays for supercapacitors. Nano Lett. **12**, 1690–2169 (2012)

Goal 13: Climate Action

Overview of the chapters:

Goal 13: Climate Action

(We must act urgently to combat the impacts of climate change)

It is evident throughout this part that biomass, its derivatives, and its application in energy production are all important factors to consider. Energy from biomass and its derivatives is expected to become a key feature of energy in the near future, in addition to fossil fuels, petroleum products, and nuclear power. In addition to damaging mankind and other organisms on the planet, overconsumption of these commercial energy resources degrades the sustainability of the ecosystem. Since the potential value of energy is of primary importance, it is imperative that we practice alternative, reusable, or renewable sources of energy. The use of fossil fuels has increased in the industrial field and for the development of post-industrial applications because of increasing affluence and population growth in our life today. Communities worldwide were working to find alternative energy generation methods and to develop new technologies for generating energy. The distribution of energy source supply can take place in the same way as traditional energy supply because a single framework of regularity is being established from multiple countries. When the European Union considers different possible energy sources in light of its strong impulses, biomass plays an essential role in the debates about energies within the context of policy, as well as in the implementation of plans that lead to the deliberate development of environmental equipment that is protected and converted into a sustainable source of energy. As a matter of fact, biomass was a significant renewable energy resource with immense potential to produce biofuels for power generation, heating, and transformation. In developed nations, bioenergy is recognized for its economic significance and potential. As a result of wastewater disposal, municipal waste disposal, and medicinal waste disposal, cities face major issues. It will be beneficial if they can be recycled or degraded biologically. Since these biowastes can be used both to generate energy and for waste management, charging them with hydrogen or other biofuels would be beneficial. There are numerous nations that have been discussing sustainability as one of the foremost topics. It remains crucial to lower modern vestiges of coal consumption to achieve sustainability. There remain insufficient outstanding fundamentals to lead to a progressively more sustainable future for renewable energy-consuming bio-body food standard. Examples of this are astral, airstream, tidal upsurge, among others. Furthermore, biowastes can be recycled or degraded biologically. Biowastes offer the potential for both energy production and waste management when hydrogen is extracted from them. We have discussed the thermochemical and biological routes of hydrogen generation in detail in this chapter. As far as generating hydrogen from biowastes is concerned, each technique has its own advantages. By utilizing biomass waste, a cleaner environment is created, and the natural system is supported to be sustainable. Biomass resources are primarily discussed in this chapter in order to generate useful energy by utilizing, recycling, and reusing them.

Utilizing Photocatalysis

Maryam Sarkhosh

Abstract Environmental pollution and water scarcity are major challenges facing humanity. One promising approach for addressing these challenges is photocatalysis, a technology that uses semiconducting materials to harness sunlight and generate reactive oxygen species (ROS) that can break down pollutants in water and air. Photocatalysis is a process in which a photocatalyst is activated by light and initiates a chemical reaction that leads to the degradation of organic pollutants, microorganisms, and inorganic contaminants. One of the major advantages of photocatalysis is that it does not produce any secondary pollutants, making it an eco-friendly technology. Additionally, the use of renewable energy sources such as sunlight as a source of energy for photocatalysis can significantly reduce the operating costs of the process. Furthermore, photocatalysis is a versatile technology that can be used for the treatment of various types of pollutants, including organic compounds, heavy metals, and pathogens. Photocatalysis has been thoroughly researched in the field of water treatment for the removal of organic contaminants, water disinfection, and the removal of heavy metals. The use of photocatalysis for water treatment has several advantages over conventional water treatment methods such as activated carbon adsorption, reverse osmosis, and chlorination. For instance, photocatalysis can achieve higher removal efficiencies for certain pollutants, and it can be used to treat a wide range of pollutants simultaneously.

1 The Photocatalytic Process

The term "photocatalyst" has a Greek root and is formed by combining two parts: "photo," meaning light, and "catalyst," which has the effect of accelerating up chemical reactions but does not undergo any permanent change and can be reused after

M. Sarkhosh (✉)
Department of Environmental Health Engineering, School of Health, Mashhad University of Medical Sciences, Mashhad, Iran
e-mail: marya.sarkhosh@yahoo.com

UNESCO-UNISA Africa Chair in Nanosciences and Nanotechnology, College of Graduate Studies, University of South Africa, Muckleneuk Ridge, PO Box 392, Pretoria 0002, South Africa

K. Kasinathan et al. (eds.), *New Technologies for Energy Transition Based on Sustainable Development Goals*, https://doi.org/10.1007/978-981-97-2527-4_19

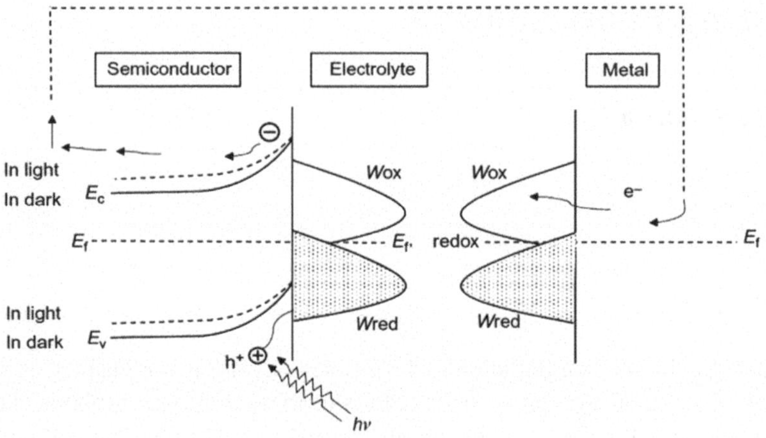

Fig. 1 Energy flow diagram of electrochemical cells

the reaction. The term "photocatalyst" is a combination of the two fields of photochemistry and catalysis. To understand photocatalytic materials, it is important to be familiar with some important terms, including the concept of bandgap. In the context of materials' electrical properties, the bandgap denotes the space that exists between the valence band and the conduction band. This gap can be characterized as the minimum energy required to elevate an electron from the valence band to the conduction band, enabling its unrestricted movement through the primary pathways of current. Conductive, semiconductive, and insulating materials can be distinguished by their different bandgaps. You can see this gap in energy diagrams (band diagrams) in Fig. 1.

When energy is absorbed, a small subset of electrons can be shifted from the valence band to the conduction band. This process leaves behind positively charged holes in the valence band for each transitioning electron. The foundation of photocatalytic reactions revolves around an interconnected sequence of events involving electrons and holes. This sequence transpires when an electron from a neighboring atom fills the void created by these vacancies, inducing a chain reaction. In the photocatalytic reactions, light energy is given to electrons in the valence band, and some of these electrons are transferred to the conduction band. These transferred electrons can then react with other atoms in the material, or with molecules in the surrounding environment, and a photocatalytic reaction occurs.

1.1 Quantum Yield of Photocatalytic Processes

The efficiency of photocatalytic processes is often referred to as "quantum yield" in scientific articles, and it shows the proportion of produced electron–hole pairs to photons that are absorbed. However, measuring the actual amount of absorbed light in photocatalytic systems is often difficult due to light scattering at the semiconductor surface, so it is usually assumed that all absorbed light is used. Also, if multiple products are generated by photocatalytic reactions, only one of them is considered. To obtain the quantum yield, a combination of all possible paths for electrons and holes must be considered. It is clear that without recombination, the quantum yield would be 100%. Recombination always happens in actual systems, and there are not the same amounts of electrons and holes there. Therefore, photocatalytic materials must undergo dimensional, technical, and surface alterations in order to perform better. The photocatalytic process involves several steps that can be summarized as follows:

- **Absorption of light:** The photocatalyst absorbs light energy, typically in the UV or visible range, to create electron–hole pairs in its structure.
- **Generation of reactive species:** The electron–hole pairs move to the photocatalyst's surface where they interact with species that have been adsorbed there, including oxygen or water, to create highly reactive species like hydroxyl radicals.
- **Adsorption of the target molecule:** The target molecule, such as a pollutant, adsorbs onto the surface of the photocatalyst.
- **Reaction with the reactive species:** The reactive species generated on the photocatalyst's surface and the adsorbed react target molecule, breaking it down into simpler and less harmful compounds, such as carbon dioxide and water.

The overall reaction can be represented by the following equation:

$$\text{photocatalyst} + \text{light energy} + \text{target molecule} \rightarrow \text{degradation products}$$

This process has many applications in environmental remediation, including water and air purification, as well as in the production of renewable fuels and chemicals.

1.2 Chemical Equations of Photocatalytic Process

In the photocatalytic process, a photocatalyst material, usually composed of metal oxides to produce electron–hole pairs, light is used to irradiate materials, such as titanium dioxide. These pairings produce reactive oxygen species (ROS), which can oxidize and break down organic contaminants. ROS includes superoxide radicals and hydrogen peroxide. The chemical reactions involved in the process can be represented by the following equations:

a. Creation of electron–hole pairs:

$$\mathrm{TiO_2}(+h\nu) \rightarrow +\mathrm{TiO_2}(h^+) + \mathrm{TiO_2}(e^-)$$

b. Formation of hydroxyl radicals:

$$TiO_2(h^+) + H_2O\,(\text{adsorbed}) \rightarrow TiO_2 + H^+ + \cdot OH\,(\text{adsorbed})$$

c. Oxidation of organic pollutants:

(1) $\cdot OH + R\text{-}H \rightarrow R \cdot + H_2O.$
(2) $R \cdot + \cdot OH \rightarrow R\text{-}OH.$

In the first equation, titanium dioxide (TiO_2) is irradiated with light, resulting that $TiO_2(h^+)$ and $TiO_2(e^-)$ are two materials that can form electron–hole pairs. Hydroxyl radicals (·OH) and hydrogen ions (H^+) can then be created when these pairs interact with water molecules, as shown in the second equation. The hydroxyl radicals are highly reactive and can oxidize organic pollutants, such as volatile organic compounds (VOCs), to less harmful substances, as shown in the third equation. Overall, the photocatalytic process can help to remove organic pollutants from air and water by using the reactive species generated by the photocatalyst material under light irradiation. In the third stage of the aforementioned reaction, a large portion of electron–hole pairs combines either in their path of motion or on the surface. Consequently, this interaction leads to the dissipation of the absorbed energy of the electron, manifesting as either heat or emitted light. Regrettably, this occurrence diminishes the overall efficiency of the photocatalyst. Surface-excited charges that have a longer lifespan enable the occurrence of oxidation and reduction reactions. This process, known as photocatalysis, can, for example, be used to break down organic pollutants in water.

1.3 Pollutant Degradation by Photocatalysis

After the absorption of sunlight (or artificial light), the degradation reaction occurs spontaneously from a thermodynamic perspective. In addition to creating a thermodynamic shortcut, a photocatalyst changes the reaction pathway, improving the reaction kinetics. The ability of pollutant degradation by photocatalysis depends on several factors:

- **Size and morphology of the photocatalyst:** A photocatalyst modifies the reaction pathway, enhancing the reaction kinetics in addition to producing a thermodynamic shortcut. The dimensions and structure of the photocatalyst exert significant influence over the available surface area, thereby directly affecting the rate of pollutant degradation. This is due to the fact that all chemical reactions occur at interfaces or surfaces. Nanoscale materials with dimensions smaller than 20 nm have been proved via research to have substantially different physical and chemical characteristics from their bigger counterparts, making them highly sought-after in this industry. Reducing the particle size increases the bandgap energy of semiconductors. However, generally speaking, as the particle size reduces, the photocatalyst material's surface area increases, enabling more absorption of contaminants and photons. In this approach, the detrimental effects of increasing bandgap energy are offset.

- **Pollutant concentration:** It is evident that the amount of pollutant that can be absorbed onto the surface of the photocatalyst material has a significant impact on the reactions because only pollutants that are absorbed onto the surface of the photocatalyst material can participate in photocatalytic reactions. Note that the concentration of the pollutant present throughout the solution does not play a role in the reaction rate.
- **Photocatalyst concentration:** The rate of pollutant degradation is heavily dependent on the concentration of the photocatalyst material. The active sites on the surface are increased by increasing the photocatalyst material concentration, which leads to the creation of more electron–hole pairs and, eventually, more degradation. However, increasing the concentration of each photocatalyst material can only improve the degradation rate to a certain extent.
- **Reaction temperature:** For the best performance, photocatalytic reactions should be carried out at an optimal temperature. Most photocatalysts exhibit optimal performance at temperatures up to 80 °C. At higher temperatures, in addition to increasing the rate of recombination, pollutant absorption on the surface of the photocatalyst material decreases, resulting in a decrease in the degradation rate.
- **Solution pH:** Changes in the pH of a solution affect the surface charges of photocatalyst particles and the potential of catalytic reactions. Therefore, this parameter is effective on the surface adsorption intensity of pollutants. For example, TiO_2 particles receive and lose protons in acidic and alkaline environments, respectively. This means that the surface structure of titanium dioxide in an acidic environment is positively charged, and in an alkaline environment, it is negatively charged. Therefore, TiO_2 nanoparticles exhibit higher oxidation activity in acidic environments. Paying attention to this parameter is essential in the design and research of photocatalysts.
- **Inorganic ions:** In most cases, polluted waters contain some inorganic ions such as manganese, iron, zinc, copper, phosphate, nitrate, sulfate, and chloride ions. These ions affect the rate of photodegradation of organic pollutants by adsorbing on the surface of the photocatalyst material.
- **Oxygen dissolved in water:** Oxygen dissolved in water plays the role of an electron acceptor in photocatalytic reactions. It captures the excited electron and prevents it from recombining with the hole. According to research, dissolved oxygen in water does not have much effect on the adsorption of pollutants on the TiO_2 photocatalyst surface.
- **Photocatalytic semiconductors:** Various semiconductor materials such as metal oxides and sulfides (TiO_2, ZnO, Fe_2O_3, WO_3, CdS, and SnO_2) can act as photocatalysts due to their unique electronic properties. These substances possess a complete valence band and an unoccupied conduction band. For the following reasons, semiconductor photocatalysts are the best choice for degrading organic pollutants.
- **Affordability:** Semiconductor photocatalysts are made from materials that are available and affordable. For example, titanium dioxide (TiO_2), which is

commonly used as a photocatalyst in many research studies, is an inexpensive and readily available material.
- **Non-toxic or low toxicity:** Semiconductor photocatalysts typically use sunlight or artificial light to create electron–hole pairs, and in this process, no toxic substances are released into the environment. Therefore, the use of these types of photocatalysts for the destruction of organic pollutants is very safe.
- **Flexibility of properties:** Semiconductor photocatalysts have the ability to adjust their properties, and by changing environmental conditions such as pH, temperature, and pressure, their properties can be altered. This feature allows these photocatalysts to be used for the destruction of organic pollutants in various conditions.

The substantial bandgap existing between the valence band and the conduction band in insulating materials hinders valence band electrons from acquiring the requisite energy for moving to the conduction band. However, a portion of the valence band overlaps with the conduction band in conductive materials, leading to a fractional displacement of electrons and conductivity in these materials. Last but not least, it is critical to comprehend the idea of electrons and holes. When an electron in a semiconducting material is excited to move into the conduction band, it leaves behind a hole in the valence band that acts like a positively charged hole. These holes can move through the material and participate in chemical reactions, similar to the way that electrons move through conductive materials. The combination of electrons and holes in semiconducting materials can result in unique chemical reactions and photocatalytic properties.

2 Applications of Nanostructured Photocatalysts

In the past decade, the technical and economic importance of photocatalysis has been significantly investigated. The improved performance is related to advancements in nanotechnology, such as the invention of nanoparticle photocatalysts, which greatly enhance the catalytic efficiency of certain materials. Numerous nanotechnologies' uses, many of which have already found commercial success, include the creation of solar hydrogen, anti-fouling, anti-microbial, and self-cleaning surfaces. Extensive research has been conducted to improve this technology and expand its potential range of applications, including anti-adhesive and anti-fingerprint coatings, soil remediation, and organic material decomposition, including microorganisms and fats. When semiconducting materials such as photocatalysts come into contact with specific light, they increase the rate of chemical reactions (e.g., organic molecule decomposition). Nanoscale catalyst particles exhibit better reactivity compared to larger or bulkier particles due to their higher surface area. Many cases are under investigation, but none seems to have the efficiency of titanium dioxide. The use of titanium dioxide requires UV light or light in the blue end of the visible spectrum. Application volumes are mainly limited to the outer surface, but even in indoor

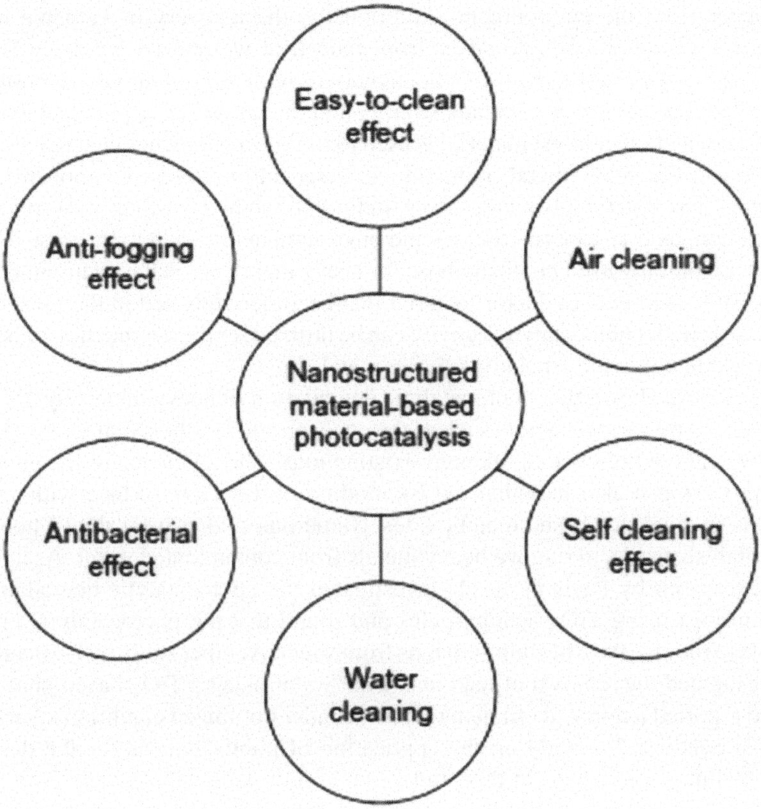

Fig. 2 Applications of photocatalysts

products such as ceramic bathroom fixtures, it is widely used despite the reduction in natural light. Additionally, research is being conducted to shift the catalytic spectrum range toward visible light. Several applications of the photocatalysis process are presented in Fig. 2.

In recent years, application programs have been developed for environmental cleanup, drinking water, and industrial and health-related fields.

2.1 Heavy Metal Removal

Heavy metals are thought to be extremely harmful, including mercury, chromium, lead, and other metals. Consequently, the elimination of these detrimental metals holds paramount significance for both human well-being and the enhancement of water quality. Heterogeneous photocatalysts have been effectively employed to eliminate heavy metals such as mercury, chromium, cadmium, lead, arsenic, nickel, and copper from the environment. Additionally, the recovery of valuable metals including gold, platinum, and silver from industrial wastewater has been accomplished through the utilization of photocatalytic reduction methods. A promising method for removing heavy metals from contaminated water is photocatalysis. In this process, a photocatalyst material is used to initiate a photochemical reaction that breaks down the heavy metal pollutants in water into harmless components. This technology has several advantages over traditional methods such as adsorption and precipitation, as it is a more efficient and environmentally friendly process. One of the most commonly used photocatalysts for heavy metal removal is titanium dioxide (TiO_2). TiO_2 is a semiconductor material that becomes activated in the presence of light, and its high photocatalytic activity can be utilized for the degradation of organic and inorganic pollutants, including heavy metals.

Studies have shown that photocatalytic degradation of heavy metals by TiO_2 can be influenced by various factors such as the concentration of the metal ions, pH, light intensity, and the presence of other co-existing ions. The efficiency of the photocatalytic process can also be enhanced by modifying the TiO_2 surface with various materials such as metals and metal oxides. Numerous studies have shown how well photocatalysis works to remove heavy metals from contaminated water. As an illustration, research by Pavia et al. [1] investigated the photocatalytic degradation of cadmium ions using TiO_2 nanoparticles and found that the photocatalytic process was able to remove 98% of cadmium ions from water. Another study by Roshani et al. [2] investigated the removal of lead ions from water using a TiO_2-based photocatalyst and reported a removal efficiency of 89% under optimized conditions. Sarkhosh et al. [3] conducted a study on the application of photocatalysis for the degradation of pollutants in water. In this study, titanium dioxide nanoparticles were used to explore the photocatalytic degradation of chromium (VI) in aqueous solution. They looked at how many factors, including pH, initial pollutant concentration, and irradiation period, affected the photocatalytic process' effectiveness. The outcomes demonstrated that the pH of the solution affected the photocatalytic degradation of chromium (VI), with pH 3 exhibiting the best efficiency. Moreover, the degradation efficiency increased with increasing irradiation time and decreasing initial concentration of the pollutant. The study also demonstrated that the photocatalytic process was effective in removing more than 90% of chromium (VI) from aqueous solutions within 120 min of irradiation.

Overall, photocatalysis has emerged as a promising technology for the removal of heavy metals from contaminated water. It has the potential to grow into an affordable and long-lasting solution for dealing with the issue of heavy metal pollution in water resources with more study and development.

2.2 *Organic Matter Degradation*

Photocatalytic processes have demonstrated the ability to transform organic compounds like alcohols, carboxylic acids, phenolic derivatives, and chlorinated aromatics into innocuous end products such as carbon dioxide, water, and basic mineral acids. Furthermore, these processes exhibit the capability to mineralize water-contaminating substances like pesticides and herbicides, including 2,4,5-trichlorophenoxyacetic acid and S-triazine herbicides, as well as addressing oil contamination in aquatic environments. This method relies on the use of photocatalysts, such as titanium dioxide (TiO_2), which generate reactive oxygen species upon exposure to light, leading to the degradation of organic compounds. The TiO_2-based photocatalyst was effective at degrading phenol, with a degradation rate of 97.9% achieved after 120 min of treatment [4]. The ZnO-based photocatalyst was able to remove over 90% of the sulfamethoxazole in just 60 min of treatment [5]. Overall, photocatalysis has proven to be a promising method for the removal of organic compounds from wastewater. By using renewable energy sources, such as sunlight, and non-toxic catalysts, such as TiO_2 and ZnO, photocatalysis offers an environmentally friendly solution for wastewater treatment.

With further research, photocatalysis could become a widely used technique for the treatment of industrial wastewater and other contaminated water sources. Up to 98% of the dye could be eliminated after 90 min of exposure thanks to the dye's exceptionally efficient photocatalytic breakdown. The study's findings demonstrated that photocatalysis can be a useful technique for removing organic contaminants from wastewater. Sarkhosh et al. [6] also studied the effect of various factors on the photocatalytic degradation of the organic dyes, including pH, catalyst loading, and initial dye concentration. The results of the study showed that the photocatalytic activity was affected by these factors, with higher catalyst loading, lower initial dye concentration, and lower pH resulting in increased degradation rates. The study also showed that the TiO_2-based photocatalytic performance was stable over several cycles of use, indicating that it has potential for practical applications in the removal of organic compounds from wastewater.

2.3 *Mineral Compounds' Removal*

Numerous mineral compounds are also susceptible to photophysical changes on catalyst surfaces, in addition to organic molecules. Decomposition presents a viable option for non-mineral entities such as bromate, chlorate, azide, halide ions, nitric acid derivatives, as well as palladium and rhodium species, along with sulfur-based species. Additionally, compounds including cyanide, thiocyanate, ammonia, nitrates, and nitrites, as well as metallic salts such as silver nitrate ($AgNO_3$) and mercury chloride (HgCl), and organic metallic compounds like methylmercury chloride (CH_3HgCl) can be efficiently eliminated from water.

2.4 Water Purification

The photocatalysis method can effectively eliminate bacteria and viruses. Heterogeneous photocatalysis has shown efficacy in eradicating various microorganisms, including *Streptococcus mutans*, *Streptococcus natos*, *Streptococcus seristos*, *Escherichia coli*, *Saccharomyces cerevisiae*, *Lactobacillus acidophilus*, and Poliovirus 1. With the escalating occurrence of algal blooms in freshwater reservoirs, the possibility of cyanobacteria contaminating these water sources can be effectively mitigated through the decomposition facilitated by a titanium dioxide catalyst. Photocatalytic disinfection sensitized by titanium dioxide had a significant effect on the destruction of Chlorococcales (green algae) with thick cell walls. The photocatalytic process involves the use of a catalyst that absorbs light and produces highly reactive species that react with contaminants, leading to their degradation. The researchers used titanium dioxide (TiO_2) as the photocatalyst and tested its efficiency in removing organic pollutants from water. The results showed that the photocatalytic process was highly effective in removing pollutants such as dyes, pesticides, and pharmaceuticals from water. The researchers tested the efficiency of a photocatalyst made from silver-doped titanium dioxide ($Ag\text{-}TiO_2$) in removing *Escherichia coli* and other bacteria from water. The photocatalytic process was highly effective in removing bacteria from water, with a removal efficiency of over 99%.

Overall, photocatalysts have great potential for water purification due to their high efficiency and environmentally friendly nature. With further research and development, photocatalytic water purification technology could become a cost-effective and sustainable solution for providing clean and safe drinking water.

2.5 Decomposition of Natural Organic Materials (DNOMs)

Humic substances (HS) are a class of naturally occurring, heterogeneous, yellowish-brown colored, high molecular weight organic compounds that are found everywhere. A portion of filtered water that is adsorbed onto XAD-8 resin, a non-ionic polymeric adsorbent, at pH 2, is another way to describe them. They are a significant source of dissolved organic carbon in surface waters (including groundwater and seawater), and they give the water its characteristic yellowish-brown hue. The amount of humic compounds varies depending on where you are. Typically, seawater contains 2–3 μg of this substance per liter. The source and age of the material exert influence over factors such as size, molecular weight, elemental composition, structure, as well as the quantity and arrangement of functional groups. It is recognized that humic compounds have an impact on the solubility and adsorption of water-repellent contaminants, water photochemistry, trace metals and toxicants, and other pollutants in aquatic settings. Humic substances are believed to serve as a nourishing sustenance for bacterial proliferation while concurrently impeding the bacterial breakdown of impurities, such as certain dyes, in natural water sources. Moreover, they exhibit the capability to form intricate complexes with heavy metals like iron, manganese, and lead, rendering their removal more challenging. Additionally, these substances facilitate the dispersion of metals within the environment, contributing to the process of metal transfer, while also exacerbating pipe corrosion. Humic compounds react with hypochlorite ions and serve as a source of methyl groups. In water treatment facilities, hypochlorite ions are utilized as a disinfectant, but they also produce trihalomethanes, haloacetic acids, and other chlorinated chemicals as byproducts, as well as nitriles (some of which can cause cancer). When humic acid reacts with chlorine, it produces more than 150 different byproducts that have recently been identified.

Overall, these investigations underscore the promising capacity of photocatalysis as a potent technique for the degradation of various organic compounds present in water, encompassing both natural organic matter and phenol. By utilizing photocatalysis, it is possible to treat organic pollutants in water without generating secondary pollution, making it a promising technology for wastewater treatment.

2.6 Sea Water Purification

Photocatalysis has also shown great potential in seawater purification. In comparison to pure water, the rate of humic compound degradation in seawater is slower. No hazardous byproducts of the breakdown process have been found. One study investigated the use of photocatalysis for seawater purification using TiO_2 nanoparticles immobilized on a membrane surface. The results showed that the photocatalytic membrane system effectively removed organic matter and bacteria from seawater with a high removal efficiency of over 99%. The researchers also observed that the photocatalytic membrane system had good stability and reusability (source: Kim and Lee [7]). Overall, photocatalysis has shown great potential for seawater purification, offering a cost-effective and environmentally friendly solution for removing contaminants from seawater.

2.7 Air Purification

With potential uses for reducing pollution in settings like homes, workplaces, factories, airplanes, and spacecraft, photocatalytic technologies are well suited for eliminating organic pollutants in the gas phase, such as VOCs. The capacity of this technique to remove pollutants is being studied in order to broaden the scope of photocatalytic processes in indoor air applications. Due to the potential of exposure to harmful pollutants present in the air, pollutant removal is crucial in indoor air applications. One of the most significant air pollutants in enclosed spaces are bioaerosols, which contain more than 60 different types of bacteria, viruses, and fungi which have been identified as known airborne diseases. Tuberculosis, legionellosis, influenza, colds, anthrax, scarlet fever, chickenpox, aspergillosis, pneumonia, meningitis, diphtheria, and yellow fever are among the illnesses that can be spread by bioaerosols. Activated carbon filters, HEPA filters, ozone, ionization, and biological protection filters are some examples of traditional indoor air filtration technologies. However, none of these methods is 100% successful.

Infectious microbes like *Bacillus anthracis* (anthrax), which can be used to make bioterrorism weapons, can be rendered inactive through photocatalytic oxidation. Gram-positive and Gram-negative bacteria that were subjected to fluorescent light radiation (like that utilized in indoor environments) while being coated with a visible light-sensitive catalyst likewise became inactive. Using AFM, FTIR, and CFU (as a function of time and by oxidative degradation of cell wall membranes), a thorough mechanism and detailed description of the photocatalytic slaughter of *E. coli* on air-coated surfaces were studied to understand the degree of inactivation kinetics of the microbe and reduction through photocatalysis. For various air pollution control applications, new photocatalytic reactors and light-based catalytic systems have been reported. These setups employ catalysts supported on polyester, carbon nanotubes, integration with other pollution control mechanisms, membrane technologies, utilization of silver for antibacterial properties in cotton fabrics to reduce airborne *E. coli*, and high-surface-area oxidized copper catalysts embedded in silica structures. From an environmental health standpoint, the anti-fungal characteristics of photocatalysts against mold fungus on hardwood surfaces within buildings were substantiated through microbial testing involving *A. niger* and UV radiation. Photocatalysts have been widely studied and employed for air purification due to their ability to generate reactive oxygen species (ROS) and degrade various pollutants under light irradiation. Titanium dioxide (TiO_2) is one of the most extensively studied photocatalysts due to its excellent photocatalytic activity, chemical stability, and low cost. Numerous studies have demonstrated that TiO_2 is an efficient way for the eradication of diverse air pollutants, encompassing sulfur dioxide (SO_2), nitrogen oxides (NOx), and volatile organic compounds (VOCs). One study conducted by Zhang et al. [8] investigated the photocatalytic removal of VOCs using TiO_2 nanoparticles under visible light irradiation. The outcomes demonstrated that TiO_2 had outstanding photocatalytic activity and could efficiently eliminate VOCs at low concentrations. Another study by Ong et al. [9] examined the use of TiO_2 for the removal of NOx in indoor air. The study found that TiO_2 was able to significantly reduce the concentration of NOx in indoor air, making it a promising candidate for indoor air purification. In addition to TiO_2, other photocatalysts such as zinc oxide (ZnO), tungsten trioxide (WO_3), and bismuth vanadate ($BiVO_4$) have also been investigated for air purification applications. For example, a study by Fang et al. [10] investigated the use of ZnO nanoparticles for the removal of SO_2 from flue gas. The findings revealed that ZnO demonstrated exceptional photocatalytic efficacy, efficiently eliminating SO_2 when exposed to visible light radiation. In summary, the photocatalytic approach is a versatile and effective decontamination technique that may inactivate a variety of dangerous microbes in a variety of situations. This is a versatile disinfection technique that is secure, non-toxic, and reasonably priced. Among other things, photocatalyst thin films on different substrates are beneficial for self-purifying surfaces, materials for medical implants, surgical instruments, surfaces in lab and hospital settings, and equipment [11–14].

References

1. D.L. Pavia, M. Sillanpää, C. Punta, Photocatalytic removal of cadmium ions using TiO_2 nanoparticles in a batch reactor. Chem. Eng. J. **405**, 126789 (2021)
2. B. Roshani, S. Sharifnia, A. Khataee, Effective photocatalytic removal of lead ions from water using TiO_2–ZnO nanocomposite as a new photocatalyst. Environ. Sci. Pollut. Res. **27**, 6106–6117 (2020)
3. M. Sarkhosh, M.R. Zare, M. Farzadkia, A. Esrafili, Photocatalytic removal of hexavalent chromium from aqueous solution by TiO_2 nanoparticles: optimization of operating parameters. J. Environ. Health Sci. Eng. **15**, 1–10 (2017)
4. L. Zhao, X. Zhang, X. Cai, X. Wang, Enhanced photocatalytic degradation of phenol by TiO_2/g-C_3N_4 composite under visible light irradiation. J. Taiwan Inst. Chem. Eng. **70**, 177–184 (2017)
5. Y. Li, Y. Xie, Y. Wang, Y. Liu, Photocatalytic removal of sulfamethoxazole by ZnO: kinetics, pathways and influencing factors. Chemosphere **200**, 309–316 (2018)
6. M. Sarkhosh, F. Ghanbari, M. Montazerzohour, A.R. Khataee, Enhanced photocatalytic degradation of organic dyes using TiO_2-based photocatalysts: a review. J. Environ. Health Sci. Eng. **15**, 10 (2017)
7. J.G. Kim, C.H. Lee, Photocatalytic membrane system for seawater purification. J. Ind. Eng. Chem. **26**, 233–239 (2015)
8. X. Zhang, Y. Li, Z. Li, M. Zhu, Y. Liu, J. Li, Visible-light-driven photocatalytic removal of low-concentration volatile organic compounds using TiO_2 nanoparticles. J. Environ. Sci. **75**, 249–256 (2019)
9. Z.X. Ong, K.W. Tan, P.E. Poh, TiO_2-based photocatalytic indoor air purification: a review. J. Environ. Chem. Eng. **8**, 104283 (2020)
10. M. Fang, X. Li, X. Li, M. Li, F. Yu, Removal of SO_2 in flue gas by ZnO nanoparticles under visible light irradiation. Environ. Sci. Pollut. Res. **28**, 11188–11197 (2021)
11. Y. Guan, M. Li, Y. Li, J. Li, X. Li, J. Li, B. Li, Highly efficient removal of organic pollutants from water using titanium dioxide-based photocatalysts: a review. J. Hazard. Mater. **407**, 124631 (2021)
12. Y. Chen, L. Yang, Y. Wu, R. Guo, J. Lu, Photocatalytic degradation of organic pollutants and inactivation of bacteria in water using Ag-TiO_2: a review. Environ. Sci. Pollut. Res. **27**, 6676–6689 (2020)
13. T. Zhang, Y. Zhou, L. Xu, X. Liu, N. Chen, S. Wu, Photocatalytic degradation of natural organic matter in drinking water using TiO_2 nanoparticles: influential factors and degradation pathway. J. Environ. Sci. **39**, 47–54 (2016)
14. Y. Li, X. Wang, J. Wang, X. Liu, C. Hu, Photocatalytic degradation of phenol in aqueous solution using TiO_2 nanoparticles: effect of H_2O_2 and operating parameters. Water Res. **148**, 246–256 (2019)

Leftover Living Cells' Derived Biomass and Microorganisms as a Source for Hydrogen Energy Production

R. Harikrishnan, M. Mani, M. Kumar, and K. Kaviyarasu

Abstract The use of biofuel like hydrogen instead of fossil fuel in various sectors such as vehicle transport, electricity generation, running industries, and powering homes can reduce greenhouse gas emissions to the atmosphere. At present form, the costs of producing biofuel are relatively higher than fossil fuels. The contributions from worldwide researchers in the experimental study of hydrogen production by various routes are essential which may help in designing the reactors for generating the hydrogen with less cost and optimizing the experimental conditions. This chapter initially discusses about the bad effects of greenhouse gas emitting fossil fuels and the importance of renewable hydrogen energies toward the hydrogen era in the introduction section. The hydrogen extraction techniques from the leftover biowaste materials into hydrogen production through biological (*fermentation, biophotolysis, and microbial electrochemical cells*) and thermochemical routes (*gasification, pyrolysis, and high-pressure aqueous*) have been discussed in a separate section. Few portions of this chapter summarize the analysis results of recently published articles focused on the title of "H" generations from the various biowaste, or leftover living cells derived biomass and microorganism biomass sources. The information such as types of techniques (*biological and thermochemical routes*), substrates, and inoculum used, hydrogen production rate, and new advancements and achievements done by the various worldwide researchers are summarized in Table. From the summarization results of the table, it is revealed that most of worldwide researchers have used

---

R. Harikrishnan (✉)
Department of Physics, St. Joseph College of Engineering, Sriperumbudur, Chennai, Tamil Nadu 602117, India
e-mail: harifiziks@gmail.com

M. Mani
Department of Physics, Kalaignar Karunanidhi Govt. Arts College, Tiruvannamalai, Tamil Nadu 606603, India

M. Kumar
Department of Physics, Arignar Anna Govt. Arts College, Cheyyar, Tamil Nadu 604407, India

K. Kaviyarasu
UNESCO-UNISA Africa Chair in Nanosciences/Nanotechnology Laboratories, College of Graduate Studies, University of South Africa (UNISA), Muckleneuk Ridge, PO Box 392, Pretoria, South Africa

K. Kasinathan et al. (eds.), *New Technologies for Energy Transition Based on Sustainable Development Goals*, https://doi.org/10.1007/978-981-97-2527-4_21

fermentation techniques for hydrogen generation. The biowastes such as agricultural residues, food waste, and wastewater were used as a main source for the hydrogen generations by worldwide researchers.

1 Introduction

Nowadays, cities face major issues due to the disposal of wastes such as wastewater, municipal waste, and medicinal waste. Recycling or degrading them biologically will be useful. Hydrogen or other biofuel generation from such biowastes will be useful since it offers both energy generation and waste management. The industries such as agriculture, educational institution, restaurants, municipalities, and hospitals dispose waste in enormous amounts, which can be used for biofuel, heat, and electricity production. There exists a larger number of organic materials to produce living cells—derived biomass materials endlessly. In addition, biomass serves as a renewable. Day-by-day, industries continuously dispose biowaste in the form of garbage, wood, and manure. The carbon content absorbed during the photosynthesis-assisted way of growing biomass is nearly the same as the carbon released from their leftover biomass, and hence, biomass acts as a carbon–neutral energy source. Currently, large amount biowood waste is available everywhere, and biofuel production using leftover wood biomass (*rather than using the biomass from the cutting of trees*) can avoid deforestation. As a developing country, India imports 80% of crude oil from various countries for fulfilling energy demand, and at the same time, 54% of electricity generation is raised due to the burning effect of coal fuel [1]. Therefore, the release of carbon emissions is highly expected in India due to the world's no. 1 population and energy demand. Under the absence of oxygen, due to anaerobic decomposition, the fossil fuels such as oil, coal, and petroleum products are formed as a result of buried living materials such as plants and animals above the layer of rocks. Living organism like plants creates energy through the photosynthesis process by intaking energy from sun radiations and storing as in the form of biomass. Meanwhile, living animals also store energy in the form of biomass by eating plants as food. Conversion of dead living materials through anaerobic decomposition into fossil fuels typically takes a longer period; for example, carbon-rich fossil fuel requires million years. The existence of carbon-rich fossil fuels as an energy source on Earth took millions of years of geological process. The term global warming is applied to climate change related to the rapidly increased temperature of our atmosphere which occurs due to the use of large amounts of fossil-based fuels in many sectors such as industrial developments, deforestation, and transport of vehicles. The extraction of energies from the various sources can be divided into two categories: (i) renewable and (ii) non-renewable energies. The extraction of renewable energies from natural sources such as sunlight and wind is constantly being replenished at a higher rate. The term renewable energy is frequently referred to as clean energy, as a promising term which does not involve in polluting our environment and atmosphere. Commonly, the release of CO_2 and greenhouse gas emissions from these sectors is the reason for climate change. Due to global warming, the world faces many issues such as intense storms, production of heat waves, wildfires, migration of animals and change of their destination pattern, rising oceans, and threatening of species in the biodiversity. Renewable energy sources such as hydro, solar, wind, nuclear power, geothermal, and biomass-derived can fulfill energy demand in the twenty-first and upcoming centuries which offers zero-emission or less emission of carbon [2]. The stored energies in the living plant part can be used in direct and indirect ways. In direct ways, burning biomass permits in the generation of heat which is used for cooking foods, lighting the environment, and electricity generation. In an indirect way, biomass can be used to produce biofuel such as ethanol, hydrogen, and biodiesel.

1.1 Biomass as a Source of Energy

In the beginning, people (*caveman*) have used biomass materials as an energy source, for example, wood fires. Historically, the first man-made biomass material was believed to be wood which is a part of trees. The biomass energy released from the wood fires was used for cooking foods, making people warm in cold countries, and in addition, it protected our ancestors from deadly animals. The energies extracted from biomass materials can be categorized into renewable and non-renewable. The term biomass is defined as a material whose constituents contain living cells, and hence, they are extracted from living things (*plants and animals*). In this era of the modernized world, much attention has been paid by worldwide researchers and engineers to the extraction of energy from various sources and storing them in energy storage devices since there exists a huge demand for the production of transport vehicles and electrical devices. Greenhouse gas emission is a common phenomenon that occurs during the process of energy extraction from the burning of biomasses and fossil fuels (*coal, oil, and gas*), which pollute our atmosphere. However, biomass materials offer a lower amount of greenhouse gas emissions. Therefore, a development in the extraction of energies from various biomass sources is much needed. Living cell-derived plant energy commonly found in the crops is used as a fuel which offers the production of high output of biomass per hectare with the intake of low energy. In general, grain biomass material can be chosen as a transportation fuel; meanwhile, the burning effect of straw mass can be used either in the form of heat or electricity. In addition, the degradation of plant biomass causes the formation of sugars, where the change of cellulose into glucose is responsible behind it. Hydrogen fuel, methane, biodiesel, and ethanol gases' production from the leftover living cell-derived biomass materials is also possible. Many biomass materials, such as corn and sugarcane, can be used for biodiesel, ethanol, and hydrogen production through the fermentation created by the microbial species. In recent decades, there has been enormous research on "biofuel extraction from the algal biomass" since the growth rate of algal metabolites is 20 times faster than the plant biomass production, and it can be easily grown in many places of our earth and easily cultivated. Biofuels can be found in three states of matter. Bioalcohols, bioethanol, and bio-oils fall under the category of liquid biofuel. Landfill gas, synthetic gas, and hydrogen are commonly found in the gaseous form. Global warming issues can be minimized with biofuel, which may (*fewer* CO_2 *gases*) or may not emit CO_2 gases during production and bioenergy from them [3].

Biofuel is a type of fuel which is produced from living cells derived biomass over a short time rather than the slow process that occurs naturally over millions of years to produce fossil fuel. This kind of fuel is commonly used to produce heat and electricity, and in addition, it can also be used for transportation purposes. The biofuels extracted from the living cells derived from biomass are considered as a non-renewable source. The use of biofuels in transportation and their production on a large-scale level are two of the best solutions for solving the global warming issue. This kind of fuel can be used in many sectors such as aviation and transportation. The most synthesized man-made biofuels are bioethanol, biodiesel, and hydrogen. There has been enormous progress in developing technology through experimental and theoretical research to produce next-generation biofuel which can be obtained from the biowaste and algal biomass. In addition, the next-generation biofuel can be the best substitution in place of petroleum and can be used in many sectors such as fuel tanks, engine fuels, and pipelines [4].

1.2 Hydrogen as a Fuel

Hydrogen is the lightest element, and it is rarely available in its pure form in earth atmosphere. During the burning effect, hydrogen interacts with the oxygen element which releases water molecule and additionally thermal energy.

$$2H_2 + O_2 \rightarrow 2H_2O\ (\text{gas}) + \text{energy}. \tag{1}$$

The released thermal energy can be used in powering the vehicles, and therefore, hydrogen can be considered as one of the best fuels. The term "*energy*" plays a role in determining the aspects such as the country's economy, infrastructure, vehicle transportation, and standard of human life. Among many other renewable energy sources, the use of hydrogen energy source is highly suitable in vehicles. Recently, the world's top automobile companies, such as Honda, Toyota, and Hyundai, are manufacturing hydrogen fuel cell-based vehicles which are commonly found in Asia, Europe, and North America. Highly educated and high-income families usually target using such hydrogen fuel cell-filled vehicles. According to the report, as of June 2018, more than 6500 vehicles (running based on hydrogen fuel) have been sold to consumers. One can expect that, by the year 2030, hydrogen fuel cell-based vehicles cost estimated to be competitive with internal combustion engine-based vehicles [5]. Among many other types of fuels, such as gasoline (49 MJ kg^{-1}), hydrogen has the highest energy density per weight which is about 142 MJ kg^{-1}. It acts as a main substance for the production of ammonia and methanol as well as it is useful for refining petroleum products; however, it requires a greater amount of hydrogen.

In order to avoid a climate crisis and to save the ecosystem of our mother planet, recently in 2015, United Nations Climate Change Conference near Paris, France, adopted the statement (*Paris Climate Accords*) that greenhouse emissions should be reached zero in 2050 (UNFCCC, Paris Agreement. United Nations, 2015). This can be achieved when all sectors all over the world reduce the usage of fossil fuels and target hydrogen as an energy carrier as well as an effective fuel.

This "H_2" fuel is commonly used as fuel and is considered one of the green renewable energy sources that may dominate futuristic transport vehicles since it offers zero carbon emission and higher efficiency. Hydrogen is considered one of the best candidates for fossil fuel which offers wide flammable and high heating capacity during combustion. The product released during the combustion of hydrogen is water which is free from pollutants, and the conversion of water into hydrogen and oxygen is also possible. In liquid form, it serves as a fuel in rockets. Due to their strong reducing properties, it is used in semiconductor processing, metal smelting, petroleum extraction, etc. "H_2" produced from the various methods can be categorized into four types: (i) green, (ii) blue, (iii) gray, and (iv) brown. If the highest level of CO_2 emissions occurs during the H_2 production (for example, gasification of fossil fuel), then the produced H_2 is called as brown hydrogen. Gray hydrogen is typically released from the H_2 production through natural gas which involves in the emissions of CO_2. Blue hydrogen is referred to as the generated hydrogen from the various plant systems containing CO_2 capture technologies. The production of green hydrogen does not involve in the production of CO_2. Therefore, the production of green H_2 production is more beneficial to our environment. Therefore, this fuel will dominate in the application of electricity generation and fuel combustion engine fuels. Literature showed that in 2019, 75 million tons of hydrogen have been produced mainly for the refining process in the chemical industry. In 2019, hydrogen production has resulted in the emission of CO_2 gases more than 800 million tons [6]. One can expect the origin of the hydrogen era when there exists an advancement in hydrogen vehicle-based technologies. The efficiency (35–50%) of energy extraction from the burning effect of hydrogen in the transport vehicle is usually higher than any vehicles running based on the hydrocarbon fuels [7]. It has been shown that hydrogen fuel cells do not involve in the emission of harmful gasses. However, during the production (reforming) of hydrogen from fossil fuels, it involves in the production of harmful CO_2 gaseous. The most abundant H element is rich in biomass materials and bonded with the other elements, and it must be extracted and stored for energy applications.

The list of possible sources for the production of H_2 is given below.

i. Fossil fuel (commonly exits in the Earth's crust in the form of coal and crude oil).
ii. Unprocessed hydrocarbons (an organic compound contains carbon and H_2 structure).
iii. Fusion-type nuclear reaction.
iv. Living cells derived biomass sources.
v. Water hydrolysis.

As of 2040, it is estimated that the use of hydrogen energy consumption in transport vehicles could save petroleum per day about 18.3 million barrels per, predicting the replacement of 150 million tons of H_2 approximately. In addition, if such a quantity of hydrogen can be produced through petroleum reforming, 11 million barrels of petroleum per day can be saved.

1.3 Biomass as a Source for H_2 Generations

It was estimated that about 11.4 billion of biomass materials/year have been produced by global farmland land and forests areas, where agricultural, pasture, timber, and other areas account for the production of biomasses with 40, 30, 18, and 12%, respectively [8]. Biomass as a source of energy is available all over the place of the Earth in two different forms: (a) woody biomass and (b) agricultural waste. Biomass can be used in numerous ways for the production of heat, hydrogen fuel, biodiesel fuel, chemicals, electricity, bioethanol, biogas, and hydrocarbons. It usually requires several processes. Biomass materials are rich in hydrogen, which can be collected separately using several techniques from various sources. The H_2 can be extracted by using several ways, including (i) thermal process, (ii) electrolysis of water, and (iii) biological methods; meanwhile, the most quantity of H_2 is extracted by steam reforming technique [9]. According to the literature, 48%, 30%, and 18% of H production were achieved by using the sources of natural gas, oil, and coal, respectively. The remaining 4% of H_2 was achieved through the water electrolysis process. Overall, the achievement of the worldwide production of H_2 is higher than 1 billion m^3/day [10].

Organic waste-based biomass is also an important source which contains stored energy, and it can be extracted in several techniques. In India, 0.5 billion metric tons of organic waste are disposed per year, and it is estimated that this waste can be used for power generation of up to 17,500 MW. Among the organic wastes, most of the wastes have been disposed from the sugar mills since 580 sugar mill factories are currently running in India, which disposes of bagasse. These bagasse wastes have the potential in generating power up to 5000 MW by cogeneration technique [1, 11]. Literature reveals that nearly 10–14% of consumption of energy by humankind is possibly done by biomass sources. The developing countries put more effort into extracting energy from biomass [12]. We focus on the H_2 generation through the biomass sources such as the residue of agricultural products, forest by-products, disposed biowaste from various industries, municipal wastes, household wastes, and wastewater.

1.3.1 Lignocellulosic Biomass

The term Lignocellulose is referred to as a dried matter of plant part, also called lignocellulosic biomass, which commonly consists of two types of carbohydrate polymers: (i) cellulose and hemicellulose and (ii) lignin (aromatic rich polymers). Plant material's cell wall uses lignin as a defense mechanism against the attack of microbial pathogens and protects the cellulose and hemicellulose. They act as disease resistance, and in addition, it is also developed to prevent the multiplication of *hypersensitive response-programmed cell death* (HR-PCD). Lignocellulosic biomass can be found in a variety of sources, including residues of agricultural and forestry products, crop material, and cellulose-rich wastes. They fall under the abundant raw materials which can be used for H_2 generation as well as biodiesel. Therefore, any biomass materials consisting of the above-mentioned carbohydrate polymer mixtures are called lignocellulosic biomass. Due to the presence of complicated structures, they are very challenging in the processing of H_2 productions, and it requires the combination of enzymes, microorganisms, and the heating process to overcome the effect of recalcitrance. These carbohydrate structures of plant cell walls contain a different form of sugar monomers which are bounded to the lignin with a covalent bond. They provide a higher potential for H_2 generation when compared with fossil fuels, and they do not involve compromising global food security. The non-edible source lignocellulosic biomass typically contains the compositions of cellulose, hemicellulose, and lignin about (38–50%), (23–32%), and (15–25%), respectively. The presence of lignin is responsible behind the rigid nature of LCB, restricting them in the process of conversion of biomass into H_2, and hence, the removal of such complicated polymers is necessary before the fermentation process. Recently, the developed nanozymes displaying enzymatic mimetic behavior replace the use of microbial enzymes. In addition, enzymatic mimetic behavior can be tuned by synthesizing the nanozymes with different sizes and morphologies. In a previous study, $CeFe_3O_4$ nanoparticles have been used in the process of degradation of lignin-complicated polymer structures from the corncob biomass and it was further used to improve H_2 generation through DF. Their result revealed that lignin-degraded corn cob has displayed the degradation of lignin which was about 43.26%. From the modified biomass, the rate and maximum cumulative H productions were achieved with 1.55 mL/h and 78.45 mL, respectively [13].

1.3.2 Straw Biomass

Straw biomass falls under the category of highest disposed agricultural materials. Annually, 7 billion tons of straw waste are distributed among them (for example, rice, wheat, and corn straw), and 900 million tons are being produced in China. The annual global rice and wheat straw are produced about 731 and 584 million tons, respectively. Apart from rice and wheat straw, the cultivation of maize is the third highest mass portion of straw biomass. Recently, rice straw [14, 15], wheat straw [16, 17], and maize straw [18] as the biomass source for hydrogen generation have been reported. The residual straw collected from the agricultural area contains C6 and C5 sugars which are mainly used for H_2 generation. However, if the residual is composed of lignin, it affects the H_2 generation. Straw biomass materials naturally contain hemicellulose, cellulose, and lignin. The plant cell wall is typically composed of cellulose. Plant cell walls are made up of complicated polymer structures, and crystallinity strongly affects enzymatic reaction for H_2 generations. Such polymerized complicated structure is raised due to the presence of *Van-Der-Waals* force acting the hydrogen and covalent bond of cellulose molecules [19].

1.3.3 Microalgae Biomass

Microbial algae are also used for H_2 generation. Food and agricultural organizations have estimated that the production of algae per hectare is about 150–600 tons/year [20]. In microalgae (dried biomass), the carbohydrates, proteins, and lipids were found approximately with the compositions (32–60%), (7–31%), and (2–13%), respectively. In addition, their cell wall is made up of cellulose, hemicellulose, pectins, and arabinogalactan protein. Hence, they are highly suitable in H_2 production due to the presence of rich in carbohydrate contents [21]. H_2 generation closer to 100 mL/gVS was achieved with the help of carbohydrate-rich microalgae biomass through the fermentation process. However, other research paper reveals the production of high-yield H_2 through microalgae composed of lower carbohydrate, and the discrepancy raised in this case is due to the different hydrolysis strategies which are used to reduce the sugars in the microalgae biomass [22] (Fig. 1).

H_2 production can be done in several ways, for example, hydrocarbon reforming from fossil fuel (which is found on Earth's crust and took millions of years to form), coal (created by the buried plant and animals' part) gasification, and partial oxidation process. However, they require a large magnitude of temperature and pressure. The biological pathway of hydrogen production can be divided into several types, which

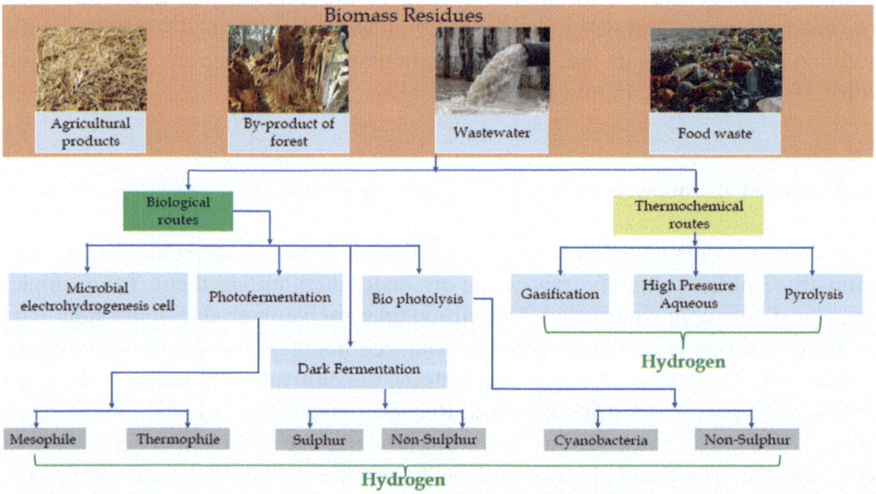

Fig. 1 Schematic diagram of H_2 generation by using leftover living cells' derived biomass through the biological and thermochemical routes

can be classified based on the different conditions, such as biomass type, light–dark environment, and types of organisms. Some types of organisms use sunlight as input energy and split the water into H_2, and such a process is called biophotolysis. A few types of organism use carbon-rich substrate and breaks them down into H_2, and they are called fermentative microbial species. Each route of H_2 production has distinct benefits as well as limitations. In the sun light-driven photosynthetic process, cyanobacteria and green algae split water molecules into H_2. However, the presence of oxygen molecules limits the hydrogenase enzyme activity, which reduces H_2 production [23].

2 Biological Routes of Hydrogen Generation

Higher energy is required during the water electrolysis and thermal processes for achieving hydrogen production. However, the biological method of producing H_2 does not require fossil fuel or electricity as an input. This method requires water, solar radiation, and biomass material as resources, and microorganism takes responsibility for producing H_2. In this technique, the rate of H_2 production is limited, and it involves in the production of reaction wastes. However, this method of H_2 production is still promising to worldwide researchers and engineers since the disposal of wastes is biodegradable, and it offers a renewable energy production strategy [24]. Microbial species such as green and blue algae are able to split H_2O into hydrogen and oxygen anions through direct and indirect photolysis routes. The photo-fermentation and photolysis of water may involve the complications such as a lower amount of hydrogen production and light dependency [1]. Therefore, H_2 production through dark fermentation is more reliable; it offers a simple process and requires moderate working conditions.

- Cyanobacteria and green algae use sunlight to split water molecules into H_2 production by direct and indirect bio photolysis.
- Photosynthetic microorganism uses sunlight to induce a fermentation process, where they break organic sources into H_2 and CO_2.
- Microorganism breaks down the organic sugars into H_2 production through dark fermentation.

2.1 Biophotolysis

H_2 generation due to photolysis can be categorized into (i) direct and (ii) indirect photolyses.

(i) *Direct Photolysis*

According to the literature, direct photolysis can be achieved only by microalgae involving photosystem II which directly induces water-splitting activity for H_2 evolution. In this technique, solar energy can be converted into hydrogen energy which is given by the mathematical relation.

$$2H_2O + \text{solar energy} \rightarrow 2H_2 + O_2. \tag{2}$$

In this typical process, the photosystem intakes energy from the photon and causes the transport of electrons to the ferredoxin, and due to this phenomenon, the reduction of water takes place. When hydrogen-producing enzyme named hydrogenase intakes electrons from the transport of reduced ferredoxin, catalytic conversion of a protein into H_2 occurs, which obeys the following relation (Fig. 2):

$$2H^- + 2FD^- \rightarrow H_2 + 2FD. \tag{3}$$

Two types of photosystems (PS I and PS II) take responsibility for H_2 generation in the presence of solar energy.

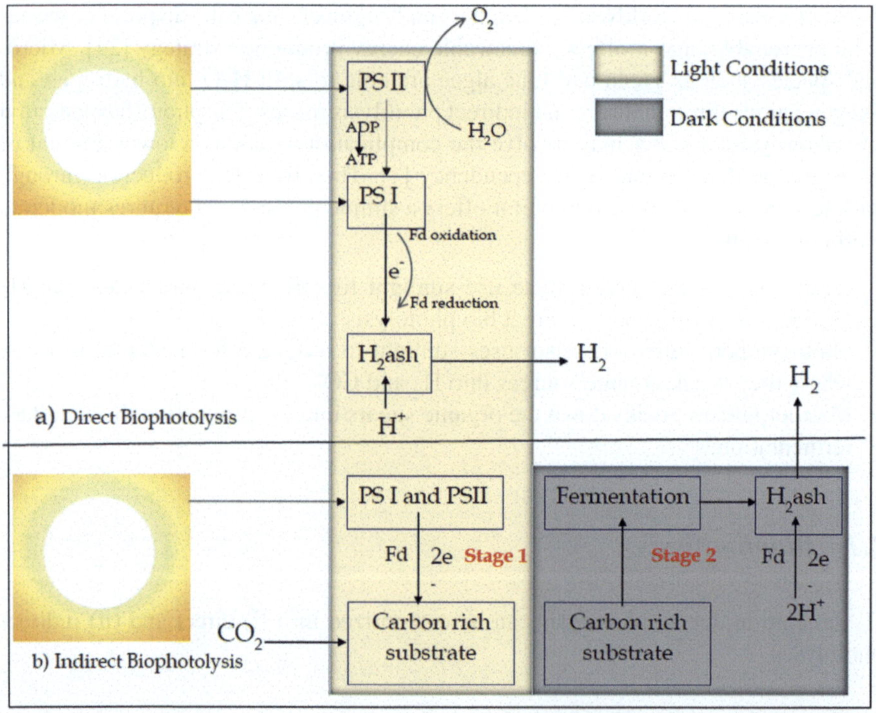

Fig. 2 Typical schematic diagram depicts the H_2 generation from the direct and indirect biophotolysis processes

(ii) *Indirect Photolysis*

In the second category, the derived electrons form the glycolysis and can enter the linear electron chain system at the PQ, and then it is possibly transferred to the (Fe–Fe)-ase. In this case, electron flow is derived by the photosystem. Hence, this indirect photolysis process is independent of the photosystem II pathway. In the case of green algae *Chlamydomonas reinhardtii*, most of the electrons transferred to the hydrogenase is derived from the photosystem II [25]. In this process, H_2 can be extracted by using the green algae, where it uses the condition of sulfur deprivation. Sulfur deprivation has been used for enhancing anaerobic conditions in the cell culture since they downregulate O_2 evolution in the photosystem II. Sulfur deprivation, or the absence of sulfur, reduces the O_2 production level and falls below the level of O_2 required for respiration during the photosynthesis process of algae. This imbalance is responsible for overall oxygen consumption by the cell culture, enabling anaerobic conditions in the photon energy-dependent algal cell cultures, where in the chloroplast, the [Fe] hydrogenase pathway of electron transport occurs, permitting the production of H_2 generations. As a consequence, the anaerobic condition is enabled in the green algae cell culture, which induces electron transport by the Fe-hydrogenase pathway, causing the production of H_2 gas. In addition, under such conditions of sulfur deprivation, bypassing the sensitivity of [Fe]-hydrogenase to an oxygen molecule, H_2 can be generated by making the release of oxygen and H_2 through separate reactions, where H_2 is released from the glycogen and starch. These cellular substances in the microbial species are grown under a CO_2 fixation process driven by photon energies. Therefore, indirect photolysis is a two-step process. The first step involves the growing of biomass in the presence of light, and the second process involves the H_2 extraction from the algal biomass (carbohydrates) under dark conditions, commonly called as a fermentative pathway of hydrogen production [26].

2.2 Microbial Electrohydrogenesis Cell (MEC)

Electrogenic microorganisms are capable of producing an electrical potential of up to 0.3 V when they release electrons and protons by intaking energy from an energy source (for example, acetic acid). External voltage can also be used to break the limitation of thermodynamic limitations, which is very useful in breaking down the non-fermentable organic sources into H_2 production. In a typical MEC, an additional external voltage causes the reduction of protons to release H_2, where bacterial activity is responsible for energy generation. In this technique, due to external voltage supply, oxidation of organic matter and reduction of the proton occur at the anode and cathode, respectively [27]. Typically, in this process, electrons and protons are released at the anode when microbial species intake energy from a source (for example, acetic acid), which are then utilized by the cathode to produce H_2; however, it sufficiently requires additional voltage in order to trigger the biohydrogen production in a sufficient way. The redox potential essentially required for initiating proton reduction is about 0.414 V. Electric potential larger than 0.11 V created by the bacterial activity is more beneficial for H_2 production at the cathode. This route gives an additional voltage from the circuits and permits higher achievement of H_2 production by bypassing the endothermic barrier created by the microbial system during the formation of fermentation and end products. The electrical potential required in this approach is found to be lower than the theoretically calculated electric potential (1.23 eV) to initiate the water electrolysis process for hydrogen generation [28].

However, the produced H_2 energy is depended on the type of substrate used. Organic substances containing lactic and acetic acid can achieve the highest percentage of H_2 production. Some biomass materials contain complicated polymer structures (for example, lignocellulosic). Hence, they make it harder for H_2 generation through the DF since thermodynamic limitation further stops the degradation of biomass materials. To overcome the difficulties mentioned above, the fermentation reactor must be integrated and optimized by introducing the microbial electrolysis cell, where it uses an external power supply and offers the decomposition of complicated polymer structures containing organic biowaste materials' end products (for example, volatile fatty acid). In addition, some reports suggested that the remaining residue products resulting from the fermentation degradation and electrochemical decomposition contain enriched essential micronutrients (for example, P, N, K, Mg, Ca, etc.). Leaving leftover products (from other techniques) may cause harmful effects such as polluting ground soil and affecting animals' health. Therefore, the residues left (biofertilizer) from the MEC reactor can be further dispersed to the ground for terrestrial plant growth [29] (Fig. 3).

The concept of microbial cells for electricity generation was first originated from the findings of the degradation of organic compounds by microbial species, leading to electricity generation, which was discovered by M. C. Potter in 1911. Later in 1931, the achievement of 35 V by the microbial fuel cell was done by B. Cohen. Electricity generation was done due to the ability of bacteria to transfer its electron from their cell interior into exterior cells, where the bacteria cause the oxidation of organic substrates at the anode. The anaerobic biological process takes a role in converting the organic substances (sugars and organic acids) into H_2. This process can be compared with the water electrolysis process. The difference between them is originated due to the catalytic activity of microbial species, which offers the further degradation of dead-end products under anaerobic conditions. In a typical microbial electrochemical cell, the anode and cathode electrodes are submerged into the electrolyte, and in addition, the electrodes are connected to an external power supply. This technique commonly uses microbial species, which is responsible for the oxidation and proton reduction process. The exoelectrogenic bacteria have the ability to transfer its exterior cell electrons into solid oxide surfaces, and hence, such kinds of bacteria are attached to the anode materials. This technique permits H_2 production with less contamination, and hence, it has more advantages than other approaches. Due to this technology, previous work impressed the H_2 production with 9.6 mol H_2/mol glucose revealing the achievement of 80% efficiency compared with the theoretically calculated values [30]. Through dark fermentation, H_2 can be produced by using carbon-rich carbohydrates as a substrate but not proteins. By combining the

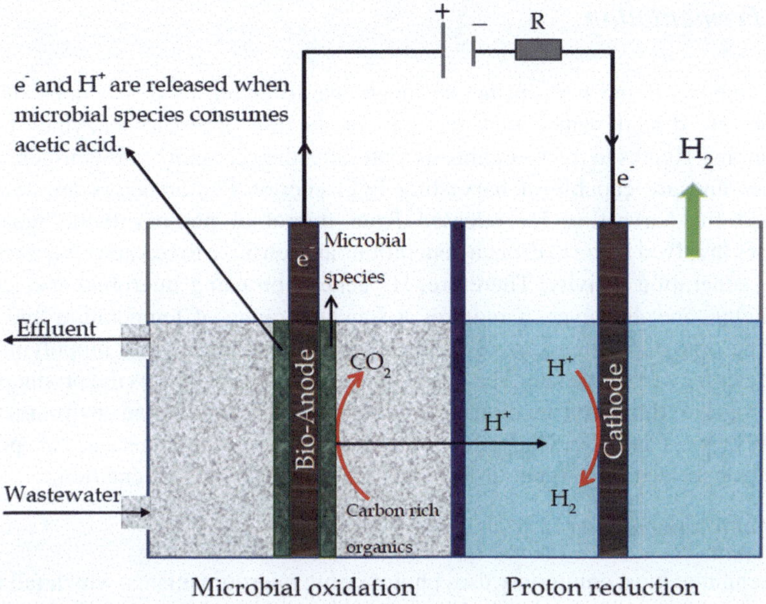

Fig. 3 Typical schematic diagram used for H_2 generation from the microbial electrochemical cells composed of electrode, wastewater, and power supply system

fermentation and input voltage supply in MEC, one can achieve H_2 from the protein. Some wastewater contains a large amount of protein, and hence, H_2 extraction from the wastewater protein is more beneficial since it also offers wastewater treatment [31].

The efficiency of H_2 generation can be lowered when the formation of excessive non-fermentative organic products occurs. Without giving additional power, microorganisms cannot achieve the conversion of dead-end products (for example, organic acetate) into biohydrogen. The formation of acetate and butyrate (dead-end products per one mole of glucose) has resulted in the production of 4 mol H_2 and 2 mol H_2, respectively, during the DF. However, adding an external power supply can result in additional hydrogen production where the reduced proton and electron fuse together at the cathode to form hydrogen. Maintaining the appropriate temperature is an important role in the electrohydrogenesis H_2 generation since the inappropriate temperature affects the wastewater or agricultural residuals, which can further affect the growth of microbial species. In addition, appropriate temperature affects the fermentation process along with the formation of other by-products. Literature reveals that increasing the temperature up to a certain point supports the metabolic activity of microbial and growth factors, which cause the increased activity of the fermentation process.

2.3 Fermentation

Photosynthesis is the main principle plants use to create food, but none of them generate H_2 directly since they do not contain the necessary enzymes for H_2 generations. Thanks to microorganisms, naturally, they contain hydrogen-generating enzymes and are capable of harvesting light energy. Photoreactors are specially designed for harvesting H_2 released from microbial photosynthetic reactions. Enzymes involved in the hydrogen generations are sensitive to oxygen which reduces the H_2 generation activity. Therefore, H_2 generation using microbial species is a challenging one; however, it requires a standard range of temperature and pressure. The first-step process involves in the growing of sugars and biopolymers by intaking energy from the sun. The fermentation process permits us the production of biohydrogen extraction from biomass sources, which is found as an environmentally friendly process and offers the reduction of alternative fossil fuel usage. This process can be categorized into (i) photo-fermentation and (ii) dark fermentation.

(i) *Photo-Fermentation*

This technique most commonly uses phototrophic microorganisms, which fall under the category of a purple non-sulfur bacterial group and is mostly adopted in the generation of hydrogen. This species degrades carbon-containing substrates (organic acids, carbohydrates, organic materials' filled wastewater) into H_2 production. Using the enzyme nitrogenase, these bacterial types convert long and short volatile fatty acids into H_2 and remnant metabolites when exposed to sun radiations. In addition, these species have the capacity to extract hydrogen from various types of organic matter. The well-known microbial species, such as *Rhodobacter sphaeroides*, *Rhodobacter capsulatus*, *Rhodovulum sulfidophilum*, and *Rhodopseudomonas palustris*, are effectively used in extracting the H_2 with high yield from the volatile fatty acids, including lactic, butyric, acidic, malic, and propionic acids through photo-fermentation. The results of the few studies concluded that H_2 production can also be done directly from the substances such as glucose and sucrose. Some studies have also demonstrated that enabling the optimized conditions such as the appropriate temperature of cell growth, light intensity, physicochemical parameters, and substrate type is crucial terms for high hydrogen yields [32]. Photosynthetic bacteria use nitrogen-deprived conditions for breaking down organic compounds through nitrogenase enzymes. Microorganism collects nitrogen from the atmosphere and converts it into ammonia ions, where nitrogenase permits the nitrogen fixation process. Such types of enzymes are found only in the cyanobacteria, non-sulfur purple species, and green sulfur bacterial medium. These bacterial media consume acetic acid and intake energy from the energy source of ATP; in addition, the transfer of electrons occurs from ferredoxin to nitrogenase enzyme, where N_2 fixation takes place. Meanwhile, in the absence of N, nitrogenase protons are converted into hydrogen. This process typically occurs in the presence of various spectral ranges of sun radiations. And the efficiency of hydrogen generation is influenced by the availability of photon energies.

Silver and gold nanorods were used in a previous study for enhancing the photosynthetic hydrogen generation by *Chlorella vulgaris*. The study found that nanoscale-level distributed iron clusters with zero valance state have been used for increasing the growth of chlorophyll and carotenoid pigments. The following reaction is adopted in the H_2 production during the photosynthetic process:

$$\text{Substrate (Malate, Lactate, and acetate)} + nH_2O \rightarrow H_2 + xCO_2. \quad (4)$$

In additions, they play a major role in the production of carbohydrate products, resulting in the growth of algal mass. The metabolism of microorganisms has been further tuned due to addition of nanoparticles, as a result the production of metabolites.

(ii) *Dark Fermentation*

Living microorganisms such as bacteria and algae interact with biomass sources, and they break down organic matter (see table) to isolate hydrogen. In this way of hydrogen production, it does not require the light environment [33]. Some microbial species have the capacity to use their biomass constituents, such as carbohydrates, proteins, and lipids, as a substrate for H_2 production. The main step involved in hydrogen production is the turning of the metabolism of glucose into pyruvate. Most commonly, electron transfer mechanism is used for H_2 production in two stages.

i. Oxidation of organic substrate into pyruvate.
ii. Conversion of pyruvate into volatile fatty compounds [24].

In general, the metabolism of anaerobic-type bacteria uses organic matter for H_2 extraction through dark fermentation. Enormous research in H_2 generation through fermentation led to a development in designing the reactor configurations and parameter optimization. However, this technique faces low-yield H_2 generations. Literature reveals that hydrogen yield with ≤ 2.6 mol/mol hexose has been achieved, which is usually lower than the theoretically calculated maximum H_2 yield with 4 mol/mol hexose. This can be explained by the reason that some portion of glucose is consumed by the microbial cultures and not metabolized in the process of H_2 generations. In most of microbial bacterial species, the turning of glucose into pyruvate occurs via the Embden–Meyerhof–Parnas pathway. The produced Acetyl-CoA through the pyruvate is metabolized in the form of acetate or butyrate, and if it is metabolized in the form of acetate, the formation of one extra mol of H_2 occurs due to reduction of NADH/mole, where NADH is formed when glycolysis turns into NAD^+. Hence through this process, overall hydrogen production with 4 mol can be achieved when one mole of glucose consumed. However, butyrate formation causes the achievement of one-half the times of H_2 yield produced from the acetate formation. However, different operational conditions may cause the formation of other products, such as ethanol, acetone, volatile fatty acids, isobutyrate, and propionate, which lead to lower H_2 yield achievements. The addition of micrometal elements such as Co, Ni, Fe, Mn, and Zn is essential for dark fermentation performing microbial, where it improves the activity of enzymes.

The thermophilic bacteria-mediated dark fermentation process for hydrogen release is one of the techniques which offers low cost, versatile, and sustainable [34]. The bacterial medium uses the carbohydrate source to generate acetic, butyric acids, and H_2 under the fermentation process. Obligate anaerobes use pyruvate ferredoxin oxidoreductase pathway for breaking down glucose into pyruvate as well as acetyl-CoA. When a conversion occurs from the pyruvate to acetyl-CoA facilitated by the hydrogenase enzyme, then hydrogen is released. Three types of hydrogenases such as (i) Ni–Fe, (ii) [Fe–Fe], and (iii) [Fe]-hydrogenase were found in the microbial cultures. The operating condition for activating these hydrogenases is different and explained by the following aspects. Under the condition of fermentation and energy conservation, the types such as [Ni–Fe]-hydrogenase and [Fe–Fe]-hydrogenase undergo H_2 production; meanwhile, [Fe]-hydrogenase plays a role in H production under fermentation only. The efficiency of H_2 production done through the [Fe–Fe] hydrogenase is higher than [Ni–Fe]-hydrogenase since [Ni–Fe]-hydrogenase requires higher energy of ATP [35]. Microbial mass production requires micronutrients such as iron, nickel, copper, and zinc. Nanoscale-level distribution of micronutrients is more beneficial due to their nature of quantum size effect, higher surface area, and higher reactivity. Among the micronutrient, iron is more beneficial since it offers microbial biomass production as well as supports in the formation of ferredoxin and hydrogenase, enabling the pyruvate ferredoxin oxidoreductase pathway. In addition, due to its conducting nature, iron nanoparticles play a role in microbial electron transfer, which is essential for improving the high yield of hydrogen generations. The magnetite nanoparticles (Fe_3O_4) usually form the electronic chains in the mixed bacterial culture and further improve the electron transport and activity of co-enzymes [36]. In addition, nanoscale-level distributed metal particles have been used for trapping and storing electrons, and through this way, the reduction of protons occurs from H^+ to H_2.

A reactor plant system for performing the biofermentation (light–dark conditions) can be easily made and operated with lost cost effect and a fast H_2 synthesis process. This technique offers a higher conversion rate of biomass cellular (for example, kitchen waste, organic wastewater) substrate into H_2 [37, 38]. Literature reported that increasing the temperature during the fermentative hydrogen production (through thermophilic conditions) is one of the proven ways for high-yield H_2 generations since higher temperature enables the reaction kinetics for accelerating the metabolism activities, and hence, production of microbial algae mass has been increased [39, 40]. The previous study further revealed that lower hydrogen production was achieved when the glucose undergoes degradation by splitting it into volatile fatty acids and soluble metabolites in the absence of iron oxide nanoparticles. However, the addition of iron oxide nanoparticles has reduced such splitting process and enhanced the production of acetate and butyric acid [41].

Hydrogen generation due to acetate production pathway

$$C_6H_{12}O_6 + 2H_2O \rightarrow 2CH_3COOH + 2CO_2 + 4H_2, \quad (5)$$

$$C_6H_{12}O_6 + 2H_2O \rightarrow CH_3CH_2CH_2COOH + 2CO_2 + 2H_2, \quad (6)$$

$$C_6H_{12}O_6 + 2H_2 \rightarrow 2CH_3CH_2COOH + 2H_2O. \quad (7)$$

In a previous study, the H_2 gas production was studied using four bottles, cassava waste pulps, and cow dung. The schematic view of experimental setup for H production is given below (Fig. 4).

(i) *Preparation of raw material*: They collected cassava waste pulps from a starch compound-producing industry. Initially, it was placed in the icebox during shipping to avoid contamination, and then, it was maintained at a temperature of 20 °C using a refrigerator until used. Meanwhile, cow dung was cultivated and handled as an inoculum. Heat shock pre-treatment was done for 30 min which permitted the reduction of hydrogen intaking microorganisms. Then acid hydrolysis was carried out with the help of additions of 1 kg pre-treated waste and 1% HCL and heated at a water path with temperatures 10 °C up to 100 °C. The sugar contents in the waste before and after acid hydrolysis were evaluated using a refractometer. The pH value of the medium was adjusted through the additions of HCL or NAOH until the required pH value, and simultaneously pH value was noted using a pH meter. Before the acid hydrolysis, the sugar

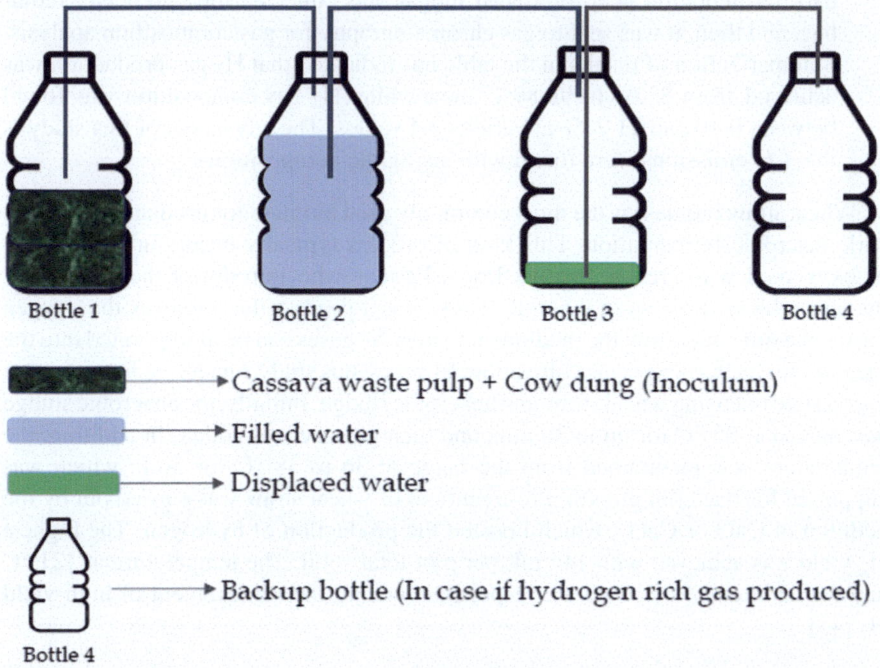

Fig. 4 Typical schematic diagram depicts the H_2 generation through fermentation which was done in a previous study by using four bottles, cassava waste pulps, and cow dung

concentration in the waste was measured as Brix. The acid hydrolysis process typically was typically done due to the production of high sugar yield as well as adjustment of pH. Therefore, acid hydrolysis has resulted in an increased sugar yield of about 9 Brix. The breaking down of each sugar in the waste generates H_2, and hence, acid hydrolysis process in a controlled way can be useful in the maximum production and production rate of H_2.

(ii) *Experimental setup for hydrogen production*: Four plastic bottles (capacity 6 L per bottle) have been used as an experimental setup for H_2 generation, followed by fermentation. To perform the fermentation process, Bottles 1 and 2 were loaded with the acid-hydrolyzed CWP and water, respectively. In Bottle 1, bacterial metabolism activity and fermentation typically take place. The filled water in Bottle 2 was responsible behind the dissolution of CO and CO_2 from Bottle 1 to achieve H_2 generation. The production of biogas (H_2 gas) has resulted in the water displacement from Bottles 2 to 3. Fourth bottle was used as a backup, at which much biogas production causes further water displacement from Bottles 3 to 4. Each bottle is tightly packed to avoid leakage and connected to the pipes. In this study, the pH values from 5.5 to 7 were chosen and tested for their effect on the H_2 generation. The most of the literature suggested that the optimum (higher) amount of H_2 was achieved at pH 6. The intracellular activity of bacterial culture is reduced at below 5.5, and hence, metabolism activity is reduced, resulting in less achievement of metabolites and H_2 production through dark fermentation [42]. Produced biogas was collected through a 1-L Tedlar bag, and then, it was sent to gas chromatography for gas composition analysis. Summarization of results in the table has indicated that H_2 gas production was achieved from 83.67 to 99.89%, meanwhile, N_2 gas composition was found between 0.10 and 11.74% at various pH values. The advantage of this study is that CO emissions were found with negligible compositions.

Wheat straw biomass is the most commonly used biomass for producing H_2 using dark anaerobic fermentation. This kind of process typically occurs in the absence of oxygen. It was French chemist Louis Pasteur who introduced the concept of the anaerobiosis process in the mid-1850s. It is a process that removes the oxygen during the time of boiling the medium, and then N_2 gases can be incorporated into the medium for bacterial species cultivation. In a previous study, anaerobic fermentation was carried out using wheat straw and anaerobic sludge. Initially, the anaerobic sludge was heated at 85 °C for up to 30 min, and then glucose was added. In addition, the temperature was maintained from the range of 36 to 38 °C for 36 h, which was supported for bacterial growth. Pre-treatment of wheat straw was carried out by the addition of CaO or CaO_2, which boosted the production of hydrogen. The highest H_2 yield was achieved with 144 mL per g of total solid. The temperature at 121 °C and the addition of 6% CaO are responsible behind the achievement of high-yield H_2 [43].

2.4 Microalgae Pathway of Hydrogen Generation

Microalgae can grow and store energy in the form of biomass by intaking energy from the sun's radiations in the presence of water and CO_2. The photoautotrophic algae groups intake carbon sources (for example, bicarbonate species) which are the results of the dissolution of CO_2. Initially, as a result of the photosynthetic process (*photosystem II*) by algae driven by the sun radiations, the splitting of the water molecule and evolution of O_2 occurs. Then, electron transfer through ferredoxin occurs from the photosystem system I to hydrogenase, where the reduced ferredoxin acts as a donor of electrons, and electron transfer is induced by light absorption of photosystem I. Hence, photosystem centers in the algae absorb energy and store it in the form of biomass. The anaerobic conditions and presence of light permit the delivering of photosynthetic electrons to [Fe–Fe]-H_2ase instead of NADPH via reduced ferredoxin. In addition, [Fe–Fe]-type hydrogenases catalysts utilize them to reduce protons to form molecular H [44].

The eukaryotic green algae, for example, *C. reinhardtii* are commonly found in fresh water and moist places. They perform the photosynthesis process, and in addition, it performs other fermentation and hydrogen production. Oxygenic photosynthesis-performing species involves in the absorption of light from the sun radiation, separation of charges, transporting of electrons, and fixation of CO_2 in the aqueous biomass. Algae initially absorbs energy from the sun radiations to extract electrons from the water molecules, and then, it was used for CO_2 fixations, where the extracted electrons interacting with the NADP reduce it into NADPH which is the source for the conversion of CO_2 into carbohydrates during the Calvin–Benson cycle. In addition, due to the process of carrying protons from the stromal centers to the lumenal portion of the thylakoids, a proton gradient is created across the thylakoid membrane, and then the enzyme named adenosine triphosphate (ATP) synthase utilizes it to the creation of ATP, and by this way, CO_2 fixation is achieved. Anaerobic conditions permit in the binding of reduced Fd with the Fe–Fe hydrogenase; subsequently, it donates electrons to initiate catalytic activity. Finally, hydrogenase combines the produced proton and electron to achieve molecular hydrogen. It has been shown that photosynthetic reduced power can be divided into two pathways: (i) aerobic condition-initiated CO_2 reduction and (ii) anaerobic condition-initiated H_2 production. The important thing is to note that ATP is required in the reduction of CO_2 and does not require in the H_2 generations [45]. Due to the presence of a large amount of chlorophyll (*photosynthetic pigment*) in microalgae, they involve in the photosynthetic process with high efficiency to synthesize biomass with larger carbohydrates. Living cells of microalgae are usually found in freshwater, seawater, and wastewater. Hence, they are easily cultivated and can be used as a source for biodiesel, bioethanol, and hydrogen generation. They possess several advantages compared to terrestrial biomass: (i) faster growth rate, (ii) do not require agricultural land to grow, (iii) absence of lignin (or absence of biopolymers with complicated structure).

3 Thermochemical Routes of Hydrogen Generation

3.1 Pyrolysis

In the pyrolysis route, H_2 can be extracted from the various sources containing hydrogen elements bonded with the carbon, known as hydrocarbons which act as a reactant. The heating process causes the decomposition of such compounds in the oxygen-absence environment. H_2 and methane acting as energy extracted from the biomass source are considered as renewable. However, H_2 extraction from the biomass source is sometimes referred as non-sustainable since in such a process, CO_2 as a side product is released. Therefore, integrating the pyrolysis reactor system with the CO_2 capturing technologies can support the production of blue H_2. However, when compared with the pyrolysis of fossil fuels, pyrolysis of biomass releases less CO_2 emissions. Thermal decomposition of biomass sources is a common phenomenon that occurs due to high temperatures, whereas thermal vibrations permit the breaking of long-chain molecules into smaller molecules. The term "*pyrolysis*" is originated from a combination of two Greek words, where pyro and lysis are called as fire and separation, respectively. In some cases of extreme pyrolysis, the formation of carbon-rich residue takes, and this process is known as carbonization. This process can be used in the gasification process for extracting the side tar products as well as for boosting syngas production. The inspirational applications of pyrolysis are the conversion of biomass sources into syngas and biochar productions and sometimes plastic wastes into usable oil products. Integrating with the pyrolysis system with the gasification system, one can achieve biomass into the rich form of syngas products. Introducing the inert gases in the pyrolysis process, chemical reactions between some molecular species can be avoided, which may cause the complicated molecular chain containing side product formation. In a previous study, thermal studies have shown that the decomposition of xylan and lignin started at a temperature of 200 °C, while the decomposition of hemicellulose takes place at temperature ranges between 220 and 315 °C. The pyrolysis temperature of the cellulose compound begins at 300 °C and reaches peak values at 344 °C. The decomposition of lignin typically occurs in a wide range of temperatures from 227 to 807 °C [46] (Fig. 5).

The use of partial oxidation in the gasification process has been carried out by mankind for 150 years. Low temperature-assisted gasification causes the release of a significant amount of hydrocarbon, while high temperature leads to the production of syngas and the minimization of hydrocarbons. In high temperatures (1 above 1000 °C) and oxygen environments, due to gasification, the formation of H_2-rich products and reduced char and tar products occur. At medium temperature (650–800 K), the decomposition of biomass materials turns into high-yield products such as liquid oils, solid forms of charcoal, and gaseous molecules. In addition, higher quality of fuels can be achieved in the pyrolysis process, when they are followed by the reforming of biogas and gasification products. For H_2 production, both gasification and pyrolysis reactor systems must be integrated with the reforming of gas stages, whereas the water–gas shift is commonly used in tuning the reformed gas product

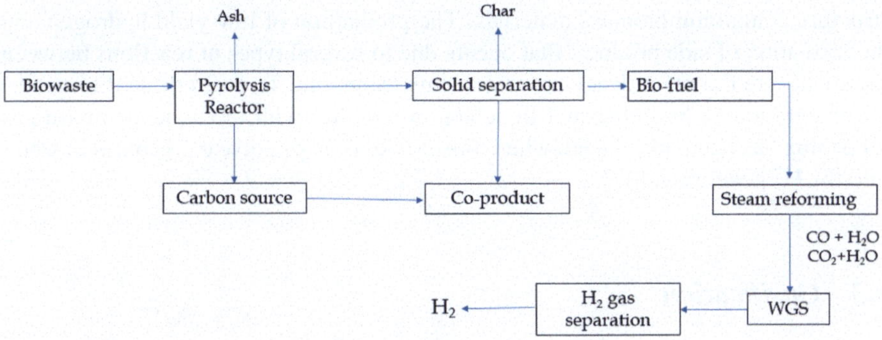

Fig. 5 Block diagram depicting the different steps involved in the hydrogen generation from the pyrolysis process

into H_2. The produced H_2 can be purified by introducing pressure swing absorption [10]. The following reaction involves in the generation of H_2 through waster gas shifting:

$$CO + H_2O \rightarrow CO_2 + H_2. \tag{8}$$

Comparative studies in a previous study revealed that an increased temperature alters the yield of hydrogen generation, which was achieved between the temperatures 250 and 800 °C. The temperature of pyrolysis is strongly depending on the type of materials used. Lignin content-rich biomass materials usually require higher temperatures, and it strongly depends on the types of catalyst used. The use of catalyst is more common in pyrolysis since it offers removable of tar products and enhances the productivity of syngas. The catalysts such as Ni, Fe, a combination of Ni–Fe/CaO, and Ni/SiC are used in the pyrolysis reactor. However, due to low cost, alumina is highly recommended in the pyrolysis of biomass [47].

3.2 Liquefaction

It is one of the potential routes for H extraction from biomass which permits the use of low temperatures (250–325 °C). However, in this route, high pressure lying around 20 MPa is used. This route of hydrogen generation can be divided into two categories: (i) direct liquefaction and hydrothermal liquefaction. This approach is more beneficial over the pyrolysis process since the drying of biowaste materials is not necessarily done, which can save input energies and is commonly used in the drying process of biowaste materials typically done in the first-step process of pyrolysis. In hydrothermal liquefaction, the temperatures and pressure ranges of (250–370 °C), and (4–22 MPa), respectively, are preferably chosen. In addition, the use of catalysts is highly recommended for breaking down the complicated polymer structure containing biomass materials. The production of low-yield hydrogen and the formation of side products that occurs due to several types of reactions between chemicals are the main disadvantages of this technique. To avoid these difficulties, the integration of hydrothermal liquefaction together with the aqueous phase gas reforming has been introduced, where this process can dissolve oxygenates in water into the H_2 generation [47].

3.3 Gasification

It is also one of the methods for H_2 production from biomass sources which typically takes place during the process of heating at high temperatures. Biomass is less expensive than fossil fuels. This technique handles the thermochemical process with 900 °C heating at low pressure and involves the addition of oxygen surroundings and partial combustion of biomass. The following phenomena, such as biomass drying, pyrolysis of the feedstock catalytic cracking, oxidation, reduction, and absorption of carbon emission, are expected during biomass gasification. The chemical products such as carbon monoxide, carbon dioxide, methane, and hydrogen components are released during the gasification. The gasification system must have the following things: (i) carbon capturing technology since the role is the production of green hydrogen, (ii) syngas purification steps, (iii) water–gas shift reaction for achieving the maximum efficiency of hydrogen generations. Permitting the water–gas shift reaction is essential since this maximizes H_2-rich production by converting the CO contents of syngas into CO_2 and H_2, which is given by the following water–gas shift exothermic reaction:

$$\mathrm{CO(syngas)} + \mathrm{H_2O} \rightarrow \mathrm{CO_2} + \mathrm{H_2}, \tag{9}$$

$$\text{Dried biowaste} + \text{Steam} \rightarrow \mathrm{H_2} + \mathrm{CO_2} + \mathrm{CO} + \mathrm{CH_4} + \text{hydrocarbon} + \text{char} \tag{10}$$

However, this technique may face the formation of by-product tar from the syngas, which must be removed with additional cost effects. Hence, this technique faces the drawback that the formation of by-product tar from the syngas, which must be removed with additional cost effects, causing the reduction of overall efficiency [48]. Recently, Wu et al. designed a solar-driven biomass gasification model consisting of a cooling, heating, and solar powdered system, which was mainly designed for optimizing the operating conditions. Biogasification system model developed by Zhang et al. [49] has offered high H_2 yield, less energy consumption, and overall increased efficiency, which was achieved with the incorporation of autothermal calcia (CaO) looping. Ishaq et al. [50] have proposed integrated thermoelectric generators supporting high-energy and exergy efficiencies. Ebrahimi and Ziabasharhag have proposed a new system consisting of a tri-generation of liquid natural gas, power, and heating technology, which offers higher efficiency. However, the above-mentioned

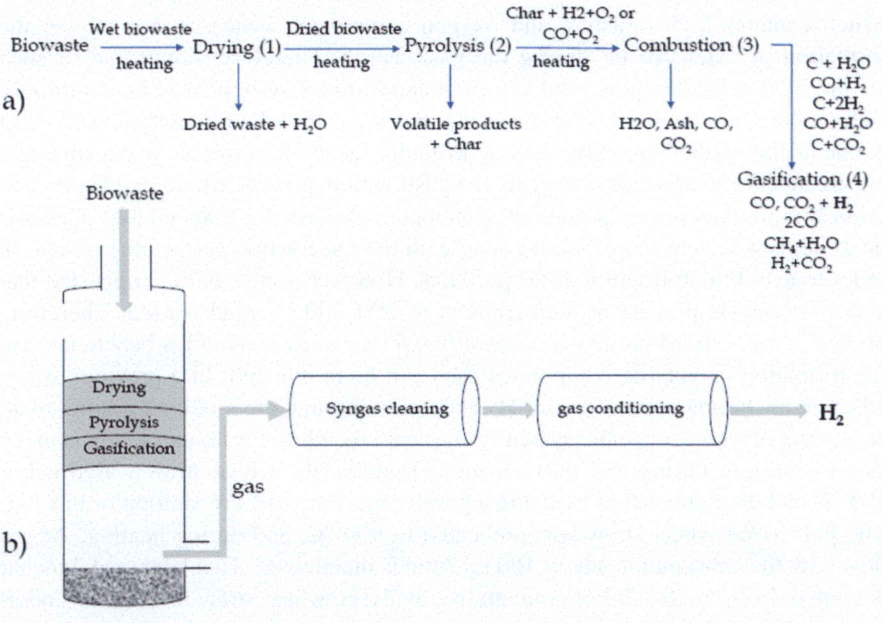

Fig. 6 **a** Steps involved in the generation of hydrogen through gasification and **b** schematic diagram with reactor system are depicted

technique is economically not stable when compared to hydrogen generation through natural gas steam reforming [51] (Fig. 6).

Under high temperatures, the biomass gasification process can be categorized into two types: (i) steam biomass gasification and (ii) air biomass gasification. In general, steam gasification offers a high yield of H_2 than air biomass gasification. To remove the moisture in the wet biomass initially, they are heated at $\geq$ 150 °C to remove the moisture in the wet biomass. The required heat was achieved for drying the biomass takes place outside the stages of the gasification process. The dried biomass is subjected into thermal decomposition with the help of pyrolysis, which can be achieved at temperatures between 200 and 650 °C, causing the formation of char and pyrolysis compounds. Later syngas is produced from the char after achieving temperatures between 700 and 1000 °C. The factors such as temperature, heat transfer rate, and steam-to-biomass ratio play a key role in improving the gasification process [53, 54].

3.4 *Tar Removal Before the Gasification*

Thermochemical technologies permit the conversion of biomass into H_2, CO, and hydrocarbons. However, some biomass materials exhibit undesirable properties, which contains high moisture and oxygen component; hence, it may trigger the formation of extensive tar during the gasification. Therefore, removable of such tar products is challenging, and the gasification plant system must be optimized. Before the gasification process, pre-treatments are required, and catalytic reforming is one of the methods to solve such a difficulty faced. Torrefaction is considered a pre-treatment process before starting the gasification process, where in this process, initially carbonization is carried out at temperatures between 200 and 320 °C. Gasification plant system must be integrated with the torrefaction performing system in order to avoid the formation of tar products. However, some studies suggested that a slow pyrolysis process at temperatures of 300–700 °C is also used. Therefore, the use of carbonized biomass (*commonly referred to as biochar*) is beneficial, and the formation of volatile compounds (tar) can be minimized. In a previous study, wheat straw biomass was used for H_2 extraction through the modified gasifier plant consisting of a tar removable system. A typical procedure for H_2 extraction is given as (i) overnight drying and pulverization: Initially, the wheat straw was dried at 105 °C and then pulverized by using a grinder machine. (ii) Preparation of biochar: The pulverized wheat straw was preheated to 900 °C, and during heating, Ar gas flow with the constant rate about 100 mL/min is maintained. Then obtained, biochar is cooled down to 20 °C. For comparative analysis, wheat straw biochar was additionally prepared by slow pyrolysis technique with the help of heating at 500 °C for 4 h. (iii) Carbonization of wheat straw: Carbonization was achieved in the same furnace, and constant rate of gas flow (100 mL/min) was introduced for 30 min before heating the raw material. (iv) Gasification of carbonized wheat straw: This process was carried out by using a two-stage fixed reactor arrangement. About 1.5 g of torrefied wheat straw (Step III) and 0.75 g of prepared wheat straw (step II) were loaded in the upper and bottom of the quartz basket, respectively. In addition, 30 mL/Min of Ar was introduced into the quartz basket to remove air molecules. After that, the upper and bottom portions were preheated to a temperature of 250 °C and 900 °C, respectively, and maintained heating for up to 10 min. Then, Ar flow was stopped, and water was introduced to form steam. In the upper and bottom stages, biomass gasification and catalytic reforming occur, respectively. When the temperature of 900 °C was reached, additionally upper stage was heated for up to 10 min. The collected liquid effluent through this process was dissolved in 20 mL of ethanol. The dissolved products in the ethanol system containing concentrated liquid organic are called as tar. In addition, the gases produced during the gasification were passed through the controllable condensing system, and then, they were collected for GC mass analysis. The results of the study revealed that the pre-treated wheat straw at 280 °C has displayed superior behavior leading to the achievement of H_2 proportion with 53% volume, while CO proportion with 31% Volume. The cumulative gas yield (maximum), H_2 yield, and syngas yield were achieved at about 112.10 mmol/g of reactants, 59.91 mmol/g reactants, and 94.10 mmol/g of reactants, respectively. The conclusion of the study suggested that the carbonization of wheat straw improved gasification and steam reforming for enhancing hydrogen production [52] (Table 1).

Table 1 Summarization of the results (comparative studies between biomass type used, techniques involved, pre-treatments, and hydrogen production achievements) from the recently published articles

S. No.	Biomass type	Species/substrate/inoculum/media	Techniques involved	Pre-treatment/NPs or other combinations additions	Processing/achievements/ advancements	References
1	Agricultural	Wheat straw	BG and SR	Carbonization (tar removal process) has been carried out to achieve higher content of H_2-rich syngas	Maximum of cumulative gas and hydrogen yield was achieved with 112.10 mmol/g of reactants and 59.91 mmol/g of reactants, respectively. Syngas yield achievement was about 94.10 mmol/g of reactants	[52]
2	Food waste	Fruits and vegetables/LAB and LU-HPB (inoculum)	Dark fermentation (lactate driven)	Digestate preheated to 90 °C for inactivating the methanogens	VHPR = 976 NmL/L-h (at pH 7, 5%) has been achieved. Hydrogen production was boosted due to the consumption of lactate rather than carbohydrate	[53]
3	Food waste	Spent coffee ground (SCP)	CO_2/steam-injected gasification	Wastes are dried under oven heating at temperatures 104–106 °C (for 24 h). Pre-treated SCP contains C with 57.74%	Gasification medium contains distinct streams; O_2/N_2, $O_2/CO_2/N_2$, and $O_2/H_2O/N_2$. Reactor wall temperature was set to 1000 °C. Hydrogen yield increased up to 69.21% against O_2/N_2 injections	[54]

(continued)

Table 1 (continued)

S. No.	Biomass type	Species/substrate/inoculum/media	Techniques involved	Pre-treatment/NPs or other combinations additions	Processing/achievements/ advancements	References
4	Agricultural biowaste	The cassava waste pulps	Fermentation	Pre-heating the waste was carried out using the oven at 80 °C (30 min) for stopping the growth of H_2 consuming species. For adjusting the PH of the medium, hydrochloric acid was added with control	Four bottles' experimental setup has been done: 1st bottle = acid hydrolyzed substrate (digestor); 2nd bottle = 6 L water; 3rd bottle = biogas; and 4th bottle = displaced water. Biohydrogen production was achieved with 12.39 ± 0.8 L (with 6.09 PH and 83.24 h of period)	[55]
5	Wastewater	Cheese whey wastewater, and brewery spent grains	DF	Filtered anaerobic sludge was used followed by the nutrition addition to improve bacterial populations	Highest production of SCCA and $BioH_2$ was achieved as 21.67 g L^{-1} and 30 g COD/L, respectively, for highest organic load placing on the reactor. The produced gas was further made to recirculate the reactor for improving SCCA production	[56]
6	Agricultural and food waste	Seed digestate (contains cattle manure, corn silage, chicken manure, olive pomace with 45, 25, 15, and 15%, respectively)	Anaerobic fermentation	Seed digestate was pre-treated at 85 °C for stopping the production of methanogens and enhancing H yield	Optimum H_2 yield was achieved as 50.4 mL per g of VS; H_2 production was 436 ± 10 mL	[57]

(continued)

Table 1 (continued)

S. No.	Biomass type	Species/substrate/inoculum/media	Techniques involved	Pre-treatment/NPs or other combinations additions	Processing/achievements/advancements	References
7	Sludge	Effluent sledge collected from balm oil mill (contains anaerobic digester)	Anaerobic fermentation	Inactivation of methanogens has been done through preheating at 80 °C (with the help of magnetic stirrer)	Fermentation medium contains glucose (10 g/L) as a major source than other constituents (sodium chloride and acetate each with 1 g/L, peptide, and yeast extract each with 3 g/L) and pH was set to 5.5. Iron nanoparticles (nutrient) were added in the medium which supported in the metabolic enzymatic growth of [Fe–Fe] hydrogenase and boosted H yield to 72.33%	[41]
8	Agricultural	Tomato plant waste ($C = 54.70\%$ in terms of elements; glucose = 24%; lignin = 21.9%; hemicellulose = 17%)	Hydrothermal gasification		Autoclave reactor was used with temperature and pressure 300–600 C and 20.0–42.4 MPa, respectively. KOH and K_2CO_3 are used as a catalyst. The biomass/water was used with 1.5 g/20 mL. Highest H_2 yield was achieved as 21.8 mol/kg of C at 600 °C. Alkali addition has resulted improved H_2	[58]

(continued)

Table 1 (continued)

S. No.	Biomass type	Species/substrate/inoculum/media	Techniques involved	Pre-treatment/NPs or other combinations additions	Processing/achievements/advancements	References
9	Agricultural	Black skin bagasse	Pyrolysis	Bagasse was crushed and heated at 105 °C up to 2 h and crushed to obtain powdered form with particle sizes ranging from 2 to 3 mm	Bio-oil formed due to microwave power for 30 min during pyrolysis. Addition of Ni as a cracking catalyst has resulted improved pyrolysis efficiency. Maximum hydrogen yield = 35.85 g/kg of biomass (under optimized condition)	[59]
10	Algal biomass	Chlorella sp./anaerobic granules	Anaerobic biomass digestion		The different concentration of biomass, and sulfur addition with 100 g/L, and 0.9% were used, respectively, for optimum conditions with 95.28 h. H yield was observed as 77.84 mL/g VS	[60]
11	Agricultural	Napier grass	Dark fermentation	NaOH has been added to delignify the lignocellulose mass for improving the glucose production	pH and mesophilic conditions with 4.5–7.0 and 34–36 C, respectively, were used on the grass biomass processed by the microflora. pH with 5.5 and mesophilic conditions favored the maximum H_2 production with 763.34 mL under dark fermentation	[61]

(continued)

Table 1 (continued)

S. No.	Biomass type	Species/substrate/inoculum/media	Techniques involved	Pre-treatment/NPs or other combinations additions	Processing/achievements/ advancements	References
12	Fatty restaurant waste	Residues of fruit, bread waste, vegetable peels, meats, grains, cooking oil, fats, diary/inoculum sludge (cultivated from the septic tank)	Fermentation microbial digestion pathway		Collection of wastes is mixed with the gray water Co-digestion of mixed waste has been by integrated three stage reactors Highest H yield was achieved in Reactor 1 (due to the presence of microbes: Mycobacterium, Proteiniclasticum, Clostridium, Planctomyces, and Acinetobacter) than 2 and 3	[62]
13	Wastewater	POME anaerobic sludge	MECs		MECs have been integrated with the PHP generator. Organic removal = 73%. H_2 production = 1.16 m^3-H_2/m^3d H_2; current production = 113 A/m^3	[63]

(continued)

Table 1 (continued)

S. No.	Biomass type	Species/substrate/inoculum/media	Techniques involved	Pre-treatment/NPs or other combinations additions	Processing/achievements/advancements	References
14	Agricultural	Activated sludge/agricultural straw waste	Dark fermentation	Activated sludge was cultivated after the addition of glucose followed by the centrifugation	In dark fermentation, mesophilic and thermophilic environment was set to 30 C and 55 C, respectively. Fermentation effluent, acetate, and other combinations were used for H generation in MECs By setting the reactor under a thermophilic condition, the degradation of lignocellulose has been done by the species proteobacteria, Thermotogota, and Coprothermo bacteria	[27]
15	Microbial biomass	Cyanobacterium A. halophytica (species) N-deprived BG11 medium (medium)	Dark fermentation		The rate of H_2 production was achieved with 55 ± 0.50 μmol per g of dry weight and hour by using the incubated medium containing glucose with 0.189 mmol C atom/L. Hydrogenase activity boosted H_2 production by reducing O_2	[64]

(continued)

Table 1 (continued)

S. No.	Biomass type	Species/substrate/inoculum/media	Techniques involved	Pre-treatment/NPs or other combinations additions	Processing/achievements/ advancements	References
16	Agricultural	Giant reed ($C = 41.4\%$ rich source)/photosynthetic bacteria (PSB) HAU-M1	PF	Pre-treatment of biomass has been carried out through the 1-butyl-3-methylimidazolium acetate addition	Pre-treated biomass exhibited higher H_2 generation with 61.7% compared with the untreated biomass	[65]
17	Forest residue	Pine bark	Gasification, WGS, and CLHG	Hydrogen generation system was integrated with the SOFC	WGS has been carried out for changing the CO_2 and CH_4 into hydrogen 92.19% efficiency of H_2 production has been achieved	[66]
18	Wastewater	Microalgae: *Chlorella vulgaris* Co-culture: domestic wastewater activated sludge	PF/AD		Phase I: 421.1 umol/L H_2 production was achieved. Phase III: 56.6 umol/L H_2 production was observed	[67]
19	Agroforestry waste	Herbaceous, hard wood, and soft wood/anaerobic sludge/*C. butyricum*	Dark fermentation	Degradation of lignin in the biomass has been carried by ammoniated fiber explosion and H_2O_2 addition for improving the efficiency of H_2 generation	CHP has been achieved with 146.65, 80.75, and 57.52 mL/g by using rice straw, poplar, and pine. *C. butyricum* has contributed to the releasing of H_2 through butyric acid production pathway	[68]

(continued)

Table 1 (continued)

S. No.	Biomass type	Species/substrate/inoculum/media	Techniques involved	Pre-treatment/NPs or other combinations additions	Processing/achievements/advancements	References
20	Wastewater	Anaerobic sludge (grown for 30 days)	DF	The derived seed sludge was cultivated from citric acid sewage. Glucose addition permitted in the growing of microorganism. Heating at 95 °C was carried out (for 90 min) to minimize methanogens and H_2 producing bacteria maximized	Firmicutes and Clostridium sensu stricto have been grown as dominant bacteria due to hydroxyapatite additions. Hydroxyapatite NPs' addition with optimal dosage 400 mg/L enhanced the higher yield H production (182.33 ± 2.41 mL per of glucose)	[69]
21	Forest residue	Rubber wood, eucalyptus wood	BG	CaO was used as a tar cracking catalyst	Syngas achievements with ratio of H/CO_2 are having the ranges between 1.8 and 2.3. The production of syngas from rubber wood and eucalyptus has been achieved as 2.5 and 2.7 kg/h, respectively, having the H_2/CO ratios of 1.8 and 1.9	[70]
22	Microbial biomass	*C. butyricum*/Clostridium sensu stricto1	DF	Addition of $NiCo_2O_4$ NP increased the substrate conversion by enabling the abundance of genes encoding, leading to higher activity of key enzymes	Addition of 400 and 800 mg/L NPs has resulted the H production yields with 259.67 and 188.14 mL/g glucose, respectively	[71]

(continued)

Table 1 (continued)

S. No.	Biomass type	Species/substrate/inoculum/media	Techniques involved	Pre-treatment/NPs or other combinations additions	Processing/achievements/advancements	References
23	Agricultural	Corn cob/lignocellulosic biomass/ *Clostridium cellulovorans*	DF	Degradation of lignin by using $CeFe_2O_3$ to form the cellulose–hemicellulose fractions	Cumulative hydrogen with 78.45 mL and H production rate with 1.55 mL/h have been achieved. About 43.26% lignin has been removed from the biomass	[13]
24	Agricultural residues	Barley straw, canola hull, oat hull	HG	Co-pelletization has been carried out	Canola hull pellet exhibited highest hydrogen gasification and selectivity, respectively, with 113 and 41%	[72]
25	Kitchen waste	Grass/leftover vegetables, and fruits/small portions of bread, coffee and tea leaves, meat scraps, and dairy residues/derived anaerobic sludge	AD–DF	Hyperthermophilic (thermophilic digestion)	H_2 gas generation has been enriched by reducing methane production under mesophilic conditions The responsible microbial species were Ruminococcaceae, Atopobiaceae, Lactobacillaceae, Petrotogaceae, and Synergistaceae	[73]
26	Microbial biomass	*Brown alga*, *S. japonica*, fermentation with mannitol and mannitol-rich seaweed *Clostridium tyrobutyricum* ATCC 25,755 (inoculum)	BA–F	Fermentation process has been increased due to addition of acetate	H_2 production has been achieved about 2.33 mol per mol of mannitol	[74]

(continued)

Table 1 (continued)

S. No.	Biomass type	Species/substrate/inoculum/media	Techniques involved	Pre-treatment/NPs or other combinations additions	Processing/achievements/advancements	References
27	Microalgae biomass	Sorghum, corn stubble, rapeseed oil, and inoculum	Anaerobic digestion process		Anaerobic digestion has been used under mesophilic conditions. H yield was achieved between 16 and 39 mL g VS^{-1}	[75]
28	Microbial biomass	Scenedesmus sp. SVMIICT1 (microalgae)	Dark fermentation	Biomass dried through air oven assisted heating at 80 °C. Dried biomass contains 24% lipids' composition which was used as a feedstock	Maximum H_2 yield was achieved about 186 mL per g of TOC Acidogenesis fermentation was responsible behind the H_2 formation	[76]
29	Agricultural	Sugarcane bagasse *Thermoanaerobacterium thermosaccharolyticum* KKU-ED1 (thermophilic cellulolytic bacterium)	Dark fermentation (thermophilic conditions)		Optimal conditions were found at pH = 7.0; $T = 55$ C; cellulose = 20 g/L Maximum H_2 yield was observed about 6113 mL H_2/L	[77]
30	Algal biomass	Cyanobacteria *Lyngbya limnetica*	Dark fermentation	$NiFe_2O_4$ addition has resulted the improved growth of fungal cellulose	Cumulative hydrogen production was achieved with 1820 mL/L	[78]

(continued)

Table 1 (continued)

S. No.	Biomass type	Species/substrate/inoculum/media	Techniques involved	Pre-treatment/NPs or other combinations additions	Processing/achievements/ advancements	References
31	Agricultural	Rice straw Klebsiella sp. WL1316	Fermentation	About 20 mg/L NiO NPs additions have resulted the improved CHP and HY with 47.29% and 37.78%, respectively	Rice straw hydrolysate has been used for fermentative H_2 production CHP = 4842.19 ± 23.43 mL/L and HY = 101.45 ± 3.32 mL/g substrate have been achieved	[79]
32	Wastewater	Methanogenic granular sludge (inoculum) Escherichia, Acinetobacter, Alcaligenes, Brevibacterium, Clostridium, and Mycobacterium	Fermentation		HY = 5.95 mM/gCOD at optimal substrate with 25 g-COD/L and HY = 4236 mM/gCOD at optimal PH = 11 and T = 35 C were achieved	[80]
33	Agricultural	Sugarcane bagasse Sewage sludge	Dark fermentation anaerobic digestion	Pyrolysis was used for preparing the biochar	Study suggested that prepared biochar boosted H_2 generation due to improved metabolism activity of microbial species Biochar prepared at 600 °C (pyrolysis temperature) displayed higher H_2 production	[81]

(continued)

Table 1 (continued)

S. No.	Biomass type	Species/substrate/inoculum/media	Techniques involved	Pre-treatment/NPs or other combinations additions	Processing/achievements/advancements	References
34	Agricultural	Corn stover/dark fermentative bacteria = Enterococcus, Sporanaerobacter, Para clostridium, and Clostridium_sensu_stricto_1; photosynthetic bacteria = HAU-M1	Dark fermentation and photosynthetic fermentation	Dark fermentation: enzymatic hydrolysates' portion of corn stover was used as a substrate. C, N, and P (micronutrient) were added (and also added in photo-fermentation). $T = 40 \pm 2$ °C and PH = 5.5 ± 0.1. Photo-fermentation: addition of 125 g of zeolites and agitation have been carried out. pH = 7.0 was set during fermentation. Temperature = 30 ± 1 °C and light intensity = 3000 ± 200 lx	Hydrogen production was achieved about 5.24 L/(L d) at 18 h period of hydraulic retention time in dark fermentation mode; meanwhile, hydrogen production rate was achieved in photo-fermentation mode with 4.60 L/(L d) at 24 h of hydraulic retention time	[82]
35	Agricultural	Corn stover (particle size < 0.25 mm). Cellulose, hemicellulose, and lignin were found with compositions 39.12%, 30.95%, and 10.73%, respectively. Photosynthetic bacteria HAU-M1	Photo-fermentation	Pyrolysis process with temperature 400 °C (at an increasing rate of 8.5 °C/min) has been carried out to prepare corn biochar. (Powdered biochar was found with C, H, N, and S element composition of 51.49, 3.23, 1.72% N, and 1.1%, respectively)	Biochar with the addition of nZVI has resulted H_2 with 286.83 ± 2.77 mL at pH 4.5. With the addition of nZVI, at PH = 5, 6, and 7, the highest hydrogen yield was obtained with 361.02 ± 10.11, 419.36 ± 10.70 and 382.67 ± 25.08 mL, respectively	[83]

(continued)

Table 1 (continued)

S. No.	Biomass type	Species/substrate/inoculum/media	Techniques involved	Pre-treatment/NPs or other combinations additions	Processing/achievements/advancements	References
36		Taihu blue algae/granular sludge (Inoculum)	MECs' anaerobic digestion fermentation		H_2 production has been achieved with 202.87 mL ($V = 0.4$)	[84]
37		Shrimp pond sediment	Anaerobic digestion dark fermentation	Dark fermentation has been carried out at $T = 39$ °C	The produced biogas contains 27.97% of hydrogen composition	[85]
38	Wastewater	Paper industry wastewater (mixture of dye, toxicity, and other contents)	MECs	Nickel, nickel–cobalt, and nickel–cobalt–phosphorous deposited on the electrode to boost H_2 production. Batch mode = $V = 0.6$, and $T = 30 \pm 2$ °C	Highest H_2 production rate was achieved with 0.16 ± 0.002 $m^3(H_2)/(m^3$ d) and 0.14 ± 0.002 $m^3(H_2)/(m^3$ d) at different treatment conditions by using nickel–cobalt–phosphorous coated electrode	[86]
39		*Agave tequilana* bagasse/granular sludge (inoculum)	Anaerobic digestion and pyrolysis	The lignocellulosic biomass contains hemicellulose = 12.44 ± 1.43%, cellulose = 55.08 ± 3.92%, and lignin = 13.12 ± 1.99%. Inoculum dried at 104∘C (for 24 h of period) and made in the form of 850 μm mesh. Acid hydrolysis and hydrolysates additions have been carried out on the *Agave tequilana*		[87]

(continued)

Table 1 (continued)

S. No.	Biomass type	Species/substrate/inoculum/media	Techniques involved	Pre-treatment/NPs or other combinations additions	Processing/achievements/advancements	References
40		Methanogens (inoculum)	MECs		The influence of temperatures ($T = 55$, 65, 75, and 85 °C) on the H_2 production has been studied by using the methanogens and sulfate reducing microbial bacteria. Among the different temperatures, highest H_2 production with 2.814 ± 0.091 M/M of glucose has been achieved at $T = 55$ C	[88]
41		Starch, sunflower oil, peptone, dog food, pig feed, sewage sludge	Dark fermentation		Peptone: H_2 production with 5.0 ml H_2/g COD; sunflower oil: 2.3 ml H_2/g COD. Starch: H_2 yield was observed about 1.55 mol H_2/mol hexose; dog food: H_2 yield = 46.5 ml H_2/g COD; pig feed: 32.1 ml H_2/g COD; sewage sludge: 9.3 ml H_2/g COD. Mixture of peptone and sunflower oil with starch has resulted reduced H_2 yield by 23% from the starch	[89]

(continued)

Table 1 (continued)

S. No.	Biomass type	Species/substrate/inoculum/media	Techniques involved	Pre-treatment/NPs or other combinations additions	Processing/achievements/ advancements	References
42		Sucrose (35% per liter) Anaerobic activated sludge	Fermentation		24.09 L H_2 production has been achieved at PH = 6.2 s	[90]
43	Biowaste	Apple and pumpkin waste/ biocatalyst: *C. kluyveri*	AF	Wastes are made in the form of paste and mixed with the bacterial culture	Achievement of 75% portion of H_2 in the released gas	[91]
44		*R. palustris*	PBC		*R. Palustris* was collected from the anaerobic conditions which contains nitrogenase protein responsible for bioH_2 production. 0.63 μmol of H_2 was achieved per hour	[92]
45	Wastewater	Pharmaceutical intermediate wastewater/*E. harbinense*	AF		H generation was initiated by ethanol-type fermentation through metabolic route of *E. harbinense*. H production was done with the maximum achievement rate of 11.2 ± 0.4 mmol/L h	[93]

(continued)

Table 1 (continued)

S. No.	Biomass type	Species/substrate/inoculum/media	Techniques involved	Pre-treatment/NPs or other combinations additions	Processing/achievements/advancements	References
46	Food waste	Fed food waste/diluted effluents (inoculum and substrate)	MECs		The generation of H_2 gas was achieved as 0.43 mL per and maximum of 3.5 mL within 8 days. Maximum H_2 efficiency has been achieved, when 40% biomass contents and 0.3 V are used	[94]
47	Wastewater	Traditional Chinese medicine wastewater	Acidogenic fermentation	Aeration pre-treatment was carried out up to 30 days followed by the addition of sucrose as a rich carbon source having the COD with 2 g/L	The presence of *C. carboxidivoran* and *C. butyricum* species is responsible, respectively, for H evolution through acetate and butyrate generation pathway. Hydrogen yield was achieved as 3.50 ± 0.09 mmol per gram of COD	[95]

(continued)

Table 1 (continued)

S. No.	Biomass type	Species/substrate/inoculum/media	Techniques involved	Pre-treatment/NPs or other combinations additions	Processing/achievements/advancements	References
48	Wastewater	Anaerobic digested sludge		Anaerobic conditions = 37 °C; reactor was filled with substrate, nutrient solution, and inoculum with 80, 10, and 10%, respectively N_2 gas passed before starting the fermentation. Fermentation has been carried out under 25 sets of batch with different concentrations of butyrate additions and with different pHs	Microbial community analysis has confirmed that the addition of butyric acid resulted in the reduction of H_2 generating bacteria (Clostridium) and supported in the growing of other bacteria which does not involve in the H generation	[96]
49	Wastewater	Heat-treated anaerobic granular sludge (inoculum)/corn (pH 13.1); brewery (pH 3.8) wastewater)	Dark fermentation	Collected sludge heated at 85 °C through water bath for 90 min for disabling the activity of methanogens	At pH = 6, maximum hydrogen production was achieved at 302 mL from the corn (nejayote) wastewater. Hydrogen production with the mixtures (corn wastewater/brewery wastewater = 60/40) was achieved with 270 mL at pH = 10.8	[97]

(continued)

Table 1 (continued)

S. No.	Biomass type	Species/substrate/inoculum/media	Techniques involved	Pre-treatment/NPs or other combinations additions	Processing/achievements/ advancements	References
50	Wastewater	Fermentation medium contains soft drink wastewater = 88% and corn steep liquor = 12%	Dark fermentation	Iron NPs have been synthesized (with average particle 8.6 nm) using lignin portion of oil palm empty fruit branches which composed of carbohydrate, proteins, fat, and humidity with 73.1%, 11.2%, 2.0% fat, and 8.7%, respectively	Optimal conditions for H_2 generation (17.67 ± 0.54 mL) were determined at 37 C and pH = 7. In additions, additions of magnetic and non-magnetic iron nanoparticles boosted H_2 generation up to 91.0 and 74.3%, respectively	[98]
51	Wastewater	Cheese whey	Thermophilic dark fermentation	Additions of sodium bicarbonate (for maintaining pH = 4.5) and micronutrient have been carried out	Hydrogen production was achieved with 149.9 molH_2 m^{-3} d^{-1} at 33 kg COD m^{-3} d^{-1}	[99]
52	Wastewater and agricultural	Agave bagasse/anaerobic granular sludge	Dark fermentation and enzymatic hydrolysate	Sludge preheated to 105 °C for 24 h to improve pore forming bacterial species	Enzymatic hydrolysates and dark fermentation have been carried Maximum volumetric H_2 was achieved with 12.9 L H_2/L-d at an OLR of 90 g COD/L-d from the enzymatic hydrolysate's mode The production of 6.3 L H_2/L-d (volumetric maximum) has been achieved mostly due to Clostridium species through dark fermentation	[100]

(continued)

Table 1 (continued)

S. No.	Biomass type	Species/substrate/inoculum/media	Techniques involved	Pre-treatment/NPs or other combinations additions	Processing/achievements/advancements	References
53	Food waste	Food waste composed of stovers, peelings, rice, grains, pasta, and vegetables	Anaerobic dark fermentation	Ni^0 added in the cultivated media	Optimum H_2 production was achieved about147 mL H_2/gVS from the added effluent with 29,765 mgCOD L^{-1}	[101]
54	Agricultural	Corn straw and acid hydrolysate	Dark fermentation		H_2 production has been discussed by three modeling systems	[102]
55	Wastewater	Molasse	Dark fermentation	The additions of zero-valent iron NPs and iron oxide NPs have been carried out to boost H_2 production and to improve COD reduction	pH = 6 has enabled boosting of H_2 through butyric acid pathway Zero-valent iron NPs and iron oxide NPs additions increased the hydrogen production by 71% and 69.4%	[103]

(continued)

Table 1 (continued)

S. No.	Biomass type	Species/substrate/inoculum/media	Techniques involved	Pre-treatment/NPs or other combinations additions	Processing/achievements/ advancements	References
56	Organic waste	Sewage sludge/potato starch; proteins by enzymatic peptone; and fats by sunflower oil	Dark fermentation	Potato starch; proteins by enzymatic peptone; and fats by sunflower oil	Firmicutes in the cultivated media dominated the dark fermentation H_2 production. The presence of Genera Ruminococcus (26–90%) and Thermoanaero bacterium. Maximum hydrogen yield with 138 mL/g volatile solids has been achieved from the starch with 60.5 of H_2 composition in the syngas (67%)	[104]
57	Wastewater	Molasse wastewater = substrate sewage sludge = inoculum	Dark fermentation	Mesophilic condition was set with 37 ± 1 °C	Most portion of metabolites is composed of acetic acid and butyric acid which were responsible for higher H_2 production. H_2 production rate was achieved with 8.08 ± 0.48 L/L/d	[105]

AD anaerobic digestion; COD Chemical oxygen demand; HG Hydrothermal gasification; SR Steam Reforming; Mx Maximum; BG Biomass gasification; AF Anaerobic Fermentation; PBC Photobiocatalysis; PF Photo fermentative; AD anaerobic digestion; CHP Cumulative hydrogen production; WGS Water Gas Shift; CLHG Chemical looping hydrogen generation; SOFC solid oxide fuel cell; PF Photo fermentation; POME Palm Oil Mill Effluent; PHP Pico hydro power; VHPR maximum volumetric hydrogen production rate; MECs Microbial Electro Chemical Cells. BA-F Brown Algae Fermentation

4 Nanoparticles and Micronutrient Addition Boosts H_2 Production

A recent study supported the information that the use of nanoparticles in algal biotechnology has permitted the improvement of photosynthetic activity by enabling nitrogen metabolism and protein levels in the algal species, which in turn supports algae growth through inducing the synthesis of carbohydrate, photosynthetic pigments (chlorophyll, carotenoids, and anthocyanin), lipids, and nitrogen [32]. Therefore the addition of nanoparticles and nanofluid accelerates the high hydrogen production yield. According to the previous study, the addition of silver nanoparticles into anaerobic batch reactors has enhanced the H_2 generation in both the acidogenesis and fermentative ways, where Clostridium butyricum and glucose was used as cell culture and substrate, respectively. They have achieved a high yield of H_2 generation with 2.48 mol/mol glucose since the silver nanoparticle presence reduced ethanol production, which in turn increased H_2 production. They further revealed that 20 nmol L^{-1} was the optimal condition for high-yield H_2 generation [106]. However, this technique permits one-third of the substrate into H_2 production; meanwhile, the remaining portion of the substrate turns into forming the by-products such as acetate, butyrate, butanol, and acetone. It has been shown that some microorganism uses nanoparticle and converts their presence to advantages in an anaerobic environment by transferring the free electrons into acceptors. The use of 5 nm gold nanoparticles has accelerated the fermentation process for high-yield H_2 production with 36.3 compared with the absence of gold nanoparticles [107]. Nanofluid-supported gasification process enhances the H_2 production with high yields. Due to the excellent thermal conductivity of nanoparticles (having particle size lesser than 100 nm), they are used as in the form of nanofluids, which replaces the base oil. Oil, ethylene glycol, and water are commonly used in gasification plant systems. Due to their poor conductivity and low viscosity properties, recently, they are replaced by nanofluids. It was Maxwell who first theorized the concept of the thermal conductivity model suggesting the addition of spherical particles in solid–liquid suspension for improving the thermal conductivity of working liquids. Recently, Aminossadati et al. [108] have analyzed the role of nanofluids by using the convection concept, and they found that the nanoparticles enhance the cooling effect. Sheikhzadeh et al. analyzed the heat transfer in the chamber [109]. The study further concluded that thermal conductivity is strongly influenced by the size of particles as well as the position of the heating source.

The thermal conductivity of water has been increased by 31.83% due to the addition of graphene support, and hence they have investigated the effectiveness of tube heat exchanger when graphene-supported nanofluid was used. Their study concluded that the efficiency of heat transfer is increased by 35.6% in the graphene-supported water when compared with the water alone. Increased Brownian motion of suspended particles in solid–liquid media is responsible for the improved heat transfer when the temperature is increased. In addition, viscosity and heat transfer coefficient are also improved as the concentration of nanoparticles is higher. Recently, researchers performed a heat exchanging mechanism in a large-scale level gasifier-based reactor by using Al_2O_3 and CuO as a source (heat exchanger) for boosting heat transfer. They found that Al_2O_3 permits more heat transfer than CuO. Due to the addition of Al_2O_3, and CuO nanoparticles, hydrogen productions have been increased by 18.5% and 8.35%, respectively [110].

4.1 Iron Nanoparticles' Additions

In addition, the efficiency of H_2 production can be increased by incorporating iron into a bacterial cultural medium. Due to high reactivity and higher volume-to-surface (*s/v*) ratio, they interact with the enzyme called [Fe–Fe]-hydrogenase of bacterial medium for high-yield H_2 generations. According to a recent study, iron nanoparticles' addition (200 mg Fe/L) with the bacterial medium has resulted in maximum fermentative hydrogen releases with 103.33 mL; meanwhile, glucose composition and hydrogen yield was achieved as 9.12 g/L and 0.79 mol H_2 per mole of glucose, respectively [41].

4.2 Nickel Ferrite and Nickel Oxide Nanoparticles

In a recent study, $NiFe_2O_4$ nanoparticles were added to bacterial cells, and their conclusion revealed that nanoparticle additions improved the activity of hydrogenase enzyme, leading to improved hydrogen generations [111]. Nanoscale-level distributed nickel–cobalt oxide particles were used to enhance hydrogen production through dark fermentation. The addition of $NiCo_2O_4$ (400 mg/L) has resulted in the achievement of a high yield with 259.67 mL/ g glucose [72]. In a previous study, the effects of the valance state of Fe and Ni nanoparticles (Fe^0, Ni^0, Fe^{2+}, and Ni^{2+}) were tested on the mesophilic hydrogen production from glucose under dark fermentation conditions, where the headstock pre-treated, anaerobic sludge was used. In the absence of metal nanoparticles' addition, they measured biohydrogen yield with 247 mL/g VS. They revealed that due to the addition of Ni^{2+}, Fe^o, and Fe^{2+} nanoparticles, increased hydrogen yield has been achieved with 55, 37, and 15%, respectively [112].

4.3 Sodium Acetate

In a previous study, an attempt was made to produce carbohydrate-rich indigenous microalgae strains such as (i) *Scenedesmus subspicatus* GY-16, (ii) *C. vulgaris* FSP-E, and (iii) *Ankistrodesmus gracilis* GY-09 which was grown by mixotrophic conditions, and it was used as a feedstock for high-yield H_2 production. The conclusion of the study suggested that higher biomass and carbohydrate production was achieved in the *C. vulgaris* FSP-E with 825.6 mg/L/d and 365.8 mg/L/d, respectively. In addition, they found that the addition of 2.0 g/L of sodium acetate has resulted in increased biomass (1022.3 mg/L/d) and carbohydrate (498.5 mg/L/d) production. They further revealed that the use of semi-batch mode in the operating photoreactor has further increased biomass and carbohydrate productivity with 1063.3 and 384.8 mg/L/d, respectively. The conclusion of their work, along with high-yield H_2 production, has confirmed the fact that carbohydrate-rich microalgae will be a potential biomass for effective H_2 generations. In semi-batch mode, the medium is loaded with urea and acetate at 560 mg/l and 2.0 g/l, respectively. In addition, light intensity and CO_2 aeration have been introduced with μmol/m^2/s and 0.2 vvm, respectively. In semi-batch mode, the sources such as carbon and nitrogen are fed to the growth of microalgae cell culture, where carbohydrate production was achieved through repetitive feeding and discharging cycles [113]. Sodium acetate contains a higher level of carbon source, and hence, it is used for inducing the heterotrophic growth factors in microalgae. They tested the concentration of sodium acetate with 1000–3000 mg/L on biomass productivity. Biomass productivity increases as increasing the concentration from 1000 to 2000 mg/L. However, the addition of 3000 mg/L has resulted in the reduction of microalgae biomass. Carbohydrates found in the microalgae are the portions of plastids or inner cell walls [114, 115]. Their cellulose lacks in the presence of lignin, and hence, they can be easily transformed into hydrogen compared with the lignocellulosic biomass materials. One important thing is to note that microalgae growth is higher than terrestrial plants.

5 Conclusion

Recycling or degrading biowastes biologically will be useful. The extraction of hydrogen from such biowastes will be useful since it offers energy production and in addition waste management. In this chapter, the biological and thermochemical routes of hydrogen generation have been discussed in a detailed way. Each technique has distinct advantages in the generation of hydrogen from the biowastes. The rate of hydrogen production strongly depends on the different factors, conditions (light, dark, pH), and substrates used. Recently published papers under the titles of biohydrogen production from waste materials have been scanned in the Scopus database and the conclusion of the published sources is summarized in the table. Researchers have used the biomass materials such as agricultural waste, food waste, restaurant waste, forest residues, wastewater, microbial biomass, and kitchen residue for hydrogen generation through techniques. The summarized result has revealed that most researchers carried out hydrogen generation through fermentation techniques. In addition, the literature survey has revealed that the addition of micronutrient, nanoparticles, and sodium acetate improved hydrogen generation by enhancing the metabolism activity of microbial species.

References

1. V. Singh, D. Das, Chapter 3—*Potential of Hydrogen Production From Biomass, Science and Engineering of Hydrogen-Based Energy Technologies* (Academic Press, 2019), 123–164
2. S.F. Ahmed, N. Rafa, M. Mofijur, I.A. Badruddin, A. Inayat, M.S. Ali, O. Farrok, T.M. Yunus Khan, Biohydrogen production from biomass sources: metabolic pathways and economic analysis. Front. Energy Res. **9**, 753878 (2021)
3. K. Ullah, M. Ahmad, V.K. Sharma, P. Lu, A. Harvey, M. Zafar, S. Sultana, C.N. Anyanwu, Algal biomass as a global source of transport fuels: overview and development perspectives. Progr. Nat. Sci. Mater. Int. **24**, 329–339 (2014)
4. M. Arshad, M.A. Zia, F.A. Shah, M. Ahmad, An overview of biofuel, in *Perspectives on Water Usage for Biofuels Production*, ed. by M. Arshad (Springer, Cham, 2018)
5. Y. Manoharan, S.E. Hosseini, B. Butler, H. Alzhahrani, B.T.F. Senior, T. Ashuri, J. Krohn, Hydrogen fuel cell vehicles. Curr. Status Future Prospect Appl. Sci. **9**, 2296 (2019). https://doi.org/10.3390/app9112296
6. International Energy Agency, *Energy Technology Perspectives* (International Energy Agency, Paris, France, 2020)
7. A.F. Ghenciu, Review of fuel processing catalysts for hydrogen production in PEM fuel cell systems. Curr. Opin. Solid State Mater. Sci.Opin. Solid State Mater. Sci. **6**, 389–399 (2002). https://doi.org/10.1016/S1359-0286(02)00108-0
8. H.L. Zhu, Y.S. Zhang, M. Materazzi et al., Co-gasification of beech-wood and polyethylene in a fluidized-bed reactor. Fuel Process. Technol. **190**, 29–37 (2019)
9. J.K. Norskov, C.H. Christensen, Toward efficient hydrogen production at surfaces. Science **312**, 1322–1323 (2006)
10. S. Venkata Mohan, A. Pandey, *Biohydrogen Production: An Introduction* (The Netherlands, Amsterdam, 2013)
11. A. Kumar, N. Kumar, P. Baredar, A. Shukla, A review on biomass energy resources, potential, conversion and policy in India. Renew. Sustain. Energy Rev. **45**, 530–539 (2005)
12. P. McKendry, Energy production from biomass (part 1): overview of biomass. Bioresour. Technol.. Technol. **83**, 37–46 (2002)
13. M. Singhvi, B.S. Kim, Green hydrogen production through consolidated bioprocessing of lignocellulosic biomass using nanobiotechnology approach. Bioresour. Technol. **365**, 128108 (2022)
14. H. Chen, J. Wu, H. Wang, Y. Zhou, B. Xiao, L. Zhou, G. Yu, M. Yang, Y. Xiong, S. Wu, Dark co-fermentation of rice straw and pig manure for biohydrogen production: effects of different inoculum pretreatments and substrate mixing ratio. Environ. Technol. **42**, 4539 (2021)
15. T. Sheng, L. Zhao, L. Gao, W. Liu, G. Wu, J. Wuf, A. Wang, Enhanced biohydrogen production from nutrient-free anaerobic fermentation medium with edible fungal pretreated rice straw. RSC Adv. **41** (2018)
16. J. Zhu, W. Song, X. Chen, S. Sun, Integrated process to produce biohydrogen from wheat straw by enzymatic saccharification and dark fermentation. Int. J. Hydrogen Energy **48**, 11153–11161 (2023)
17. T.A. Shah, S. Ali, A. Afzal et al., Simultaneous pretreatment and biohydrogen production from wheat straw by newly isolated Ligninolytic Bacillus Sp. strains with two-stage batch fermentation system. Bioenerg. Res. **11**, 835–849 (2018)
18. H. Bao, X. Zhang, H. Su, L. Li, Z. Lv, X. Zhang, Study on the hydrogen production ability of high-efficiency bacteria and synergistic fermentation of maize straw by a combination of strains, 16 (2019)
19. M. Yadav, K. Paritosh, V. Vivekanand, Lignocellulose to bio-hydrogen: an overview on recent developments. Int. J. Hydrogen Energy **45**, 18195–18210 (2020)
20. L. Cao, I.K.M. Yu, X. Xiong, D.C.W. Tsang, S. Zhang, J.H. Clark, C. Hu, Y.H. Ng, J. Shang, Y.S. Ok, Biorenewable hydrogen production through biomass gasification: a review and future prospects. Environ. Res. **186**, 109547 (2020)

21. E.S. Biris-Dorhoi, D. Michiu, C.R. Pop, A.M. Rotar, M. Tofana, O.L. Pop, S.A.S. Socaci, A.C. Farcas, Macroalgae-A sustainable source of chemical compounds with biological activities. Nutrients **12**, 3085 (2020)
22. C.-H. Liu, C.-Y. Chang, C.-L. Cheng, D.-J. Lee, J.-S. Chang, Fermentative hydrogen production by *Clostridium butyricum* CGS5 using carbohydrate-rich microalgal biomass as feedstock. Int. J. Hydrogen Energy **37**, 15458–15464 (2012)
23. P. Sivagurunathan, G. Kumar, A. Pugazhendhi, G. Zhen, T. Kobayashi, K. Xu, *Biohydrogen Production from Wastewaters* (InTech, 2017)
24. C. Ding, K.L. Yang, J. He, *Biological and Fermentative Production of Hydrogen* (Woodhead Publishing, 2016), 303–333
25. G. Buitrón, J. Carrillo-Reyes, M. Morales, C. Faraloni, G. Torzillo, *Chapter 9: Biohydrogen Production from Microalgae* (Woodhead Publishing, 2017), 209–234
26. L. Zhang, T. Happe, A. Melis, Biochemical and morphological characterization of sulfur deprived and H_2-producing *Chlamydomonas reinhardtii* (green alga). Planta **214**, 552–561 (2002)
27. F. Ndayisenga, Z. Yu, B. Wang, G. Wu, H. Zhang, I.A. Phulpoto, J. Zhao, J. Yang, Thermophilic-operating environment promotes hydrogen-producing microbial growth in a lignocellulose-fed DF-MEC system for enhanced biohydrogen evolution. Process Safety Environ. Protect. **167**, 213–224 (2022)
28. S. Venkata Mohan, A. Pandey, *Chapter 1—Biohydrogen Production: An Introduction* (2013), 1–24
29. D.M. Le, H.R. Sørensen, N.O. Knudsen et al., Biorefining of wheat straw: accounting for the distribution of mineral elements in pretreated biomass by an extended pretreatment-severity equation. Biotechnol. Biofuels. Biofuels **7**, 141 (2014)
30. P. Bakonyi, G. Kumar, L. Koók, G. Tóth, T. Rózsenberszki, K. Bélafi-Bakó, N. Nemestóthy, Microbial electrohydrogenesis linked to dark fermentation as integrated application for enhanced biohydrogen production: a review on process characteristics, experiences and lessons. Biores. Technol.. Technol. **251**, 381–389 (2018)
31. L. Lu, D. Xing, T. Xie, N. Ren, B.E. Logan, Hydrogen production from proteins via electrohydrogenesis in microbial electrolysis cells. Biosens. Bioelectron.. Bioelectron. **25**, 2690–2695 (2010)
32. S. Bosu, N. Rajamohan, Influence of nanomaterials in biohydrogen production through photo fermentation and photolysis—review on applications and mechanism. Int. J. Hydrog. Energy (2022)
33. M. Kamran, *Chapter 8—Bioenergy, Renewable Energy Conversion Systems* (Academic Press, 2021), 243–264
34. J.F. Soares, T.C. Confortin, I. Todero, F.D. Mayer, M.A. Mazutti, Dark fermentative biohydrogen production from lignocellulosic biomass: technological challenges and future prospects. Renew. Sustain. Energy Rev. **117**, 109484 (2020)
35. P. Mishra, S. Krishnan, S. Rana, L. Singh, M. Sakinah, Z. Ab, Outlook of fermentative hydrogen production techniques: an overview of dark, photo and integrated dark-photo fermentative approach to biomass. Energy Strat. Rev. **24** (2019)
36. M. Nasr, A. Tawfik, S. Ookawara, M. Suzuki, S. Kumari, F. Bux, Continuous biohydrogen production from starch wastewater via sequential dark-photo fermentation with emphasize on maghemite nanoparticles. J. Ind. Eng. Chem. **21** (2015)
37. T. Laurinavichene, B.F. Belokopytov, K.S. Laurinavichius, D.N. Tekucheva, M. Seibert, A.A. Tsygankov, Towards the integration of dark−and photo−fermentative waste treatment. Potato as substrate for sequential dark fermentation and light-driven H_2 production. Int. J. Hydrogen Energy **35**, 8536−8543 (2010)
38. A. Pandey, P. Sinha, A. Pandey, Hydrogen production by sequential dark and photofermentation using wet biomass hydrolysate of *Spirulina platensis*: response surface methodological approach. Int. J. Hydrogen Energy **46**, 7137–7146 (2021)
39. N.S. Engliman, P.M. Abdul, S. Wu, Influence of iron (II) oxide nanoparticle on biohydrogen production in thermophilic mixed fermentation, Int. J. Hydrogen Energy **42** (2017)

40. B.S. Zainal, M.A. Ahmad, M. Danaee, N. Jamadon, N.S. Mohd, S. Ibrahim, Integrated system technology of pome treatment for biohydrogen and biomethane production in Malaysia. Appl. Sci. **10** (2020)
41. A.H.A. Aziz, N.S. Engliman, M.F. Mansor, P.M. Abdul, S.N. Arisht, N.S. Jamali, M.F. Tiang, Synergistic enhancement of biohydrogen production by supplementing with green synthesized magnetic iron nanoparticles using thermophilic mixed bacteria culture. Int. J. Hydrog. Energy **47**, 40683–40695 (2022)
42. R. Łukajtis, I. Hołowacz, K. Kucharska, M. Glinka, P. Rybarczyk, A. Przyjazny, M. Kamiński, Hydrogen production from biomass using dark fermentation. Renew. Sustain. Energy Rev. **91**, 665–694 (2018)
43. J. Zhang, C. Kong, M. Yang, L. Zang, ACS Omega **5**, 9151–9161 (2020)
44. J.H. Hwang, H.C. Kim, J.A. Choi et al., Photoautotrophic hydrogen production by eukaryotic microalgae under aerobic conditions. Nat. Commun.Commun. **5**, 3234 (2014)
45. U.M.N. Murthy, M.L. Ghirardi, *Algal Hydrogen Production* (Academic Press, 2013), 66–70
46. Z. Hui, L. Yanqiu, M. Aihong, C. Shen, L. Qinghai, Z. Yanguo, A novel method for kinetics analysis of pyrolysis of hemicellulose, cellulose, and lignin in TGA and macro-TGA. RSC Adv. **5**, 26509–26516 (2015)
47. L. Jara-Cobos, M. Abril-González, V. Pinos-Vélez, Production of hydrogen from lignocellulosic biomass: a review of technologies. Catalysts **13**, 766 (2023)
48. S. Kaskun, An overview of hydrogen-rich gas production from biomass by using thermal technologies. IOP Conf. Ser. Earth Environ. Sci. **614**, 012010 (2020)
49. S. Zhang, N. Gao, C. Quan, F. Wang, C. Wu, Autothermal CaO looping biomass gasification to increase process energy efficiency and reduce ash sintering. Fuel **277**, 118199 (2020)
50. H. Ishaq, S. Islam, I. Dincer, B.S. Yilbas, Development and performance investigation of a biomass gasification based integrated system with thermoelectric generators. J. Clean. Prod. **256**, 120625 (2020)
51. A. Ebrahimi, M. Ziabasharhagh, Energy and exergy analyses of a novel integrated process configuration for tri-generation heat, power and liquefied natural gas based on biomass gasification. Energy Convers. Manag. **209**, 112624 (2020)
52. G. Kong, K. Wang, X. Zhang, J. Li, L. Han, X. Zhang, Torrefaction/carbonization-enhanced gasification-steam reforming of biomass for promoting hydrogen-enriched syngas production and tar elimination over gasification biochars. Bioresour. Technol. **363**, 127960 (2022)
53. L.J. Martínez-Mendoza, R. Lebrero, R. Muñoz, O. García-Depraect, Influence of key operational parameters on biohydrogen production from fruit and vegetable waste via lactate-driven dark fermentation. Bioresour. Technol. **364**, 128070 (2022)
54. J.P. Rodrigues, G.F. Ghesti, E.A. Silveira, G.C. Lamas, R. Ferreira, M. Costa, Waste-to-hydrogen via CO_2/steam-enhanced gasification of spent coffee ground. Cleaner Chem. Eng. **4**, 100082 (2022)
55. A.J.F.T. Balbin, H.A. Gonzales, L.A.M. Bornillo, J.H.A. Tagupa, R.H. Jaro, A.L. Ido, R.O. Arazo, High biohydrogen yield from fresh cassava waste pulps via gas dissolution in a dark fermentative environment. Clean. Waste Syst. **3**, 100039 (2022)
56. O. Sarkar, U. Rova, P. Christakopoulosa, L. Matsakas, Green hydrogen and platform chemicals production from acidogenic conversion of brewery spent grains co-fermented with cheese whey wastewater: adding value to acidogenic CO_2, Sustainable. Energy Fuels **6**, 778–790 (2022)
57. M.-Y. Shen, M. Torre, C.-Y. Chu, P. Tratzi, M. Carnevale, F. Gallucci, V. Paolini, F. Petracchini, Green biohydrogen production in a co-digestion process from mixture of high carbohydrate food waste and cattle/chicken manure digestate. Int. J. Hydrogen Energy **47**, 40696–40703 (2022)
58. N.Ü. Cengiz, E. Yildirir, M. Sağlam, M. Yüksel, L. Ballice, Hydrogen and methane production from tomato processing plant waste by hydrothermal gasification. J. Supercrit. Fluids **190**, 105751 (2022)
59. Z. Zhang, K. Huang, C. Mao, J. Huang, Q. Xu, L. Liao, R. Wang, S. Chen, P. Li, C. Zhang, Microwave assisted catalytic pyrolysis of bagasse to produce hydrogen. Int. J. Hydrog. Energy **47**, 35626–35634 (2022)

60. P. Sharma, M. Sivaramakrishnaiah, B. Deepanraj, R. Saravanan, M. Venkateswar Reddy, A novel optimization approach for biohydrogen production using algal biomass. Int. J. Hydrog. Energy (2022)
61. D. Balakrishnan, N. Manmai, S. Ponnambalam, Y. Unpaprom, C. Chaichompoo, R. Ramaraj, Optimized model of fermentable sugar production from Napier grass for biohydrogen generation via dark fermentation. Int. J. Hydrogen Energy **48**, 21152–21160 (2023)
62. N.S. Alhajeri, M. Eraky, M.A. Qyyum, A. Tawfik, Eco-friendly fermentation module for maximization of hydrogen harvesting from fatty restaurant waste diluted with grey water. Int. J. Hydrog. Energy (2022)
63. A. Kadier, R. Singh et al., A novel pico-hydro power (PHP)-Microbial electrolysis cell (MEC) coupled system for sustainable hydrogen production during palm oil mill effluent (POME) wastewater treatment. Int. J. Hydrogen Energy **48**, 21066–21087 (2023)
64. N. Chinchusak, A. Incharoensakdi, S. Phunpruch, Enhancement of dark fermentative hydrogen production in nitrogen-deprived halotolerant unicellular cyanobacterium *Aphanothece halophytica* by treatment with reducing agents. Biomass Bioenerg.Bioenerg. **167**, 106624 (2022)
65. Q. Zhang, X. Shui, M.K. Awasthi, T. Zhang, J. Yang, H. Zhang, Z. Chen, C. Zou, D. Jiang, 1-Butyl-3-methylimidazolium acetate pretreatment of giant reed triggering yield improvement of biohydrogen production via photo-fermentation. Bioresour. Technol. **64**, 128068 (2022)
66. M. Rajabi, M. Mehrpooya, A. Sami, Biomass fueled chemical looping hydrogen generation, high temperature solar thermal and thermochemical energy storage hybrid system. J. Energy Storage **55**, 105657 (2022)
67. M.A. Javed, A.A. Hassan, RSC Adv. **12**, 29785–29792 (2022)
68. Z. Cai, W. Zhang, J. Zhang, J. Zhang, D. Ji, W. Gao, ACS Omega **7**, 21433–21443 (2022)
69. H. Mo, N. Wang, Z. Ma, J. Zhang, J. Zhang, L. Wang, W. Dong, L. Zang, ACS Omega **7**, 10550−10558 (2022)
70. N.L. Slatter, B. Vichanpol, J. Natakaranakul, K. Wattanavichien, P. Suchamalawong, K. Hashimoto, N. Tsubaki, T. Vitidsant, W. Charusiri, ACS Omega **7**, 44951−44961 (2022)
71. Z. Li, J. Wang, K. Tian, C. Zhou, Y. Pei, J. Zhang, L. Zang, ACS Omega **7**, 41594–41605 (2022)
72. T.R. Sarker, S. Nanda, A.K. Dalai, Hydrogen-rich gas production from hydrothermal gasification of fuel pellets obtained from co-pelletization of agricultural residues. Int. J. Hydrog. Energy (2022)
73. P. Liczbinski, S. Borowski, W. Cieciura-Włoch, Anaerobic co-digestion of kitchen waste with hyperthermophilically pretreated grass for biohydrogen and biomethane production. Biores. Technol.. Technol. **364**, 128053 (2022)
74. W. Li, L. Lu, C. Cheng, N. Ren, S.-T. Yang, M. Liu, Biohydrogen production from brown algae fermentation: relationship between substrate reduction degree and hydrogen production. Biores. Technol.. Technol. **364**, 128069 (2022)
75. J.F. González, A.I. Parralejo, J. González, A. Álvarez, E. Sabio, Optimization of the production and quality of biogas in the anaerobic digestion of different types of biomass in a batch laboratory biodigester and pilot plant: numerical modeling, kinetic study and hydrogen potential. Int. J. Hydrog. Energy **47**, 39386–39403 (2022)
76. H. Kopperi, R. Katakojwala, S. Venkata Mohan, Catalytic hydrothermal liquefaction of Scenedesmus sp. biomass integrated with dark-fermentation: biocrude and low-carbon fuel production in a biorefinery approach. Sustain. Energy Fuels **6**, 1499–1511 (2022)
77. K. Saripan, A. Reungsang, S. Sittijunda, Co-production of hydrogen and ethanol by *Thermoanaerobacterium thermosaccharolyticum* KKU-ED1 from alpha-cellulose and cellulose fraction of sugarcane bagasse. Bioresour. Technol. Rep. **15**, 100759 (2021)
78. N. Srivastava, A. Hussain, D. Kushwaha, S. Haque, P.K. Mishra, V.K. Gupta, M. Srivastava, Nickel ferrite nanoparticles induced improved fungal cellulase production using residual algal biomass and subsequent hydrogen production following dark fermentation. Fuel **304**, 121391 (2021)

79. Q. Zhang, S. Xu, Y. Li, P. Ding, Y. Zhang, P. Zhao, Green-synthesized nickel oxide nanoparticles enhances biohydrogen production of Klebsiella sp. WL1316 using lignocellulosic hydrolysate and its regulatory mechanism. Fuel **305**, 121585 (2021)
80. B.A. Ayil-Gutiérrez, M.C. Tamayo-Ordoñez et al., Study of factors involved in the behavior of biofilms formed by biohydrogen-producing microflora identified by molecular biology using dairy wastewater. Int. J. Hydrog. Energy **46**, 25965–25973 (2021)
81. J. Bu, H.-L. Wei, Y.-T. Wang, J.-R. Cheng, M.-J. Zhu, Biochar boosts dark fermentative H_2 production from sugarcane bagasse by selective enrichment/colonization of functional bacteria and enhancing extracellular electron transfer. Water Res. **202**, 117440 (2021)
82. Y. Li, Z. Zhang, D. Jiang, Y. Jing, C. Lu, H. Zhang, Q. Zhang, Continuous dark and photo biohydrogen production in a baffled bioreactor and electrons distribution analysis. Bioresour. Technol. **337**, 125440 (2021)
83. S. Zhu, X. Yang, Z. Zhang, H. Zhang, Y. Li, Y. Zhang, Q. Zhang, Tolerance of photo-fermentative biohydrogen production system amended with biochar and nanoscale zero-valent iron to acidic environment. Biores. Technol.. Technol. **338**, 125512 (2021)
84. H. Wu, H. Wang, Y. Zhang, G. Antonopoulou, I. Ntaikou, G. Lyberatos, Q. Yan, In situ biogas upgrading via cathodic biohydrogen using mitigated ammonia nitrogen during the anaerobic digestion of Taihu blue algae in an integrated bioelectrochemical system (BES). Bioresour. Technol. **341**, 125902 (2021)
85. K. Khanthong, C.W. Purnomo, W. Daosud, Y. Laoong-u-thai, Microbial diversity of marine shrimp pond sediment and its variability due to the effect of immobilized media in biohydrogen and biohythane production. J. Environ. Chem. Eng. **9**, 106166 (2021)
86. A.K. Chaurasia, R. Shankar, P. Mondal, Effects of nickle, nickle-cobalt and nickle-cobalt-phosphorus nanocatalysts for enhancing biohydrogen production in microbial electrolysis cells using paper industry wastewater. J. Environ. Manage. **298**, 113542 (2021)
87. E. Emilia Rios-Del Toro, H. Chi, V. González-Álvarez, H.O. Méndez-Acosta, J. Arreola-Vargas, H. Liu, Coupling the biochemical and thermochemical biorefinery platforms to enhance energy and product recovery from *Agave tequilana* bagasse. Appl. Energy **299**, 117293 (2021)
88. N.K. Singh, R. Singh, Modeling and statistical analysis of heat-shocked sulfate-reducers and methanogens rich consortiums for hydrogen and methane production in a bio-electrochemical cell. Int. J. Hydrog. Energy **46**, 25819–25831 (2021)
89. Yu.V. Litti, A.A. Kovalev, D.A. Kovalev, I.V. Katraeva, S.N. Parshina, E.A. Zhuravleva, E.A. Botchkova, Characteristics of the process of biohydrogen production from simple and complex substrates with different biopolymer composition. Int. J. Hydrogen Energy **46**, 26289–26297 (2021)
90. B. Jabbari, E. Jalilnejad, K. Ghasemzadeh, A. Iulianelli, Modeling and optimization of a membrane gas separation based bioreactor plant for biohydrogen production by CFD–RSM combined method. J. Water Process Eng. **43**, 102288 (2021)
91. M. Venkateswar Reddy, Y.-C. Chang, Production of biofuel precursor molecules (monocarboxylic acids, biohydrogen) from apple and pumpkin waste through an anaerobic fermentation process. J. Mater. Chem. A **9**, 19788–19795 (2021)
92. Z. Wang, D. Gao, H. Genga, C. Xing, Enhancing hydrogen production by photobiocatalysis through *Rhodopseudomonas palustris* coupled with conjugated polymers. RSC Adv. **11**, 33714 (2021)
93. R. Liu, Y. Lin, X. Ye, J. Hu, G. Xu, Y. Li, Improved hydrogen production from pharmaceutical intermediate wastewater in an anaerobic maifanite-immobilized sludge reactor. RSC Adv. **11**, 33714–33722 (2021)
94. A. Segundo-Aguilar, L.V. González-Gutiérrez, V.C. Payá, J. Feliu, G. Buitrón, B. Cercado, Energy and economic advantages of simultaneous hydrogen and biogas production in microbial electrolysis cells as a function of the applied voltage and biomass content. Sustain. Energy Fuels **5**, 2003–2017 (2021)
95. C. Sun, T. Sheng, L. Lia, L. Yanga, Biohydrogen production from traditional Chinese medicine wastewater in anaerobic packed bed reactor system. RSC Adv. **11**(11), 5601–5608 (2021)

96. Y. Chen, Y. Yin, J. Wang, Influence of butyrate on fermentative hydrogen production and microbial community analysis. Int. J. Hydrogen Energy **46**, 26825–26833 (2021)
97. Y.A. Del Angel-Acosta, L.H. Alvarez, R.B. Garcia-Reyes, J. Carrillo-Reyes, A. Garcia-Gonzalez, E.R. Meza-Escalante, Co-digestion of corn (nejayote) and brewery wastewater at different ratios and pH conditions for biohydrogen production. Int. J. Hydrogen Energy **46**, 27422–27430 (2021)
98. J.R.N. Junior, L.A.Z. Torres, A.B.P. Medeiros, A.L. Woiciechowski, W.J. Martinez-Burgos, C.R. Soccol, Enhancement of biohydrogen production in industrial wastewaters with vinasse pond consortium using lignin-mediated iron nanoparticles. Int. J. Hydrog. Energy **46**, 27431–27443 (2021)
99. G. Lovato, I.M.G. Augusto, A.D.N.F. Júnior, R. Albanez, S.M. Ratusznei, C. Etchebehere, M. Zaiat, J.A.D. Rodrigues, Reactor start-up strategy as key for high and stable hydrogen production from cheese whey thermophilic dark fermentation. Int. J. Hydrog. Energy **46**, 27364–27379 (2021)
100. R. Palomo-Briones, J.J. Montoya-Rosales, E. Razo-Flores, Advances towards the understanding of microbial communities in dark fermentation of enzymatic hydrolysates: diversity, structure and hydrogen production performance. Int. J. Hydrog. Energy **46**, 27459–27472 (2021)
101. M.S. Akca, O. Bostancı, A.K. Aydin, I. Koyuncu, M. Altinbas, BioH_2 production from food waste by anaerobic membrane bioreactor. Int. J. Hydrog. Energy **46**, 27941–27955 (2021)
102. Y. Wang, M. Tang, J. Ling, Y. Wang, Y. Liu, H. Jin, J. He, Y. Sun, Modeling biohydrogen production using different data driven approaches. Int. J. Hydrogen Energy **46**, 29822–29833 (2021)
103. S.N. Malik, S.K. Rena, Enhancement effect of zero-valent iron nanoparticle and iron oxide nanoparticles on dark fermentative hydrogen production from molasses-based distillery wastewater. Int. J. Hydrog. Energy **46**, 29812–29821 (2021)
104. Yu.V. Litti, D.A. Kovalev, A.A. Kovalev, A.Y. Merkel, A.V. Vishnyakova, Y.I. Russkova, A.N. Nozhevnikova, Auto-selection of microorganisms of sewage sludge used as an inoculum for fermentative hydrogen production from different substrates. Int. J. Hydrogen Energy **46**, 29834–29845 (2021)
105. T. Tang, Y. Chen, M. Liu, Y. Zhang, Z. Yu, Biohydrogen production, sludge granulation, and microbial community in an anaerobic inner cycle biohydrogen production (AICHP) reactor at different hydraulic retention times. Int. J. Hydrogen Energy **46**, 30300–30309 (2021)
106. W. Zhao, Y. Zhang, B. Du, D. Wei, Q. Wei, Y. Zhao, Enhancement effect of silver nanoparticles on fermentative biohydrogen production using mixed bacteria. Biores. Technol.. Technol. **142**, 240–245 (2013)
107. S.M. Aminossadati, B. Ghasemi, Natural convection cooling of a localised heat source at the bottom of a nanofluid-filled enclosure. Eur. J. Mech. B Fluid **28** (2009)
108. G.A. Sheikhzadeh, A. Arefmanesh, M.H. Kheirkhah, R. Abdollahi, Natural convection of Cu-water nanofluid in a cavity with partially active side walls. Eur. J. Mech. B Fluid **30**, 166–176 (2011)
109. J. Zhou, M.A. Ali, T. Hai, K. Sharma, K.H. Aziz, F.Q.A. Alyousuf, K.T. Almoalimi, S.F. Almojil, A.I. Almohana, A.F. Alali, Enhanced hydrogen generation in a combined hybrid cycle using aluminum and cooper oxide nanomaterial based on biomass and vanadium chloride cycle: optimization based on deep learning techniques and environmental appraisal. Int. J. Hydrog. Energy (2022)
110. N. Fahoul, M.H. Sayadi, M.R. Rezaei, S. Homaeigohar, Nickel ferrite nanoparticles catalyzed dark fermentation of dairy wastewater for biohydrogen production. Bioresour. Technol. Rep. **19**, 101153 (2022)
111. M. Taherdanak, H. Zilouei, K. Karimi, The effects of Fe^0 and Ni^0 nanoparticles versus Fe^{2+} and Ni^{2+} ions on dark hydrogen fermentation. Int. J. Hydrog. EnergyHydrog. Energy **41**, 167–173 (2016)
112. C.-Y. Chen, H.-Y. Chang, J.-S. Chang, Producing carbohydrate-rich microalgal biomass grown under mixotrophic conditions as feedstock for biohydrogen production. Int. J. Hydrogen Energy **41**(7), 4413–4420 (2016)

113. J.E. Ramirez-Morales, E. Tapia-Venegas, J. Toledo-Alarcon, G. Ruiz-Filippi, Simultaneous production and separation of biohydrogen in mixed culture systems by continuous dark fermentation. Water Sci. Technol. **71** (2015)
114. H. Abdulla, Enhancement of rice straw composting by lignocellulolytic actinomycete strains. Int. J. Agri. Biol. **9** (2007)
115. F. Ndayisenga, Z. Yu, J. Zheng, B. Wang, H. Liang, I.A. Phulpoto, T. Habiyakare, D. Zhou, Microbial electrohydrogenesis cell and dark fermentation integrated system enhances biohydrogen production from lignocellulosic agricultural wastes: substrate pretreatment towards optimization. Renew. Sustain. Energy Rev. **145**, 111078 (2021)

Compostable Wastes Generated from Biomass

R. Ramesh, A. Saran, G. Damodaran, A. Lavanya, A. Periya Nayagi Shilpa, F. Liakath Ali Khan, and K. Kaviyarasu

Abstract Biomass in general can be defined as the organic matters which are produced by the plants of both the terrestrial and aquatic means. Due to the chemical reaction by the bioactive microorganisms, the huge mass of the wastes generated in the environment has been processed into biomass wastes. Being the source of power biomass plays a vital role in the field of the energy sector. The whole world is looking for an alternative way of producing energy from sources that are available as raw materials to generate different forms of energy. According to a recent study, fossil fuels will get depleted in a few decades due to overconsumption, and also a lot of practical difficulties are there in the case of nuclear-based energy sources. All over the world millions and trillion tons of waste are generated per day from various fields including agricultural waste, food waste, industrial waste, wastes produced from forests, animal waste, poultry droppings, sewage waste, etc. All these wastes can be utilized into useful energy using different conversion technologies including fermentation, anaerobic digestion, pyrolysis, combustion, transesterification, gasification, thermochemical conversion, etc. In this chapter, the waste generated from biomass and the process involved in the conversion of biomass into useful energy are depicted. Also, the by-products generated from biomass, like biodiesel, biogas, and other products and their role as a biofuel are discussed.

R. Ramesh (✉) · A. Saran · G. Damodaran · A. Lavanya · A. Periya Nayagi Shilpa
Department of Physics, Sacred Heart College (Autonomous), Tirupattur, Tamil Nadu 635601, India
e-mail: rramesh@shctpt.edu

R. Ramesh · F. Liakath Ali Khan
Department of Physics, Islamiah College (Autonomous), Vaniyambadi, Tamil Nadu 635752, India

K. Kaviyarasu
UNESCO-UNISA Africa Chair in Nanosciences/Nanotechnology Laboratories, College of Graduate Studies, University of South Africa (UNISA), Muckleneuk Ridge, PO Box 392, Pretoria, South Africa

K. Kasinathan et al. (eds.), *New Technologies for Energy Transition Based on Sustainable Development Goals*, https://doi.org/10.1007/978-981-97-2527-4_22

1 Introduction

Among the most widely used energy resource, "*biomass*" plays a major role in recent days. The concern about the disastrous effects was increasing consumption of fossil fuels like the greenhouse effect and climatic changes lead the effect on human health. This current article explains the sources which are available in biomass, as well as its characteristics and the chemical make of biomass-related products [1–3]. The explanation of various changing technologies, like physicochemical conversion, thermochemical, and biochemical conversions was done, and their related materials were evaluated. Next, the biomass global position versus renewable resource energy was examined. Furthermore, from the standpoint of environmental and economics, the production of energy which is derived from biomass was examined. Finally, the discussion was done on the market of global energy carriers from the derived energy of biomass which acts as the barrier to increasing their shares [4–7]. In the manufacturing of eco-friendly upgraded chemicals and biofuels, the feedstocks from the agricultural wastes were used as the biorefinery rather than lignocelluloses, hence its energy security was increased. As a result, a possible application for producing green fuels like bio-butanol and bioethanol avoids the food which was extracted from the biomass renewable energy versus fuel conundrum giving the effectiveness of cost as the positive result [8]. Taking the use of various agricultural opportunities biorefinery produces bioenergy by using biomass and experimenting with various biomass conversion strategies. The increasing affluence and population increment resulted in public health on degradation in recent days of our life, and owning mostly the usage of fossil fuels was increasing in the industrial field and for the development of postindustrial applications. To get alternative methods and to make development for generating energy worldwide communities were working [9–12]. An international framework for renewable energy was being attempted as a significant organization to promote renewable energy, hence the distribution of energy source supply can be done in the way of traditional energy supply. In the strong impulse of the European Union, on considering the different possible energy sources, biomass plays a crucial role in the debates regarding energies within the context of policy, in the execution of plans which leads to the deliberated development of the protected environmental equipment and transformation of saving. Indeed, the substantial renewable energy resource was taken from the potential of biomass, with enormous potential for the generation of biofuels for the transformation, power, and heat. The economic and potential significance of bioenergy is acknowledged in developed nations [13–15]. Bioenergy promotion was increased as part of agricultural policies, biomass significantly creates economic-related opportunities in rural areas, it also includes increasing the income of farmers from the various and many productive usages of lands; it reduces the migration to countryside-city which leads to the protection of rural culture and our environment. To maintain the release of greenhouse gas (GHG) in the atmosphere and their effects from the primary stage of consumption of fossil fuels in 1988, an scientific community called International Panel on Climate Change (IPCC) was invented as the huge level commitments are brought to the problem of climatic changes in the countries and new bodies. The argument was ongoing in tandem regarding the use and application of renewable energy resources, biomass plays a major role to securing energy and giving green development. On the latest invention on the usage of biomass, potential ways for optimizing the generation of liquid, energy, and heat biofuels in the affordable plentiful naturally available resources have been devised [16–18]. As biomass was a biogenetic product during the burning process carbon dioxide will be emitted, it will not affect the carbon dioxide in the atmosphere and also it is 100% of a renewable resource. The plants require CO_2 for the growth, and function of metabolic activities and the process of degradation of other plants occur due to the release of CO_2 in the environment. Biomass is any organic material that results directly or indirectly from the photosynthesis process. Biomass is the composition of trees, algae, shrubs, and crops that is derived from plants and occurs in nature. The variability in the usage, origin, and their components can change the entire concept of biomass [19–21]. The biomass products are derived from fossil resources and petrochemicals except for plastics which consist of mixed organic components. In the world, an enormous quantity of biomass was stored in many regions from forests to the sea. In our world in approximation, there are 1.8 trillion land reserves and 4 billion tons of aquatic reserves biomass are there. The output capacity of the entire biomass potential on Earth is larger than 80 times of global annual energy consumption which is 33,000 EJ. Through the expenditure of wood, shrub, plant waste, and animal waste up to 50% of biomass was generated in some of the developing countries, and the biomass energy production in developed countries was 11% of total energy production [22]. Biomass takes only 14% of primary global energy or around 56 million TJ/year, whereas biomass energy was mostly used in recent days. Furthermore, the consumption of this resource is not distributed uniformly around the globe [23–25].

1.1 Biomass Waste

The term "*biomass waste*" refers to a broad category of materials, including forestry residues, agricultural wastes, processing waste from other food products, and trash from the processing of fruit [26]. If that kind of garbage is not properly disposed of, it might result in major health or environmental issues. Therefore, it is crucial to design an eco-friendly and efficient approach for utilizing and managing different types of biomass waste. Currently, biomass waste is viewed as a low-value resource that is often underutilized. However, its value as a tool for producing outcomes with added value is now more widely acknowledged [27].

1.2 Classification of Biomass

Biomass has a remarkably diversified chemical makeup. For instance, cereals and bovine manure are primarily made of starch, among plant leftovers, cellulose and hemicellulose make up the majority, and lignin in variable amounts. Various biomass substrates have been revealed in investigations to contain significant amounts of small organic components, which may affect the treatment processes [28]. The following is a list of them 4–61% of proteins, 4–60% of nucleic acid, and 2–40% of lipids in alga. In certain straws and flax, there are 2–4% acetyls trees such as sorghum grass, eucalyptus, and pine which contain uronic acids in amounts between one and four percent. There is an estimate that up to 10% of protein can be found in plants, including maize, whitewood, reeds, pine, and Betula utilizes [29]. Due to their significant diversity, amount, and compositional differences, biomass cannot be classified in a unified way since there are myriad ways to classify them, depending on their goals and scope as shown in Fig. 1. The origin, function, and end products of biomass are typically classified into two categories. Biomass types in nature are classified based on their ecology or type of plant [30].

1.2.1 Using Biomass as a Feedstock and Categorizing It

There are numerous groups of biomasses based on the former classification:

- Mixtures of biomass;
- Biomass derived from wood or woody plants;
- Biomass derived from herbs;

Fig. 1 Types of biomass wastes

- Biomass from aquaculture;
- Waste biomass from animals and humans.

1.3 *Wood and Woody Biomass*

Woody biomass consists largely of lignin and sugars, which make up most of its composition. In this category, you will typically find substances that are directly burnable (or gasified) and can be converted into energy through several processes, such as woody shrub wood, bark, leaves, and roots [31]. For energy and fuel production, biomass is usually derived from four main sources:

- Residues from production;
- Timber residues that are not merchantable;
- Wood wastes after consumption;
- Wastes are generated by urban and agricultural activities.

Currently, among the renewable energy resources woody biomass plays a major role. In 2010, woody biomass was estimated to have produced 30 EJ of energy, with 16 EJ being generated for domestic use and 14 EJ being generated for industrial use [32].

1.3.1 Phytoplankton Biomass

Solid biofuels are classified based on the European standard EN 14961-1, which defines herbaceous biomass as follows. "*Byproducts of food processing industries like cereal straw and grains with non-woody stems make up herbaceous biomass* [33]. *It is derived from plants with non-woody stems that die back each year*".

1.3.2 Aquatic Biomass

As aquatic biomass, emerging plants, macroalgae, and microalgae all belong to this category: (i) As fast as 60 m can be grown, macroalgae are multicellular creatures. Hydrocolloids and food are commonly extracted from them. (ii) Microalgae is a microscopic organism made up of diatoms, green algae, and golden algae [34]. An algae that has a single cell and is brown can be up to a few millimeters in size. As the world's largest source of biomass, they contribute a great deal to aquatic microflora. The amount of green algae in freshwater is particularly high. It is also possible for these algae to produce oils, but starch is their primary product. In the same way that diatoms produce both oils and carbohydrates, golden algae do the same. (iii) Marshes and swamps are ideal habitats for emerging plants. A total of more than 55,000 species of bacteria can be found in nature, including 100,000 different species of brackish water, freshwater, and terrestrial algae. Using algae is beneficial because they produce metabolites and compounds from sunlight, water, CO_2, and algae biomass as a result [35]. The biodiesel in the third generation was produced by biomass in the aquatic region, biodiesel will not compete with edible crops, and it can be able to produce a large amount per hectare of biomass than the crops produced in the land. However, commercial biofuels must also overcome significant economic and technological obstacles. The level of irradiation should also be considered among other factors. The concentrations of CO_2 and O_2 in the water, temperature, pH, salinity, and nutrient content all affect the performance of aquatic biomass [36].

1.3.3 Waste Biomass from Animals and Humans

Human excrement, animal manure, bones, and meat meals are common sources. Before stricter regulations in the environment on health, odor issues, and pollution have taken effect to begin from the wastes which are already used, and lands used in agriculture or as fertilizer. To turn this waste into valuable products, the major practical technique is digestion in an anaerobic manner. By-products of the bioenergy process, such as biogas, can be used directly to heat water and rooms and to drive engines and turbines [37].

1.4 Biomass Conversion Technologies

Typically, biomass energy conversion involves the following processes:

- A thermochemical conversion
- The biochemical conversion processes
- Conversions that involve physicochemical processes.

1.4.1 Biomass Is Converted by Thermochemical Means

During thermochemical conversion processes, heat and chemical reactions are applied to produce energy. Thermochemical conversion occurs through four processes: combustion, gasification, liquefaction, and pyrolysis [38].

1.4.2 Burning Process

The combustion of biomass generates approximately 90% of renewable energy. In combustion plants, biomass can be used from different sources like animal dung, wood, shells of hardest vegetables, rice husks, and dry leaves [39–41]. When biomass and oxygen were burned with heat, CO_2, a hot atmosphere, and water vapor are produced.

Using an equation from below, you get:

$$\text{Combustion of biomass} + O_2 \rightarrow CO_2 + H_2O + \text{Heating}. \quad (1)$$

Biomass combustion can be approximated by the following chemical equation:

$$\text{CH1.44O0.66} + 1.03\,O_2 \rightarrow CO_2 + 0.72\,H_2O + \text{Heat}. \quad (2)$$

Although the thermal energy produced from the average amount of biomass is 20 MJ/kg, several variables can affect the amount of heat created. During combustion in the present atmosphere, biomass was burned, creating chemical energy that can be converted into mechanical and electrical energy. The process of combination is therefore a chemical reaction of exothermic process [42]. Within combustion chambers, combustion occurs at temperatures between 800 and 1000 °C. Dry biomass should have a moisture content of $< 50\%$. The combination of plants in biomass that use as a fuel is extracted from the woody plants and forest leftovers, typically producing from 20 to 50 MW of energy with the efficiency of electricity ranging between 25 and 30%. Electricity from biomass has been produced on a measure of 50–80 MW with the 30–40% of its needs through fluidized bed systems, sophisticated gas cleaning, and further innovations [43].

1.4.3 Lignocellulose Materials Are Pretreated in Various Ways

(a) *Mechanically*

As a result of the milling, biomass is reduced to a smaller size using methods and machinery. Biomass breakdown—Using ultrasonic equipment and methods to get molecules in small sizes was achieved by breaking down the biomass. The high-water pressures and its temperature—Hydrolysis of the lignin, hemicellulose, and cellulose by using high pressure and temperature water. The weak acid—After spraying weak acid on raw materials, the mixture is heated between 160 and 220 °C for a brief period. Concentrated acids—Aqueous solutions of strong concentrate acids (such as H_2SO_4 and HCl) were used in the hydrolyzed lignin, hemicellulose, and cellulose [44].

(b) *Chemically*

Calcium (or sodium) hydroxide combined with ammonia is warned to create alkaline hydrolysis that eliminates some hemicellulose and reduces lignin levels [45]. Using an organic solvent or a mixture of an organic solvent and water, lignin is removed before cellulose is enzymatically hydrolyzed. H_2O_2, ozone, oxygen, or air can be used as oxidative-oxidizing chemicals for the delignification of cellulose.

(c) *The Role of Temperature*

Liquids with organic cations and inorganic anions (RTIL)—Chemically defined as cations and anions, these liquids can be used to separate lignin and improve the accessibility of cellulose without utilizing acidic or alkaline solutions. Expansion of steam—As a result of the decompression process associated with the explosion of biomass, lignin, and hemicellulose are broken apart and destroyed. AFEX—A high-pressure and high-temperature reactor uses liquid ammonia [46]. Thereafter, the lignin and hemicellulose concentrations decrease, and the cellulose crystallizes, because of the drastic reduction in pressure.

(d) *Chemicals/Mechanicals*

Carbon dioxide—A batch reactor containing biomass is injected with high-pressure CO_2. Due to the pressure reduction, the subsequent hydrolysis is improved by explosive decompression. The mechanical process—Adding alkali to lignocellulosic biomass during mechanical pretreatment is commonly called the alkaline pretreatment process [47]. Additionally, the insoluble cellulose was added to the hemicellulose, lignin, and other soluble substances.

(e) *Biological*

Hydrolysis by microorganisms—Microbes (like white and brown fungi) break down hemicellulose and lignin during a reaction batch [48].

1.5 The Pyrolysis Processes

The pyrolysis process produces liquids (bio-oil), solids (charcoal), and gaseous products (combustible gases) by partially burning biomass at temperatures of about 500 °C. As biomass-producing gases are heated, their volatile parts can be vaporized, and the vapors will liquefy and become liquids. It is possible to use the liquid fuel produced by this process for heating and electricity generation later on [49]. As well as liquid fuels, pyrolysis also produces charcoal, gas, and numerous other combustible products. The liquid fraction of the pyrolysis process, or bio-oil, is currently the main product of pyrolysis, which may be used as fuel in cogenerators. In comparison with other biofuels, bio-oil can be produced at a lower cost, produce neutral CO_2 emissions, and, depending on the refinement process, can have a high energy density [50].

1.6 *The Gasification Processes*

Syngas is the combination of primary gases like carbon monoxide (CO), hydrogen (H_2), and nitrogen (N_2) that is produced by gasifying solid carbon sources (solid biomass). A variety of goods can be made using syngas, including methanol, ammonia, synthetic gasoline, and others, as an energy source or as the raw material for a variety of materials in the refinery industries and petrochemical industries [51]. The synthesis of the gas method was first improved in the first half of the twentieth century. It is important to note that there were different methods of treatment and procedure choices, but they all share some similarities. Despite this, the most common method involves huge temperatures, the matrix form of the solid was partially oxidized, leading to syngas containing mostly H_2 and CO, with a significant amount of CO_2 and CH_4. Gas is produced in many combustible and noncombustible ratios, depending upon the biomass type and the condition of operation [52]. Natural gas calorific value is typically between 10 and 50% of the generated gas, resulting in a calorific value of 4.5–6 MJ/m^3.

1.6.1 The Exothermic Phase of Oxidation

Biomass gasification requires an oxidation phase to reach the required operating temperature for endothermic processes. A reaction that takes place without oxygen results in oxidation. As shown in the below equations, oxidation takes place during the following processes [53].

$$\text{A coal combustion reaction : C} + \text{Oxygen} \rightarrow \text{carbon dioxide} \quad \Delta H = -394\ \text{kJ/mol} \tag{3}$$

$$\text{Partially oxidized : C} + \frac{1}{2}\,\text{Oxygen energy} \rightarrow \text{CO} \quad \Delta H = -111\ \text{kJ/mol} \tag{4}$$

$$\text{Hydrogen combustion : Hydrogen} + \frac{1}{2}\text{Oxygen} \rightarrow \text{Water} \quad \Delta H = -242\ \text{kJ/mol.} \tag{5}$$

At this stage, the thermal energy was obtained by the primary objectives which are required for the whole process. The product combustion, for instance, was composed of water, carbon dioxide, and other compounds.

1.6.2 The Drying Process (Endothermic Stage)

A temperature above 100 °C is used to remove moisture from biomass during the drying stage, after which it is heated to 150 °C to finish the process. In this phase, raw material moisture content determines the amount of heat required, which is typically obtained from the previous oxidation phase [54].

1.6.3 The Endothermic Stage of Pyrolysis

In pyrolysis, carbonaceous carbon mass is decomposed anaerobically by heat. A liquid fraction known as tars are produced at relatively low temperatures by condensing complex organic substances. A solid fraction is produced by "*char*" and a gaseous fraction is produced by the combination of light hydrocarbons and incondensable gases such as carbon dioxide, hydrogen, and oxygen [55].

1.6.4 Endothermic Reduction (Reduction)

Pyrolysis and oxidation produce a mixture of gas and coal. The phase of reduction (reactions on the phase of gas) changes this composition. After the by-products of the earlier processes react in a final step, syngas is produced [56].

1.6.5 The Liquefaction Processes

At huge pressure ranging from 10 to 25 MPa and temperatures of range 280 to 370 °C, biomass liquefaction takes place [57]. Additionally, a liquid granulate that resembles crude oil is produced along with a variety of by-products from solid, gaseous, and aqueous products.

2 Conversion of Biomass Biochemically

The biochemical conversion process breaks down biomass into readily available carbohydrates like fuel liquid, biogas, and their related products. There are two biochemical processes most commonly employed, anaerobic digestion and fermentation [58].

2.1 Digestion in Anaerobic Conditions

Biological processes such as anaerobic digestion produce renewable energy as well as assist with the proper management of waste. This process consists of four fundamental phases: methanogenesis, acidogenesis, hydrolysis, and acetogenesis. It is made possible by various chemical reactions in the whole process in the oxygen-free environment by microorganisms. During this process, different kinds of chemical

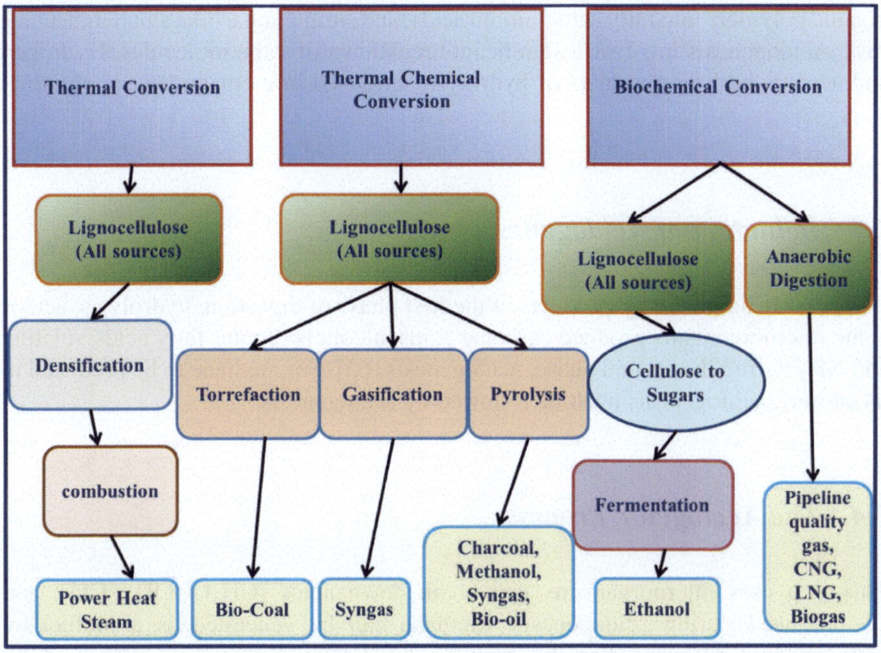

Fig. 2 Biomass conversion technologies

reactions will take part through normal pathways of metabolism [59–61]. A mixture of CO_2 and methane, and the other gases related to it, is produced by reducing complex organic macromolecules into simpler ones. A variety of feedstocks are majorly used in this kind of process, including animal manure, agricultural leftovers, MSW, and sewage sludge. A simplified representation of the process of anaerobic digestion is seen in Fig. 2. As a result, the energy content of the biogas produced is 20–40% lower than that of the raw materials used in the process. A moisture content of 80–90% is best for the procedure when dealing with organic waste [62]. In addition to being able to use the biogas directly in gas turbines and ignition engines, this method produces a valuable product. Total conversion efficiency is 21% with this technique. Thermal energy can be collected from engines and turbines using an exchanger. Anaerobic digestion is divided into the following phases [63].

2.2 *The Hydrolysis Process*

Biomass, a long-chain organic polymer containing lipids, carbohydrates, and proteins, is consumed first, followed by hydrolysis, a process that breaks down organic polymers into fatty acids, amino acids, and simple sugar-like small molecules [64]. Acidogenesis involves a significant breakdown of large molecules. Hydrogen and acetate, both by-products of hydrolysis, are used in methanogenesis, the final step.

2.3 *Induction of Acidogenesis*

By breaking down the by-products of the first phase of digestion, hydrolysis, acidogenic microorganisms produce carbonic acids, alcohols, lighter fatty acids volatile, and NH_3 during the second phase, acidogenesis [65]. For methane to be produced in its entirety, acidogenesis must be followed by acetogenesis.

2.4 *The Acetogenic Process*

This step uses microorganisms that break down acids (CH_3COOH), CO_2, and H_2 produced during acidogenesis. Methane can be generated as a biofuel by methanogenic archaea after the biomass breakdown process is completed by acetogens [66].

2.5 *The Methane Cycle*

The final stage of anaerobic digestion produces methane, as shown in Eqs. 6 and 7 or hydrogenotrophic methanogenesis. It is produced primarily from CH_3COOH and CO_2, the products of acetogenesis [67].

3 Fermentation Processes

Anaerobic microbes, primarily yeasts, fermented organic materials into ethanol and carbon dioxide during the fermentation of simple sugars (*hexoses and pentoses*). Microorganisms that are used in this process include the Saccharomyces cerevisiae, and feedstocks used in this kind of process include sugars, starches, and lignocellulosic materials [68]. In addition to feedstock, temperatures, pHs, inoculum, and fermentation periods, several variables influence process quality and yields.

$$\text{Sugars} \rightarrow \text{Ethanol} + \text{carbon dioxide} + \text{residues}$$

$$C_6H_{12}O_6(\text{hexoses}) \rightarrow 2C_2H_5OH + 2CO_2 \tag{6}$$

$$3C_5H_{10}O_5(\text{pentoses}) \rightarrow 5C_2H_5OH + 5CO_2 \tag{7}$$

Carbohydrates can be converted into ethanol through a variety of metabolic processes depending on their starting substrate. This involves (a) carbohydrate metabolism and glycolysis through the Embden-Meyerhof pathway (EMP) and (b) pentose metabolism through the pentose phosphate pathway (PPP). There is a more rapid rate of hexose conversion compared with pentose conversion [69].

3.1 Biomass Conversion by Physicochemical Reactions

As biomass is physiochemically converted into high-density biofuels are created. Different kinds of oil from vegetables and fats from animals can be changed into biodiesel through the process of esterification and/or in the transesterification process. First-generation biodiesel is generally made from rapeseed oil and sunflower oil [70]. Tomato oil and rice bran oil are the two main oils from vegetables that were used in the biodiesel in first generation. Waste cooking oil (WCO), as well as microbial oils, such as algal oil, could be used in the production of biodiesel in the second and third generations [71].

$$\text{Fuel oil} + \text{short chain alcohol} \rightarrow \text{Distilled biodiesel} + \text{glycerol} \tag{8}$$

A low temperature (50–70 °C) and tamp pressure are required for the transesterification of triglycerides. By adding glycerol to the produced product, neutralization and purification take place.

4 Sources of Biomass

In addition to livestock farms and sewage, algae and aquatic crops are the most significant sources of biomass. There are also a variety of other sources, including agricultural remnants (such as shavings and sawdust produced by factories that process wood), animal residues, and animal feces. Waste streams from anthropogenic activities and municipal solid waste (MSW) only fall under the biomass category if they cannot be further processed [72]. Forests, oceans, and other habitats contain large amounts of biomass. There are approximately 1.8 trillion tons and 4 billion tons of sources in the biomass reserves on land and in water, according to several sources. According to the World Energy Council, the biomass available in the world has an output of energy of about 33,000 EJ, which is greater than 80 times the global consumption of total energy (36). A current estimate suggests that biomass provides approximately 56 million TJ (1230 Mtoe) of primary energy per year—about 14% of the world's energy. Additionally, how this resource is utilized throughout the world differs widely. It is estimated that up to 50% of the whole energy requirements in some of the improving countries come from burning wood, shrubs, and animal waste, while about 11% of the total energy produced in developed countries comes from energy of biomass. Approximately 3.2 million TJ (70 Mtoe) of the United States energy needs are supplied from biomass every year, or 3% of its consumption of whole energy [73]. In the European Union, biomass represents 3.5% of the annual supply of energy (about 40 Mtoe/yr). However, few countries, such as Austria, Sweden, and Finland, produce more biomass energy than the European average, at 18%, 17%, and 13%.

5 Energy from Biofuels

Plant biomass (stalks, trunks, stems, and leaves) is being used in biofuel production to reduce oil dependence and lessen environmental impact. For biomass to be converted into biofuels, it must be broken down into sugars by biotechnology as shown in Fig. 3. As lignocellulosic biomass is a major source of fuel ethanol production today, wood chips, grasses, corn stalks, and other biomass are more difficult to convert into sugar than cereals (corn, wheat, etc.). The use of cellulosic ethanol is therefore recommended to help meet our energy needs. Feedstocks as well as finished products (such as butanol) can be used in a variety of ways [74]. A lot of potential is believed to exist in butanol. For food production, algae are not competing with arable land. Algae can be used to produce a variety of products, such as biodiesel and hydrogen. The creation of hydrogen or even electricity from solar energy and water by artificial photosynthesis could provide a virtually limitless supply of energy in the long run. Biotechnology, particularly synthetic biology, can play a significant role in boosting production and promoting the use of sustainable bioenergy if advanced solar to biomass conversion techniques are developed, socioeconomic and environmental issues are considered when designing technological solutions, and next-generation feedstocks are developed. Types of biofuels and their generations: Biofuels are fuels extracted from biomass, currently alive organisms, and their by-products from metabolism, such as cow dung. As opposed to nuclear energy, coal and petrol, they are renewable sources of energy [75]. The type of technology used, and the biomass feedstock used by biofuels can help categorize them into "generations". In addition to feedstocks and production technologies, a description related to the first and second generations of biofuel principles was given.

5.1 First-Generation Biofuels

A first (1st)-generation biofuel is an energy source that is made up of food crops (such as sugarcane or rice) and other food-based feedstocks (such as cooking oil). Currently,

Fig. 3 Energy from biomass to biofuels

there is a considerable amount of biofuel production techniques and their products in the market are well-defined [76]. In the first generation of biofuel, bioethanol, biodiesel, and biogas are the most important. The process of bioethanol production involves the fermentation of starch to sugar and the biomass from the sugar (such as cereal crops like corn, maize). If the fuel specification is followed, that gasoline can be blended with it in any proportion of 10% (US), either in pure form or in pure form with specially designed vehicles. Bioethanol and isobutylene are used to make ethyl-tertiary butyl-ether (ETBE). A mixture of up to 15% can be made with gasoline. Currently, it contributes the most biofuels to the European market. Transesterification (reaction of triglycerides with acids or methanol and alkali) is the process of making biodiesel (FAME) from vegetable oils like rapeseed, soya, palm fruit, or other oils [77]. A pure form of this fuel is available for use in vehicles that are specially adapted, or it can be blended with the automotive biodiesel to a maximum of 5% (up to 30% if it is a captive fleet fuel). In landfills, for instance, manure and other humid biomass materials (such as food waste) are anaerobically treated, and the resultant gas is converted to biomethane for use in natural gas vehicles and feed-in to the natural gas grid. Biogas can be produced not only from crops, and it can be produced from various biomass products, which leads to confusion over whether it is referred to in the first or second generation. Nonetheless, this paper considers biogas to be part of the first generation, given the maturity and wide usage of the technology that produces it, as do many other experts in the field [78]. These fuels are generated by using plant parts that are easily extracted, such as corn kernels with a high content of starch, grains, and sugar in canes, or oilseeds. By-products left over from vegetable oil production, such as press cake or glycerin, as well as dried distillers' grains with solvents, are typically used for feed or chemicals. The use of by-products for process energy is one of the integrated concepts being developed in the R&D stage. As of 2006, there was a worldwide production of nearly 50 billion liters (approximately 39.5 million tons) of first-generation biofuel, including 5.4 million tons of biodiesel. In Europe, ETBE plants could produce approximately 100 tons per day. A total of 2.3 million were produced in 2005 (39). Biofuels have a high carbon content in the first generation [79].

5.2 Second-Generation Biofuels

A wider variety of feedstocks can be used to manufacture future biofuels than the first generation, which is primarily made of non-food crops. A plant's whole biomass like straw of wheat, grass, wood, or waste streams that which has rich in amount of cellulose and lignin, can be used [80]. In part of the process of breaking down this biomass, two main conversion pathways are considered: (i) hydrolysis of lignocellulose (*by chemical or biochemical means*) into sugar, which is made by fermentation to create alcohol. This technology is known as "cellulosic bioethanol" and is currently under development; (ii) pyrolysis and gasification of lignocellulose into raw gas and oil using thermochemical processes (*high temperatures*). Carbon monoxide and hydrogen form the main components of synthesis gas (*syngas*) after the resultant gas is treated and conditioned. In addition to being able to be changed from liquid to gaseous fuels, that gas can also be converted into another kind of gas and also as the fuel [81]. Synthetic biofuels are produced through this route. Bioethanol and Fischer–Tropsch fuels are the most promising liquid synthetic biofuels. Biogas and dimethyl ether (DME) are synthetic gaseous biofuels, both of which can be used as natural gas substitutes. Additionally, hydrogen can be produced from the conditioned and cleaned product gas. Biofuels can also be produced from biomass using pyrolysis or hydrothermal treatment (e.g., *HTU diesel*) by deoxygenation. Compared with the first generation of biofuels, here there is a lower GHG content. Food crops that have been left over are used to make them [82].

5.3 Third-Generation Biofuels

The feed stocks from the biotechnological invention are essential to third-generation biofuels. Lignin, cellulose, and hemicellulose, the structural components of the cells of plants, are engineered to be manipulated in response to a particular task required of them. Plant scientists, for instance, are seeking to develop plants/trees which can be able to grow naturally, but the cell walls were triggered to give modification in their strength, so the sugars can be released more easily [83]. By using functionally engineered enzymes to break down plants with special properties, third-generation biofuels create synergies between these types of interventions and processing steps. In particular, this second generation of biofuels has only been explored gradually.

As CO_2 is emitted and sequestered, it is carbon neutral. Microorganisms are used to produce this product. Currently, aquatic microorganisms are being used as feedstocks for third-generation biofuels. In the development of third-generation biofuels, microalgae were leading the way. Carbon is absorbed more efficiently by microalgae, and there are high lipid contents. Microalgae are also easy to grow and can be harvested in a short period. In comparison with soybeans, microalgae have a lipid content that is 25–200 times higher. As well as making biodiesel out of microalgae oil, making alcohol from its carbohydrate, and recycling nitrogen and phosphorus into fertilizer, microalgae oil can be used as fertilizer [84].

6 Derived Biomass from Various Sources

6.1 A Willow That Is Dedicated to Storing Energy (Salix viminalis)

Increasingly, willow is being used to generate energy. Trees and shrubs (*which serve as energy generators most commonly*) are two types of trees in nature. This soil is highly resistant to contamination and expands rapidly. To get renewable energy sources willow was also used in 3–5 years because of its rapid growth cycle [85]. It generates minimal nitrogen, chlorine, and sulfur emissions into the atmosphere when its wood is burned, unlike coal. Heavy metals from the soil are not accumulated in it, so it is suitable for energy applications. A soil purification method based on energy-dedicated willow is not recommended. In dry climates where chlorine concentrations may exceed lethal levels low soil purification remains a problem when excess nitrogen is eliminated satisfactorily. A modest domestic wastewater disposal system could be purified using willow [86]. Sludge from the treatment of sewage is used to produce nutrients for water purification, while biomass is used to produce biofuel. It is almost impossible to find a better fertilizer for willow plantations than household wastewater. Plant nutrient requirements and waste material concentrations are compensated for in some cases by the soil. Compared with industrial wastewater, household wastewater is less contaminated with heavy metals, which may limit microbial growth in soil and plant roots. Reductions in garbage consumption result from both processes. Researchers found that using gravel alone can efficiently reduce and eliminate BOD, nitrogen, phosphorus, and potassium in *Salix viminalis*, which grows on hydroponic gravel [87]. It increases nitrogen and phosphorus uptake by 12%, phosphorus uptake by 5%, and potassium uptake by 18% compared with the plant alone.

6.2 *Sorghum moench (Sweet Sorghum)*

In terms of energy use, it is one of the most promising plants. Aside from its high photosynthetic efficiency, it also requires little or no fertilizer, is easy to adapt to a variety of climates and soils, and has a high surface efficiency. It takes 3–5 months for seedlings to grow in warm temperatures [88]. The high sugar content in the juices of sorghum makes it an attractive plant for bioethanol production since it can easily be fermented and converted to biofuel.

6.3 *Elaeis guineensis (Oil Palm)*

Asia (*particularly Malaysia*) is a particularly popular place to find this plant, but Europe is less common than those described above. South Africa is the home of this plant. Edible oil is manufactured from its fruits. It is now time for the palm to be exploited. After planting, it takes 3–12 years for the plant to reach its full potential. By the 25th year, the harvest has dropped to the lowest level. There are many tropical countries where oil palms can be grown, so they offer more potential for biomass production in tropical areas [89]. Cellulose, hemicelluloses, and lignin make up most of the oil palm biomass. An oily substance surrounds the seed of the palm fruit. As a result of the extraction process, oil is extracted from the fruit's cellulose mass. While seed oil can be used to make soap, the oil can also be eaten.

6.4 *The Waste Generated by Agriculture*

It is common for agricultural production processes to generate large amounts of useless biomass. Based on the agricultural branch from which the waste originated, agricultural wastes can be classified. Four categories of waste can be identified, including plantation waste, food processing waste, breeding waste, and slaughter waste [90]. The major sources of manure are sludge, whey, molasses, dairy production waste, straw, stems, post-slaughtering sediments, diluted manure, and slurry. Agricultural wastes are costly to collect and transport from the point of collection to their ultimate use and storage, which adds to the difficulties of economically viable energy production from agricultural waste. As a fuel source for combustion, vaporization, and fermentation, agricultural biomass can be directly utilized in the industrial sector. The vaporization process is extremely advantageous due to the scarcity of gas and liquid fuels. Warmer climates as well as temperate climates can benefit from agricultural biomass [91]. As one of the world's largest palm oil producers, Malaysia uses industrial waste to produce palm oil. In addition to processing waste into animal feed and fertilizer, attempts are being made to convert biofuel (*bioethanol*) into electric energy.

6.5 *Wastes of Food and Agriculture Industry*

Agricultural and food industry wastes can be used to generate renewable energy (peelings, seeds, mill cakes produced by juice and beer producers, distillery stillage, whey, mushroom substrates, fish waste, butcher waste, etc.) due to their rapid expansion. The University of Warmia and Mazury in Olsztyn has a Chair of the Foundation of Safety that is concerned about this issue [92]. The use of methane fermentation wastes from the production of sugar beet pulp is currently being researched. It takes years for beet pulp to devolve into a waste product, once considered a valuable resource. Willows that are devoted to energy are being fertilized with them.

6.6 *Methane Fermentation and Biogas from the Wastes*

An oxygen-less environment is ideal for fermentation in which microorganisms break down organic matter. Among the products of fermentation is biogas, which consists of primarily CO_2 and methane. Production yield was boosted through multi-substrate fermentation. Liquid and solid waste are often stabilized by fermentation methods [93]. The generation of biogas using agricultural waste, animal dung, and industrial and urban waste has been hailed as a possible alternative energy source by many countries. 2.1 kWh are produced by one cubic meter of biogas and 2.9 kWh are generated by one cubic meter of biogas. As early as the 1990s, biogas was recognized to have multiple benefits and could be produced using a variety of biomass sources. Manufacturing of this type also serves as a means of utilization. There are various ways to dispose of food industry waste, including sugar wastewater. In addition to hydrogen and methane, biogas also contains carbon dioxide [94]. The production of hydrogen is the subject of extensive research, and it is being described as the fuel of the future. Due to their high carbohydrate concentrations with comparatively low nitrogen content, the sugar industry and cellulose production wastes are among the sources of biohydrogen. It is also important to consider plant watering waste. Nitrogen levels decrease in plants, though wastes remain a rich source of hydrogen. Fermentation wastewater can be converted into hydrogen through an intriguing process. First, hydrogen is produced in the fermentation process, and then methane is produced in the second phase. Food industry wastes, which are more important for recycling can be used as substrates in some algae, like *Laminaria japonica* [95].

7 Sustainable Development

It is stated that sustainability is an activity that is capable of meeting current needs without jeopardizing the ability to meet the needs of future generations. The primary motivation for encouraging bioenergy research and development is the potential of bioenergy to replace conventional fuels, electricity, and heat sources as a more sustainable source. An aspect of sustainability that combines economics, environmental aspects, and social aspects is sustainability [96]. Assessments of sustainability are based on indicators and life-cycle assessments. Small-scale operations (for example, a farm or biorefinery) can be considered sustainable as well as regional, national, or global companies (for instance, soybean biodiesel or cellulosic ethanol made from mixed feedstock). On focusing on the single operation, the implications along the supply chain were considered as sustainability to the assessments. In applying land-use efficiency factors in the production of feedstock, spatial footprints can be used to determine the sustainability of biorefineries [97]. A system's ability to maintain itself over time, whether in the short term or the long term, is called its "*viability*". One of its most prominent features is the ability to remain profitable for long periods. In addition to economic factors, environmental considerations play an important role in viability. It is necessary to preserve soil quality to produce profitable feedstocks, and public support is necessary for bioenergy systems in general.

A system's viability is considered alongside sustainability, which measures how much it contributes to a larger system's capacity to meet future demands—a region, a country, or the entire world. For bioenergy systems to be sustainable, several factors have to be taken into consideration, such as the quality of the soil and quality of water, gas in the greenhouse (GHG) balance, the quality of air, productivity, and biodiversity. There are overlapping socio and economic concerns beyond profitability, and social acceptability, which include employment, growth, trade on an international level, energy security, and the resource occurring in nature accounting [98]. The viability of a bioenergy system can be evaluated through indicators. Statistics that measure human activities' potential or realized effects on the environment, society, or economy of concern are considered to be a sustainability indicator. These indicators may be based on management practices (such as fertilizer use) or their effects (such as soil or water nutrient levels). Sustainable biofuels, as well as the council on sustainable biomass production (CSBP), provide indicators based on management practices. Indicators of effect can be used to analyze the sustainability of a bioenergy business or pathway to provide empirical support for management-based indicators. Indicators must reflect the entire supply chain to the maximum extent possible. Identifying low-cost feedstock production locations and locations close to markets, these indications can help make decisions like picking a conversion technology [99–101].

8 Environment and Ecology

A food chain is composed of five levels of trophic status, which are the bulk or weight of living tissue in an environment:

- Photosynthesizing green plants are the producers
- Producers are eaten by herbivores

- Herbivores are eaten by predators
- Those who eat other predators are called top predators.

A pyramid-shaped distribution of biomass is observed in terrestrial ecosystems, with mass decreasing gradually from producers to apex predators. In the waterways the biomass was distributed in the shape of a dipyramid, which is narrowing from both the top and bottom, appearing green or blue-green, while others with a pyramid-shaped biomass distribution are clear [102]. As a result of biomass, the community's energy and resource distribution is reflected. Carbon is recycled from the atmosphere by plant biomass, which binds and emits CO_2 during growth. Since carbon dioxide is recycled into new growth by replanting biomass harvest, there was no increase in the carbon in the atmosphere. As "*old*" biomass is burned to produce fossil fuels, it is converted into "*new*" carbon dioxide, increasing the effect of greenhouse, and reducing non-renewable resources. To comply with environmental regulations such as the pure air act amendments in the USA, the biomass may be blended with the existing power plants like coal due to its low sulfur content [103]. Carbon dioxide and particulate emissions are significantly increased by large-scale biomass burning or clearing land with fires.

8.1 *Energy from Biomass*

Fuel energy derived from biological sources is called bioenergy. Plants, grass clippings, leaves, manure, rice hulls, peanut shells, sugarcane, corn hulls and shells, peanut shells, rice hulls, corn shells, corn cobs, and rice are examples of organic matter used as biomass energy sources. Biomass energy is still the dominant source of energy in poor countries, which is obtained primarily from wood, crop leftovers (called bagasse), and manure [104]. There are several woodstoves in these areas, for example, which provide heat for the homes and cooks. There is an estimated 2740 Quads worth of biomass produced worldwide each year [1 Quad = 1560 British thermal units (BTUs)], which is eight times the whole amount of the energy consumed locally (340 Quads). The biomass energy source is therefore an important renewable energy source [105]. With innovations in harvesting methods and processes of various techniques, short-rotation forestry, and much more efficient stoves and boilers, biomass may provide a huge amount of the world's total energy needs. The introduction of biomass power technology and new energy crops (see Economics, environmental) is expected to offer several economic benefits, particularly for rural populations [106].

8.2 Indicative Estimate

Inventory procedures are often used to gather elaborate details related to the resources of biomass. From above, this is extremely difficult to determine the amount of biomass. New remote sensing instruments, such as the sensor for imaging the vegetables using laser, can still provide insight into forest places and other vegetation production types in the three dimensions [107]. It calculates, for example, the height of a vertical column of vegetation and the number of branches and leaves. As trees and vegetation grow, above-ground biomass increases. For example, Saguaro cacti are directly influenced by their height, which is reflected in their biomass above ground. Biomass estimation is typically performed using regression methods. Regression techniques are used to determine an organism's weight or mass by selecting individuals based on an appropriate selection process [108]. By using computer-based image analysis as an explanation variable, a new method for predicting plant biomass can be developed. Studies show that predicted crown area alone is responsible for sapling mass and seeding trees there was a variation of up to 97%. An organism's growth or size can be measured and studied through cabinetry, which refers to a process of gathering data and generating biomass relationships. A fractal network theory of branching tubes is presented in, which encompasses allometric scaling laws related to multivariate allometry. In most cases, biomass data display heteroscedasticity, which means that their error variance does not remain the same over time. To provide accurate parameter estimates, weighted regression analysis is used. It is time-consuming to physically gather biomass data. Taking direct measurements may be too time-consuming and expensive for big creatures, like trees. A realistic replacement was sub-sampling (*see sampling on the environment*). It proposes two methods for selecting sample paths that will yield an unbiased estimation of wood biomass: randomized branch sampling and significance sampling [109].

There are two major periods in most biomass inventories. First, numerous organisms are selected for study and their traits are measured. Set area sample plots are typically used to group organisms into clusters. There is no measurement of biomass for these organisms. We measure biomass and properties proportional to the tiny organism's sample in the second phase as we did in the first phase. Based on the second-phase sample, the regression function is developed. A function (*or functions*) is used to estimate the average biomass per unit area for the first phase sample [110]. If biomass regressions have already been produced, no second-phase sample is needed. However, the formula used to generate the regression function is predicated on the basic assumption that the population being inventoried and the population used to generate the regression function are comparable. Combining the error from the first- and second-phase plots of the sample can be accomplished using Cunia's technique. To implement this approach, the estimators must be based on the following formula:

$$\hat{W} = b_1 z_1 + b_2 z_2 + \cdots + b_m z_m = b' z. \tag{9}$$

To calculated a sample plot of the vector's statistics information in the z, whereas b is the biomass regression function in the vector coefficient. The variance of $\hat{w}$ can be calculated in the following way

$$S_{\hat{w}\hat{w}} = b' S_{zz} b + z' S_{bb} z, \tag{10}$$

where S_{zz} and S_{bb} are the covariance matrices of z and b. The covariance matrices of z and b can be found by using S_{zz} and S_{bb}. A sample plot error variance component is represented by the first term and a biomass regression variance component is represented by the second term. A detailed review of aquatic microbial biomass is provided, as are details on a review of biomass.

9 Categorization of Wastes Arising from Agriculture and Industry

There are several organic wastes found in agro-industrial wastes (slurry from animal excreta, manures from farmyards, mushroom compost, soiled water, silage effluent, and so on) that, recycling can either be completed through integrated waste utilization or by reintroducing them to their natural habitats through the use of clean technology. Plants waste, animal waste, and natural trash are classified as three different types of waste. Specifically, the rubber, seafood, palm oil, and biochemical industries discuss the sources and challenges of agro-industrial wastes. Both physical and biological methods are usually available for recycling agro-industrial wastes since they usually are free of non-natural materials [111].

9.1 Component Characteristics of Agro Wastes

Biofuel production requires the modification of agricultural leftovers according to their unique properties, unlike wood, which has a limited range of properties that can influence their selection as raw materials. There are many characteristics associated with crop agro-residue, and they vary depending on the type of crop and the environment where the crop is planted. To utilize agro-leftovers for manufacturing various goods and developing various technologies, it is essential to understand and characterize their physical, thermal, and chemical properties. Agricultural waste can be used to manufacture a few items with just a few information requirements. Information such as, for example, must be gathered [112–116].

- Characteristics of agriculture-related waste, including bulk density, angle of response, value of moisture condition, and equilibrium in the moisture value (EMV).

- Temperatures for softening and fusion of ash, as well as high and low heating values.
- Determine the concentration of volatile matter and ash, as well as carbon fixation from the proximal analysis.
- A chemical analysis, such as measuring the agro-residue C, H, N, alkali metals, silica, and so on, to determine chemical compositions.
- Statistical analysis using thermogravimetric methods to determine weight loss rates based on temperature.

9.2 Utilization of Agro-waste Biomass

The composition of agricultural waste differs greatly in the form of lignin, cellulose, and hemicellulose. Following pretreatment, the three factors, along with the amount and variety of sugars produced, are the most important factors to consider. Additionally, sugar quality and quantity play a role in the kind of product that can produce [117]. A pretreatment with the alkaline solution can result in the synthesis of xylans from hemicellulose, for instance. Xylans can produce xylooligosaccharides, considered a prebiotic when enzymes (*Xylanases*) are involved. Through microorganisms, glucose can also be produced from agricultural leftovers after cellulose has been converted into ethanol. Combustion gasification, liquefaction, and pyrolysis can all be used to treat agro-waste bio-waste immediately after it is generated. As well as biological processing, microbial processing, and fermentation, chemical processing is also available [118]. Based on pretreatment methods, agro-waste biomass is the primary composition of cellulose (40–50%), hemicellulose (20–30%), and lignin (10–25%). These are then hydrolyzed by enzymatic means either using cellulose cellulases or hemicellulases to convert cellulose and hemicellulose into monomer sugars. During pretreatment, cellulose–hemicellulose–lignin networks are disrupted, increasing carbohydrates' accessibility for fermentation and hydrolysis. It is possible to improve the overall economics of agro-waste biomass by converting lignin into valuable products [119]. The depolymerization of the lignin or the hydrodeoxygenation of the lignin with hydrogen-donating solvents can be used as a gasoline additive or phenolic building block chemical. To produce biochemicals from agro-waste, the integration of thermochemical and biochemicals is essential. There are a few building block chemicals used in the synthesis of chemicals from organic matter and completed products in the industry of petrochemical, whereas biorefineries use biomass to produce similar types of building blocks for organic chemicals and polymers to meet societal demands [120, 121].

9.3 The Circular Bioeconomy of Agro-waste

Sustainable biorefineries can be created using agricultural waste, which is studied in the field of bioeconomy. In this process, forestry and agricultural waste are converted into energy and chemicals [122]. A significant step forward has been made in the management of carbon emissions and the migration of greenhouse gases by the transition from petroleum biorefineries to waste biorefineries. Throughout this chapter, a circular bioeconomy is presented, which involves recycling, reuse, and remanufacturing as the viable and the effectiveness of cost was alternated to disposing of waste. The economic circulation was related to the use of agro-waste, which is considered to be the biorefinery strength of fundamentals [123]. A continuous supply of raw materials throughout the year is ensured by the extensive availability of feedstock, thereby reducing total production costs. Instead of being used directly as fuel, the value-added products were produced from the by-products of biorefineries things. A waste biorefinery represents a cost-effective and progressive waste management system in addition to environmental and energy sustainability. It is possible to define circular agriculture as a system of using agricultural wastes for bio-derived products, such as biofuel or other fuel sources, in any country on Earth [124]. It is possible to propose the take-make-waste concept or approach to replace the existing linear economy based on the take-make-waste concept as shown in Fig. 4. Agricultural agriculture can reduce the number of external inputs through the generation of take-waste. Also, the eliminating discharge of wastewater from the source of water at the surface and also from the underground on the environment leads to the reduction of negative impact and closing nutrient loops, are all methods to produce agricultural products with no discharges. Most of India's crude petroleum oil comes from extremely expensive imports [124]. Aside from fossil fuels, vehicle, and traffic emissions can exacerbate air pollution during the day due to ever-increasing emissions from vehicles and traffic in urban areas, especially if fossil fuels are burned. The production and use of bioethanol produced from agricultural waste could impact air quality more than fossil fuel use in this context. Using biofuel along with fossil fuel has been found to reduce CO_2 emissions (up to 50% or more) and other harmful emissions around the world [125]. A 20% ethanol blend with conventional fuel has been implemented in many countries around the world, which can reduce carbon dioxide emissions by 10 million tons by 2022. In India or other developing countries, blending biofuels with fossil fuels makes the reduction of carbon dioxide emissions by 33% by 2030, it is compared with the current days or the level of previous years. Compressed natural gas (CNG) originated from methane gas (whose original volume is reduced to 1% or less during compression) is the primary component of compressed natural gas (CNG). In the current era, India consumes compressed natural gas (CNG) about 40 million metric tons (MMT). There is a compound annual growth rate (CAGR) of 15% projected for this amount of CNG. 65% of it is imported; the need for CNG is increasing daily (Fig. 4) [126, 127].

Compressed biogas (CBG) is now being promoted by the Indian government as an alternative to compressed natural gas (CNG). It reduces crude oil consumption

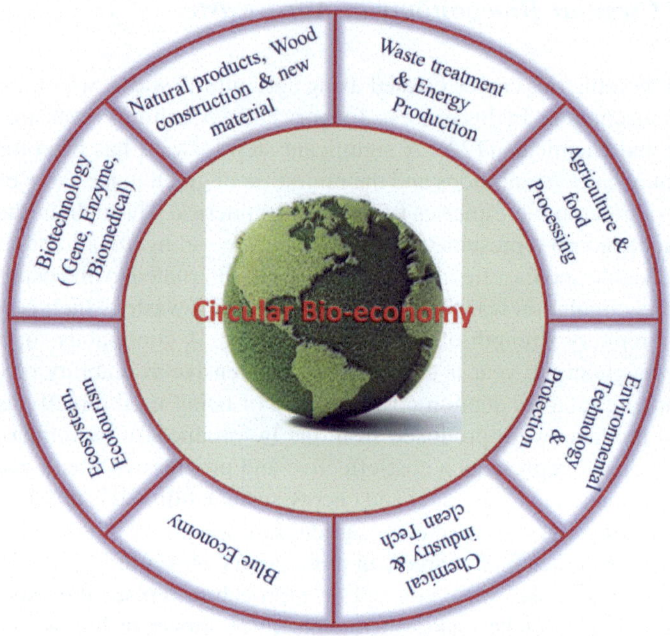

Fig. 4 Circular bioeconomy of bio wastes

and the price of fossil fuel imports (saving almost 15–20% overall) by partially replacing petrol and diesel with CBG fuel. A blend of ethanol with gasoline fuels or oils will save India about 12,000 crores on its oil import costs by quadrupling industrial bioethanol production. Using agricultural waste, as well as convincing locals, is the only way to promote biofuel production. Three important sectors are being supported by India's efforts to develop biofuels [59]. In addition, the pollution in the environment is reduced as well, resulting in huge savings in import costs. Additionally, this work can help farm communities increase employment opportunities and return on investment [128].

10 Summary and Conclusion

The whole chapter highlights the importance of biomass, its derivatives, and its application in energy production is portrayed. Apart from commercial energy sources of energy like fossil fuels, petroleum products, and nuclear power, energy from biomass and its derivatives seeks an interesting key feature shortly. Overconsumption of these commercial energy resources deploys the sustainability of the ecosystem and creates harmful effects and pollution to mankind and other organisms existing on the Earth. We need to practice alternative and reusable or renewable sources as the potential value of energy is the primary importance. The wastes generated in the environment become a hazardous problem due to the dumping of wastes. Newer technologies are utilized nowadays for the management of solid and liquid wastes and converting them into useful energy resources like biogas, biodiesel, ethanol, and other related liquid fuels will compensate for the energy need of human society in many sectors. Also, the utilization of wastes generated from biomass leads to a clean environment and supports the ecosystem to develop the sustainability of the natural system that is maintained. The whole chapter is mainly focusing on the concept of usage, recycling, and reuse of biomass resources into useful energy.

References

1. S.V. Vassilev, D.J. Baxter, L.K. Andersen, C.G. Vassileva, T.J. Morgan, Anoverview of the organic and inorganic phase composition of biomass. Fuel **94**, 133 (2012)
2. A. Demirbas, Use of algae as biofuel sources. Energy Convers. Manage. **51**(12), 2738–2749 (2010)
3. Y. Tamaki, G. Mazza, Measurement of structural carbohydrates, lignins, and micro-components of straw and shives: effects of extractives, particle size and crop species. Ind. Crops Prod. **31**(3), 534–541 (2010)
4. G.W. Huber, S. Iborra, A. Corma, Synthesis of transportation fuels from biomass: chemistry, catalysts, and engineering. Chem. Rev. **106**(9), 4044–4098 (2006)
5. P. Lauri, P. Havlík, G. Kindermann, N. Forsell, H. Böttcher, M. Obersteiner, Woody biomass energy potential in 2050. Energy Policy **66**, 19–31 (2014)
6. A. Dibenedetto, The potential of aquatic biomass for CO_2-enhanced fixation and energy production. Greenhouse Gases Sci. Technol. **1**(1), 58–71 (2011)
7. S.V. Vassilev, C.G. Vassileva, Composition, properties and challenges of algae biomass for biofuel application: an overview. Fuel **181**, 1–33 (2016)
8. V. Lebaka, Potential bioresources as future sources of biofuels production: an overview, in *Biofuel Technologies: Recent Developments*. ed. by V.K. Gupta, M.G. Tuohy (Springer, Berlin, 2013), pp. 223–258
9. M. Canakci, J.H. Van Gerpen, Biodiesel production via acid catalysis. Trans. ASAE **42**(5), 1203–1210 (1999)
10. N.D. Kaushika, K.S. Reddy, K. Kaushik, *Biomass Energy and Power Systems. Sustainable Energy and the Environment: A Clean Technology Approach*, vol. 9 (Springer, Cham, 2016), pp. 123–137
11. J. Chen, C. Li, Z. Ristovski, A. Milic, Y. Gu, M.S. Islam, S. Wang, J. Hao, H. Zhang, C. He, H. Guo, H. Fu, B. Miljevic, L. Morawska, P. Thai, Y.F. Lam, G. Pereira, A. Ding, X. Huang, U.C. Dumka, A review of biomass burning: emissions and impacts on air quality, health and climate in China. Sci. Total Environ. **579**, 1000–1034 (2017)
12. Y. Sun, J. Cheng, Hydrolysis of lignocellulosic materials for ethanol production: a review. Bioresour. Technol. **83**(1), 1–11 (2002)
13. M. Imai, K. Ikari, I. Suzuki, High-performance hydrolysis of cellulose using mixed cellulase species and ultrasonication pretreatment. Biochem. Eng. J. **17**(2), 79–83 (2004)
14. F. Teymouri, L. Laureano-Perez, H. Alizadeh, B.E. Dale, Optimization of the ammonia fiber explosion (AFEX) treatment parameters for enzymatic hydrolysis of corn stover. Bioresour. Technol. **96**(18), 2014–2018 (2005)
15. P. McKendry, Energy production from biomass (part 2): conversion technologies. Bioresour. Technol. **83**(1), 47–54 (2002)

16. M.R. Rahimpour, Z. Arab Aboosadi, A.H. Jahanmiri, Synthesis gas production in a novel hydrogen and oxygen perm-selective membranes tri-reformer for methanol production. J. Nat. Gas Sci. Eng. **9**, 149–159 (2012)
17. A. Molino, S. Chianese, D. Musmarra, Biomass gasification technology: the state of the art overview. J. Energy Chem. **25**(1), 10–25 (2016)
18. M. Akia, F. Yazdani, E. Motaee, D. Han, H. Arandiyan, A review on conversion of biomass to biofuel by nanocatalysts. Biofuel Res. J. **1**(1), 16–25 (2014)
19. J. Zhang, X. Weng, Y. Han, W. Li, Z. Gan, J. Gu, Effect of supercritical water on the stability and activity of alkaline carbonate catalysts in coal gasification. J. Energy Chem. **22**(3), 459–467 (2013)
20. D. López Barreiro, W. Prins, F. Ronsse, W. Brilman, Hydrothermal liquefaction (HTL) of microalgae for biofuel production: state of the art review and future prospects. Biomass Bioenergy **53**, 113–127 (2013)
21. S. Mahalaxmi, C. Williford, *Biochemical conversion of biomass to fuels. Handbook of climate change mitigation and adaptation* (Springer, Berlin, 2014), pp.965–999
22. A.R.K. Gollakota, N. Kishore, S. Gu, A review on hydrothermal liquefaction of biomass. Renew. Sust. Energy Rev. **81**, 1378–1392 (2018)
23. M.J. Taherzadeh, K. Karimi, Pretreatment of lignocellulosic wastes to improve ethanol and biogas production: a review. Int. J. Mol. Sci. **9**(9), 1621–1651 (2008)
24. H. Fukuda, A. Kondo, H. Noda, Biodiesel fuel production by trans esterification of oils. J. Biosci. Bioeng. **92**(5), 405–416 (2001)
25. D.Y.C. Leung, X. Wu, M.K.H. Leung, A review on biodiesel production using catalyzed trans esterification. Appl. Energy **87**(4), 1083–1095 (2010)
26. G. Habert, Y. Bouzidi, C. Chen, A. Jullien, Development of a depletion indicator for natural resources used in concrete. Resour. Conserv. Recycl. **54**(6), 364–376 (2010)
27. A. Jama, W. Nowak, (*Salix viminalis* L.) in purifying sewage sludge treated soils. Pol. J. Agron. **9**, 3–6 (2012)
28. M. Marmiroli, B.H. Robinson, B.E. Clothier, N.S. Bolan, N. Marmiroli, R. Schulin, Effect of dairy effluent on the biomass, transpiration, and elemental composition of *Salix kinuyanagi* kimura. Biomass Bioenergy **37**, 282–288 (2012)
29. C. Mant, J. Peterkin, E. May, J. Butler, A feasibility study of *Salix viminalis* gravel hydroponic system to renovate primary settled wastewater. Bioresour. Technol. **90**, 19–25 (2003)
30. L. Matsakas, P. Christakopoulos, Fermentation of liquefacted hydrothermally pretreated sweet sorghum bagasse to ethanol at high-solids content. Bioresour. Technol. **127**, 202–208 (2013)
31. L. Wang, Z. Luo, A. Shabazi, Optimization of simultaneous saccharification and fermentation for the production of ethanol from sweet sorghum (*Sorghum bicolor*) bagasse using response surface methodology. Indus. Crops Prod. **42**, 280–291 (2013)
32. T.L. Kelly Yong, K.T. Lee, A.R. Mohamad, S. Bhatia, Potential of hydrogen from oil palm biomass as a source of renewable energy worldwide. Energy Policy. **35**, 5692–5701 (2007)
33. S.H. Shuit, K.T. Tan, K.T. Lee, A.H. Kamarruddin, Oil palm biomass as a sustainable energy source: a Malaysian case study. Energy **34**, 1225–1235 (2009)
34. A. Demirbas, Importance of biomass energy sources for Turkey. Energy Policy **36**, 834–842 (2007)
35. S.N. Dodic, T. Zelenovic Vasiljevic, R.M. Maric, A.J.R. Kosanovic, J.M. Dodic, S.D. Popov, Possibilities of application of waste wood biomass as an energy source in Vojvodina. Renew. Sustain. Energy Rev. **16**, 2355–2360 (2012)
36. E.J. Olguin, H.W. Doelle, G. Mercado, Resource recovery through recycling of sugar processing by-products and residuals. Resour. Conserv. Recycl. **15**, 85–94 (1995)
37. G. Lustrato, G. Alfano, G. Ranalli, Bio-hydrogen and bio-methane co-production by sequential two-phases dark fermentation from agro-industrial wastes (IMERA). Environ. Eng. Manag. J. **11**, S76 (2012)
38. K. Zieminski, M. Frac, Methane fermentation process as an anaerobic digestion of biomass: transformations, stages and microorganisms. Afr. J. Biotechnol. **11**(18), 4127–4139 (2012)

39. V.N. Gunaseelan, Anaerobic digestion of biomass for methane production: a review. Biomass Bioenergy **13**, 83–114 (1997)
40. I.K. Kapdan, F. Kargi, Bio-hydrogen production from waste materials. Enzyme Microb. Technol. **38**, 569–582 (2006)
41. K.W. Jung, D.H. Kim, H.S. Shin, Continuous fermentative hydrogen and methane production from *Laminaria japonica* using a two-stage fermentation system with recycling of methane fermented effluent. Int. J. Hydrogen Energy **37**, 15648–15657 (2012)
42. S.H. Kim, H.C. Cheon, C.Y. Lee, Enhancement of hydrogen production by recycling of meth-anogenic effluent in two-phase fermentation of food waste. Int. J. Hydrogen Energy **37**, 13777–13782 (2012)
43. T. Sayara, R. Basheer-Salimia, F. Hawamde, A. Sánchez, Recycling of organic wastes through composting: process performance and compost application in agriculture. Agronomy **10**, 1838 (2020)
44. M. Patchaye, B. Sundarkrishnan, S. Tamilselvan, N. Sakthivel, *Microbial Management of Organic Waste in Agro Ecosystem: Microorganisms for Green Revolution* (Springer, Singapore, 2018), pp. 45–73
45. M. Balat, H. Balat, C. Oz, Progress in bioethanol processing. Prog. Energy Combust Sci. **34**, 551–573 (2008)
46. U. De Corato, I. De Bari, E. Viola, M. Pugliese, Assessing the main opportunities of integrated bio refining from agro-bioenergy co/by-products and agro industrial residues into high-value added products associated to some emerging markets: a review. Renew. Sustain. Energy Rev. **88**, 326–346 (2018)
47. S. Jayakumar, M.M. Yusoff, M.H. Rahim, G.P. Maniam, N. Govindan, The prospect of microalgal biodiesel using agro-industrial and industrial wastes in Malaysia. Renew. Sustain. Energ. Rev. **72**, 33–47 (2017)
48. R. Agrawal, A. Verma, R.R. Singhania, S. Varjani, C. Di Dong, A.K. Patel, Current understanding of the inhibition factors and their mechanism of action for the lignocellulosic biomass hydrolysis. Bioresour. Technol. **27**, 125042 (2021)
49. M.K. Islam, H. Wang, S. Rehman, C. Dong, H.Y. Hsu, C.S.K. Lin, S.Y. Leu, Sustainability metrics of pretreatment processes in a waste derived lignocellulosic biomass biorefinery. Biores. Technol. **298**, 122558 (2020)
50. H.H. Ibrahim, P.E. Bilsborrow, A.N. Phan, Intensification of pre-treatment and fractionation of agricultural residues. Chem. Eng. Process **159**, 108231 (2021)
51. H.N. Özbek, D. Koçak Yanık, S. Fadıloğlu, F. Göğüş, Effect of microwave-assisted alkali pre-treatment on fractionation of pistachio shell and enzymatic hydrolysis of cellulose-rich residues. J. Chem. Technol. Biotechnol. **96**, 521–531 (2020)
52. D. Halda, M.K. Purkait, Lignocellulosic conversion into value-added products: a review. Process Biochem. **89**, 110–133 (2020)
53. V.I. Popa, Biomass for fuels and biomaterials, in *Biomass as Renewable Raw Material to Obtain Bio Products of High-Tech Value.* (Elsevier, New York, 2018), pp. 1–37
54. Z. Luo, Q. Qian, H. Sun, Q. Wei, J. Zhou, K. Wang, Lignin-first bio refinery for converting lignocellulosic biomass into fuels and chemicals. Energies **16**, 125 (2023)
55. S. Quereshi, T.K. Naiya, A. Mandal, S. Dutta, Residual sugarcane bagasse conversion in India: current status, technologies, and policies. Biomass Convers. Biorefinery **12**, 3687–3709 (2020)
56. S. Takkellapati, T. Li, M.A. Gonzalez, An overview of biorefinery-derived platform chemicals from a cellulose and hemicellulose biorefinery. Clean. Technol. Environ. Policy **20**, 1615–1630 (2018)
57. G. Fiorentino, M. Ripa, S. Ulgiati, Chemicals from biomass: technological versus environmental feasibility. A review. Biofuel Bioprod. Biorefinery **11**, 195–214 (2017)
58. H.Y. Leong, C.K. Chang, K.S. Khoo, K.W. Chew, S.R. Chia, J.W. Lim, J.-S. Chang, P. Show, Waste biorefinery towards a sustainable circular bioeconomy: a solution to global issues. Biotechnol. Biofuels **14**, 1–5 (2021)
59. A.T. Ubando, C.B. Felix, W.H. Chen, Biorefineries in circular bioeconomy: a comprehensive review. Bioresour. Technol. **299**, 122585 (2020)

60. M.S. Venkata, S.K. Butti, K. Amulya, S. Dahiya, J.A. Modestra, Waste biorefnery: a new paradigm for a sustainable bioelectro economy. Trends Biotechnol. **34**, 852–855 (2016)
61. K.M. Zech, K. Meisel, A. Brosowski, L.V. Toft, F. Müller-Langer, Environmental and economic assessment of the Inbicon lignocellulosic ethanol technology. Appl. Energy **171**, 347–356 (2016)
62. A. Roussos, N. Misailidis, A. Koulouris, F. Zimbardi, D. Petrides, A feasibility study of cellulosic isobutanol production—process simulation and economic analysis. Processes **7**, 667 (2019)
63. V. Venkatramanan, S. Shah, A.K. Rai, R. Prasad, Nexus between crop residue burning, bioeconomy and sustainable development goals over North-Western India. Front. Energy Res. **8**, 614212 (2020)
64. T. Ronzon, A.I. Sanjuán, Friends or foes? A compatibility assessment of bioeconomy-related sustainable development goals for European policy coherence. J. Clean Prod. **254**, 119832 (2020)
65. N.R. Baral, A. Shah, Comparative techno-economic analysis of steam explosion, dilute sulfuric acid, ammonia fiber explosion and biological pretreatments of corn stover. Bioresour. Technol. **232**, 331–343 (2017)
66. R. Banerjee, S.P.J. Kumar, N. Mehendale, S. Sevda, V.K. Garlapati, Intervention of microfluidics in biofuel and bioenergy sectors: technological considerations and future prospects. Renew. Sustain. Energy Rev. **101**, 548–558 (2019)
67. R.A. Sheldon, Green chemistry, catalysis and valorization of waste biomass. J. Mol. Catal. A Chem. **422**, 3–12 (2016)
68. K. Kohli, R. Prajapati, B.K. Sharma, Bio-based chemicals from renewable biomass for integrated bio refneries. Energies **12**, 1–40 (2019)
69. S.P.J. Kumar, V.K. Garlapati, P.J.K. Samudrala, A. Dash, P. Scholz, R. Banerjee, Sustainable green solvents and techniques for lipid extraction from microalgae: a review. Algal Res. **21**, 138–147 (2017)
70. G. Kaur, A.K. Srivastava, S. Chand, Advances in biotechnological production of 1,3-propanediol. Biochem. Eng. J. **64**, 106–118 (2012)
71. S. Roy, D. Banerjee, M. Dutta, D. Das, Metabolically redirected bio-hydrogen pathway integrated with bio methanation for improved gaseous energy recovery. Fuel **158**, 471–478 (2015)
72. A. Rao, A. Sathiavelu, S. Mythili, Genetic engineering in bio butanol production and tolerance. Braz. Arch. Biol. Technol. **59**, 1–14 (2016)
73. M. Hans, S. Kumar, A.K. Chandel, I. Polikarpov, A review on bioprocessing of paddy straw to ethanol using simultaneous saccharification and fermentation. Process Biochem. **85**, 125–134 (2019)
74. H. Mattila, J. Kuuskeri, T. Lundell, Single-step, single organism bioethanol production and bioconversion of lignocellulose waste materials by phlebioid fungal species. Bioresour. Technol. **225**, 254–261 (2017)
75. S. Shrestha, X. Fonoll, S. Kumar Khanal, L. Raskin, Biological strategies for enhanced hydrolysis of lignocellulosic biomass during anaerobic digestion: current status and future perspectives. Bioresour. Technol. **245**, 1245–1257 (2017)
76. S.P.J. Kumar, R. Banerjee, Enhanced lipid extraction from oleaginous yeast biomass using ultrasound assisted extraction: a greener and scalable process. Ultrason. Sonochem. **52**, 25–32 (2019)
77. H.K. Sharma, C. Xu, W. Qin, Biological pretreatment of lignocellulosic biomass for biofuels and bioproducts: an overview. Waste Biomass Valoriz. **10**, 235–251 (2019)
78. S. Saha, M. Arshad, M. Zubair, A. Ullah, Keratin as a biopolymer: extraction from waste biomass and applications, in *Springer Series on Polymer and Composite Materials*. ed. by S. Sharma, A. Kumar (Springer, Cham, 2019), pp. 163–185
79. M. Yamabhai, S. Sak-Ubol, W. Srila, D. Haltrich, Mannan biotechnology: from biofuels to health. Crit. Rev. Biotechnol. **36**, 32–42 (2016)

80. S.M. Wunderlich, N.M. Martinez, Conserving natural resources through food loss reduction: production and consumption stages of the food supply chain. Int. Soil Water Conserv. Res. **6**, 331–339 (2018)
81. A. Sridhar, A. Kapoor, P.S. Kumar, M. Ponnuchamy, S. Balasubramanian, S. Prabhakar, Conversion of food waste to energy: a focus on sustainability and life cycle assessment. Fuel **302**, 121069 (2021)
82. S. Dahiya, A.N. Kumar, J. Shanthi Sravan, S. Chatterjee, O. Sarkar, S.V. Mohan, Food waste bio refinery: sustainable strategy for circular bioeconomy. Bioresour. Technol. **248**(2018), 2–12 (2018)
83. M. Perwez, R. Ahmad, M. Sardar, A reusable multipurpose magnetic nano biocatalyst for industrial applications. Int. J. Biol. Macromol. **103**, 16–24 (2017)
84. K. Periyasamy, L. Santhalembi, G. Mortha, M. Aurousseau, A. Boyer, S. Subramanian, Bioconversion of lignocellulosic biomass to fermentable sugars by immobilized magnetic cellulolytic enzyme cocktails. Langmuir **43**, 6546–6555 (2018)
85. G. Baskar, R. Naveen Kumar, X. Heronimus Melvin, R. Aiswarya, S. Soumya, Sesbania aculeata biomass hydrolysis using magnetic nano biocomposite of cellulase for bioethanol production. Renew. Energy **98**, 23–28 (2016)
86. A.M. Abdalla, S. Hossain, O.B. Nisfindy, A.T. Azad, M. Dawood, A.K. Azad, Hydrogen production, storage, transportation and key challenges with applications: a review. Energy Convers Manag. **165**, 602–627 (2018)
87. M. Aziz, T. Oda, T. Kashiwagi, Advanced energy harvesting from macroalgae-innovative integration of drying, gasification and combined cycle. Energies **7**, 8217–8235 (2014)
88. J.C. Santamarta, F. Jarabo, J. Rodríguez-Martín, M. Paz Arraiza, J.V. López, Analysis and potential of use of biomass energy in Canary Islands, Spain. IERI Proc. **8**, 136–141 (2014)
89. E. Toklu, Biomass energy potential and utilization in Turkey. Renew. Energy **107**, 235–244 (2017)
90. A. Paiano, G. Lagioia, Energy potential from residual biomass towards meeting the EU renewable energyand climate targets. Ital. Case Energy Policy. **91**, 61–173 (2016)
91. T. Masseck, R. Valkenburg, E. Ouden, Roadmaps for energy (R4E): a systemic approach to the renewable energy transition of smart communities. Energy Proc. **115**, 90–396 (2017)
92. A. Darmawan, M.W. Ajiwibowo, K. Yoshikawa, M. Aziz, K. Tokimatsu, Energy-efficient recovery of black liquor through gasification and syngas chemical looping. Appl. Energy **219**, 290–298 (2018)
93. S. Sharma, S. Basu, N.P. Shetti, T.M. Aminabhavi, Waste-to energy nexus for circular economy and environmental protection: recent trends in hydrogen energy. Sci. Total Environ. **713**, 136633 (2020)
94. T. Lepage, M. Kammoun, Q. Schmetz, A. Richel, Biomass-to hydrogen: a review of main routes production, processes evaluation and techno-economical assessment. Biomass Bioenergy **144**, 105920 (2021)
95. Y.-C. Chen, Effects of urbanization on municipal solid waste composition. Waste Manag. **79**, 828–836 (2018)
96. T.V. Ramachandra, H.A. Bharath, G. Kulkarni, S.S. Han, Municipal solid waste: generation, composition and GHG emissions in Bangalore, India. Renew. Sustain. Energy Rev. **82**, 1122–1136 (2018)
97. S. Jayalakshmi, K. Joseph, V. Sukumaran, Bio hydrogen generation from kitchen waste in an inclined plug flow reactor. Int. J. Hydrogen Energy **34**, 8854–8858 (2009)
98. M.H. Liang, J.G. Jiang, Advancing oleaginous microorganisms to produce lipid via metabolic engineering technology. Prog. Lipid Res. **52**, 395–408 (2013)
99. A. Banerjee, C. Banerjee, S. Negi, J.S. Chang, P. Shukla, Improvements in algal lipid production: a systems biology and gene editing approach. Crit. Rev. Biotechnol. **38**, 369–385 (2018)
100. A. Nawaz, F. Aamir, R. Huang, I.U. Haq, F. Wu, M. Munir, R. Chaudhary, A. Rafique, K. Jiang, Co-production of biohydrogen and biomethane utilizing halophytic biomass *Atriplex crassifolia* by two-stage anaerobic fermentation process. Front. Chem. **11**, 1233494 (2023)

101. C. Ferroni, G. Varchi, Keratin-based nanoparticles as drug delivery carriers. Appl. Sci. **11**, 9417 (2021)
102. M.O. Andreae, P.J. Crutzen, Atmospheric aerosols: biogeochemical sources and role in atmospheric chemistry. Science **276**, 1052–1058 (1997)
103. J. Schmidt, V. Gass, E. Schmid, Land use changes, greenhouse gas emissions and fossil fuel substitution of biofuels compared to bioelectricity production for electric cars in Austria. Biomass Bioenergy **35**, 4060–4074 (2011)
104. S.H. Lamlom, R.A. Savidge, A reassessment of carbon content in wood: variation within and between 41 North American species. Biomass Bioenergy **25**, 381–388 (2003)
105. P.R. Adler, S.J.D. Grosso, W.J. Parton, Life-cycle assessment of net greenhouse gas flux for bioenergy cropping systems. Ecol. Appl. **17**, 675–691 (2007)
106. B. Fischer, A. Pigneri, Potential for electrification from biomass gasification in Vanuatu. Energy **36**(2011), 1640–1651 (2011)
107. J.P. Lange, Lignocellulose conversion: an introduction to chemistry, process and economics. Biofuels Bioprod. Bioref. **1**, 39–48 (2007)
108. P.R. Adler, M.A. Sanderson, A.A. Boateng, P.J. Weimer, H.-J.G. Jung, Biomass yield and biofuel quality of switch grass harvested in fall or spring. Agron. J. **98**, 1518–1525 (2006)
109. L.G. Angelini, L. Ceccarini, N. Nassi o Di Nasso, E. Bonari, Long-term evaluation of biomass production and quality of two cardoon (*Cynara cardunculus* L.) cultivars for energy use. Biomass Bioenergy **33**, 810–816 (2009)
110. T. Beringer, W. Lucht, S. Schaphoff, Bioenergy production potential of global biomass plantations under environmental and agricultural constraints. GCB Bioenergy **3**, 299–312 (2011)
111. A.V. Bridgwater, Review of fast pyrolysis of biomass and product upgrading. Biomass Bioenergy **38**, 68–94 (2012)
112. P. Parthasarathy, K.S. Narayanan, Hydrogen production from steam gasification of biomass: influence of process parameters on hydrogen yield. A review. Renew. Energ. **66**, 570–579 (2014)
113. J. He, W. Zhang, Techno-economic evaluation of thermo-chemical biomass-to ethanol. Appl. Energy **88**, 1224–1232 (2012)
114. X. Cai, X. Zhang, D. Wang, Land availability for biofuel production. Environ. Sci. Technol. **45**, 334–339 (2010)
115. S.C. Davis, F.G. Dohleman, S.P. Long, The global potential for Agave as a biofuel feedstock. GCB Bioenergy **3**, 68–78 (2011)
116. L.L. Escamilla-Trevino, Potential of plants from the genus Agave as bioenergy crops. BioEnerg Res. **5**, 1–9 (2012)
117. J.E. Fargione, R.J. Plevin, J.D. Hill, The ecological impact of biofuels. Ann. Rev. Ecol. Evol. Syst. **41**, 351–377 (2010)
118. M. Patel, X. Zhang, A. Kumar, Techno-economic and life cycle assessment on lignocellulosic biomass thermochemical conversion technologies: a review. Renew. Sust. Energ. Rev. **53**, 1486–1499 (2016)
119. J. Rogers, J. Brammer, Estimation of the production cost of fast pyrolysis bio-oil. Biomass Bioenergy **36**, 208–217 (2012)
120. S. Shabangu, D. Woolf, E.M. Fisher, L.T. Angenent, J. Lehmann, Techno economic assessment of biomass slow pyrolysis into different bio char and methanol concepts. Fuel **117**, 42–748 (2014)
121. C. Somerville, H. Youngs, C. Taylor, S.C. Davis, S.P. Long, Feedstocks for lignocellulosic biofuels. Science **329**, 790–792 (2010)
122. K.P. Vogel, J.J. Brejda, D.T. Walters, D.R. Buxton, Switchgrass biomass production in the Midwest USA. Agron. J. **94**, 413–420 (2002)
123. R. Dungani, M. Karina, A. Sulaeman, D. Hermawan, A. Hadiyane, Agricultural waste fibers towards sustainability and advanced utilization: a review. Asian J. Plant Sci. **15**, 42–55 (2016)
124. D. Xiao, Z. Yu, S. Qing, S. Du, H. Xiao, Development of agricultural waste/recycled plastic/waste oil bio-composite wallpaper based on two-phase dye and liquefaction filling technology. Environ. Sci. Pollut. Res. **27**, 2599–2621 (2020)

125. V. Venkatramanan, S. Shah, R. Prasad, Sustainable bioeconomy: pathways to sustainable development goals. Singapore Nat. **10**, 978–981 (2021)
126. M.S. Venkata, P. Chiranjeevi, S. Dahiya, N. Kumar, Waste derived bioeconomy in India: a perspective. New Biotechnol. **40**, 60–69 (2018)
127. S.B. Ummalyma, D. Sahoo, A. Pandey, Resource recovery through bioremediation of wastewaters and waste carbon by microalgae: a circular bioeconomy approach. Environ. Sci. Pollut. Res. **28**, 58837–58856 (2021)
128. F. Savin, The economy that runs on waste: accumulation in the circular city. J. Environ. Policy Plan **21**, 675–691 (2019)

Goal 15: Life on Land

Overview of the chapters:

Goal 15: Life on Land

(Biological diversity must be conserved, restored, and used sustainably through agricultural practices)

It is estimated that the world's population will reach nearly ~9 billion by 2050, in the current scenario. A great deal of energy consumption occurs due to the overconsumption of food, power, and everything else. The people's need and demand are changing, and we are able to meet all of our needs both at a low cost and of high quality. Due to various factors, including biotic and abiotic stresses, climate change across the globe, and increasing population in a country, essential needs have been impacted by agricultural productivity in a country. Recently, nanotechnology has been increasingly applied to farming, particularly in the area of genetically modified crops. Our agricultural sector, farms, trees, and blooms can be protected and developed more effectively by researchers who developed fresh ideas. As far as promoting farming is concerned, it is favorable compared to its objectives. The chapter discusses the detrimental effects of seed germination, plant growth, and crop yield on crops. Population growth leads to a greater demand for agriculture-related products in the world as a result of the existing situation. Food products and other essentials require an interesting factor due to the overconsumption of natural resources. Due to its ability to reach the targeted site easily, it is widely used by researchers today. Food production and quality are improved using nanotechnology in agriculture. Pesticides, growth factors, and herbicides are emitted by nanotechnology to increase crop production. It has been possible to develop pesticides using synthetic polymers and silica nanoparticles. Consequently, the soil is not damaged while crops are produced in high yields. Aside from preventing bacterial and fungal infections, it can also prevent the growth of weeds. An overview of nanomaterials and nanotechnology's application in improving crop production will be presented in this chapter.

Utilization of Nanoparticles in Crop Phytotechnology and Pharmaceuticals

Meenakshi Kaniyur Chandrasekaran, Manikandan Vani Raju, Rathi Muthaiyan Ahalliya, Meenakshi Sundari Rajendran, and Valan Arasu Mariadhas

Abstract Nanotechnology is a fast-developing area with huge potential in many human-serving industries. Nanomaterials created and produced using the aid of nanoscale have a wide range of uses including farming and health care. The utilization of nanomaterials in agriculture helps in managing the climatic conditions. Nanofertilizers and nanopesticides support in improving the yield of crops and they are safe. Considering a focus on a range of food, agricultural, and ecological issues, nanosensors are precise, efficient, and economical. The application of cutting-edge nanotechnological techniques in crop phytonanotechnology is crucial for ensuring more secure and consistent cultivation of food. In the field of medicine, nanotechnology helps in diagnosing and treating many diseases. By using nanovaccine, bacterial, viral infections, and malaria can be treated. Compared with other kinds of nanoparticles, the nanosponges are harmless and sustainable at higher degrees. By altering treatment intake, lowering digestion, extending their biological lifespan, and lowering toxic effects, nanoliposomes are being utilized to increase the therapeutic value of new or old pharmaceuticals. Hence, the usage of nanotechnology in agriculture helps in increasing the yield of crops, whereas in medicine it helps to treat various diseases.

1 Introduction

In recent days, nanotechnology is the most advanced technology in medicine. It is used in the diagnosis and treatment of many diseases. It also has an important application in the field of molecular imaging and drug delivery systems. Nowadays, it has been used by many researchers as it can reach the targeted site easily [1]. Nanotechnology is used in agriculture to improve the production and quality of foods. By using nanotechnology, they can increase the production of crops by ejecting pesticides,

M. K. Chandrasekaran · M. V. Raju · R. M. Ahalliya (✉) · M. S. Rajendran
Department of Biochemistry, Karpagam Academy of Higher Education, Coimbatore, Tamil Nadu 641021, India
e-mail: rathi.muthaiyan@kahedu.edu.in

V. A. Mariadhas
Department of Botany and Microbiology, King Saud University, Riyadh, Saudi Arabia

K. Kasinathan et al. (eds.), *New Technologies for Energy Transition Based on Sustainable Development Goals*, https://doi.org/10.1007/978-981-97-2527-4_23

growth factors, and herbicides. Polymeric and silica-based nanoparticles have been used in the development of pesticides. This results in a high yield of crops without causing any damage to the soil. Metal nanoparticles like zinc oxide, copper oxide, and titanium oxide can be used in the prevention of pesticides. It can also prevent the plants from bacterial and fungal infections [2]. Pesticides can be categorized depending on their chemistry, which includes inorganic as well as organic elements. These consist of manmade steroids, antimicrobial agents, carbamates, formamidines, thiocyanates, organotins, the organochlorines, phosphates of carbamates, dinitrophenol, and phosphates. Chemical properties, among other things, play a role in influencing the quantity of herbicides in the ground after treatment [3]. The incorporation of pesticide in methods like encapsulate and restricted release mechanisms is an important matter. Numerous studies have emphasized the value of nanomaterials and tiny capsules for use in agriculture. Insect particles with sizes from 200 to 400 can dissolve equally in aqueous or fatty acids to create tiny particles. These oils may be merely incorporated into gels that contain lotions, and fluids in order to accomplish numerous applications, such as safeguarding against damage, therapy, or eliminating harvested material. The use of materials combined with silica constitutes a component of the latest desirable controlled liberation of farming nutrients [4].

Carbon nanotubes, lipids, and polymers are the mainly used nanomaterials in agriculture. Generally, plants require both micro and macro nutrients for their growth. They can obtain these nutrients from fertilizers. Biofertilizers can be used instead of chemical fertilizers because biofertilizers are safe for the environment and soil. Nanofertilizers can also be used in the growth of plants. These nanofertilizers can be categorized into micro- and macronutrient nanofertilizers. When these nanofertilizers are incorporated with biofertilizers, it is known as nano–biofertilizer [5]. Nanotechnology has been used in seed germination to increase the yield of crops [6]. Nanofertilizers can increase the quality of nutrition in plants and decrease the stress caused by climatic changes. Even the use of a little amount of nanofertilizers can increase crop yield. They can also help in the growth and germination of seeds and increase nitrogen metabolism [7]. Pesticides distribution can be managed with the use of nanopesticides, which also allow for greater impacts to be produced with a smaller pharmacological dose. The main ingredient for the initial class of nanopesticides, which include small metal particles like aluminum and zinc, can be either synthetic nanotubes or natural nanostructures. The development and utilization of nanopesticides attempts to increase a pesticide's effectiveness and longevity while also lowering the concentration in active chemicals and minimizing or eliminating any possible hazards [8].

Elevating the utilization of independent nanosensors placed around fields of crops together with a global positioning system for exact time as well as thorough crop surveys are two of the primary duties of nanotech gadgets are advancement that leads to improved agriculture practices by providing excellent and crucial data, such as the ideal time for establishing and harvesting crops. These devices are the main steps toward the complete computerization of the crop and report the ambient temperature once every minute [9]. Because of the ability to raise the yield of crops and enhance sensitivity to both biological and synthetic problems by turning metallic into green

NPs using extracts from plants, phytonanotechnology is becoming more and more popular.

Nanomedicine has a wide range of applications in healthcare. It helps in the diagnosis, treatment, and prevention of many diseases. The field of medicine has a lot of promise for using nanotechnologies, including in imaging and diagnostic instruments, systems for delivering drugs, tissue-engineered designs, implants, and pharmaceutical treatments. It also has recently developed treatments for diseases like cancer, viral infections, diabetes, neurological disorders, and cardiac diseases [10]. Compared to conventional medications that affect the whole organism, nanovaccines are a novel class of vaccinations made of nanomaterials that concentrate on the place in the human body where an illness or disease first manifested itself. The concept that numerous elements that determine vaccine efficacy can be logically built, optimized independently, and integrated into one mechanism that induces an effective immune response serves as the basis for using nanoconstructs as vaccine vehicles. Individual components can communicate in a complementing or competitive manner, but ultimately they will function together to influence the intended antibodies [11]. In addition to these newly developed tiny materials, nanosponges have been investigated for specific and persistent drug delivery for cancer treatment due to their adequate permeation, digestion, bio compatibility, accessibility, and durability. These have demonstrated desirable qualities that make them appropriate for biological applications, including a high degree of biocompatibility, decomposition, and little cellular toxicity. In the case of nanosponges, the medication and cyclodextrin component create an incorporation compound. In contrast, the medication component is absorbed or retained within the permeable nanomaterials of non-inclusive compound nanosponges. Cyclodextrin-based nanostructures are being extensively investigated for use in pharmaceutical delivery, cleanup, monitoring, and reactions within nanosponges [12]. Whenever an energized substance less than a tiny pores penetrates an opening in a delicate membrane order, it can cause an alteration that can be detected using nanopore technologies. Consequently, small amounts of protein, RNA, DNA, etc. may be detected and analyzed using nanopore technique. Introducing the charges of a cathode and an anode to the fluid on the leading and backward sides of the cell membrane, correspondingly, produces nanopore technology [13].

Nanoliposomes are the recent advances in the field of drug delivery. Nowadays, the usage of nanoliposomes is common because it has high biocompatibility and it can control the circulation of drugs through the bloodstream [14]. A particle core, a biocompatible exterior layer, and a connecting molecule for improved bioactivity are necessary components for nanoparticles utilized for drug delivery. Prior to drug delivery, some of the nanovectors were altered. This alteration includes covering of ligands through antibodies, folic acids, and peptides. Ligands bind with nanoparticles so that it can determine the target site easily and improve specificity [15]. The application of nanotechnology in agriculture includes, nanofungicides, managing the plant virus and nanoinsecticide [16]. In the field of medicine, nanotechnology can be used in treating wounds, diagnosing diseases, drug delivery and enhancing the effectiveness of medicine [17]. In this chapter, the use of nanotechnology in medicine and agriculture has been discussed. Figure 1 implies that nanotechnology has various

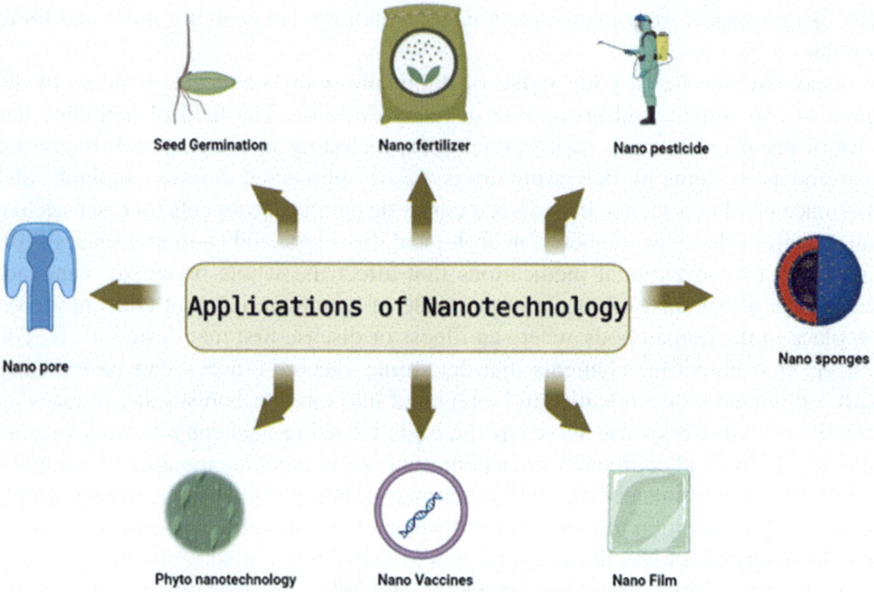

Fig. 1 Application of nanotechnology in agriculture and medicine

applications in the field of agriculture and medicine. Nanofertilizers, nanopesticides, seed germination has major importance in the crop yielding and development of plants. Nanovaccines, nanosponges, nanopores, and nanofilms play a major role in the treatment and diagnosis of many diseases.

2 Nanotechnology in Agriculture

Nowadays, agriculture mainly depends on pesticides and fertilizers. This may affect the growth of agriculture. The usage of nanotechnology in agriculture can increase the advancement and help in the regeneration of crops [18].

2.1 Seed Germination

The biological process of seed germination is complex and important, and it depends on the soil's properties, genetic makeup, and environmental conditions. Recent studies show that the metal nanoparticles like Zinc oxide, gold and Silicon dioxide have higher activity in seed germination [19]. The study of [20] reveals that nanoparticles like ZnO, SiO_2, and TiO_2 have a major effect in seed germination activity. By improving the ability of seeds to absorb water, nonmetallic NPs like multi-walled carbon nanotubes (MWCNTs) may promote seed germination in several crops. Seed germination is associated with antioxidant enzyme activities along with water and oxygen content or consumption rates [21].

2.2 *Nanofertilizers*

Nanofertilizers are nanomaterials that function as transporters/additives to provide vital nutrients or they can act as nutrients themselves. Nanofertilizers can enhance the yield of crops by producing a large amount of nutrients and support in the development of agriculture [22]. They are nonpoisonous and less hazardous to both people and the surroundings than usual fertilizers. Biofertilizers were used in the development of productivity of soil in the presence of microorganisms by producing growth factors, adjusting ambient nitrogen and dissolving phosphorus. The mixture of nano-based structures and biofertilizers are known as nanobiofertilizers [23]. Nutrients are slowly delivered using nanofertilizers, which may help to enhance the effectiveness of nutrient utilization without any correlated side effects. These nanofertilizers are made to give nutrients gradually over an extended amount of time and to significantly limit nutrient loss, assuring environmental safety. Micronutrients are components that are required by the plant in detecting/low amounts. They are important for the plants to regulate their metabolic activities. Boron, Zinc, Iron, manganese are some of the commonly used metallic nanoparticles in the production of micronutrient nanofertilizers. The macronutrient nanofertilizers are made up of more than one macronutrients that have been enclosed in a specific nanomaterial. A particular amount of macronutrients, like K, Mg, Ca, S, P, and N were incorporated with nanomaterials for their delivery to crops [24].

2.3 *Nanopesticides*

The determination of nanosized biological pesticides must take into account a number of crucial factors, including their ease of preparation, lack of toxicity, effectiveness toward an extensive variety of pests, economic viability, safety for non-target organisms and the environment, lack of accumulation in the food chain, fragrance, lack of residues, lack of impact on food quality, texture, and flavor, and ease of application. Similar to this, in order to create nano–bio-pesticides, silver salt is combined with biological molecules that have pesticidal capabilities as sealing inhibitory agents [25]. Nanoformulations for pesticide products provide various advantages, including enhanced duration, solubility, and kinesis; lesser effective components; minimal damage to organisms that are not targeted, lowering resistance development; and shielding toward early degradation. Polymers, which are strongly bound together by covalent bonds, have become the major components of nanoparticles and nanocapsule production, making them stable and more resistant than liposomes. Chitosan is widely used in agriculture for its antimicrobial activity against a variety of pathogens. It has antibacterial and insecticidal effects along with triggering defense mechanisms in plants [26]. It is not enough to analyze the particles and establish their positive interaction with the active substance found in the pesticide or another medication in nanoparticulate systems. To promote balanced and ecological growth of the atmosphere, it is also necessary to examine the harmful effects for both target and non-target creatures. A toxicology profile is required whenever a therapeutic agent or herbicide is introduced to the sector [27].

2.4 Nanosensors

In particular, a light sensor is made up of a recognizing element that is targeted for the individual persistent pesticidal component and may communicate with its additional component, the receiver, which is used to provide the output for the attachment of specific contaminants to the monitor. The recognition components, which include host recognizers, antibodies, enzymes, and molecularly imprinted polymers are capturing the focus of researchers as a means of increasing the diagnostic efficacy of any sensor [28]. Due to their diagnostic actions, immunochromatographic techniques have also been described for their role in tracking agricultural systems. The incorporation of nanomaterials into microbes causes the formation of clusters that surround the microbe as a result of magnetic forces between the negatively charged portions of the cell walls of bacteria and the cationic component of (CTBr) cetyltrimethylammonium bromide [29]. Groundwater content, quality of the soil, and nutritional status of crops are being tracked through monitoring devices for agricultural precision. Additionally, detectors for insecticides, weedkillers, and repellents in the fluids or soil have been created; they are helpful for boosting harvesting and promoting the safety of both food as well as water. These devices detect the soil's environment, humidity, and nutritional content, offering crucial data for managing crops [30].

2.5 Phytonanotechnology

Plants deal with high temperatures since their proliferation, growth, and outcome are all affected. High temperatures, elevated salt content, flooding, and other changes in the environment with high levels of contaminants are regarded to be some of the primary impacts of global climate change on crop maturation and growth. The enhanced adaptation of crops requires a multifaceted strategy, including hormone management, crop metabolic system stimulation, stress gene activity, prevention of water deficiency stress, and regulation of contaminants transport and absorption. The incorporation of nanomaterials to the progress of plants is essentially determined by the dose, material, quantity, and physical and chemical attributes of NMs

[31]. The contamination of the soil has been increased over the past years by sudden growth and commercialization. Phytonanotechnology is one of the best strategies for resolving or purifying harmful pollutants such as poisonous heavy metals (HMs) through several channels, such as reducing the excessive amount of reactive oxygen species and cellular damage produced by HMs, decreasing the levels in crop residues, and repressing heaviest elements-triggered production associated with the metal transporter-associated genes in crops for consumption.

Although excessive usage of nanoparticles, including fertile soil and herbicides, may contaminate ecosystems (soil and aquatic) by releasing from cultivation or treatments. Excessive concentrations of nanoparticles have a major effect on gene activity and may trigger oxidative stress, which leads to rupture of membranes, breakdown of electrolytes, and reduced luminescence supply in crops [32]. Furthermore, NPs treatment improved crop resistance to frost stress by preserving a level of carotenoid and chlorophyll accumulations, stimulating ascorbate peroxidase and catalase, and increasing gene production of pigment- and Rubisco-binding proteins. Due to their small size and increased surface contact area, nanoliposomes have been approved for specific delivery of antimicrobial drug nutrients, vitamins and minerals, and vitamins. Several factors, including bacterial interface pH, heat, divalent ion and microbial barrier properties, have been discovered to affect the fusing of liposomes with spores. The lipid organization of the microbial membrane is crucial for the interaction of liposomes with the bacterium. The application of nanotechnology to nutrition conservation has shown positive results in reducing spoilage. The post-harvest storage of foods derived from crops has made extensive use of nanoliposomes, coatings that are consumable, and additional varieties of nanoparticles [33]. Nanocomposites were commonly employed for nutrition processing as well as conservation because of their fungicidal or microbiological characteristics. The antibacterial property possessed by Ag nanoparticles incorporates being significantly utilized for subsequent harvest sanitation due to its destruction of product deterioration microbes. Silver nanoparticle is an excellent choice for lateral retention partly because of its distinctive characteristics, including low volatility, stability at high temperatures, and toxicity against numerous types of microorganisms [34].

2.6 Antimicrobial Polymers

Polymer interfaces that are in contact with surroundings are also prone to contamination by microorganisms carried by the wind or found in the ground. Additionally, conditions like humidity and temperature encourage the growth of microfilms, which are challenging to remove. Maintaining cleanliness for such surfaces is the optimal way to avoid fungal infections, but for long-term uses, more measures are required to make this achievable. Those methods typically use toxic, unwanted chemicals such hypochlorite, oxidized water, and compounds that produce reactive oxygen species.

The administration of particular tiny biocidal compounds additionally shows excellent efficacy in eliminating bacteria, but significant concerns exist over their instability and accumulating harm. Additionally, all of these therapies require consistent dosing to preserve their effectiveness, which also aids in the emergence of resistance to them [35].

3 Nanotechnology in Medicine

In recent days, nanotechnology has been used in the field of medicine to treat and diagnose many diseases. Apart from this, nanotechnology is very important in the drug delivery system as it can target the diseased tissues. Natural nanoparticles are made from nucleic acids, polymers, carbohydrates, lipids, and proteins that can be produced synthetically or biologically. Due to their chemical nature, they are normally harmless and exhibit little immunological reactions [36].

3.1 Nanovaccines

Many advantages provided by nanotechnology can be used to improve the efficacy of vaccination compositions. The capacity to distribute immunostimulatory adjuvants along with antigenic components is one benefit of formulations based on nanoparticles. A nanoparticulate material could include adjuvant substances directly. Because of their distinct physicochemical properties, such as biomimicking property, size and shape, nanovaccines are excellent vaccine candidates. Due to this, they have a flexible delivery mechanism for antigen and adjuvants [37]. It is feasible to rapidly resist immunity due to the presence of numerous proteins found in cell membranes and antibodies on the outermost layer of nanovaccines. Nanovaccines can be used in the treatment of bacterial infection, HIV, malaria, foot, and viral diseases [38]. Liposomes, virosomes, dendrimers, and micelles can be used as nanovaccines to conquer the authentic vaccines. Nanovaccines have the potential to be helpful in the development of immuno-therapeutic compositions to treat human tumors. Nanocarriers are the most significant component of nanovaccines because they transport subunit vaccinations comprising adjuvants and antigens for distribution. Various nanocarriers have been utilized for immunotherapy and cancer vaccinations in which they can be classified into synthetic, biogenic and semi-synthetic nanocarriers.

Table 1 shows that VLP can act against viral infections and vault protein can produce flagellin scaffold. Whereas Au NP that are coated in bacterial membranes can increase the immunity against bacterial infections and Avian retrovirus can help in intracellular delivery of protein [38]. In recent times, tissue-generated small molecules featuring the capability to trigger reliable cytotoxic responses have

Table 1 Components of nanovaccines and its application

S. No.	Nanovaccine components	Application
1	VLP that are modified genetically	Protecting against anti-viral infection
2	Hollow vault protein	Flagellin scaffold
3	Membrane of bacteria coated with Au nanoparticles	To produce antibacterial immunity
4	Avian retrovirus	Intracellular protein delivery

obtained extensive consideration in tumor therapies. Nanoparticle-based immunizations were demonstrated to provide various advantages to component silver nanoparticles. These are made through simply incorporating silver inside the nanoparticles or coating the outermost layer of the nanoparticles with Ag. Because nanoparticles defend against Ag disintegration and cause prolonged Ag absorption and prolonged absorption, it can elicit stronger immune reactions than dissolving Ag-based immunizations.

3.2 *Nanosponges*

Cyclodextrin-based nanosponges with distinct qualities like biological compatibility, porosity architectures, controlled-release conduct, and improved absorption are being researched as effective transporters of drugs and therapeutic substances in the management of a variety of disorders particularly tumor and cancer. The number of linking agents used can alter the permeability and dimension of nanosponges; usually, rising the volume of a linker utilized leads to nanosponges with lesser diameters and larger permeation. Such nanosystems were resistive to synthetic solvents and have a stable temperature (possibly 300 °C), therefore being ideally suited for an array of nanoformulations [39]. The consistency of materials (like unsteady substances) and the soluble nature of therapeutics and drug compounds can both be enhanced by using cyclodextrin tiny particles with variable lipid-based pores and a hydraulic matrix. In addition, nanosponges with advantages for specific absorption were developed to increase the dissolution of docetaxel in water-based environments [40]. Versatile auto-catabolizing DNA enzyme nanosponges were created exhibiting suppression of genes with regulation and focused treatment distribution. For specific administration into tumor cells, the created DNAzyme nanosponges might be loaded with diverse sequential substance configurations. Auto-assembled nanosponges encoded with DNA containing diverse cancer cell-fastening compounds have been developed for facilitating the highly effective diffusion of tumor-explicit drugs. As the outcome of the combined administration of both doxorubicin and oligonucleotides with antisense targeting miRNA-21, they induced combinatorial cytotoxic therapy. DNA nanosponges might be capable to enhance anticancer treatment by altering the cellular apoptosis-related aspect of genes that they cause. Adaptive resilience and biological compatibility have been taken into consideration when designing active DNA small particles for customizable cancers and DNAzyme-based regulation of genes with high effectiveness precise delivery [41]. Ultrasonic-mediated dispersion solvent distillation approach was used to create ethyl cellulose nanosponges for the intended administration of withaferin-A with cytotoxic features. With a higher rate of cell death compared to the unbound ribociclib, this nanosystem demonstrated strong toxicity over the cell lines of breast cancer MCF-7 and MDA-MB-231, providing potential substrates for localized drug administration for aggressive tumors [42].

3.3 Nanopores

The primary natural micro pores tend to produce excellent outcomes due to their simple construction, a lot of adaptability, and reproducibility, which enable consistent present monitoring. To properly discover particular molecules inside an incredibly heterogeneous or essentially complicated liquid is a few of the current goals of the nanopore group. They indicated how gene alteration may end in accurate molecule localization and showed that just one molecule can be detected among a diverse collection of molecules for additional study of their biologic carriers [43]. Nanopore innovation facilitates the detection of further gene targets, keeping on top of the development of new pathogens. Through improvement of the interface orientation inside the nanopore container, they increased foundation sensing by increasing data collecting speeds. Physical and clinical imaging focus greatly on relationships between drugs and gene sequences. Plenty of the present methods are unreliable or entail molecular labeling, which can occasionally cause a biological molecule's original configuration to be disrupted. A label-free technique with excellent specificity at femtomolar amounts of the chemicals is possible with nanopore monitoring. The characterization of specific RNA and pharmaceutical interactions utilizing organic or artificial pores has formed the subject to multiple investigations [44]. There is a huge opportunity for metallic nanopores to be used as individual chemical biosensors. The inclusion of metallized nanopores treated with homocysteine was successful in sensing sugar [45]. While producing nanoparticles or nanoporous particles that carry medications for lung carcinoma therapy, knowledge of hydrodynamic features such as dimensions, form, and the charge on their surfaces is important. Artificial nanoporous compounds are a distinct family of nanoporous substances which indicate extremely flexible modalities of drug administration with the ability to combine detecting and medicinal qualities. In combination with lipid and polymer related nanoparticles, they may assist to tackle pharmaceutical delivery problems due to their durability, distinctive chemical and physical qualities, adjustable exterior features, flexible artificial techniques, and crystalline structure with the least amount of immediate or long-term adverse reactions. Additionally, when compared with biological nanoparticles, the development of permeability and fine-tuning of the pore capacity is significantly simpler with artificial particles [46]. Massive macromolecules like protein, siRNA, mRNA, or DNA are difficult to package in non-permeable or microporous drug transport vehicles since they are huge compounds. These substances are typically incorporated or affixed to surfaces using non-permeable transporters as coating modifiers. Enzymes that exist in the living surroundings may readily dissolve these corresponding compounds. MSNs found with substantial porosity are useful for enclosing and transporting macromolecules which are resistant to cell walls, especially those used in the treatment of cancer. For a more efficient and better treatment with radiation, permeable particles of platinum are being manufactured. Because of its great functionality, affordable price, and fewer side effects with little immunity, natural substances tend to be utilized in pharmaceutical delivery systems more than inorganic particles This section reviews the most presently studied permeable biological particles and their current advancement [47].

3.4 Nanoliposomes

In addition to surgical treatment for removing the tumor, chemotherapy, radiotherapy, vaccinations, and endocrine and specialized medicines are also alternatives for cancer. The majority of people receiving treatment possess operation along with either radiation treatments or medications. Chemotherapy is the most prevalent aspect of therapy. Actually, the majority of standard chemotherapy has an insufficient ability to treat cancer, whether due to their struggle to reach appropriate doses in the tissues being targeted or due to their cytotoxicity and toxicity on healthy cells and systems. The majority of chemotherapy drugs work by inhibiting cell replication in order to stop malignant cells from dividing uncontrollably. Because normal cells remain constantly growing, antineoplastics cannot distinguish across benign and cancerous cells, which might have significant adverse reactions throughout chemotherapy. Among the essential features that set tumor tissue apart is the requirement for capillary activation in order for tissues to develop quickly and excessively. This leads to defective design and the formation of an extremely permeable network ranges from 10 to 800 nm. The increased penetration and adhesion response (EPR) is a special pathology that results from inadequate lymphatic circulation. In addition, as a result of inadequate lymphatic flow, tiny size systems penetrate the tumor cell through the holes and passively diffuse to the tumor area. In the tumor as in comparison to normal tissue, levels of this particular type of transport can be as much as ten times greater in this manner. Certain cell surface receptor subtypes are frequently overexpressed by tumor cells, which supports their unchecked proliferation. Because this substance (peptide, protein, or antibody) detects the binding site, the structure of the nanosystem can be manipulated on its outermost layer or designed to promote selective development in tumor tissue [48]. In addition to its ecological and scientific benefits, nanoliposomes are now regarded as a highly efficient drug delivery technology. A number of clinical investigations demonstrate that nanoliposomes are excellent choices for a variety of forms of delivery, including for the administration of painkillers and anti-allergic pharmaceuticals as well as cytotoxic, antimicrobial, and antibiotic agents. Benefits of nanoliposome-based compositions over non–nanoliposome-based formulations for drug delivery via by mouth, topical, and injectable routes. To boost their mode of practice, nanoliposomes are additionally coupled to different therapeutic approaches. The volume of substance confined is lower than that needed lacking protection as a result of greater rigidity and precision. While utilizing premium pharmacological chemicals, this might be beneficial [49]. By increasing pharmacological drugs' dispersion and accessibility, laboratory, and animal strength, and capacity to avoid collisions with various particles, nanoliposomes can improve the effectiveness of therapeutic agents. Additionally, nanoliposomes can inhibit the dissolution of an isolated medication, allowing for prolonged interaction with the area where it acts and a better outcome. Typically, lipophilic medications are included in the layer of phospholipid while medication that is hydrophilic can be absorbed in the watery segment. As a means to ensure just the infected cells are impacted by the treatment, aggressively targeting liposomes travel straight to the specified body parts or tissues and distribute the medication for an extensive amount of duration [50]. Usage of liposomes in chemotherapy in the HCC tumor habitat contains a wide variety of cells, notably lymphoid fibroblasts, vascular endothelial HCC cells, and other stem cells. Both the growth of tumors and how well they respond to anticancer therapy are significantly influenced by these cells and outside networks that generate. Considering the key element of HCC, cells with this condition overstate a variety of proteins and cellular receptors, including ASGP-R, GA-R, and the transferrin receptor. This accumulation separates cells with HCC from different kinds of tissue and may function as an objective for liposomes utilized in cures for HCC.

3.5 Nanofilms

The biggest issue in studying biomaterials may be managing biological function. Communication with cells is known to be affected by a variety of material characteristics, including isoelectric point, hydrophobic nature, terrain and physical rigidity. Furthermore, organisms may respond to biologically active substances offered by an element. To create the best material for a particular biological reaction every single one of these qualities ought to be separately controllable. Biological activity and structural stiffness are frequently challenging to separate in practice. A major instance arises from nanofilm biomaterials developed via the LbL technique, where stiffness is often regulated by chemically binding of the polymerization matrix after post-assembly, and biological action is typically provided by macromolecules that are incorporated in the coating or that are absorbed by the outer layer. Biomolecule loading is typically restricted to the film interface whenever linkage occurs before bioactivation, and implanted proteins can grow unavailable to interacting fibroblasts [51]. Technologies for biological imaging are being enhanced by applying nanoscale film-like particles. Nanotechnology is a perfect contender for tumor imaging due to their highly tiny size particularly given the likelihood that they will collect in tumor sites. High-contrast pictures are produced because of nanoparticle's greater laser performance contrasted to that achieved by organic dyes. Finally, nanomaterials can cling to tumor cells because of their large surface-area-to-volume proportion, which makes it simpler to identify and imaging them. Research to this point has demonstrated that nanoscale tiny films might be utilized to enhance ultrasound procedures, PET, and fluorescence detection methods. Experiments have demonstrated that carbon oxide-containing coatings helpful for biological visualization can be created using the successively (LbL) approach. A track-etched (TE) membrane with high hole sizes has a preferentially compacted extracellular matrix (ECM) nanofilm on it. Through the use of scanning electron microscopy (SEM), the film was able to visualize connections between cells and the transfer of individual cells through the outermost layer of membrane while also providing an environment for the study of diverse intracellular interactions. Finally, the various studies suggest the prospect of creating versatile nanotubes that operate as theranostic agents as well as improving clinical imaging. The use of one compound as a marker that glows and a medicinal ingredient is demonstrated in this paper. The development of nanoparticle thin films for that possible multifunctional usage may be the subject of additional study [52].

4 Conclusion

From the last few years, the application of nanotechnology has developed significantly, particularly in the area of the farming industry. Researchers created fresh ideas for enhancing the protection of our agricultural sectors, farms, trees, and blossoms while also maximizing their development. It is favorable compared to the goal it set out to achieve, in order to promote farming. Improving the dissolution, incorporation, accessibility, and controlled release of therapeutics was a major motivation for the utilization of nanotechnology in medicine. Although the development of pharmaceuticals is associated with uncertainty and the search for substances with pharmacological activity in nature is less popular currently than it was a half-century ago, it has become routine to use nanotechnology to increase the potency of known typical bioactive molecules. Thus, the nanotechnology and medicine advancements, improvements in illness diagnostics, as well as the capacity to combine medical care and diagnosis are becoming realities.

References

1. S. Bayda, M. Adeel, T. Tuccinardi, M. Cordani, F. Rizzolio, The history of nanoscience and nanotechnology: from chemical-physical applications to nanomedicine. Molecules **25**(1), 112 (2019)
2. X. He, H. Deng, H.-M. Hwang, The current application of nanotechnology in food and agriculture. J. Food Drug Anal. **27**(1), 1–21 (2019)
3. I. Mubeen, M.F.B. Mfarrej, Z. Razaq, S. Iqbal, S.A.H. Naqvi, F. Hakim, W.F.A. Mosa, M.M.A. Moustafa, Y. Fang, B. Li, Nanopesticides in comparison with agrochemicals: outlook and future prospects for sustainable agriculture. Plant Physiol. Biochem.Biochem. **198**, 107670 (2023)
4. M. Asif, S. Islam, M.A. Malik, Z.M. Dar, A. Masood, S. Shafi, B. Rashid, S. Sidique, *Nano Pesticides Application in Agriculture and their Impact on Environment* (IntechOpen eBooks, 2021)
5. S. Arora, G. Murmu, K. Mukherjee, S.K. Saha, D. Maity, A comprehensive overview of nanotechnology in sustainable agriculture. J. Biotechnol.Biotechnol. **355**, 21–41 (2022)
6. M.A. Khan, T. Khan, Z.-U.-R. Mashwani, M. Riaz, N. Ullah, H. Ali, A. Nadhman, Plant cell nanomaterials interaction: growth, physiology and secondary metabolism, in *Comprehensive Analytical Chemistry* (2019), pp. 23–54
7. M.T. El-Saadony, A.S. ALmoshadak, M.E. Shafi, N.M. Albaqami, A.M. Saad, A.M. El-Tahan, E.-S.M. Desoky, A.S.M. Elnahal, A. Almakas, T.A.A. El-Mageed, A.E. Taha, A.S. Elrys, A.M. Helmy, Vital roles of sustainable nano-fertilizers in improving plant quality and quantity-an updated review. Saud. J. Biol. Sci. **28**(12), 7349–7359 (2021)
8. A. Bratovcic, W.M. Hikal, H.A.H.S.-A. Ahl, K. Tkachenko, R.S. Baeshen, A. Sabra, H. Sany, Nanopesticides and nanofertilizers and agricultural development: scopes, advances and applications. Open J. Ecol. **11**(04), 301–316 (2021)
9. M. Kaushal, S.P. Wani, *Nanosensors: Frontiers in Precision Agriculture* (Springer eBooks, 2017), pp. 279–291
10. S. Sim, N.K. Wong, Nanotechnology and its use in imaging and drug delivery (review). Biomed. Rep. **14**(5) (2021)
11. M. Agarwal, Role of nanovaccine in immunotherapy. J. Cell Sci. Ther. (2015)
12. S. Iravani, R.S. Varma, Nanosponges for drug delivery and cancer therapy: recent advances. Nanomaterials **12**(14), 2440 (2022)
13. B. Lin, J. Hui, H. Mao, Nanopore technology and its applications in gene sequencing. Biosensors **11**(7), 214 (2021)
14. A. Hatami, A. Heydarinasab, A. Akbarzadehkhiyavi, F.P. Shariati, An introduction to nanotechnology and drug delivery. Chem. Methodol. **5**(2), 153–165 (2021)
15. S. Anjum, S. Ishaque, H. Fatima, W. Farooq, C. Hano, B.H. Abbasi, I. Anjum, Emerging applications of nanotechnology in healthcare systems: grand challenges and perspectives. Pharmaceuticals **14**(8), 707 (2021)
16. A.Y. Ghidan, T.M.A. Antary, *Applications of Nanotechnology in Agriculture* (IntechOpen eBooks, 2020)
17. A. Haleem, M. Javaid, R.P. Singh, S. Rab, R. Suman, Applications of nanotechnology in medical field: a brief review. Glob. Health J., 2414–6447 (2023)
18. I. Sanzari, A. Leone, A. Ambrosone, Nanotechnology in plant Science: to make a long story short. Front. Bioeng. Biotechnol. **7** (2019)
19. N. Maroufpoor, M. Mousavi, M. Hatami, A. Rasoulnia, B.A. Lajayer, Mechanisms involved in stimulatory and toxicity effects of nanomaterials on seed germination and early seedling growth, in *Advances in Phytonanotechnology*. ed. by M. Ghorbanpour, S.H. Wani (Academic Press, Cambridge, MA, 2019), pp.153–181
20. D.K. Verma, S. Patel, K.S. Kushwah, Effects of nanoparticles on seed germination, growth, phytotoxicity and crop improvement. Agric. Rev. **1**, 1–11 (2020)
21. C. Liu, H. Zhou, J. Zhou, The applications of nanotechnology in crop production. Molecules **26**(23), 7070 (2021)

22. M. Kah, R.S. Kookana, A. Gogos, T.D. Bucheli, A critical evaluation of nanopesticides and nanofertilizers against their conventional analogues. Nat. Nanotechnol.Nanotechnol. **13**(8), 677–684 (2018)
23. R. Kumari, D.P. Singh, Nano-biofertilizer: an emerging eco-friendly approach for sustainable agriculture. Proc. Nat. Acad. Sci. India Sect. B Biol. Sci. **90**, 733–741 (2020)
24. A. Nongbet, A.K. Mishra, Y.K. Mohanta, S. Mahanta, M.K. Ray, M. Khan, K.-H. Baek, I. Chakrabartty, Nanofertilizers: a smart and sustainable attribute to modern agriculture. Plants **11**(19), 2587 (2022)
25. B. Deka, A. Babu, C. Baruah, M. Barthakur, Nanopesticides: a systematic review of their prospects with special reference to tea pest management. Front. Nutrit. **8** (2021)
26. R. Kumar, N. Kumar, V.D. Rajput, S. Mandzhieva, T. Minkina, B.S. Saharan, D. Kumar, P.K. Sadh, J.S. Duhan, Advances in biopolymeric nanopesticides: a new eco-friendly/eco-protective perspective in precision agriculture. Nanomaterials **12**(22), 3964 (2022)
27. D. Abdollahdokht, G. Asadikaram, M. Abolhassani, H. Pourghadamyari, M. Abbasi-Jorjandi, S. Faramarz, M.H. Nematollahi, Pesticide exposure and related health problems among farmworkers' children: a case-control study in southeast Iran. Environ. Sci. Pollut. Res.Pollut. Res. **28**(40), 57216–57231 (2021)
28. X. Yan, H. Li, X. Su, Review of optical sensors for pesticides. Trends Anal. Chem. **103**, 1–20 (2018)
29. P. Sharma, V. Pandey, M.M.M. Sharma, A. Patra, B.M. Singh, S. Mehta, A. Husen, A review on biosensors and nanosensors application in agroecosystems. Nanoscale Res. Lett. **16**(1) (2021)
30. E. Voke, R.L. Pinals, N.S. Goh, M.P. Landry, In planta nanosensors: understanding biocorona formation for functional design. ACS Sensors **6**(8), 2802–2814 (2021)
31. Y. Ding, Y. Shi, S. Yang, Molecular regulation of plant responses to environmental temperatures. Mol. Plant **13**(4), 544–564 (2020)
32. V. Dappe, S. Dumez, F. Bernard, B. Hanoune, D. Cuny, C. Dumat, S. Sobanska, The role of epicuticular waxes on foliar metal transfer and phytotoxicity in edible vegetables: case of *Brassica oleracea* species exposed to manufactured particles. Environ. Sci. Pollut. Res.Pollut. Res. **26**(20), 20092–20106 (2018)
33. D. Bhuyan, G.W. Greene, R.K. Das, Prospects and application of nanobiotechnology in food preservation: molecular perspectives. Crit. Rev. Biotechnol.Biotechnol. **39**(6), 759–778 (2019)
34. M. Jiang, Y. Song, M.K. Kanwar, G.J. Ahammed, S. Shao, J. Zhou, Phytonanotechnology applications in modern agriculture. J. Nanobiotechnol. **19**(1) (2021)
35. M. Alvarez-Paino, A. Muñoz-Bonilla, M. Fernández-García, Antimicrobial polymers in the nano-world. Nanomaterials **7**(2), 48 (2017)
36. S. Sindhwani, W.C.W. Chan, Nanotechnology for modern medicine: next step towards clinical translation. J. Intern. Med. **290**(3), 486–498 (2021)
37. L. Deng, T. Mohan, T.Z. Chang, G.X. Gonzalez, Y. Wang, Y.-M. Kwon, S.-M. Kang, R.W. Compans, J.A. Champion, B.-Z. Wang, Double-layered protein nanoparticles induce broad protection against divergent influenza A viruses. Nat. Commun. **9**(1) (2018)
38. V. Vijayan, A. Mohapatra, S. Uthaman, I.-K. Park, Recent advances in nanovaccines using biomimetic immunomodulatory materials. Pharmaceutics **11**(10), 534 (2019)
39. D. Lembo, F. Trotta, R. Cavalli, Cyclodextrin-based nanosponges as vehicles for antiviral drugs: challenges and perspectives. Nanomedicine **13**(5), 477–480 (2018)
40. S.S.B. Rizvi, N. Akhtar, M.U. Minhas, A. Mahmood, K.U. Khan, Synthesis and characterization of carboxymethyl chitosan nanosponges with cyclodextrin blends for drug solubility improvement. Gels **8**(1), 55 (2022)
41. D. Luo, X. Lin, Y. Zhao, J. Hu, F. Mo, G. Song, Z. Zou, F. Wang, X. Liu, A dynamic DNA nanosponge for triggered amplification of gene-photodynamic modulation. Chem. Sci. **13**(18), 5155–5163 (2022)
42. M.M. Ahmed, F. Fatima, A.S. Alali, M.A. Kalam, K. Alhazzani, S. Bhatia, S. Alshehri, M.M. Ghoneim, Ribociclib-loaded ethylcellulose-based nanosponges: formulation, physicochemical characterization, and cytotoxic potential against breast cancer. Adsorpt. Sci. Technol.. Sci. Technol. **2022**, 1–11 (2022)

43. Z. Fang, L. Liu, Y. Wang, D. Xi, S. Zhang, Unambiguous discrimination of multiple protein biomarkers by nanopore sensing with double-stranded DNA-based probes. Anal. Chem. **92**(2), 1730–1737 (2019)
44. K.J. Freedman, L.M. Otto, A.P. Ivanov, A. Barik, S.H. Oh, J.B. Edel, Nanopore sensing at ultra-low concentrations using single-molecule dielectrophoretic trapping. Nat. Commun. **7**(1) (2016)
45. L.S. Lastra, V. Sharma, N. Farajpour, M. Nguyen, K.J. Freedman, Nanodiagnostics: a review of the medical capabilities of nanopores. Nanomed. Nanotechnol. Biol. Med. **37**, 102425 (2021)
46. D. Radhakrishnan, S. Mohanan, G. Choi, J.-H. Choy, S. Tiburcius, H.T. Trinh, S. Bolan, N.M. Verrills, P.S. Tanwar, A. Karakoti, A. Vinu, The emergence of nanoporous materials in lung cancer therapy. Sci. Technol. Adv. Mater. **23**(1), 225–274 (2022)
47. J. Kashima, R. Kitadai, Y. Okuma, Molecular and morphological profiling of lung cancer: a foundation for "next-generation" pathologists and oncologists. Cancers **11**(5), 599 (2019)
48. R. Gilabert-Oriol, G. Ryan, A.W.Y. Leung, N. Firmino, K.L. Bennewith, M.B. Bally, Liposomal formulations to modulate the tumour microenvironment and antitumour immune response. Int. J. Mol. Sci. **19**(10), 2922 (2018)
49. K.M. Aguilar-Pérez, J.I. Avilés-Castrillo, D.I. Medina, R. Parra-Saldívar, H.M.N. Iqbal, Insight into nanoliposomes as smart nanocarriers for greening the twenty-first century biomedical settings. Front. Bioeng. Biotechnol. **8** (2020)
50. S. Nomani, J. Govinda, Nanoliposome: an alternative approach for drug delivery system. Int. J. Adv. Pharm. Med. Bioallied Sci. **2016**, 1–10 (2016)
51. P.R. Van Tassel, Nanotechnology in medicine: nanofilm biomaterials. Yale J. Biol. Med. **86**(4), 527–536 (2013)
52. B. Choi, J.-W. Choi, H. Jin, H.-R. Sim, J.H. Park, T.-E. Park, J.H. Kang, Condensed ECM-based nanofilms on highly permeable PET membranes for robust cell-to-cell communications with improved optical clarity. Biofabrication **13**(4), 045020 (2021)

Nanotechnology for Improved Productivity of Agricultural Products

R. Ramesh, A. Lavanya, A. Periya Nayagi Shilpa, K. Rafath Fathima, and F. Liakath Ali Khan

Abstract In the chapter, the adverse effects of nanomaterials, nanotechnology in seed germination, plant growth, crop yield, and increased crop production are focused. The existing situation in the world is the rapid growth of population leads to the demand for agriculture-related products. Due to the overconsumption of natural resources, food products, and other essential items needs an interesting factor. Internal and external factors like water, salinity, drought, temperature, mineral and salt stress, nutrition availability of the soil source, nutrition inside the seeds, seed dormancy, and harmonic response to the external stimuli decide the germination and growth factor of any crop. Many reports revealed the application of organic and inorganic fertilizers and chemical fertilizers has improved crop production and as well as yield. Nowadays, innovation in science and technology urges the research community to implement it in the field of crops to achieve maximum yield. In this connection, nanoparticles including metal nanoparticles, metal oxides, nanocomposites, nanorods, nanofibers, quantum dots, and CNTS have been used as a nanofertilizer, and there is a tremendous increase in crop production. Also, nanotechnology can be used as a nano–biosensor, and nanopesticide in preventing disease, crop protection, and other related issues. This chapter mainly focuses on the technologies that are used for seed priming for efficient seed germination, methods of incorporating nanotechnology into the plant such as soil amendment, foliar application, and stem injection have been discussed. Also, the efficacy of various NPs and post-harvest methods is discussed in detail.

R. Ramesh (✉) · A. Lavanya · A. Periya Nayagi Shilpa · K. Rafath Fathima
Department of Physics, Sacred Heart College (Autonomous), Tirupattur, Tamil Nadu 635601, India
e-mail: rramesh@shctpt.edu

R. Ramesh · F. Liakath Ali Khan
Department of Physics, Islamiah College (Autonomous), Vaniyambadi, Tamil Nadu 635752, India

K. Kasinathan et al. (eds.), *New Technologies for Energy Transition Based on Sustainable Development Goals*, https://doi.org/10.1007/978-981-97-2527-4_24

1 Introduction

In this current scenario, there is a fast enlargement in inhabitants in the world, and at the end of 2050, it is expected to be nearly six billion [1]. Need of the people and the demand for the consumption of energy in various forms including, sources used as food materials, power demand, and everything is changed due to the over consumption, and we are in a situation to meet all our needs with low cost and as well as with good quality [2]. Essential needs are based on the productivity of agricultural products in a country and nowadays the productivity of crops is getting decreased because of various factors including biotic and abiotic stress, change in the climate across the globe [3], and the unavailability of water resources and these are all the roadblock in the development of the crops and to overcome all these we need innovations in the farming of agricultural products. Nanotechnology can revolutionize agricultural properties and food products. Nanomaterials can be synthesized by many methods like physical, mechanical, chemical, and green approach methods and each of these methods has its effect in engineering the desired nanoparticles. Metal nanoparticles, metal oxides, nanocomposites, nanopolymers, nanoemulsions, silicon NPs, and carbon-based NPs have adverse effects on the improvement of plant growth and increased crop productivity [4]. The application of nanotechnology helps the farmer in terms of improving the health of the soil and thereby the retention of water and the available nutrients will enhance the production of the crops and reduce the usage of too much chemical fertilizer and reduce the water intake ability. Chemical fertilizers somehow boosted the yields of the crops however due to poor efficiency their usage is being reduced, chemical fertilizers suppress the uptake and translocation of the major nutrients including nitrogen, phosphorus, and potassium [5–10]. These macronutrients and micronutrients like manganese, copper, boron, zinc, iron, etc. are being lost and not at all fully utilized by the plant. Recent findings in nanotechnology were being implemented in overcoming this issue and developing strategies for the slow and sustained release of the nutrients in the form of nanofertilizers, and it reduces the loss of mobile nutrients. Nanofertilizer can be applied to plants in two ways, one is it provides nutrients by itself, and the other way is to add/load with other nutrients and thereby reducing the production cost of the crops. In this chapter, the effect of nanomaterials and nanotechnology in improving the productivity of crops and other related effects are discussed in detail. As per the literature, nanomaterials straightly influence the effectiveness of the loam microorganisms and hence promoting the physicochemical reaction and essential quality of the loam and thereby the signal transmission of the soil microbes as the carriers of the nanomaterials which could affect the plant growth in many ways [11–18]. Reports reveal that the implication of nanotechnology enhanced the delivery of nutrients and improve soil fertility by inducing the enzymatic activity of the soil through the interaction of the nanoparticles with rhizospheric bacteria and fixing the problems for smart bioaccumulation and translocation of the elements from the soil to the plant [19, 20].

2 Improvement of the Soil Texture with the Aid of Nanoparticles

Soil plays a vital role in promoting seed germination and crop development, naturally, the soil has some minerals, and its availability will vary from one type of soil to another. Organic carbon is naturally occurring in the soil and is noticed to undergo a considerable outcome on soil standard, function, and healthiness [21]. The interaction of the soil and plant mainly be conditional on the microbial interaction and biochemical progress, and it may be intensified by nanomaterials the outcome may be the improvement of the soil fertility and soil structure, maintenance, and the cycling of nutrients, productivity and fertility are mainly dependent on the interaction of the microorganisms and soil nutrients. Due to the usage of chemical fertilizers, the soil biodiversity is being damaged and with the new inventions, nanotechnology can enlarge the present horticultural practice and facilitate sustainable agriculture thereby reducing the input wastes [22, 23]. Nanotechnologies could improve sand health and restore the degraded soil texture by formulating the engineered nanoparticles and adapting according to the climate-resilient cultivation procedure as shown in Fig. 1. There is various stress which affects soil health including salt, acidity, dryness, suboptimal root zone temperature, availability of nutrients, pH value, and sufficient soil usefulness are essential for yield productivity, and they can obstruct plant production. Application of nanomaterials induces nutrient cycling and phytostimulation in the plants via the root-associated microorganisms by the ecological chemistry cycling of macronutrients and vitamins, minerals of micronutrients, and other important minerals for the improvement progress of the plants [24, 25].

The treatment of soil microbiome with nanotechnology may put down the huge inputs of insecticides by enhancing the strength of the soil and itself will fight against infestation and disease-causing pathogens and insects. The microbes which are essential for the supply of nutrition to the host herbs are endomycorrhizal fungi and rhizobium, a azospirillum is diazotrophic bacteria that evolve in root bump. Other microbes like *Pseudomonas* sp. and *Pantoea* sp. mobilize the nutrients and can change the phosphate into soluble forms. Nanotechnology plays the role in intervening by safeguarding reactive oxygen species homeostasis and enhances the plant's capability to keep out sodium and conserve potassium, escalate the nitric oxide preparation plays an important role in the manufacture of fertilizers and increase the amylase movements to grade up the dispersible sugars contented to keep the ion stability and osmotic homeostasis to avoid the adverse discharge [26].

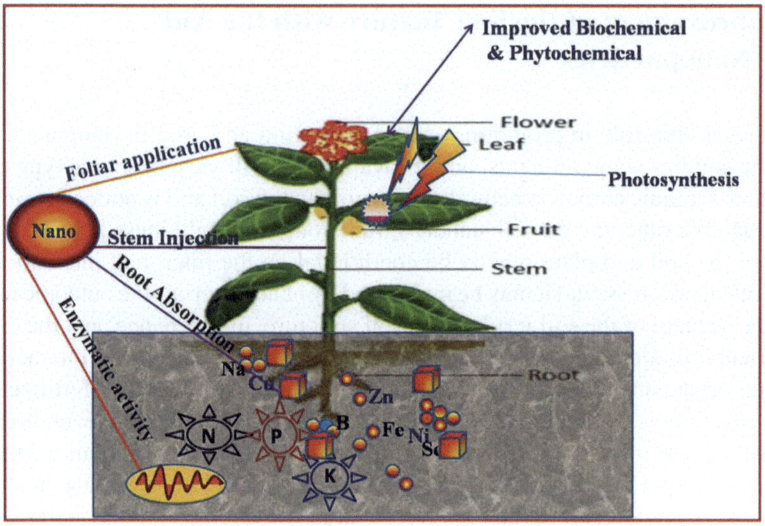

Fig. 1 Schematic representation of the application of nanoparticles in plant growth

3 Methods Adopted for Enhancement of Crop Production and Yield

3.1 Seed Priming Technology

To achieve good crop production and high yield first of all the good quality of seeds is to be selected for the cultivation of crops. In the traditional method, the seeds have been collected at the time of harvest and the collected seeds have been dried in direct sunlight and stored in a chamber that provides an optimum temperature which is made up of wood or dried muddy chambers. In modern days, those methods are not advisable due to increased temperature, seasonal, and climatic changes. The productivity of any crop directly depends on seed quality [27–30]. Environmental factors such as temperature, drought, moisture salinity, etc. affect seed germination and seedling development. Speedy and uniform sapling development leads to fortunate growth and development of the deep root system. Due to variations in the ecosystem and unfavorable conditions the level of bioactive compounds and phytocompounds like alkaloids, flavonoids, carotenoids, phenols, and proteins were affected and reduced due to various factors. To avoid such loss an appropriate method of seed priming is the essential one. Seed priming affects the fast and tedious formation of seeds and enhances plant growth [31]. It also improves the hormones of the damaged or aged seeds by increasing the plant's forbearance to living and nonliving exhaust. Seed priming is seeds that are presoaked in nanoparticle solution creating synchronized and fast germination, improved water usage, and efficacy of nutrient uptake and maturity of plant species. Nanopriming improves the seed germination, seedling rate, vigor, and dry weight through specific catabolic action that is energized throughout the premature stage of formation [32]. According to a report seed priming created physiological strength and resistance against biotic and abiotic stresses. In general plant growth includes three phases/stages, the first part is the imbibition stage in which the arrival of water into the seed by the action of absorbing. The second stage is the stimulation in which the beginning of the biochemical process, the humidity of seeds influences the managed imbibing and stimulation of pre-inception anabolism [33–35]. The third phase is the growth which demonstrates the physiological development of all the parts of the plant which causes the sugar utilization by the embryo and growing of the radical. Various methods of seed priming technology are available, hydropriming is the method of soaking the seeds in water, osmopriming is a method of soaking in an osmotic solution, and halopriming is another method of soaking in inorganic salt solutions to supply nutrients. Harmopriming is a technique in which the seed is soaked in growth regulators; Matrix priming is another method that utilizes low and high temperatures. Biopriming is the method of soaking in organic matter and nanopriming is the soaking of the seeds in nanomaterials [36]. The achievement of priming mainly belongs to the composition of interrelated elements like water, plant, priming therapy, period, climate condition, and healthiness of the grain. Reports revealed that nanopriming is an appropriate method in supplying nutrients and can penetrate the seed's coat and enlarge the intake efficiency of the water which augments the germination and growth process [37].

3.2 Fraternization of Nanoparticles with Soil

Many nanoparticles have been designed and directly applied to the soil to improve the quality of the soil. Naturally occurring nanocomposites like nanoclays starch nanocomposites, and nanoporous zeolites were applied to the soil and increased the water-holding capacity of the soil and were helpful in rain-fed crops [38]. Nanoparticles have been applied to the contaminated lands and recovery has been made. Despite the supply of nutrients or carrier of minerals nanoparticles improved the soil features. Nanomaterials like nanozeolites, oxides, metal oxides, carbon nanotubes, nanofibers, enzymes bimetallic nanoparticles have been used to remove the pollutants in the soil through the process of nanoremediation which acts as a catalyst and reduce the containments and it is a cheaper method and provide faster remediation. The various pollutants like petroleum products, pathogens, heavy metals, and organic pollutants can be removed by nanotechnology. One of the fungi arbuscular mycorrhizal is a member of the microbiota which plays a vital role the carbon, nitrogen, and phosphorous cycling, and reports revealed that various NPs have a positive effect on microbial community composition and enhancing soil texture [39].

3.3 Foliar Application of Nanoparticles

Foliar insemination has become an essential tool to control micronutrient insufficiencies in plants and has shown that it may improve crop production and feature. However, the uses of manure in a foliar the way could be restricted by the production, the inception, and the particle size. It helps to achieve the optimal performance and development of the plant species continually and exactly [40]. Due to the deduction in size of the nanoparticles, the increase in exterior area increases and hence the exposure area with the crop will be improved. Plants will absorb the applied nanoparticles with dynamics which is an added advantage over the other sources of insemination [41]. The particles lesser than 100 nm provide important nutrients for plants and are considered as nanofertilizers. The foliar type of application of nanoparticles is very effective in many ways and the doses can be reduced than the root and soil absorption method, and wastage of nutrients can be reduced. A study reported that the advantages of zinc oxide nanoparticles in low concentration have an adverse outcome on the production of seeds, growth of the crops, chlorophyll level, carotenoid content, and increase in yield. One report revealed that the application of nanoparticles in high concentrations affected crop development and later it affected the DNA and protein synthesis. Many reports revealed the application of the nanoparticles as a foliar method including chitosan had supportive results on photorespiration pigment concentration and increased crop production [42].

4 Efficacy of Various Nanoparticles in Plant Growth and Yield

4.1 Zinc-Based Nanoparticles

The divalent cation is absorbed by roots as zinc (Zn). Numerous stimulants, including alcohol dehydrogenase, which converts acetaldehyde to ethanol, are activated by zinc. Activates the hydration of carbon dioxide into bicarbonate by carbonic anhydrase (CA). While the altered metabolism of the auxin hormone is reflected in the soil's lack of zinc. Again, the availability of zinc may be determined by various elements, for example, the pH of the soil, the existence of other macro and micronutrients, soil microbes, and so forth. They combine to make and break down the nutrients [43]. Zinc solubility in soil and conception decreases simultaneously as soil pH rises, according to several studies. Higher levels of phosphorus (P) in the soil may cause Zn deficiency in some crops. Rice and maize plants grown in soil deficient in Zn frequently exhibit symptoms such as stunted enlargement and tiny brown dots on their leaves. Seed priming with Zn consequently increased plant production and interconnected feature in common bean. Seddigh, Khoshgoftarmanesh, and Ghasemi demonstrated, for instance, that seed preparation with Zinc is more efficient for wheat plants than applying zinc sulfate to the soil. They also talked about how seed priming with these newly processed manures can be a good substitute for using Zn-sulfate in the soil commercially these days [44]. Furthermore, others reported that prepared plantain seeds with zinc concentrations ranging from 0 to 2.5% enlarged yield and bract area over unprimed seeds.

4.2 Boron Nanoparticles

The cell wall's structural integrity is dependent on boron (B). Boron, on the other hand, is important for mitosis (cell division) and prolongation. For instance, the suppression of each mitosis refers to cell division and prolongation in major and minor roots is one of the most rapid responses to boron shortage [45]. The roots appear stubby and bushy because of this. Necrosis of the meristem follows the inhibition of cell division in the young leaves and shoot apex. Boron is also known to encourage the elongation and germination of pollen tubes. Boron, on the other hand, influences the growth of shoot meristems, and one of the most common signs of boron insufficiency is condensed tendril, which gives the plant a feathery or rosette looks, and expanded stems, which cause the celery clutter known as "*stem crackle*". The disarray known as "*heart decompose*" occurs in storage roots like sugar beets when boron deficiency causes the demise of dividing cells in the growth region. Under boron insufficiency, crop harvest is harshly decreased due to extreme interruption in various Boron embracing anabolic processes like the inhalation of carbohydrate, protein, and indole ethanoic acid, carbolic acid metabolism, membrane character and activity, and cell wall [46].

Boron also plays numerous roles and is linked to several methods, such as cell dividing, the metabolism of carbohydrates and nitrogen, blossoming and fruiting, the utilization of calcium, and water, being a catalyst for some reactions, and being resistant to infections. Boron deficiency interrupts blossoming and fruiting and issues low yields, including grain and fruit that is deformed or discolored [47]. The signs and syndromes of Boron insufficiency may change by species, including B insufficiency results in emptied seeds, also known as the "*empty heart*" in soybeans and groundnuts, whereas black gram seeds may exhibit no deficiency in B, despite a decrease in grain yield. In addition, rice seeds primed in a 0.5% B solution failed to germinate. The concentration of the solution is the most important aspect when priming with B. while priming rice with 0.001 and 0.1% B solutions enhanced root entrenched. Seed coating with B, for instance, is more suitable to use and worthwhile steps of applying B. However, the solution's quality is crucial and must be carried out and improved before crop application. Therefore, seed coating with boron increases development, early sapling growth, leaves area, shoot length, crop height, fruiting, and capsule harvest while simultaneously reducing the number of days required to flower, grain weight, and grain production in various crops by 50% [48].

4.3 Molybdenum-Based NPs

While molybdenum (MoO_2) is a metal, its characteristics are like the nonmetals. It mostly happens as the molybdate ion MoO_2 in aqueous solution. Among all known micronutrients, molybdenum essential is among the lowest. Molybdenum is included in nitrate absorption and nitrogen infatuation, two processes that are crucial to plants [49]. Nitrate from the soil is engrossed and absorbed by the Mo-containing nitrate-reducing enzyme. A lack of Mo in plants causes nitrates to accumulate in the leaves and not be absorbed into proteins. Mo is linked to nitrogen addiction in root lumps involving bacteria, in leguminous plants, to attach air nitrogen. As a result, legumes deficient in Mo hurt soil microorganisms' ability to fix nitrogen, leading to N deficiency. Mo deficiency causes whiptail in cauliflower, oats blue chaff, and citrus yellow spot, all of which are symptoms of a lack of molybdenum, which is influenced by the catabolism of nitrogen in crops. Molybdenum is a most important micronutrient that can be put in seed priming in very less concentrations to significantly boost development, nodulation, nitrogen absorption, nutrient intake, dry particle accumulation, and crop production [50].

According to some studies, applying Mo through seed medication is superior to applying it through the soil. In Mo-deficiency soil, preparing the seeds of subterranean clover in 0.1 or 1 percent solution of sodium molybdate was found to be uniformly effectual for soil application. According to a larger rate of Molybdenum and Nitrogen content in seeds than equal to soil applications. In fact that similar bean seeds coated with sodium molybdate increased bump, dehydrated matter aggregation, fixation of nitrogen, and yield. Furthermore, some of the studies show that whereas an eight-hour examination with prepared chickpea seeds in a 0.5-g, sodium molybdate solution produced an increase of 27% growth in culturing the pot experiment and 20% in the field examination when contrasted to soil application. The effectiveness of seed preparation can be improved by including Rhizobium culture in the priming solution along with Mo. It was discovered that the preparation of green gram seeds in rhizobia improved manganese in sodium molybdate [51].

4.4 Manganese-Based NPs

It is an essential part of nitrogen anabolism, photosynthesis, and the formation of a few chemical compounds required for plant metabolism. Manganese acts as a co-founder for several enzymes, importantly decarboxylase which is an enzyme that catalyzes the removal of carbon dioxide from a compound, and hydroxylase, which take part in an important role in the inhaling carbon cycle. As a mannoprotein, manganese is the role of the oxygen-releasing complex involved in photosystem II, where it assembles a charge throughout water oxidation. Manganese inadequacy can be common in a few regions dependent on soil quality, season, and plant species. Insufficient mostly depends on the less pH of the soil and higher organic molecules content. It can lead to severe chlorosis of leaf veins as a discolor and deform legume endocarp [52]. Under conditions of acute Manganese deficiency, premature leaf drop due to brown necrotic spots on the leaves. Some cereal leaves indicate Mn deficiency by making their leaves appear white and gray spots. Manganese has many essential roles in plant physical growth, importantly in the photosynthetic and inhaling process, so insufficiency of this micronutrient will reduce production. Germination and growth physiology and growth, leaf area, average height and root length, seed production, and seed manganese content were essentially increased when grains of different crops were lined with manganese sulfate. The usual growth and improvement of seedlings and their vitality index are very highly related to the mineral nutrients existing in the seed repository tissue. Less grain manganese content causes poor stamina of wheat plants and less yield at harvest [53]. While in some studies, it is observed that slow development and poor sapling growth with low manganese present in barley and lupine beans. For example, some report shows that safflower seeds were prepared with the help of manganese sulfate (0.004 mM) and germination physiology was developed. When conducting experiments with soybean (botanical name is glycine max L.) beads were prepared with 5- and 15-mM Mn solutions by using manganese sulfate for twelve hours at 20 °C among five layers of moisture strainer paper and a considerable increase in manganese content in the beads was observed up to 20 times.

4.5 Copper-Based Nanoparticles

Copper functions as a cofactor for various oxidant enzymes. These contain photosynthetic electron messenger plastocyanin; Cytochrome c oxidase, a terminal oxidase mitochondrial respiratory enzyme; and vitamin C oxidase. Superoxide dismutase: Another important copper-containing enzyme that detoxifies superoxide radicals (O_2). This participates in carbon incorporation and nitrogen metabolism. Its insufficiency leads to strong development delays. Copper also provides cellular strength as well as being involved in lignin biosynthesis. It also protects against wilting. Leaf chlorosis, growing retardation, and stem and branch death are common symptoms of copper deficiency [54]. Numerous grains (*per cone*), increase in number of grains, increase in grain production and biomass. It was obtained because of the inoculation of seeds with copper sulfate. Grain dressing is not required complex equipment; it is lighter, faster, cheaper, and more environmentally friendly than most water-efficient technologies, providing farmers with an extensively attractive differently for faster production growth, better settling, and yield. Seed priming methods can use different osmotic agents with different properties and efficiency levels [55].

4.6 Cobalt-Based NPs

Cobalt is useful for certain plants, but important for others. For example, in pulses, Cobalt is needed for nitrogen addiction. Cobalt influence plant catabolism and growth. It is an important constituent of numerous stimulants and coenzymes. It is essential accomplishment in slowing down leaves senescence, inhibiting ethene biosynthetic, and stimulant of alkaloid biosynthesis. Given that it increases growing, pod production, molt rate, grain weight, yield index, building assimilation, Femininity, yield, tuber, and panicle length were observed in different seeds [56]. Crops were grown with standard concentrations of cobalt nitrate or sulfate. Pattanajak and others reported that seed prepared by combining the microelement (0.16 mg g1 Molybdenum as sodium molybdate and 0.008 mg g1 Cobalt as cobalt chloride) and rhizobium importantly increased nodulation, N addiction, intake of nutrients, growth of plant and production. This outcome of seed preparation in various plant plants with cobalt as a leading solution is contributed to the article in pattypan squash (*Cucurbita pepo* L.) reported by Ambika, Manonmani, and Deepika., seed preparation with cobalt sulfate is mainly enhanced in dry particles assemblage, femininity, and berry production differentiated with water dipped seeds [6]. In adding up, coating with $CoSO_4$ stimulated autogenous ethylene synthesis throughout premature sapling steps (14 days after sprinkling), which continued until floweret inception (thirty days after sowing).

4.7 Nickel-Based NPs

Nickel is comparatively new to the various essential nutrients. It is a plenteous metal component and is easily immersed by the roots. There is no limit to the amount of aluminum in plant cells and tissues. It typically ranges between 0.05 and 5.0 mg kg^1 in parched weight. External nickel claim basic plants are unclear but may be involved in nitrogen mobilization at the time of seed development. Nickel is also used as a constituent of two compounds namely urea amidohydrolase and hydrogenase [57]. Urease activates it hydrolyzes urea to ammonia and carbon dioxide and is widespread in the plant domain. The major impact of nickel insufficiency in soybean plants (i.e., Glycine max) is a decrease in urea amidohydrolase function. The anabolic importance of urea amidohydrolase is not confirmed, but in leaves, the nitrogen of certain legumes is a domain of urea-based constituents known as ureides, like allantoic acid and citrulline malate. When urea breaks down, urea is formed. Toxicity content in nickel-inadequacy plants. It is considered that most plants have it, but not all plants contain it. You will need to use urea amidohydrolase and nickel. Hydrogenase is one of the most important enzymes in certain nitrogen-addicting plants. Hydrogenase liable for the recovery of hydrogen is used in the nitrogen fixation activity. Nickel insufficiency leads to a reduced hydrogenase process in soybean buds, and a decrease in nitrogen fixation efficiency is expected [58].

4.8 Role of Iron-Based NPs in Plant Growth

Iron belongs to the catalytic group. It plays two main roles in the plants. Iron has a higher rate of solubility. Iron takes part a crucial role in the growth of the plant. The decoration of chloroplast structure and reduction of chlorophyll can cause iron deficiency in plants. Iron is available in three covalent forms or two covalent forms with a valency of three or two [59]. Catalase and peroxidase are some of the oxidizing enzymes which is the branch of iron. As iron act as an oxidoreductase, it is very essential for synthesizing chlorophyll in plants. The cytochromes are hemoproteins and non-heme iron–sulfur proteins are the two most important oxidoreductase. For the development of chlorophyll, iron is the most fundamental thing it is used to transport and circulate elements in the plant's bloodstream. The slow movement of iron and the disability of removing the oldest iron can lead to the deficiency of chlorosis, this can be seen in the veins of the newly budded leaf at the earlier stage [60]. Chlorosis can spread if there is not enough amount of chlorophyll present in the plant then the deficiency becomes severe, and the very little leaf turns white. In the application of agriculture, researchers are developing several nanoparticles related to iron all over the world. The chlorophyll rate was increased in leaves when the experiment was done with iron nanoparticles with glycine max in the effect of the greenhouse. Similarly, iron nanoparticles were used in the black-eyed peas plant; it shows the result as the number of pods was increased in that plant and it gives a large amount of iron content 34%; and chlorophyll was increased up to 10% and leaves are compared to the control at 500 ppm. Iron nanoparticles are also used in soybeans to prevent chlorosis in that plant [61].

4.9 Sodium-Based NPs

Sodium is necessary for fixing the pathway of carbon in many plants we use the Crassulacean acid metabolism and 4-carbon atom. Carboxylation is done in the C4 and CAM pathways for regenerating the phosphoenolpyruvate substrate. Less amount of sodium ion leads to the formation of C3 at a high rate. If the plants have sodium deficiency the plants will not bloom and become necrotic and chlorinated. The fractional growth of the cell can be done by expanding the cell and sodium can be provoked [62]. For the enzymes which use nitrogen, sodium bicarbonate acts as a powerful donor of electrons, like nitrogenase and nitrite reductase in *Vigna mungo*. Potassium has osmotic activity, and it acts as a solubilizer. The seeds are primed with sodium sulfate, when seeds are at the nearest stage of germination it is compared with one without bait, priming is done by synthesizing the DNA and RNA inside the seed in the pre-germination process. After this, the seed partially hydrates, and the pre-germination metabolism continues. Hence, the outcome shows that, in final germination, the percentage of unprimed seeds was not noticeably different from it is common to have this condition, which is primed with sodium sulfite in the same way the other report has done. In the germination model versus control, we can get the constant result by doing various penetrations on fusing the grains into corn kernels [63].

4.9.1 Nanoporous Zeolites

In most cases the plant will take the minimum amount of nutrition from the fertilizers, to get more nutrients at the large level the fertilizers are released slowly to the plant with the help of nanoporous zeolite. Using the ball-milling technique, the zeolite has been prepared as nano-sized particles [64]. Then, the centrifuge is done for 1 g of nano-sized zeolite and 8 h it is stirred with calcium sulfate solution of 1.5 M and 50 ml next, it was filtered, and distilled water is used for washing. The synthesis is done in a maintained ratio of 1:10 that is air-dried solid: liquid. Tough nitrogen fertilizers have huge amounts of solubility which causes much damage to the plants and the atmosphere also the nitrogen with urea releases at a faster rate thus the plant takes only less amount of nutrients. So, we can use nanoporous zeolite. The zeolite nanoparticle is mixed with the urea and used as a fertilizer, it gives a very controlled release of fertilizers, and the plant can take more nutrients as uptake. Zeolite has the capability of bringing the soil's moister from dry land it helps the nutrients to flow into the soil. To save the soil nature, aluminum zeolite was used because it is highly penetrable [65].

4.9.2 Titanium Oxide and Its Effect on Plant Growth

The best consent of enzymic activity in plants can be produced by titanium oxide nanoparticles, it also helps to increase the growth of the plant. The process of chemical fixation in nitrogen with air and increased activity of photosynthesis is take place in the presence of sunlight. Titanium oxide nanoparticles are used in the soybean (Glycine max) for rising the activity of nitrate reductase in the rhizosphere, this process is done by boosting the nitrogen fixation. Quick growth and germination can be done with the composition of titanium oxide and silicon oxide [66]. Titanium oxide nanoparticles are also used in the spinach plant (*Spinacia oleracea*) for the treatment of seed and the foliar spray technique is done. In the 250 ppm of titanium oxide nanoparticles, the spinach seeds were socked under the light for up to 48 h, then germination takes place in the seed. The demonstration is taking place in the conservatory for 35 days. Once a week, the titanium oxide nanoparticles were sprayed to the plant. The weight of the plant is reduced by using titanium oxide nanoparticles and also the rate of nitrogen is controlled up to 23%, the total content of chlorophyll and protein are also reduced, were chlorophyll—34% and protein—34%. Also, some research says that the growth can be increased at 50 ppm in less concentration [67].

4.9.3 Carbon Nanoparticles

To test the positive and negative effects on the growth of the plants at different concentrations carbon nanotubes (CNT) are used. Carbon nanotubes also act as growth boosters for different plants like tomatoes, soybean, cabbage, carrot, rye, rapeseed, and com. Multi-walled carbon nanotubes are used in sarson, ray grass, and maize. In some plants like cucumber, onion, and cannas, the root length is increased when the seed is germinated at 200 ppm, the good elongation of roots is observed by this process [68]. The use of single-walled nanotubes in seed germination produces the best crop production and gives a good impact on plant growth in lettuce, cucumber, and radish, this is obtained at different levels of concentrations. The intake of water and need for usage is increased in tomato plants by using the carbon nanotubes on the surface change with protein in the expression of the water channel. By using the carbon nanotubes in the tomato plant, the yield is increased up to factor at 50 ppm, and the concentration was also controlled. The use of carbon nanotubes in the tomato makes the plant grow faster by the seed germination process and the capacity of absorbing the moister is increased [69].

4.9.4 Silver Nanoparticles

Silver is a well-known antibacterial metal. Silver can be in ionic form and in nanoparticle form [70]. The quiescent enzymic system has unicellular microorganism silver that can kill it. Silver nanoparticles act as pesticides for several plants like *Magnaporthe grisea*, phone, water molds, *M. grisea*, collectrichum, gloeosporioides, *Sclerotinia sclerotiorum*, *Rhizoctonia solani*, *Fusarium culmorum*. Silica nanoparticles and silver nanoparticles were sprayed on the Erysiphaceae of the infected gourd leaves and noticed that after 3 days of doing this application, the silver–silica alloys protected the powdery mildew (Erysiphaceae). The pathogen in the rose Erysiphaceae can be treated with the help of nanocolloids of silver which gives an antifungal effect to the plant [71, 72]. In the same manner, the pathogen of oak mushrooms in Korea is also treated with silver nanoparticles. The fungal effects like mycelium and germination of the conidial can be determined by using silver nanoparticles. The rise of blast and spotted disease in plants can be prohibited using Ag nanoparticles to a large extent. Some studies show that the investigation of the increase in the rate of antifungal activity in nanoparticles and albicans, triazole phone glomerate, and Trichoderma in the biological synthesis of silver nanoparticles [73, 74].

4.9.5 Silica Nanoparticles

Silica nanoparticles give a competent result against the insects in plants such as mustard weevil (phaeton weevil), coconut tick (*Dermanyssus gallinae*), and whitefly (*Trialeurodes vaporariorum*). The physical activity of a plant and its growth essentially needs the silica element in the form of silica gel [75]. In the infected plant, the stress resistance is increased with the help of a silica inducer. The silica alone cannot act as an antibacterial agent but when it is mixed with the silver nanoparticle it acts as the best antibacterial and antifungal agent. The plant which grows with fungal infection can be treated with silver–silica nanoparticles at the concentration range of 10 ppm, and the plant with bacterial infections can be treated with silver–silicon nanoparticles at the concentration range of 100 ppm [76, 77]. Thus, the 100% of control in fungal and bacterial infection in the plants was demonstrated using Ag–Si nanoparticles. The transport of target delivery in DNA and interested gene and the chemicals in the mesoporous structure, the other form of silica is governable. In cucumber and pansy plants, the phytotoxicity was 3200 ppm as shown by the nanoparticles, and in the powdery mildew of gourds. The surface area is increased in well-disciplined pores which have mesoporous nanoparticles in them [78]. To get effectivity in the delivery system the validamycin material which is an insecticide is loaded with PHSN (porous hollow silica nanoparticles). These porous hollow silica nanoparticles deliver the necessary amount of pesticide. To kill the insects in plants physisorption is done in the cuticle lipid of the insect by absorbing the nanosilicate. Thus, nanosilicate acts as a good insecticide against the insects in plants [79].

5 Nanopesticides and Their Effect on Preventing Diseases

Synthetic herbicides can lead to atmospheric pollution. To overcome pollution-related issues nanoagrochemical plays a major role as a replacement for synthetic herbicides. The nanomaterials used as a pesticide are called nanopesticides. In recent days, plant disease-causing 20–40% of the waste of crop production, to increase the yield we need to avoid pesticides, which leads to biotic stress [80]. Instead of pesticides, we can use nanopesticides which act as the global-friendly tool. The nanopesticides have antibacterial, insecticide, and antifungal properties or the synthetic molecules were enclosed by the developed nanoparticles which let go into the medium unhurried and in a constant manner. The dacoity and hexaconazole are enclosed with chitosan nanoparticles get the chitosan-related nanofungicides it is ranging between 2 and 168 nm. These anti-fungicide nanoparticles were used to treat *Ganoderma boninense* (a disease in the plant) which causes phototoxic effects in the oil palm crops, free fungicides were compared to the photosynthetic rate, seedling height, chlorophyll content, and root elongation [81]. To control banded blight of rice with a fungicide validamycin, silica nanomaterials were used as part of the pesticide treatment. To overcome the oomycetes, a herbicide called metalaxyl was used to surround silica nanomaterials ranging in size from 80 to 15 nm. Metalaxyl was wrapped with silica nanomaterial in order to provide the herbicide in a controlled manner. And also, the metal and metal oxide nanoparticles can lead to a great effect as fungicides, it can be either applied directly as nanoparticles for treating herbicides or in the composite's formation process.

6 Nano-biosensor

Nanotechnology improved biosensors by using the properties of the nanomaterials and our ability to deliver quick, accurate, and cost-effective information on field examination. Soil moisture, nutrients, fertilizers, and plant pathogen are determined using a nanosensor. Nanosensors play an important role in agriculture. The primary step in nanosensors is the coupling of specific sensing applied directly through nanomaterial. Soil moisture sensing nanosensors are developed by using various methods [82]. Based on electrical properties, certain nanoparticles are used to sense soil humidity and dielectric properties changes based on the absorption of water. For example, polyethylene terephthalate substrate is used for coating silver nanoparticles to get the correct moisture sensor and is highly stable. Nanosensor as optical moisture has several uses in parameters of optical on the absorption of water. For example, SU8 polymers are coated by using graphene oxide films producing optical humidity sensors with high response, fast and broad range. Nutrient-based nanosensors are used to indicate the efficiency of fertilizers and provide information in real-time when and where the response or action is needed [83]. The existing targets of nutrient nanofertilizers are sodium content, soil saltness, carbon content, and residue of nitrate. Cysteamine-processed gold-based nanoparticles are used to record nitrates at large affinity, permitting a quick and highly sensitive colorimetric assay. For insecticide nanosensors several nanoparticles are used such as carbon nanotubes, which are the hardest material, quantum dots are one-dimensional materials, gold nanoparticles, and nanocomposites. Different insecticide nanosensors belong to the immobilized enzyme of pesticide. For example, atrazine is detected by a nanosensor, it is composed of a tyrosinase enzyme immobile on a TiO_2 nanotube for detection of several subpart-per-trillion levels. Plant pests and insects are identified with the help of pathogen nanosensors. Pathogen or parasite nanosensors are made using quantum dots nanomaterials [84]. Regions of pathogens are detected using quantum dot FRET-based nanoparticles and show effulgent capacity. It is used to detect beet necrotic yellow virus, rhizomania, and detect specific diseases of sugar beet. This sensing material is made by using quantum dots intact with glutathione S-transferase antibodies specific to microbe polymyxa beta and GST-contacted dye rhodamine. The developed nanosensor permits the detection of pathogens by a decrease in fluorescence observation. Single-stranded DNA is linked with altered quantum dots for pathogen nanosensors. Further, Cy5 fluorescent dyestuff is linked with another DNA single strand [85]. In both probes, if pathogen DNA is present, its hydrolysis in the surrounding region and increases the fluorescence emission. Few nanosensors are small and various steps are still needed for processing. Sometimes lengthy turnaround times (24–48 h) and laboratory facility maintenance are required. New developments are generally expected to receive simpler and more convenient sensors [86].

7 Post-harvest Method

In the food delivery chain packaging of food plays a major role. We need to pack the remaining food which was not consumed after the preparation. The microorganisms, chemical formation, dirt, and contaminants can be avoided by covering the food in a fully packed way. While storing and delivering the food packing/covering that can maintain the food quality [87]. The packing must be eco-friendly, low expensive, easy to carry, give protection, it must be a re-useable product, easy to dispose of, and passive. To get all these properties, polymers were used as a packing material, which gives safe packing. To create this kind of polymers the cost effect is high, and it is less required. To overcome these defects biopolymers were used, which have the property of low resistance, less air and moisture protection, and fragility. And also, it can come across the problem like low thermal spin problems. The preservation of fruits and vegetables is more difficult because oxygen supply only decides their lifespan [88]. To increase nutrient in the fruit and vegetables and keep it fresh nanoparticles were discovered. For storing milk, packaging foods, juice storage and to increase other agriculture products lifespan nanotechnology was used, it is a very safe method. To maintain the quality and characteristics of food packaging, nanocomposite materials were used. Nanocomposites has a wide range of application in food and agriculture packaging [89].

The nanocomposite structure has the repeat spacing between the phases or one, two, or three dimensions of one of the phases less than 100 nm. Hence, nanocomposites are called the multiphase solid material [90]. Nanocomposites were prepared from nano-sized metals or ceramic materials which is more strong and harder than the other bulk materials. The thermal and gas hurdle characteristic of nanocomposites is used in the post-harvest method to expand the shelf life and it helps in food and food transport storage [91]. In the water vapor dispersion, in favor compounds, carbon dioxide and oxygen nanocomposites created a high barrier. To get standard storage conditions the new material polymer nanocomposites was introduced. The diffusing of an inert nanoscale into a polymeric was obtained by the crew. Some inorganic substances, chitosan nanoparticles or chitin, nanosilicate platelets (vide infra), clay, starch nanocrystal, silica nanoparticles, graphene, and carbon nanotubes can act as a filling material. In the comparison between the polymer nanocomposites and the conventional polymer, poly nanocomposites have a high aspect ratio [92–94].

8 Summary and Conclusion

In this chapter, the effect of nanomaterials, nanotechnology, and its role in the process of seed priming, seed germination, plant growth, crop protection, and production of crops to provide a maximum yield is discussed in various aspects including nanopriming of the seed materials like bio priming, holopriming, coating of the seeds using core–shell nanoparticles. Encapsulation of fertilizers for sustained release and controlled release of the available macro and micronutrients plays a vital role as the major stage of plant growth and production. The efficacy of the various nanoparticles including sodium, potassium, nitrogen, boron, cobalt, copper, silver-based, gold NPs, silica-based, CNTs, quantum dots, grapheme-based NPs, application of nanomaterials as a biosensor, pesticide, and fungicide was reported. Once the plant attains its maturity the proper technology of post-harvest is very important, and the packaging of the agricultural and food products needs a special type of designed NPs. Overall, the chapter will provide a depth knowledge of the application of nanomaterials and nanotechnology in the improvement of the crop production.

References

1. B. Ahmed, A. Rizvi, A. Zaidi, M.S. Khan, J. Musarrat, Understanding the phyto-interaction of heavy metal oxide bulk and nanoparticles: evaluation of seed germination, growth, bioaccumulation, and metallothionein production. RSC Adv. **9**, 4210–4225 (2019)
2. A. Jaberzadeh, P. Moaveni, H.R. Moghadam, H. Ahedi, Influence of bulk and nanoparticles titanium foliar application on some agronomic traits, seedgluten and starch contents of wheat subjected to water deficit stress. Notulae Botanicae Horti Agrobotanici Cluj-Napoca. **41**, 201–207 (2013)
3. R. Mohammadi, R. Maali-Amiri, A. Abbasi, Effect of TiO_2 nanoparticles on chickpea response to cold stress. Biolo. Trace Elem. Res. **152**, 403–410 (2013)
4. J. Faraji, A. Sepehri, Exogenous nitric oxide improves the protective effects of TiO_2 nanoparticles on growth, antioxidant system, and photosynthetic performance of wheat seedlings under drought stress. J. Soil Sci. Plant Nutr.Nutr. **20**, 703–714 (2020)
5. M. Askary, S.M. Talebi, F. Amini, A.D. Bangan, Effects of iron nanoparticles on *Mentha piperita* L. under salinity stress. Biologija **63**, 65–75 (2017)
6. S. Torabian, M. Zahedi, A.H. Khoshgoftar, Effects of foliar spray of nano-particles of $FeSO_4$ on the growth and ion content of sunflower under saline condition. J. Plant Nutr.Nutr. **40**, 615–623 (2017)
7. H. Moradbeygi, R. Jamei, R. Heidari, R. Darvishzadeh, Investigating the enzymatic and non-enzymatic antioxidant defense by applying ironoxide nanoparticles in *Dracocephalum moldavica* L. plant under salinity stress. Sci. Horticult. **272**, 109537 (2020)
8. M. Adrees, Z.S. Khan, S. Ali, M. Hafeez, S. Khalid, M.Z. Ur Rehman, Simultaneous mitigation of cadmium and drought stress in wheat by soil application of iron nano particles. Chemosphere **238**, 124681 (2020)
9. M. Iqbal, N.I. Raja, M. Hussain, M. Ejaz, F. Yasmeen, Effect of silver nanoparticles on growth of wheat under heat stress. Ira. J. Sci. Technol. Transac. A Sci. **43**, 387–395 (2019)
10. A.K.S. Mohamed, M.F. Qayyum, A.M. Abdel-Hadi, R.A. Rehman, S. Ali, M. Rizwan, Interactive effect of salinity and silver nanoparticles on photosynthetic and biochemical parameters of wheat. Arch. Agro. Soil Sci. **63**, 1736–1747 (2017)
11. S.S. Hojjat, M. Kamyab, The effect of silver nanoparticle on fenugreek seed germination under salinity levels. Russ. Agricult. Sci. **43**, 61–65 (2017)
12. R. Grillo, B.D. Mattos, D.R. Antunes, M.M.L. Forini, F.A. Monikh, O.J. Rojas, Foliage adhesion and interactions with particulate delivery systems for plant nanobionics and intelligent agriculture. Nano Today **37**, 101078 (2021)
13. R.M. Hussain, The effect of phosphorus in nitrogen fixation in legumes. J. Agric. Res. Technol. **5**, 1–3 (2017)
14. I. Lavicoli, V. Leso, D.H. Beezhold, A.A. Shvedova, Nanotechnology in agriculture: opportunities, toxicological implications and occupational risks. Toxicol. Appl. Pharmacol.. Appl. Pharmacol. **329**, 96–111 (2017)

15. H. Singh, A. Sharma, S.K. Bhardwaj, S.K. Arya, N. Bhardwaj, M. Khatri, Recent advances in the applications of nano-agrochemicals for sustainable agricultural development. Environ Sci Process Impacts **23**, 213–239 (2021)
16. N.M. Marzouk, H.A. Abd-Alrahman, A.M.M. ElTanahy, S.H. Mahmoud, Impact of foliar spraying of nano micronutrient fertilizers on the growth, yield, physical quality, and nutritional value of two snap bean cultivars in sandy soils. Bull. Nat. Res. Centre **43**, 84 (2019)
17. A.S. Soliman, M. Hassan, F. Abou-Elella, A.H. Hanafy Ahmed, S.A. El-Feky, Effect of nano and molecular phosphorus fertilizers on growth and chemical composition of Baobab (*Adansonia digitata* L.). J. Plant Sci. **11**, 52–60 (2016)
18. L. Xiong, P. Wang, M.N. Hunter, P. Kopittke, Bioavailability and movement of phosphorus applied as hydroxyapatite nanoparticles (HA-NPs) in soils. Environ. Sci. Nano **5**, 2888–2898 (2018)
19. H.Y. Yoon, J.G. Lee, L. Degli Esposti, M. Iafisco, P.J. Kim, S.G. Shin, J.R. Jeon, A. Adamiano, Synergistic release of crop nutrients and stimulants from hydroxyapatite nanoparticles functionalized with humic substances: toward a multifunctional nanofertilizer. ACS Omega **5**, 6598–6610 (2020)
20. R. Lal, Soil health and carbon management. Food Energy Secur. **5**, 212–222 (2016)
21. V.D. Rajput, T. Minkina, S.K. Upadhyay, A. Kumari, A. Ranjan, S. Mandzhieva, S. Sushkova, R.K. Singh, K.K. Verma, Nanotechnology in the restoration of polluted soil. Nanomaterials **12**, 769 (2022)
22. R.K. Maiti, P. Satya, Research advances in major cereal crops for adaptation to abiotic stresses. GM Crop. Food. **5**, 259–279 (2014)
23. M. Kalwani, H. Chakdar, A. Srivastava, S. Pabbi, P. Shukla, Effects of nanofertilizers on soil and plant associated microbial communities: emerging trends and perspectives. Chemosphere **287**, 132107 (2022)
24. S.T. Khan, S.F. Adil, M.R. Shaik, H.Z. Alkhathlan, M. Khan, M. Khan, Engineered nanomaterials in soil: their impact on soil microbiome and plant health. Plants **11**, 109 (2022)
25. Y. Wang, Z. Fang, Y. Kang, E.P. Tsang, Immobilization and phytotoxicity of chromium in contaminated soil remediated by CMC-stabilized nZVI. J. Hazard. Mater. **275**, 230–237 (2014)
26. V.D. Rajput, T. Minkina, M. Feizi, A. Kumari, M. Khan, S. Mandzhieva, S. Sushkova, H. El-Ramady, K.K. Verma, A. Singh, Effects of silicon and silicon-based nanoparticles on rhizosphere microbiome, plant stress and growth. Biology **10**, 791 (2021)
27. H. Tian, M. Kah, K. Kariman, Are nanoparticles a threat to mycorrhizal and rhizobial symbioses? A critical review. Front. Microbiol.Microbiol. **10**, 1660 (2019)
28. S. Compant, A. Samad, H. Faist, A. Sessitsch, A review on the plant microbiome: ecology, functions, and emerging trends in microbial application. J. Adv. Res. **19**, 29–37 (2019)
29. T.C.G. Kibbey, K.A. Strevett, The effect of nanoparticles on soil and rhizosphere bacteria and plant growth in lettuce seedlings. Chemosphere **221**, 703–707 (2019)
30. F. Eivazi, Z. Afrasiabi, E. Jose, Effects of silver nanoparticles on the activities of soil enzymes involved in carbon and nutrient cycling. Pedosphere **28**, 209–214 (2018)
31. M. Amde, J.F. Liu, Z.Q. Tan, D. Bekana, Transformation and bioavailability of metal oxide nanoparticles in aquatic and terrestrial environments. A review. Environ. Pollut. Pollut. **230**, 250–267 (2017)
32. A.H. Ahkami, R. Allen White, P.P. Handakumbura, C. Jansson, Rhizosphere engineering: enhancing sustainable plant ecosystem productivity. Rhizosphere **3**, 233–243 (2017)
33. M. Djanaguiraman, N. Belliraj, S.H. Bossmann, P.V.V. Prasad, High-temperature stress alleviation by selenium nanoparticle treatment in grain sorghum. ACS Omega **3**, 2479–2491 (2018)
34. H. Etesami, H. Fatemi, M. Rizwan, Interactions of nanoparticles and salinity stress at physiological, biochemical and molecular levels in plants: a review. Ecotoxicol. Environ. Saf.. Environ. Saf. **225**, 112769 (2021)
35. F. Zulfiqar, M. Ashraf, Nanoparticles potentially mediate salt stress tolerance in plants. Plant Physiol. Biochem.Biochem. **160**, 257–268 (2021)

36. A. Kumari, R. Kaur, An insight into drought stress and signal transduction of Abscisic acid. Plant Sci. Today. **5**, 72–80 (2018)
37. G.N. Rameshaiah, J. Pallavi, S. Shabnam, Nano fertilizers and nano sensors—an attempt for developing smart agriculture. Int. J. Eng. Res. Gener. Sci. **3**, 314–320 (2015)
38. E.E. Elemike, I.M. Uzoh, D.C. Onwudiwe, O.O. Babalola, The role of nanotechnology in the fortification of plant nutrients and improvement of crop production. Appl. Sci. **9**, 499 (2019)
39. W.F. Abobatta, Nanotechnology application in agriculture. Acta Sci. Agric. **2**, 99–102 (2018)
40. N. Dasgupta, S. Ranjan, D. Mundekkad, C. Ramalingam, R. Shanker, A. Kumar, Nanotechnology in agro-food: from field to plate. Food Res. Int. **69**, 381–400 (2015)
41. R. Sadeghi, R.J. Rodriguez, Y. Yao, J.L. Kokini, Advances in nanotechnology as they pertain to food and agriculture: benefits and risks. Annu. Rev. Food Sci. Technol. **8**, 467–492 (2017)
42. R. Ravichandran, Nanotechnology applications in food and food processing: Innovative green approaches, opportunities and uncertainties for global market. Int. J. Green Nanotechnol. Phys. Chem. **1**, 72–96 (2010)
43. S.M. Jafari, D.J. Mc Clements, Nanotechnology approaches for increasing nutrient bioavailability. Adv. Food Nutr. Res. **81**, 1–30 (2017)
44. F. Mustafa, S. Andreescu, Nanotechnology-based approaches for food sensing and packaging applications. RSC Adv. **10**, 19309–19336 (2020)
45. K. Pathakoti, M. Manubolu, H.-M. Hwang, Nanostructures: current uses and future applications in food science. J. Food Drug Anal. **25**, 245–253 (2017)
46. R. Sivarethinamohan, S. Sujatha, Unlocking the potentials of using nanotechnology to stabilize agriculture and food production. AIP Conf. Proc. **2327**, 20022 (2021)
47. S.C. Mali, S. Raj, R. Trivedi, Nanotechnology a novel approach to enhance crop productivity. Biochem. Biophys. Rep. **24**, 100821 (2020)
48. K. Neme, A. Nafady, S. Uddin, Y.B. Tola, Application of nanotechnology in agriculture, postharvest loss reduction and food processing: food security implication and challenges. Heliyon **7**, e08539 (2021)
49. C. Griesche, A.J. Baeumner, Biosensors to support sustainable agriculture and food safety. Trends Anal. Chem. **128**, 115906 (2020)
50. M.M. Othman, Effect of organic fertilizers and foliar application of some stimulants on barley plants under saline condition. J. Soil Sci. Agric. Eng. **12**(4), 279–287 (2021)
51. D.A. Ghazi, Response of hot pepper plants (*Capsicum frutescens* L.) to compost and some foliar application treatments. J. Soil Sci. Agric. Eng. **11**, 641–646 (2020)
52. A.E. El-Sonbaty, Effect of foliar application of NPK nanoparticle fertilization and soil addition of mineral NPK fertilization as combined treatments on soil fertility and wheat plant performance. J. Soil Sci. Agric. Eng. **12**(8), 567–572 (2021)
53. S. Kumar, G. Bhanjana, A. Sharma, N. Dilbaghi, M.C. Sidhu, K.H. Kim, Development of nanoformulation approaches for the control of weeds. Sci. Total. Environ. **586**, 1272–1278 (2017)
54. W.H. Elmer, J.C. White, The use of metallic oxide nanoparticles to enhance growth of tomatoes and eggplants in disease infested soil or soilless medium. Environ. Sci. Nano **3**, 1072–1079 (2016)
55. P. Vanathi, P. Rajiv, R. Sivaraj, Synthesis and characterization of Eichhornia-mediated copper oxide nanoparticles and assessing their antifungal activity against plant pathogens. Bull. Mater. Sci. **39**, 1165–1170 (2016)
56. M. Rai, A. Ingle, Role of nanotechnology in agriculture with special reference to management of insect pests. Appl. Microbiol. Biotechnol.Microbiol. Biotechnol. **94**, 287–293 (2012)
57. S.F. Peteu, F. Oancea, O.A. Sicuia, F. Constantinescu, S. Dinu, Responsive polymers for crop protection. Polymers **2**, 229–251 (2010)
58. P. Shukla, P. Chaurasia, K. Younis, O.S. Qadri, S.A. Faridi, G. Srivastava, Nanotechnology in sustainable agriculture: studies from seed priming to post-harvest management. Nanotechnol. Environ. Eng. **4**, 11 (2019)
59. V. Rajkumar, C. Gunasekaran, C.A. Paul, J. Dharmaraj, Development of encapsulated peppermint essential oil in chitosan nanoparticles: characterization and biological efficacy against stored-grain pest control. Pestic. Biochem. Physiol.. Biochem. Physiol. **170**, 104679 (2020)

60. K. Dangi, A.K. Verma, Efficient & eco-friendly smart nano-pesticides: emerging prospects for agriculture. Mater. Today Proc. **45**, 3819–3824 (2021)
61. B.P. Baker, T.A. Green, A.J. Loker, Biological control and integrated pest management in organic and conventional systems. Biol. Control **140**, 104095 (2020)
62. E. Thomine, J. Mumford, A. Rusch, N. Desneux, Using crop diversity to lower pesticide use: socio-ecological approaches. Sci. Total. Environ. **804**, 150156 (2021)
63. S. Kumar, M. Nehra, N. Dilbaghi, G. Marrazza, A.A. Hassan, K.-H. Kim, Nano-based smart pesticide formulations: emerging opportunities for agriculture. J. Control. Release **294**, 131–153 (2019)
64. A. Galal, Exogenous application of zinc mitigates the deleterious effects in eggplant grown under salinity stress. J. Plant Nutr.Nutr. **42**, 915–927 (2019)
65. M. Usman, M. Farooq, A. Wakeel, A. Nawaz, S.A. Cheema, H. Rehman, M. Sanaullah, Nanotechnology in agriculture: current status, challenges and future opportunities. Sci. Total Environ., 137778 (2020)
66. R.M. Kumari, N. Sharma, G. Arya, S. Nimesh, Recent progress in applied nanomaterials. Learn. Mater. Biosci., 33–64 (2019)
67. K.P. Chandrika, A. Singh, M.K. Tumma, P. Yadav, Nanotechnology prospects and constraints in agriculture. Environ. Chem. Sustain. World, 159–186 (2018)
68. M. Kah, R.S. Kookana, A. Gogos, T.D. Bucheli, A critical evaluation of nanopesticides and nanofertilizers against their conventional analogues. Nat. Nanotechnol.Nanotechnol. **13**, 677–684 (2018)
69. R.C. Choudhary, R.V. Kumaraswamy, S. Kumari, S.S. Sharma, A. Pal, R. Raliya, P. Biswas, V. Saharan, Zinc encapsulated chitosan nanoparticle to promote maize crop yield. Int. J. Biol. Macromol.Macromol. **127**, 126–135 (2019)
70. M. Singh, S. Shinjini, S. Prasad, I.S. Gambhir, Nanotechnology in medicine and antibacterial effect of silver nanoparticles. Digest. J. Nanomater. Biostruct. **33**, 115–122 (2008)
71. X.F. Zhang, Z.G. Liu, W. Shen, S. Gurunathan, Silver nanoparticles: synthesis, characterization, properties, applications, and therapeutic approaches. Int. J. Mol. Sci. **17**, 1534 (2016)
72. B. Zewde, A. Ambaye, J. Stubbs Iii, D. Raghavan, A review of stabilized silver nanoparticles—synthesis, biological properties, characterization, and potential areas of applications. Nanomedicine **4**, 1–4 (2016)
73. S. Wang, F. Wang, S. Gao, Foliar application with nano-silicon alleviates cd toxicity in rice seedlings. Environ. Sci. Pollut. Res. Int.Pollut. Res. Int. **22**, 2837–2845 (2015)
74. K. Vijayaraghavan, T. Ashok Kumar, Plant-mediated biosynthesis of metallic nanoparticles: a review of literature, factors affecting synthesis, characterization techniques and applications. J. Environ. Chem. Eng. **5**(5), 4866–4883 (2017)
75. M.J. Hodson, P.J. White, A. Mead, M.R. Broadley, Phylogenetic variation in the silicon composition of plants. Ann. Bot. **96**, 1027–1046 (2005)
76. R. Suriyaprabha, G. Karunakaran, K. Kavitha, R. Yuvakkumar, V. Rajendran, N. Kannan, Application of silica nanoparticles in maize to enhance fungal resistance. IET Nanobiotechnol.Nanobiotechnol. **8**, 133–137 (2014)
77. P.V. Ramouthar, P.M. Caldwell, S.A. McFarlane, Effect of silicon on the severity of brown rust of sugarcane in South Africa. Eur. J. Plant Pathol.Pathol. **145**, 53–60 (2016)
78. G. Guerriero, J.-F. Hausman, S. Legay, Silicon and the plant extracellular matrix. Front. Plant Sci. **7**, 463 (2016)
79. D.K. Tripathi, S. Singh, V.P. Singh, S.M. Prasad, N.K. Dubey, D.K. Chauhan, Silicon nanoparticles more effectively alleviated UV-B stress than silicon in wheat (*Triticum aestivum*) seedlings. Plant Physiol. Biochem.Biochem. **110**, 70–81 (2017)
80. H. Kosaka, T. Aikawa, N. Ogura, K. Tabata, T. Kiyohara, Pine wilt disease caused by the pine wood nematode: the induced resistance of pine trees by the A virulent isolates of nematode. Eur. J. Plant Pathol.Pathol. **107**, 667–675 (2001)
81. D. Seaman, Trends in the formulation of pesticides—an overview. Pest Manag. Sci. **29**, 437–449 (2010)

82. M. Nuruzzaman, M.M. Rahman, Y. Liu, R. Naidu, Nano encapsulation, nano-guard for pesticides: a new window for safe application. J. Agric. Food Chem. **64**, 1447–1483 (2016)
83. M. Kah, H. Walch, T. Hofmann, Environmental fate of nanopesticides: durability, sorption and photodegradation of nanoformulated clothianidin. Environ. Sci. Nano **5**(4), 882–889 (2018)
84. R. Kumar, K.K. Nair, M. Alam, R. Gogoi, P.K. Singh, C. Srivastava, M. Gopal, A. Goswami, Development and quality control of nano hexaconazole as an effective fungicide and its biosafety studies on soil nitrifiers. J. Nanosci. Nanotechnol. **15**, 1350–1356 (2015)
85. B. Singh, D.K. Sharma, A. Gupta, A study towards release dynamics of thiram fungicide from starch-alginate beads to control environmental and health hazards. J. Hazard. Mater. **161**, 208–216 (2009)
86. S. Anusuya, M. Sathiyabama, Foliar application of β-D-glucan nanoparticles to control rhizome rot disease of turmeric. Int. J. Biol. Macromol.Macromol. **72**, 1205–1212 (2015)
87. P. Kaur, R. Thakur, A. Choudhary, An in vitro study of the antifungal activity of silver/chitosan nanoformulations against important seed-borne pathogens. Int. J. Sci. Technol. Res. **1**, 83–86 (2012)
88. M. Malerba, R. Cerana, Recent applications of chitin-and chitosan-based polymers in plants. Polymers **11**, 839 (2019)
89. R. Nair, S.H. Varghese, B.G. Nair, T. Maekawa, Y. Yoshida, D.S. Kumar, Nanoparticulate material delivery to plants. **179**, 154–163 (2010)
90. S. Kumar, G. Bhanjana, A. Sharma, M.C. Sidhu, N. Dilbaghi, Synthesis, characterization and on-field evaluation of pesticide-loaded sodium alginate nanoparticles. Carbohydr. Polym.. Polym. **101**, 1061–1067 (2014)
91. B.A. Magnuson, T.S. Jonaitis, J.W. Card, A brief review of the occurrence, use, and safety of food-related nanomaterials. J. Food Sci. **76**, 126–133 (2011)
92. S.M. lfadul, A.A. Elneshwy, Use of nanotechnology in food processing, packaging and safety—review. Afr. J. Food Agric. Nutr. Dev. **10**, 2719–2739 (2010)
93. J. Brockgreitens, A. Abbas, Responsive food packaging: recent progress and technological prospects. Compr. Rev. Food. Sci. Food Saf. **15**, 3–15 (2016)
94. N. Sozer, J.L. Kokini, Nanotechnology and its applications in the food sector. Trends Biotechnol. **27**, 82–89 (2009)

SPRINGER NATURE

GPSR Compliance

The European Union's (EU) General Product Safety Regulation (GPSR) is a set of rules that requires consumer products to be safe and our obligations to ensure this.

If you have any concerns about our products, you can contact us on ProductSafety@springernature.com

In case Publisher is established outside the EU, the EU authorized representative is:

Springer Nature Customer Service Center GmbH
Europaplatz 3
69115 Heidelberg, Germany